£10·75

Lecture Notes in Physics

Edited by J. Ehlers, München, K. Hepp, Zürich
R. Kippenhahn, München, H. A. Weidenmüller, Heidelberg
and J. Zittartz, Köln
Managing Editor: W. Beiglböck, Heidelberg

81

Malcolm H. MacGregor

The Nature of the Elementary Particle

Springer-Verlag
Berlin Heidelberg New York 1978

Author
Malcolm H. MacGregor
Lawrence Livermore Laboratory
University of California
P.O. Box 808
Livermore, CA 94550/USA

ISBN 3-540-08857-1 Springer-Verlag Berlin Heidelberg New York
ISBN 0-387-08857-1 Springer-Verlag New York Heidelberg Berlin

Printed in Germany

Printing and binding: Beltz Offsetdruck, Hemsbach/Bergstr.
2153/3140-543210

"The need is for a whole revolution of thought, which can only be carried through by nonconformists."

Werner Heisenberg

"Let us recall the guiding principles for constructing a new theory. The first principle is Simplicity."

Herbert Pietschmann

FOREWORD TO THE READER

The present book describes an approach to elementary particle structure that is not the usual one. In these studies we systematically explore the experimental data on elementary particles, and we attempt to keep our preconceptions about these particles to a minimum. Using a somewhat classical viewpoint, and stressing the connections with other areas of physics, we arrive at conclusions which are different from currently-accepted notions about these particles. Some of the ideas discussed in these studies are still in their formative stages, and the readers of the book are invited to send any comments that they may have on this work to the author. It is hoped that the present volume will serve as a stimulus for further research along these lines, and that comments and extensions of the work can someday be incorporated into a revised and expanded version of this book.

All of the material that is presented here has been published in The Physical Review and other standard scientific journals. Thus this material has been refereed by experts in the field. Also, the author has given invited seminars and colloquia on this work at more than twenty major universities in the United States and Canada. Hence these ideas have been subjected to public scrutiny, and they are, in spite of their somewhat revolutionary nature, physically plausible concepts. The correctness of these concepts is of course another question, and the main purpose of the present book is to demonstrate in detail the agreement between these concepts and the experimental data. In particular, we find accurate agreement between the present elementary particle model and the so-called New Particles, which were discovered after this model had already been formulated and published. From a phenomenological viewpoint, we find a rather close correspondence between elementary particle physics and certain aspects of nuclear physics, and the correspondence has ramifications which are important for both of these areas of physics. Also, some of our most striking results are obtained by making a straightforward extrapolation of standard relativity theory and applying it to rotating systems. It should be kept in mind that elementary particle physics is not a domain unto itself, but is just a segment of the field of science which embraces nuclear, atomic, and molecular physics, and the same universal laws must govern all of these systems.

This book has been written with a twofold purpose. The first purpose is to present these studies of the elementary particle in a manner that will be of interest to the specialist in elementary particle physics, because the ideas that we develop here are novel, central to the main problems of particle physics, and in some respects at variance with our usual ways of thinking about these particles. The second purpose is to make these discussions complete enough that they can be read and understood by the scientist or would-be scientist who is not a specialist in elementary particle physics. In line with this second purpose, we have included an appendix, Appendix A, which gives a brief overview of the subject of elementary particle physics, and which, in particular, introduces the standard elementary particle terminology. This overview is important for the reader who is not completely familiar with the subject, and it is also of some help to the worker in the field in that it outlines the facets of elementary particle physics about which we are mainly concerned.

The author would like to take this opportunity to express his appreciation to Robert Howerton, to the late Simon Pasternack, to Glenn Fellows, to Dirk Brinkman, to members of the Particle Data Group, and to the members of his family, who in various ways have been of assistance in this project during the past several years.

Livermore, California, 1978

TABLE OF CONTENTS

Page

Page

APPENDICES

LISTS OF FIGURES AND TABLES

FIGURE | Page

FIGURE | Page

PROLOGUE: THE WAVE-PARTICLE DUALITY OF THE ELEMENTARY PARTICLE

In seeking to ascertain the nature of the elementary particle, we must right at the outset come to grips with one of its most puzzling features — namely, the wave-particle duality of the elementary particle. This wave-particle duality was first clearly recognized in the case of the photon, but it has subsequently been shown to apply also to electrons, nucleons, and atoms. The wave properties of the photon are exemplified in the fact that a single photon has a region of coherence which extends for a distance of a meter or more in all directions. The particle properties of the photon emerge most clearly from Compton scattering, in which a high-energy photon makes a billiard-ball type of collision with an electron, and hence delivers its energy and momentum into a volume that has dimensions on the order of 10^{-11} cm (the Compton wavelength of the electron). Thus the "wave region" of an energetic photon is roughly thirteen orders of magnitude larger than the "particle region" of the photon. An optical-frequency photon is about 10^6 times larger than an MeV photon, so that if we assume a wave-field extent of about a meter, the wave-to-particle dimensionality ratio for an optical photon is roughly 10^7.

In dealing with the photon, we must take into account both its wave aspects and its particle aspects. But since these two aspects differ dimensionally by many orders of magnitude, they are in a practical sense separate entities, so that we can discuss the wave properties of the photon without having to consider its particle properties, and vice versa. A free electron is in many respects similar to a photon; it has wave properties that have been observed to extend over a distance of more than 10^{-3} cm, and yet it has particle interactions which indicate a size of 10^{-11} to 10^{-13} cm (depending on whether we choose the Compton radius or the classical electron radius). Thus a free electron has a wave-to-particle dimensionality ratio of roughly 10^9. A bound electron has wave properties that extend over the dimensions of an atom — roughly 10^{-8} cm — , and a free slowly-moving proton or neutron also has a wave field of this general magnitude. Thus the wave-to-particle ratios for these particles are roughly 10^4 to 10^5, which are much smaller than for photons or free electrons, but which are still large enough that the wave and particle aspects of these particles can be rather cleanly separated. However, when we now come to the cases of atomic nuclei and of massive elementary particles (particles which contain substates), we encounter a completely different scale of interaction distances. An atomic nucleus is a bound-state cluster of nucleons — protons and neutrons. From the systematics of the nuclear shell model, we know that the protons and neutrons inside a nucleus exhibit wave properties; and since the dimensions of atomic nuclei are on the order of 1 or 2 times 10^{-12} cm, the wave properties of bound nucleons occur over dimensions of this magnitude. But the nucleon itself is known to have a spatial extent (diameter) of about 2 times 10^{-13} cm. Thus in atomic nuclei the wave aspects of a nucleon are only about ten times as large as its particle aspects. As a consequence, the wave and particle properties of a bound nucleon are mixed together, and they cannot be cleanly separated. When we come to the internal structure of the nucleon itself, this problem assumes its most critical form. Both protons and neutrons have root-mean-square (rms) electromagnetic radii of 0.8 fermi (0.8×10^{-13} cm). If we assume that a nucleon is composed of three equal-mass substates (which are denoted as "quarks"), and if each quark has a mass that is about 1/3 of the nucleon mass, then the Compton radius $R = \hbar/mc$ of a quark is 0.6 fermi. If we cluster together three quarks, each with a radius of 0.6 fermi, then a simple calculation shows that the overall rms radius of this three-quark structure is just about 0.8 fermi. Thus on the inside of a nucleon we have a system of quarks whose "wave dimensions" are essentially equal to their "particle dimensions", so that the wave and particle aspects of these quarks are intermingled in an inseparable manner.

We can summarize the above discussion in the following table:

Particle	Wave dimensions	Particle dimensions	Dimensional ratio
free photon	~ 1 meter	$\sim 10^{-5}$ to 10^{-11} cm	$\sim 10^{7}$ to 10^{13}
free electron	$\sim 10^{-3}$ cm	$\sim 10^{-12}$ cm	$\sim 10^{9}$
bound electron	$\sim 10^{-8}$ cm	$\sim 10^{-12}$ cm	$\sim 10^{4}$
free nucleon	$\sim 10^{-8}$ cm	$\sim 2 \times 10^{-13}$ cm	$\sim 10^{5}$
bound nucleon	$\sim 2 \times 10^{-12}$ cm	$\sim 2 \times 10^{-13}$ cm	~ 10
bound quark	$\sim 2 \times 10^{-13}$ cm	$\sim 1 \times 10^{-13}$ cm	~ 2

The dimensionality ratios shown in this table are only approximate (for example, they depend on the energies of the particles), but they serve to illustrate the following point: in the domain of the atomic nucleus, and to an even greater extent in the domain on the "inside" of an elementary particle, the wave and particle aspects of the constituent substates are of comparable dimensions. This means that the "particle" properties of these substates will interfere with and modify the "wave" properties of the substates in an essential manner. In Chapter 20 we illustrate this result for the case of atomic nuclei. In the case of the elementary particle, the properties of the quark substates are not well known, but it is apparent that the "geometry" of a quark will strongly influence the quark wave functions.

The relevance of this discussion to the reader is that the great majority of the currently-studied elementary particle models are based on the use of SU(3)-related symmetry groups as a means for classifying elementary particle states. These symmetry groups[†] relate mainly to wave properties, so that the "wave" aspects of the elementary particle are emphasized. However, in the present studies we will instead focus on the "particle" aspects of the elementary particle — that is, on its spectroscopic properties rather than on its interactive properties. These two approaches to elementary particle structure are so different that they appear almost to be separate topics. But it is clear that both the "wave" and the "particle" aspects of the elementary particle must be brought into agreement. Thus the results that we obtain here and the results that are obtained from SU(3) and other wave-related approaches must ultimately be confronted with one another, so that a common area of agreement can be delineated. Unfortunately, the studies in the present book are limited almost entirely to developing and presenting the "particle" point of view, and we give only a few intercomparisons of the present results with those of SU(3). Also, we focus our attention mainly on spectroscopic experimental data, and we thus ignore whole classes of experiments — such as deeply-inelastic electron scattering — which are highly relevant to the structure of the elementary particle. The present studies are obviously incomplete, but they serve to focus attention on aspects of the elementary particle which are not emphasized in the approaches based on unitary symmetry.

With this prologue, we now move on to the main body of the text, and to our investigation of The Nature of the Elementary Particle.

[†]A good basic discussion of SU(3) and related symmetries, and of quark models, is given in B. T. Feld's book Models of Elementary Particles, Blaisdell, Waltham (1969).

PART I. MASS QUANTIZATION: THE SEARCH FOR THE BASIS STATES

"We may have big surprises ahead of us. In particular, the spectrum of masses of the building blocks remains utterly baffling."

J. D. Bjorken, Summary Talk: Neutrino '77
Elbrus, USSR, June 18 - 24 (1977)

INTRODUCTION AND SUMMARY

What is the nature of the elementary particle? This is the question that has dominated scientific thought for at least the past 2500 years. And it is an especially exciting question today, because physicists now have the tools which should provide the answer. Indeed, the answer may already lie before us, buried in the wealth of experimental elementary particle data which has been accumulated during the past few decades.

The speculations of the ancients on the subject of the elementary particle — the ultimate building block of matter — were necessarily limited by the fact that the world of microscopic phenomena lay beyond the range of their vision; and it is a demonstrable fact that the great truths in the natural sciences have emerged from observation rather than introspection. However, the explosion of scientific progress which has occurred during the past two centuries has now given us the techniques for making detailed measurements, if not actual observations, in the realm of the very small. As scientists have refined their experimental techniques, we have moved from studies of molecules to studies of atoms, and then to studies of atomic nuclei. The final step in this progression is the elementary particle, which in some sense represents the bed rock of the universe. It seems intuitively plausible that if the physical systems we study get progressively simpler, then the physics itself should also get simpler — or at least more clear-cut. Quantum mechanical effects, which are characterized by Planck's constant h, become progressively more important as the mass of the system is decreased. Since elementary particles represent the lightest physical systems, they are the most strongly affected by quantum mechanical considerations, and they should provide the most basic information about the ultimate nature of quantum mechanics. Now quantum mechanics, as the name suggests, is the mechanics of systems whose variables are quantized. This quantization appears very strikingly in the angular momenta of molecules, atoms, nuclei, and elementary particles, and it appears in the energy values of photon wave packets. When we get down to the level of the elementary particle, do we find that masses are also quantized? From a cursory inspection of the mass values for the almost two hundred elementary particles that have been observed to date, it is not immediately obvious that this mass quantization exists. However, this plethora of elementary particles suggests that the so-called "elementary particle" must itself be a composite structure, and perhaps it is composed of elementary particle "building blocks" which have quantized masses. This is the subject that we investigate in Part I of the present book.

As a guide to the reader, we now briefly summarize the contents of Chapters 1 through 9, which form Part I of the book. In this search for a fundamental set of elementary particle building blocks, our procedure is to make a direct study of the experimental data on elementary particles. Do these data reveal the pattern of their own existence? In an attempt to answer this question, we start with the simplest approach: assuming that the two hundred elementary particle states which have been identified to date (see Appendix B) are composite structures formed from a set of basis states, we first explore the possibility that these basis states have small binding energies (a few percent), since the binding energies in all other areas of physics are known to be small. This approach leads us to consider the mass quantum $M \simeq 70$ MeV as an elementary particle basis state, so that we can reproduce the lowest-mass meson resonance, the 140 MeV pi meson, as the state-antistate pair $\pi = M\overline{M}$. When we extend this result to include the other meson and baryon resonances, we find to our astonishment that the search for required basis states is ended: we can accurately reproduce the entire spectrum of meson and bayon resonances by using just the quantum M, provided that we allow it to appear in both spinless and relativistically-spinning configurations. The lowest-mass meson state has provided the decisive clue!

The identification of the mass quantum $M = 70$ MeV as the fundamental building block for all particles which are more massive than the electron raises an immediate question. This question is not, as you might think, to explain how it is that such a simple model can reproduce the observed elementary particle states: we will demonstrate this result in detail later on. The question is rather to explain why it is that such an elementary model hasn't been thoroughly explored years ago. To answer this question, we must consider the zeitgeist, the intellectual climate, which was prevalent in physics in the early 1930's, when the field of modern elementary particle physics came into being. This zeitgeist was (and still is) of such a nature as to preclude serious attempts at constructing simple representational models for elementary particles. The present book is written in an effort to alter this situation. The discussion of this zeitgeist in elementary particle physics constitutes Chapter 1 of the book.

In Chapter 2 we pose a number of fundamental questions about the elementary particle. As Neils Bohr reminds us, we can't really ask pertinent questions until we know the answers. Thus, by including Chapter 2 in this book, we are promising to give answers, or at least partial answers, to the questions that we ask. The answers to these questions emerge from the step-by-step analysis of the experimental data that is contained in the subsequent chapters of the book. The questions listed in Chapter 2 are useful in that they serve as both the motivation and the guide for this analysis.

Chapter 3 describes the basic ansatz of the present work – the assumption that the elementary particle is a composite structure, formed from a subset of quark states. Since no one has been able to break apart an elementary particle even with the most powerful accelerators, the composite nature of the elementary particle is not immediately obvious. However, the interrelationships that can be traced among the various particles seem to point quite clearly to this composite nature. The task of tracing through these particle interrelationships is in fact the principal subject matter of the book.

Chapter 4 is a discussion of the decisive conclusion which underlies all of the present work – the conclusion that the quark basis states in the elementary particle are held together with small binding energies – a few percent. The workings of the quark model have always suggested this conclusion, and the regularities which occur in the experimental data lead directly to this same result.

There is one definitive test that can be made to investigate quark binding energies. If quark binding energies are only a few percent, then the existence of a quark state $M \simeq 70$ MeV is unequivocally required in order to reproduce the 140 MeV pi meson. Furthermore, if the spinless 494 MeV $K^{\pm}$ meson is composed of seven quanta M, as we will show, then the quantum M must itself be spinless. A spinless 70 MeV mass quantum M has not been observed experimentally, and it is not a feature of conventional quark models, which include only spinning quarks. Thus the argument that quark binding energies are small devolves into the argument that the spinless 70 MeV mass quantum M must exist. Chapter 5 is a summary of evidence based on elementary particle mass values which points to the existence of the quantum M. This evidence is obtained from both the meson and the baryon resonances. Interestingly, some of the clearest evidence comes from the systematics of the "New Particles". The New Particles are a series of relatively-long-lived high-mass ($\geq$1800 MeV) meson resonances that were unexpectedly discovered quite recently, after most of the modern-day concepts of the elementary particle (including the present one) had already been formulated and published, and they thus provide an interesting a posteriori test of these concepts. The validity of any theoretical model is determined, in the long run, by the results of experiments, and not by the opinions of theorists, which is why the surprise appearance of these very-narrow-width high-mass New Particles has had such an impact on the elementary particle scientific community (as is evidenced by the prompt awarding of a Nobel prize to the discoverers of these particles).

As we discuss in Chapter 6, there is one completely independent type of evidence which points to the existence of the quantum M. In Chapter 6 we present arguments which indicate that the special-relativistic mass increase which is observed in a moving particle applies to rotational motion as well as to translational motion; and we also show that in the laboratory frame of reference, both special relativity and general relativity give rise to the same set of equations when applied to rotating systems (which is a result that has been noted in connection with certain Mössbauer experiments on rotating systems). Then, applying these relativistic equations to a sphere of matter that is rotating at the full relativistic limit (with its equator moving at the velocity of light), we show that <u>the rotating sphere is half again as massive as it was at rest</u>. Thus a spinless 70 MeV quantum M, when set into full-relativistic rotation, becomes a spinning 105 MeV quantum, and the association of this quantum with the spinning 105 MeV muon seems unmistakable. Furthermore, by calculating the moment of inertia of this relativistically-spinning sphere about its axis of rotation, and by setting the radius of the sphere equal to its Compton radius, $R = \hbar/mc$, we obtain $1/2\ \hbar$ as the calculated spin angular momentum of the sphere, which is equal to the observed spin of the muon.

This line of inquiry is continued into Chapter 7, where we demonstrate that the main spectroscopic properties of the muon, and also its de Broglie wavelength, are reproduced by this model of a relativistically-spinning sphere. Historically, the muon has come to occupy a position as an "unwanted" elementary particle. Initially hailed as the mediator of the nuclear force, as predicted by Yukawa, it was soon discovered to interact only weakly (electromagnetically), whereas the nuclear force is a strong (hadronic) interaction. Thus the muon, which used to be called the mu meson, is not a meson state at all; it appears to be nothing more or less than a heavy electron. But the muon does (embarrassingly) exist, and it is, after all, the lowest-mass heavy-particle state (where "heavy particles" are defined here to mean all particles more massive than the electron); furthermore, it has a lifetime that, as we will show, is phenomenologically related to the lifetimes of the other unstable heavy particles. Hence the muon must be of fundamental importance in the overall scheme of heavy-particle resonances. Since present-day theories of these heavy particles have no place to include the muon, the relationship we have established here between the spinless 70 MeV mass quantum M and the spinning 105 MeV muon μ is of more than passing interest.

In Chapter 8 we move on to a study of spinning hadronic quarks, where it turns out that the key piece of information is provided by the example of the nucleon.

A number of phenomenological arguments suggest that the nucleon is composed of three approximately-equal-mass spin 1/2 quarks. Thus, under our assumption of small quark binding energies, each spin 1/2 "nucleon quark" must have a mass that is about 1/3 of the nucleon mass, or roughly 315 MeV. If these spinning nucleon quarks follow the same relativistic mass relationship that we established above for the sphere and the muon, then a "nonspinning" nucleon quark would have a mass that is 2/3 of this value, or 210 MeV. Now 210 MeV is just three times the mass of the quantum M, which suggests that the spin 1/2 nucleon quark is composed of three quanta M in a relativistically-spinning configuration. In Chapter 8 we summarize experimental evidence from both the baryon and meson resonances which delineates the existence and the properties of this spinning quark. For clarity, we denote this spinning nucleon quark as the spinor S in order to distinguish it from the more conventional nonstrange quarks of Gell-Mann and Zweig. There are many similarities between the spinor S and the corresponding Gell-Mann and Zweig quark, but there are also essential differences. In particular, the spinor S does not carry the fractional baryon quantum number which characterizes the Gell-Mann and Zweig quark; instead, it carries the integral baryon quantum number which characterizes the Fermi and Yang quark. The spinor S occurs in baryon resonances in the form of invariant Fermi-Yang triplets, $S\bar{S}S$, and the conservation of these spinor triplets is precisely equivalent to the conservation of the baryon quantum number B.

Chapter 9, the final chapter in Part I, contains a discussion of the basic long-lived N, Λ, Σ, Ξ, and Ω baryon and hyperon states, and it completes our development of the light-quark set of basis states.

Chapters 1 through 9 constitute the derivation and general description of the elementary particle basis states. What we obtain from these chapters is the strongest possible mass quantization condition: all elementary particles which are more massive than the electron can be formed as spinless and spinning combinations of the mass quantum $M \simeq 70$ MeV. (It is of phenomenological interest to note that the ratio of the electron mass to the $M = 70$ MeV mass is almost precisely equal to the dimensionless fine structure constant $\alpha = e^2/\hbar c \simeq 1/137$.)

In Chapters 10 - 14 we move from mass quantization to angular momentum quantization. Specifically, we decompose the total angular momentum of an elementary particle into its spin and orbital components, and we find that these have different quantization uncertainties. In Chapters 15 - 17 we face up to the problem of reproducing the observed spectrum of baryon and meson resonances, including the high-mass New Particle resonances. Chapters 18 - 20 extend and amplify these results.

Since this approach to elementary particle structure is not a familiar one, we will endeavor in each section to outline the goals that we are seeking to achieve.

Many of the results that are obtained here may seem to be very simple, which in fact they are, but this does not mean that they are necessarily trivial. Indeed, the most important concepts are usually the simplest ones. The results that we obtain from these studies are in many respects incomplete, but they hopefully constitute a step in the right direction.

"The aim of scientific inquiry is not to reveal absolute truth but to discover more useful ways of thinking about things."

Andrew Weil

CHAPTER 1. THE ELEMENTARY PARTICLE: VICTIM OF AN HISTORICAL ACCIDENT

"It is the customary fate of new truths to begin as heresies and end as superstitions."

T. H. Huxley

Modern elementary particle physics had its inception in the early 1930's with the invention of the first high energy accelerator and the discovery of the neutron. This occurred right after a furious quarter of a century of activity in atomic physics, during which the concept of quantum mechanics was born, formulated in detail, and then crystallized. Thus these quantum mechanical ideas which had so recently been created in the area of atomic physics were taken over and applied to the new field of particle physics. Hence we must first understand the historical preconceptions that the particle physicists inherited from the atomic physicists before we can understand why it is that certain theoretical approaches have been attempted, or have not been attempted, in the area of elementary particle physics.

Modern atomic physics had its genesis with the Rutherford discovery of the massive and point-like atomic nucleus, and with the subsequent Bohr discovery that electrons appear to move in orbits around the atomic nucleus. These conceptual breakthroughs stimulated a great deal of research into the spectroscopy of the atom. The analogy between the $1/r^2$ coulomb force in the atom and the $1/r^2$ gravitational force in astronomy led to the natural assumption that electrons move in elliptical orbits around the nucleus, and Sommerfeld's explanation of the atomic fine structure splitting as arising from special-relativistic corrections to the electron orbits seemed to confirm this viewpoint. However, spectroscopic difficulties soon began to arise. The discovery by Goudsmit and Uhlenbeck of electron spin and an associated electron magnetic moment led to an atomic spin-orbit correction term that is comparable to and apparently extraneous to the Sommerfeld relativistic correction term. Also, attempts at combining two elliptical electron orbits so as to obtain the proper ground-state energy for the helium atom were unsuccessful. The helium ground-state problem is of course an example of the unsolved three-body problem in physics, and the calculational failure here was probably as decisive as any other factor in the eventual change from classical to quantum methods for dealing with atomic spectroscopy.

At about the time when these difficulties with classical atomic spectroscopy were beginning to be recognized, in the middle 1920's, another avenue of approach began to suggest itself. The quantization of the electron orbits in an atom implied that a wave-like phenomenon might be applicable to these orbits. Also, de Broglie's suggestion that the electron has a wave-particle duality similar to the wave-particle duality of the photon was soon confirmed by Davisson and Germer, and independently by Thomson. The electron wave-length experiments, the wave-mechanics approach in treating atomic structure, and formal rules such as the Heisenberg Uncertainty Principle and the Pauli Exclusion Principle, all appeared during this period as quantum concepts or results which seemed to transcend anything that one could obtain by using a classical approach to atomic spectroscopy.

The development of quantum mechanics had a decisive effect upon conventional atomic spectroscopy — it essentially destroyed it. In 1926 Born introduced the idea that the square of the wave function of an electron orbital should be interpreted as a position density and not as a charge density, and the further step was soon taken of moving from position densities to momentum densities — from coordinate space representations to momentum space representations. The position and the momentum of a particle became functions that had to be interpreted statistically, and the important quantities for quantum mechanics became the "observables" — the properties of a particle that could actually be measured. Now, in an atom the stationary states are not "observables". For example, there is no practical way in which the frequency of rotation of an electron moving in a bound state in hydrogen can be measured; the only thing that can be observed is a transition from one orbit to another. Thus atomic transitions, from a quantum point of view, are more fundamental than atomic stationary states; the transitions can be observed, but the stationary states cannot. The quantum-mechanical idea of regarding transitions as being more fundamental than stationary states crystallized at about the same time as that at which the "stationary-state spectroscopists" were encountering what appeared to be insuperable difficulties in contructing suitable classical electron orbitals. The result was that physicists essentially gave up on classical atomic spectroscopy — that is, they abandoned attempts at constructing visualizable models of atomic stationary states, and they concentrated henceforth on formal wave-mechanical methods for reproducing stationary-state energies and occupation numbers — methods which do not require complete spectroscopic descriptions of the states being calculated. Thus, for example, we now have formal rules for the ways in which electrons fill atomic orbits and couple to one another vectorially, but we have no "models" with which to explain the workings of these rules.

It was just at the time when the quantum approach to atomic structure was firmly decided upon that the field of modern elementary particle physics made its appearance. The discovery of the neutron and the invention of high energy particle accelerators in the early 1930's opened up the domain of subatomic particles. Since quantum mechanics was invented to account for atomic spectroscopy and not subatomic spectroscopy, one of the first points to determine in elementary particle spectroscopy was the applicability of quantum concepts to subatomic particles. A link between quantum mechanics and subatomic particles had of course already been established in the case of the electron. A further link was provided by Yukawa's conjecture that a massive particle is the quantum of the nuclear force field just as the photon is the quantum of the electromagnetic field, and by the subsequent discovery of the strongly-interacting pi meson with a mass value that is appropriate to account for the range of the nuclear force. Much later, a wave-particle duality for the neutron similar to that of the electron was verified experimentally. However, the decisive step in our present-day concept of the elementary particle came earlier than either the discovery of the pi meson or the discovery of the wave properties of the neutron; it occurred in 1928 with the Dirac formulation of the electron. Relativistic Dirac equations provide an accurate description of the dynamical properties of the electron. They also lead to a picture for the ground state of the electron which shows it as part of a complicated system that includes an infinite sea of virtual electron-positron pairs. Thus the ground state of an electron becomes no more simple or fundamental than the excited (dynamical) states of the electron, and the important physical reality becomes the entire ensemble of possible electron states, and in particular the ensemble of transitions that can occur between these states. Hence we arrive at a parallel situation between atomic physics and elementary particle physics. In each case, the symmetry properties of the observable transitions have become the dominant focal point for theoretical investigation, transcending in importance the properties of the stationary states.

The spin and the magnetic moment of the electron appear naturally in the Dirac equations, and their interactions are accurately accounted-for. But the Dirac equations, spectroscopically speaking, tell us nothing about the geometry of the electron. The generally-accepted assumption that the electron is point-like leads to conceptual difficulties in accounting for both the large spin value and the large magnetic moment of the electron. However, for the historical reasons summarized here, physicists today do not tend to worry about these difficulties, any more than they worry about providing a detailed description of the S-state motion of the electron cloud which presumably surrounds the proton in the ground state of the hydrogen atom.

An obvious limitation in studying the atom and the electron, as in atomic physics, is that the forces involved are entirely electromagnetic. Thus the quantum theory for these systems is based solely on electromagnetic interactions. One of the main reasons for extending these studies to include the strongly-interacting <u>hadronic</u> particles — the proton and neutron, and the families of meson and baryon resonances — is the hope that in processes involving strong (hadronic) interactions we can probe more deeply into the basic quantization process. In point of fact, this hope may have been realized. The raison d' etre for the present book is to describe in mathematical detail the empirical fact that the strongly-interacting hadronic particles admit a very simple but accurate spectroscopic description — a description which is a curious mixture of some clearly quantum concepts with some essentially classical concepts, and a description in which the quantization of particle masses occurs in a clear and unmistakable manner.

The experience of quantum mechanics to the contrary, it has always been the hope of physicists that the stable ground states of physical systems should in some sense be simpler or more symmetric than the excited states of these systems, and that individual elementary particles should be simpler than their bound-state aggregates. This viewpoint has not received any encouragement from present-day elementary particle theories, which are essentially all based on formal symmetry and field-theoretic approaches to particle structure. These theories are expressed in terms of formidable mathematical models, with the complexity of the mathematics apparently far outstripping the complexity of the particles themselves. However, if we abandon this formal theoretical approach, and if we start from scratch and use the experimental data rather than theoretical preconceptions as our guide, we are led to a detailed microscopic model which indicates that the elementary particle may in fact be a reasonably simple entity. Under normal circumstances, this direct use of the experimental data would seem to be a natural starting point for anyone attempting to formulate a theory of elementary particles. But the impact of atomic physics on elementary particle physics was so immediate and so powerful that the formal quantum mechanical approach to particle structure was accepted without question right from the outset. As a result, elementary particle physicists violated the Golden Rule of Scientific Research: try the simple solutions first, and go on to more complicated solutions only if the simple solutions can't be made to work.

Since the present book is devoted to the experimental elementary particle data and their interpretation, references to current theoretical ideas will be mentioned only incidentally, and mainly in the next three chapters, which serve as an introduction to some of the spectroscopic problems with which we will be struggling in this book. In Chapter 2 we pose several questions about the elementary particle, and the remainder of the book consists of our attempts to obtain answers to these and other questions by making a careful study of the experimental data. The experimental data appear to tell their own story in a rather straightforward manner, so that the present discussion is essentially self-contained.

References for Chapter 1.

The quotation by T. H. Huxley at the beginning of Chapter 1 is taken from A. P. French, Principles of Modern Physics, Wiley, New York (1958). Some of the historical comments in this chapter are taken from the author's notes of a lecture "The Development of Concepts in the History of Quantum Theory" given by Werner Heisenberg at the University of California at Berkeley on May 2, 1973. For an interesting article about the early period of modern physics, see S. A. Goudsmit and G. E. Uhlenbeck, "It Might as well be Spin", Physics Today, June 1976, pages 40-48. As a student at the University of Michigan, the present author was fortunate enough to have Professor Uhlenbeck as the lecturer for much of his graduate work in physics.

The quotation by Andrew Weil at the end of the Introduction and Summary is taken from his book The Natural Mind, Houghton (1972).

The quotation by Werner Heisenberg on the Frontispiece is taken from an article by Andre Mercier in Foundations of Physics, Plenum Publishing Corporation, New York, Volume I, page 285 (1971). The quotation by Herbert Pietschmann on the Frontispiece is taken from his Inaugural Lecture as an Honorary Member of the Hungarian Physical Society, delivered in Budapest, May 31, 1976.

CHAPTER 2. WHAT IS THE NATURE OF THE ELEMENTARY PARTICLE?

2.1 What can we Learn about Elementary Particles?

> "Physics is mathematical not because we know so much about the physical world, but because we know so little: it is only its mathematical properties that we can discover."
>
> Bertrand Russell

The question about the ultimate nature of the elementary particle, perhaps more than any other question, has dominated investigations in the natural sciences for many centuries; it is really the fundamental question in elementary particle physics. The scientific tools available for answering this question have increased enormously in the past century, and in particular in the past two decades. And yet, at the time of the writing of this book, the answer seems almost as obscure to us as it was to the ancient Greeks.

The line of inquiry about the ultimate nature of the elementary particle can be traced back as far as Democritus, who speculated that if we keep on dividing matter into smaller and smaller pieces, we will finally reach an indivisible particle, the atom, which constitutes the basic building block of the universe. Aided by the tools of modern science, we have sharpened and refined this idea to the point where we now have the molecular domain of the biologist and chemist, the atomic domain of the atomic physicist, the subatomic domain of the nuclear physicist, and the subnuclear domain of the elementary particle physicist. We know that molecules are formed as geometrically-well-defined clusters of atoms; that atoms are formed as atomic nuclei surrounded by clouds of electrons; and that atomic nuclei are formed as reasonably-well-defined clusters of nucleons – protons and neutrons. Of what, then, are nucleons formed? More specifically, can we understand the nucleon in terms that suffice for the nucleus, the atom, and the molecule? Or have we reached a realm with the elementary particle in which a whole new set of realities comes into play – a set of realities that requires a new language and a new way of thinking?

As we discussed in Chapter 1, the development of the field of atomic physics necessitated a jump in our thinking beyond the ideas of classical mechanics – a jump into the domain of quantum mechanics. And contemporary elementary particle physicists have implicitly posited a further jump once the domain of the elementary particle is reached, with the nature of this jump depending on the physicist in question. Thus, for example, we have explanations for the ultimate reality of the nucleon which picture it as (1) a pole in a complex energy-momentum plane, or (2) a location in an SU(3) diagram, and/or (3) a collection of nucleon substates — denoted as nucleon "quarks". Furthermore, in the quark model of the nucleon, the nucleon quarks, which have never been observed directly, are now assumed to occur in a variety of forms, denoted colloquially as "flavors", none of which have any classical significance. Hence, to at least the great majority of present-day elementary particle physicists, the nature of the elementary particle is quite different from that of the atomic nucleus, the atom, or the molecule; these physicists believe that the ultimate realities of the elementary particle involve considerations which are unimportant at the level of the nucleus or the atom, so that we cannot learn the essential facts about the elementary particle by extrapolating from our experience with the nucleus and the atom. However, even if this viewpoint is correct, which it certainly is in some approximation, there may nevertheless be some aspects of the elementary particle that can be understood in terms of concepts which also apply in other areas of physics.

The main obstacle to our understanding of the elementary particle is the fact that these particles are too small to be observed in any direct manner. All we can do is to perform various experiments on them — for example, by smashing them against one another — and then devise mathematical models which reproduce the results of these experiments. This is the situation characterized by Bertrand Russell in the quotation at the beginning of this section. With this viewpoint in mind, let us ask the following question about the ultimate nature of the elementary particle:

> Does the elementary particle have properties that can be understood in terms of more-or-less classical concepts?

Now, when we go to examine the properties of the elementary particle, it becomes apparent that these fall into two general categories — spectroscopic and dynamical. The spectroscopic properties are the properties of a single isolated particle — its mass, charge, spin angular momentum, and so on; the dynamical properties are delineated by the interactions which occur when two or more particles are brought together. With respect to the properties of the elementary particle, the lessons that were carried over from the domain of atomic physics, as described in Chapter 1, are that a complete spectroscopic description of the elementary particle may not be possible, and that the main emphasis should be placed on dynamical rather than on spectroscopic studies. Present-day elementary particle physicists have taken these lessons to heart. Thus, for example, elementary particle theorists attempt to calculate mass values, but they do not attempt to calculate spin angular momentum values;

and elementary particle experimentalists, especially in recent years, have tended to measure dynamical rather than spectroscopic properties of elementary particles, although the discovery of the New Particles (see Chapter 17) has at least temporarily reversed this trend. In contrast to most of the present-day workers in the field, we will concentrate in this book almost exclusively upon the spectroscopic aspects of the elementary particle. Thus we rephrase the above question about the elementary particle as follows:

> Does the elementary particle have spectroscopic properties that can be understood in terms of more-or-less classical concepts?

Keeping Bertrand Russell's dictum in mind, we recast this question in still a slightly different form:

> Can a mathematic model be devised that reproduces the spectroscopic properties of the elementary particle?

We would naturally assume that this question has been thoroughly explored. However, for the reasons described in Chapter 1, and as a perusal of the contemporary literature on the elementary particle clearly indicates, this question has not been dealt with to any extent on the microscopic level, at least in modern times. By analyzing the experimental elementary particle data in a step-by-step manner, we find that the answer to this question is yes; we can devise a mathematical model that reproduces the main spectroscopic properties of the elementary particle. This model, furthermore, is a model that can be visualized in more-or-less classical terms. But is this model consistent with the dynamical information that we have about the elementary particle? Unfortunately, this is a question about which we will have little to say in the present book, and it is a question that must eventually be answered before we can assess the true significance of the model. But as far as we have been able to ascertain, this mathematical model does not appear to conflict with experiment, although it does conflict with the preconceptions of many physicists. The virtues of this model are that it reproduces spectroscopic properties which have not been forthcoming from any other approach to elementary particle structure, and that it blends together both classical and quantum concepts in a very intriguing manner. Thus it should be at least a useful step forward in our attempts to understand the ultimate nature of the elementary particle.

Our conclusion in this section is that the domain of the elementary particle, while it contains some undeniably unique features, is nevertheless more closely related to other areas of physics than we might suspect from a study of the elementary particle models which are currently in vogue. Of course, we have not yet described the mathematical model upon which we base this conclusion. Before proceeding with the development of this model, which is a task that occupies most of the book, we list in the next section a number of specific questions about elementary particles.

2.2 Some Spectroscopic Questions about Elementary Particles

> "A problem cannot be properly formulated before its solution is known, for questions and answers have to be in complete reciprocal agreement."
>
> Neils Bohr

In the discussion of the preceding section, we stated that the emphasis in this book will be on spectroscopic rather than dynamical aspects of the elementary particle. However, we did not go into any details as to the kind of spectroscopic information we are seeking. In order to clarify this point, we list in the present section a series of spectroscopic questions about the elementary particle. These questions are not comprehensive, but they are central to the problem at hand; and, in particular, they are questions for which we obtain answers, or at least partial answers, in the present studies. As Neils Bohr points out in the quotation cited above, we don't really know what questions to ask until we have some clues as to the answers. And, in fact, there is little point in discussing a question unless we have some hope of answering it. This is the reason that, in spite of the voluminous contemporary literature on the elementary particle, there is very little discussion of the questions that are raised here; present-day formal approaches to elementary particle structure are not designed to deal with this kind of specific spectroscopic information.

The spectroscopic questions that we wish to consider are the following:

(1) Is the "elementary particle" elementary or composite?

(2) If the elementary particle is composite, what are the binding energies and the basis states?

(3) Does the muon fit in with these basis states?

(4) Does the relativistic mass increase that is observed in translational motion also apply to rotational motion?

(5) Does the spin of an elementary particle arise from the rotation of an extended mass distribution?

(6) Does the magnetic moment of an elementary particle arise from the rotation of an extended charge distribution?

(7) Does the concept of orbital angular momentum apply to an elementary particle?
(8) Does the baryon quantum number have a physical interpretation?
(9) Does the strangeness quantum number have a physical interpretation?

Our purpose in writing down these questions is twofold: first, it gives us an opportunity to briefly review some of the contemporary ideas about these questions; and second, it serves to define just what it is we are searching for in our perusal of the experimental data. Some of these questions, such as Numbers (8) and (9), are of a qualitative nature; their answers are based on physical insight. Other questions, such as Numbers (5) and (6), are of a quantititive nature; to answer these questions, we must produce calculations that match the experimental data.

We now briefly consider each of these questions in turn.

Question (1). Is the "elementary particle" elementary or composite?

This question is discussed in Chapter 3. Elementary particle physicists are of two schools of thought here, with the majority now favoring the idea that the elementary particle is composite. From the present studies, we conclude that the elementary particle is definitely composite – formed from a basic set of substates. In accordance with common usage, we will sometimes denote these substates as quarks, although the quarks that we obtain here are not the same as the quarks of Gell-Mann and Zweig.

Question (2). If the elementary particle is composite, what are the binding energies and the basis states?

This question is discussed in Chapters 4, 5, 8, and 9. It is really the question which forms the essence of the present studies. In contrast to the opinion of most workers in the field, we conclude from an examination of the experimental data that quark binding energies are small (less than 5%). Consequently, since the binding energies serve to delineate the basis states, we end up with a set of quark states (Chapters 5 and 8) which are considerably different from the conventional quarks of Gell-Mann and Zweig. We will often refer to this new set of quarks as <u>light-quark basis states</u>.

Question (3). Does the muon fit in with these basis states?

This question is discussed in Chapter 7. The answer, spectroscopically speaking, is that the muon does fit in with these light quark basis states. This result is important because the muon, as the lightest of the "heavy particles" (those particles which are more massive than the electron), is of undeniable importance in the scheme of things, and yet present-day theories of elementary particles have no explanation for its existence.

Question (4). Does the relativistic mass increase that is observed in translational motion apply also to rotational motion?

This question is discussed in Chapter 6. The present-day elementary particle theorists to whom the writer has talked personally, and this includes a great many people, believe that it does not apply, at least in any way that is important to them. However, classical textbooks on the subject indicate that it does apply, both from the viewpoint of special relativity and from that of general relativity, and the experimental data substantiate this viewpoint. Phenomenologically, this relativistic result is indispensable for the formulation of the light-quark model: it gives correctly-calculated values for spectroscopic quantities such as spins and magnetic moments; it relates spinning and nonspinning basis states; and it brings the muon into the mass spectrum of the hadron resonances.

Question (5). Does the spin of an elementary particle arise from the rotation of an extended mass distribution?

Question (6). Does the magnetic moment of an elementary particle arise from the rotation of an extended charge distribution?

These two questions, which are closely related to one another, are also closely related to Question (4). By taking the simplest model we can devise for an elementary particle – a rotating sphere – , and by applying the equations of relativity to the rotational motion, we obtain the spin angular momentum and the magnetic moment as directly calculated quantities (Chapters 6, 7, 18 and 19). The main point of interest here to the reader is that these are original results; this simplest of all models does not appear to have been discussed anywhere in the published literature on elementary particles. And it is a result that we were led to from a study of the experimental data.

Question (7). Does the concept of orbital angular momentum apply to an elementary particle?

This question is discussed in Chapters 11 - 13. Many elementary particle physicists feel that orbital angular momentum as applied to elementary particles is not a particularly well-defined concept. However, by studying mass and lifetime systematics with the aid of the light-quark basis states, we are able to decompose the total angular momentum of a hadron resonance into its spin and orbital components.

Question (8). Does the baryon quantum number have a physical interpretation?

In the light-quark model, the answer is yes. Each baryon resonance is characterized by the presence of an invariant spinor triplet $S\bar{S}S$ (see Chapters 8 and 15), which carries the baryon quantum number B, and which does not appear in meson resonances (see Chapters 16 and 17), except in the form of $S\bar{S}S \cdot \bar{S}S\bar{S}$ pairs in some of the high-mass New Particle $N\bar{N}$ resonances.

Question (9). Does the strangeness quantum number have a physical interpretation?

In the light-quark model, the answer again is yes. Strangeness is related to the particle-antiparticle symmetry or asymmetry of the light-quark substates, and it is thereby related in an intriguing and useful manner to the binding energies of these substates: strange substates and matching antistates have 4% binding energies, whereas nonstrange excitations or substates have $\sim$0% binding energies.

We now go on to Chapter 3, where the first of these nine questions is considered in more detail.

References for Chapter 2.

The quotation by Bertrand Russell at the beginning of Sec. 2.1 is taken from Arthur Koestler, The Act of Creation, Macmillan, London (1971), page 251. The quotation by Neils Bohr at the beginning of Sec. 2.2 is from André Mercier in Foundations of Physics, Plenum Publishing Corporation, New York, Volume I, page 285 (1971). References to the quark model of Gell-Mann and Zweig are given at the end of Chapter 4.

CHAPTER 3. IS THE "ELEMENTARY PARTICLE" ELEMENTARY OR COMPOSITE?

"The distinguishing characteristic of these new lectures ... is the unequivocal adoption of nuclear democracy as a guiding principle. ... It is too soon to say with absolute certainty that there are no aristocrats among nuclear particles ... . My standpoint here, however, is that every nuclear particle should receive equal treatment under the law."

G. F. Chew

"The striking successes of the quark model, in predicting the static properties of the elementary particles and many of the dynamical features of their interactions, suggests that the possibility of the existence of quarks must be given very serious consideration."

B. T. Feld

In this chapter we ask if the elementary particle is elementary or composite; and in the next chapter, having concluded that it is composite, we inquire as to the binding energies of the substates. The answers that we obtain to these two questions, when weighed against the background of contemporary thought, are sharply contrasting: we argue in the present chapter for the composite nature of the elementary particle, which is a result that is in agreement with the notions of the majority of the present-day elementary particle physicists; however, we argue in the next chapter that the fundamental elementary particle building blocks (quarks) have very small binding energies (less than 5%), which is in marked disagreement with the notions of most particle physicists.

Is the so-called "elementary particle" really elementary, or is it in fact a composite structure made up of an even more elementary set of "elementary particle building blocks"? In particular, the "elementary particle zoo" now contains almost two hundred different particles, whose properties are described in Appendix B. Is each of these particles a separate "fundamental" entity, or are these hundred-odd particles formed as various combinations of a small subset of basis states? If we could knock elementary particles apart into their constituents, then we would of course answer this question directly. However, the most determined efforts of physicists have thus far failed to break apart an elementary particle. Elementary particles do decay into other elementary particles, but they do so at their own predetermined rates, and in a manner that does not reveal their internal substructure (except in some of the recently-discovered New Particle decays).

In discussing the question of the "elementality" of the elementary particle, the situation that occurs in the area of nuclear physics serves as a useful example. In nuclear physics, which is the study of the atomic nucleus, more than 1100 different atomic nuclei have been identified experimentally. All of these nuclei are composed of just two building blocks — the proton and the neutron. Since protons and neutrons have small binding energies in nuclei (less than 1%), the protons and neutrons preserve their essential single-particle identities even when bound together. Thus the atomic nucleus is clearly a composite structure, with protons and neutrons constituting the fundamental nuclear building blocks. In addition to the 1100 or so standard atomic nuclei, short-lived "hypernuclei" can also be formed by adding other building blocks, such as lambda particles, to a nucleus, but these hypernuclei clearly are composite structures of the same type as we have just described. We will have more to say about the subject of nuclear physics in Chapter 20.

One feature of atomic nuclei which reveals their composite nature is that they can be broken apart by an impact which exceeds the binding energy. However, in the case of elementary particles we have not been able to break apart a proton, or a neutron, or a lambda hyperon, no matter how large an impact we apply. Thus we have not been able to observe their substates, if any, directly. As a consequence, contemporary theoretical elementary particle physicists have formed into two camps — those who believe that the elementary particle is indeed "elementary", and those who believe that the hundred-odd elementary particle states are formed as various combinations of a set of quark substates. The physicists in the former camp are the dispersion relation theorists, who insist that the essential reality of an elementary particle is its manifestation as a mathematical singularity in a complex energy-momentum plane. In this viewpoint, if there are a hundred elementary particles, then there are a hundred singularities, and all of these singularities are on an equal footing. In fact, only by studying the entire assemblage of the hundred singularities can we learn all that there is to learn about any one of them. This is the concept that is known popularly as "elementary particle democracy". The dispersion relation theorists argue that the elementary particle is indeed elementary.

At the opposite end of the theoretical spectrum are the quark model theorists. These workers believe that the elementary particles are formed as combinations of quarks and antiquarks. The original quark model of Gell-Mann and Zweig featured only three strongly-interacting quarks, so that this early quark model was an economical way in which to represent elementary particle states. However, as the number of observed particles increased, the number of required quarks also increased, and now as many as 18 different quarks are postulated in order to accurately account for the observed spectrum of elementary particles. Thus much of the simplicity of the original quark hypothesis has been lost.

As we stated at the beginning of this discussion, the question of the composite nature of the elementary particle would have been resolved if we had been able to break elementary particles apart into their basic constituents. However, the fact that we have not been able to accomplish this doesn't necessarily mean that the basic constituents don't exist. Quarks may have some kind of special property which keeps them bound together in a particle even during violent collisions with other particles, and papers on the problem of quark confinement now abound in the literature. Resorting once again to Bertrand Russell's dictum at the beginning of Chapter 2, we can test for the compositeness of the elementary particle by seeing whether dispersion relations or quark models yield the most useful mathematical expressions for reproducing the properties of the elementary particle. A voluminous contemporary literature, which we will not discuss here, exists on this subject. The unexpected appearance of the New Particle resonances (Chapter 17) has probably tipped the scales in the direction of the quark modelists, since the dispersion relation approach has not been applied with much success to these high-mass narrow-width resonances.

In the present studies, we assume as an ansatz that the elementary particle is composite, and we use the experimental data as our guide in unraveling the mystery of the unseen basis states. What we find is in some respects the simplest of all quark models; we find that the mass quantum M = 70 MeV is the building block for all particles more massive than the electron. However, the quantum M appears in four distinctive spinless basis state forms (1M, 3M, 4M, and 7M) and in two relativistically-spinning forms (as the muon: $\mu \equiv (M)_{spinning}$; and as the spinor S: $S \equiv (3M)_{spinning}$), and some of these basis states occur with different combinations of particle and antiparticle substates, so that this new quark model in fact has considerable complexity. Before discussing this light-quark model in any detail, we first discuss the crucial step which led to the formulation of the model — namely, the conclusion that we reach about quark binding energies, which is the subject of the next chapter.

References for Chapter 3.

The quotation by G. F. Chew at the beginning of the chapter is from the book Strong-Interaction Physics by M. Jacob and G. F. Chew, W. A. Benjamin, Inc., publishers, New York (1964), page 104. The quotation by B. T. Feld at the beginning of the chapter is from his book Models of Elementary Particles, Blaisdell Publishing Company, Waltham (1969), page 361. References to the quark model of Gell-Mann and Zweig are given at the end of Chapter 4. The proliferation of quarks in recent times is discussed, for example, by J. D. Bjorken in the SLAC report SLAC-PUB-1668, which is a symposium summary of the International Symposium on Lepton and Photon Interactions, Stanford University, Stanford, California, August, 1975. For an interesting review of contemporary quark models, see Y. Nambu, "The Confinement of Quarks", Scientific American, November, 1976, page 48. Also see D. B. Lichtenberg, J. G. Willis, and J. T. Kiehl, Physical Review Letters 39, 1592 (1977), "Interpretation of the $\Upsilon(9.5)$ as Evidence for Another Quark".

CHAPTER 4. WHAT IS THE BINDING ENERGY OF THE QUARK?

"The agreement of quark predictions with experiment could be described as spectacular, and yet virtually every prediction really makes sense only if quarks are not too heavy or if the binding is not too strong."

Jerrold Franklin
Physical Review, Vol. 172, p. 1816 (1968)

Suppose we assume, as we discussed in Chapter 3, that the almost two hundred elementary particles which have been observed to date (Appendix B) are composite structures formed as combinations of a basis set of quark substates. What then are the binding energies that hold these quarks together? This is really the decisive question to answer, for the binding energies determine the range of possible quark masses.

It turns out, as we discuss later on in this chapter, that there is a straightforward empirical test which helps to pin down the binding energy of the quark. However, before discussing this test, it is useful to consider binding energies in other areas of physics, and also to summarize some of the current ideas about quark binding energies and the ramifications that are associated with these ideas.

In the domain of atomic physics, where electrons are bound to a massive central core, electron binding energies amount to only a tiny fraction of the total energy of the system. In the domain of nuclear physics, where protons and neutrons are bound together to form the massive central core of the atom, binding energies are much larger and are measured in millions of electron volts (MeV). But these nuclear binding energies still amount in all cases to less than 1% of the total energy of the system. Now if we assume that protons and neutrons are each formed of three quarks, as seems phenomenologically correct, then what shall we assume for the binding energies of these quarks? In order to fully appreciate the difficulties that are inherent in this question, we must consider not only the nucleon, but also the pi meson, the lightest strongly-interacting particle, and we must consider possible forms that the quark model can take.

One of the earliest forms of the quark model was set forth by Fermi and Yang[1] in 1949, long before the name "quark" was applied to these models. Fermi and Yang assumed that the nucleon is the prototype for the quark, so that the quark Q and the nucleon N both have the same quantum numbers. They then constructed the nucleon N and the antinucleon $\bar{N}$ as the quark-triplet bound states

$$N = Q\bar{Q}Q \qquad \text{and} \qquad \bar{N} = \bar{Q}Q\bar{Q} \ , \tag{4.1}$$

and they constructed the pi meson as the quark-antiquark bound state

$$\pi = Q\bar{Q} \ . \tag{4.2}$$

Now the mass of the nucleon N is 939 MeV[†] (which in the Fermi-Yang model can be taken for the sake of argument as the mass of the quark Q), and the mass of the charged pion is 140 MeV. Thus if we picture the pion as being a $Q\bar{Q}$ bound state, then the quark-antiquark binding energy in the pion is 1878 MeV - 140 MeV = 1738 MeV, or 93% of the total mass! With such a strong binding energy, the theoretical calculation of quark bound-state masses becomes a practical impossibility.

In the later quark model of Gell-Mann and Zweig[2], the pion is again constructed as the quark-antiquark bound state

$$\pi = Q\bar{Q} \ , \tag{4.2}$$

but the nucleon and antinucleon are now constructed as the quark-triplet bound states

$$N = QQQ \qquad \text{and} \qquad \bar{N} = \bar{Q}\bar{Q}\bar{Q} \ . \tag{4.3}$$

This formulation leads to fractional baryon quantum numbers for the quark Q, but it has the useful consequence for theorists that it accounts for the non-appearance of certain SU(3) groups of particles. In the Gell-Mann and Zweig model, the quark masses are not specified. However, in order to reproduce the mass of the nucleon, the nucleon quark Q must have a mass that is at least 1/3 of the nucleon mass, or 313 MeV. Since the pion is formed with quarks of this same mass, the $Q\bar{Q}$ binding energy in the pion must be at least 616 MeV - 140 MeV = 476 MeV, or 77% of the total mass. This large a binding energy again leads to an intractable theory from the standpoint of calculating absolute mass values, although the model can be used to deduce ratios of mass values. In a later version of the Gell-Mann and Zweig model, the quarks Q are supplemented by "gluons", which are field quanta that bind the quarks together. The

---

[†]As is mentioned in Appendix A, we use units in which the velocity of light c is set equal to unity, so that the terms <u>mass</u> and <u>energy</u> can be used interchangeably.

gluons can then be the principal carriers of the mass values, so that the "bare" quarks Q, in a field theory context, can have very small mass values. However, this formulation just passes the mass problem on from the quarks to the gluons. From our present phenomenological point of view, gluons, if they exist, should be regarded as additional quark states.

If we could observe quarks Q directly, then we would of course know their masses. But quarks have never been observed. One explanation that has been advanced for the nonobservability of single quarks is that they are very massive objects when unbound, so that high energies are required in order to unbind a pion or nucleon. However, experiments with very high energy accelerators have now shown that if this explanation is correct, then the quark mass must be at least 25 GeV. If this is true, then the binding energy of a quark-antiquark pair in a pion is at least 99.7% of the total mass! Since physicists do not really believe that quarks can be this massive, it now seems apparent that something other than the mass values is responsible for the fact that single quarks are not observed. This remains as an unsolved problem in elementary particle physics.

The problem of the quark masses leads to an important conceptual difficulty with respect to conventional quark models. Quark model theorists have implicity assumed that quark masses are large, and therefore quark binding energies are very strong. However, the phenomenological results which have been obtained by using the quark model indicate just the opposite. As the quotation by Franklin at the beginning of this chapter points out, the fact that quark quantum numbers and quark scattering amplitudes are additive suggests that quarks behave as more-or-less independent entities, which means that they must be weakly bound together. For example, in the Gell-Mann and Zweig quark model the quark magnetic moments are assumed to be additive. This additivity implies that anomalous magnetic moments must be small. But according to conventional notions, small anomalous magnetic moments are associated with small virtual-pion fields and hence with small quark binding energies. For strongly-interacting quarks that have large binding energies, anomalous magnetic moments will be large, which is in contradiction to the workings of the quark model. Now, it can be argued that quark charges retain accurate ratios under strong interactions because the charge is related to a conserved vector current; but this mechanism is not available for the quark magnetic moments. Thus the assumption of additivity for quark magnetic moments can only be interpreted as meaning that quarks have small binding energies and therefore small masses.

The same logical connection between additivity and small binding energies occurs with respect to high-energy quark scattering amplitudes, as was noted by B. T. Feld in his book Models of Elementary Particles[3]: "The success of the factorizability model (of quarks) seems to imply that, at least for scatterings involving appreciable momentum transfer, the individual quark-rearrangement probabilities (amplitudes) may be thought of as being statistically independent, so that a given reaction requires the coincidence of independent quark deflections through the appropriate angle for all of the quarks involved. In a very crude sense, this model would seem to imply relatively weak quark binding energies and, correspondingly, relatively small quark masses."

We thus have evidence from the workings of the quark model itself that quarks may in fact be relatively light objects which are held together by small binding energies. This of course raises the problem again as to why an elementary particle can't be broken apart into its basic quark constituents. But this is a problem that occurs even if quarks are assumed to be very massive, since we know that the proton is stable even at ultrahigh cosmic ray energies — energies which are large enough to exceed any reasonable assumption about quark binding energies. Thus the stability of the proton must reside in some property which is apart from our usual concept of binding energies. If this is true, then the stability of the proton imposes no restrictions *per se* on quark binding energies, and hence on quark masses.

From the standpoint of the Gell-Mann and Zweig quark model, the ratios of the quark masses constitute a problem. In its original form, the G-Z model features two nonstrange quarks, Q_u and Q_d, and one strange quark, Q_s. Gell-Mann and others[4] have argued that the strange-to-nonstrange quark mass ratio is $m_s/m_u \sim 25$ (where we have $m_u \simeq m_d$). However, other workers[5] place this mass ratio at about 5 - 7; and recent estimates[6], which are based on the assumption of weak quark binding energies, range as low as $m_s/m_u \simeq 1.3$. (In the present studies, we conclude that baryon strangeness quantum numbers are carried by spinless excitation quanta, as described in Chapter 9, so that we do not use the Q_u, Q_d, Q_s set of spinning quark basis states.)

Our discussion of quark masses and quark binding energies has thus far been on the level of theoretical speculation. However, as we mentioned at the beginning of this chapter, there is a direct test of the experimental data that can be made to resolve this matter. If the quark model is to make any sense, then the lowest-mass meson state, the pi meson, must be formed as a single quark-antiquark pair, $\pi = Q\bar{Q}$. This is the form in which the pion appears in both the Fermi-Yang[1] and Gell-Mann and Zweig[2] quark models (Eq. 4.2). If quark binding energies are weak, then each quark Q in the pion must have a mass which is approximately one half of the pion mass, or about 70 MeV. Thus the mass quantum $M \simeq 70$ MeV should be a basic building block for elementary particles; that is, we have the causal relationship

$$(\pi \equiv M\bar{M} = 140 \text{ MeV}) + (\text{weak binding energies}) \rightarrow (M = \bar{M} \simeq 70 \text{ MeV}). \qquad (4.4)$$

Hence the occurrence of small quark binding energies mandates the existence of the quantum M ≃ 70 MeV, and conversely.

The mass quantum M has never been observed directly, and it is not a postulated quantity in either the Fermi-Yang or Gell-Mann and Zweig quark models. However, as we describe in Chapter 5, there are abundant indications in the mass spectra of the elementary particles for the existence of mass intervals which are in units or multiples of 70 MeV. Furthermore, in Chapters 6 and 7 we show that the spinless mass quantum M = 70 MeV can be related to the spinning μ = 105 MeV muon by the equations of both special and general relativity. Thus, although the mass quantum M = 70 MeV has not been observed directly, it can be observed indirectly in the form of the muon μ. As an experimental indication of a close relationship between the pion and the muon, and hence also between the quantum M and the muon, we note that the decay mode of the charged pion is into a muon plus a neutrino:

$$\pi^{\pm} \rightarrow \mu^{\pm} + \nu , \qquad (4.5)$$

where $\pi^{\pm} \equiv M^{\pm}M^{o}$, and where the quantum M^{o} annihilates in the decay process, with the quantum $M^{\pm}$ being converted into a spinning muon $\mu^{\pm}$ as a spinning neutrino ν is simultaneously created (the $\mu^{\pm}$ and the ν are both spin 1/2 particles).

If we accept the fact that a mass quantum M ≃ 70 MeV does exist as a basic quark state for elementary particles, this implies that quark binding energies must be small — a few percent. But can we pinpoint these binding energies more precisely? It turns out that we can, and we briefly outline the results here so that they will serve as a guide for following the discussions in Chapters 5 and 8. The derivation of these results, which involves a definition of the light-quark set of basis states, is implicitly contained in the hadron systematics that we develop in the remainder of the book.

An examination of the way in which 70 MeV mass quanta M combine together in hadron resonances indicates that there are four basic spinless combinations:

$$\begin{array}{llll} 1 \equiv M \simeq 70 \text{ MeV}; & & 3 \equiv 3M \simeq 210 \text{ MeV}; \\ 4 \equiv 4M \simeq 280 \text{ MeV}; & & 7 \equiv 7M \simeq 490 \text{ MeV}. \end{array} \qquad (4.6)$$

Now the quantum M, as we have seen in Eq. (4.4), occurs in both particle (M) and antiparticle ($\bar{M}$) forms. Since the basis states 3, 4, and 7 in (4.6) each contain several subquanta M, there is some freedom as to the way in which the particle and antiparticle substates are assigned. Phenomenologically, the basis states 3 and 4 occur in two distinct configurations: the first is an <u>unmixed</u> configuration, which we write as

$$3 \equiv MMM, \quad \bar{3} \equiv \bar{M}\bar{M}\bar{M}, \quad 4 \equiv MMMM, \quad \bar{4} \equiv \bar{M}\bar{M}\bar{M}\bar{M}; \tag{4.7}$$

the second is a mixed configuration, which we write as

$$\dot{3} \equiv M\bar{M}M \text{ or } \bar{M}M\bar{M}, \quad \dot{4} \equiv M\bar{M}M\bar{M}. \tag{4.8}$$

It seems intuitively plausible that binding energies between matching unmixed basis states and antistates (e.g., $3\bar{3}$ or $4\bar{4}$) will be much stronger than binding energies which involve mixed basis states (e.g., $3\dot{3}$ or $\dot{3}\dot{3}$), and this turns out to be empirically correct. Phenomenologically, two distinct light-quark binding energies can be deduced from the experimental data — a 4% binding energy which occurs between matching state-antistate pairs, and an essentially 0% binding energy (with the mass values chosen here) which occurs for states that have mixed configurations. State-state pairs and antistate-antistate pairs do not bind together hadronically.

There is another striking empirical result which emerges from these basis state configurations. The unmixed basis states 3, $\bar{3}$, 4, and $\bar{4}$ in Eq. (4.7) are "strange"; that is, they carry the strangeness quantum numbers +1, -1, +1, and -1, respectively. The mixed basis states in Eq. (4.8), on the other hand, are "nonstrange"; they carry the strangeness quantum number zero. By combining this result with the results shown in the preceding paragraph, we can empirically relate strangeness quantum numbers to basis-state binding energies: strange basis states bind with 4% binding energies, and nonstrange basis states bind with ~0% binding energies. This is a relationship that can be clearly observed in the experimental hadron data. The systematics of hadron binding energies is discussed in detail in Appendix C.

The basis state 7 ≡ 7M in Eq. (4.6) is the K meson, which appears experimentally in three distinct forms — as the $K^{\pm}$, the K^o_L, and the K^o_S kaons; the K^o_L and K^o_S are long-lived and short-lived states of the neutral kaon (see Chapter 10), and K^o_L mesons can be transformed into K^o_S mesons by a scattering process which evidently rearranges the substates M.

The light-quark basis states 1, 3, 4, and 7 described above are all spinless. But a spin 1/2 basis state is also required so that we can reproduce the spin 1/2 nucleon. The extension to spinning basis states is made in Chapter 8, where it is seen that the binding energies for spinning quarks are similar to those for spinless quarks. The three quarks in the nucleon [see Eq. (4.1)] have 4% binding energies, so that each of these spinning substates (which we denote here as spinors S) has a mass of about 330 MeV. As was mentioned in the Introduction and Summary, the spinor S appears to be a relativistically-spinning form of the spinless quantum 3 ≡ 3M; that is,

$$S \equiv (3)_{\text{spinning}} . \tag{4.9}$$

In direct analogy to the spinless basis states 3, $\bar{3}$, and $\dot{3}$, the spinor S occurs in both unmixed (S and $\bar{S}$) and mixed ($\dot{S}$) configurations, which have 4% and ∿0% binding energies, respectively. Thus the 4% and 0% binding energy dichotomy occurs for all of the light-quark basis states — both spinless and spinning. Spinors S appear in baryon resonances in the form of unmixed Fermi-Yang triplets: e.g., $N = S\bar{S}S$ and $\bar{N} = \bar{S}S\bar{S}$; and they appear in meson resonances in the form of mixed spinor pairs: e.g., $\dot{S}\dot{S}$ and $\dot{S}\dot{S}\dot{S}\dot{S}$.

In answer to the question "What is the binding energy of the quark?", we conclude, as will be documented in detail in the remainder of this book, that quark binding energies are small, less than 5%, and that they in fact occur with two distinct values — B. E. ≃ 4% and B. E. ≃ 0%, as outlined above. In assessing the validity of this conclusion, there is an historical note that is of interest here. Atomic binding energies are very small, as we mentioned at the beginning of this chapter. Furthermore, nuclear binding energies, which are larger and are associated with the "strong interaction", are still always less that 1%. Hence if we postulate a quark binding energy of 4%, as we have just done, we are invoking a binding energy that is much stronger than binding energies which occur in any known area of physics. Now, a phenomenological approach is on a much firmer foundation if it is based on observed quantities rather than on merely-speculated quantities. Thus it becomes highly desirable that we find some direct evidence among the hadron resonances for a binding energy as large as 4%. At the time that this conclusion about quark binding energies was first published by the author, in the Coral Gables Proceedings of 1971,[7] there was no experimental evidence for a binding energy of this magnitude. However, the existence of a $\bar{p}n$ bound state resonance was subsequently reported in a Syracuse experiment,[8] and the binding energy of this $\bar{p}n$ bound state turned out to be 4.4 ± 0.1%! Thus we now have direct experimental evidence for a binding energy as large as 4%, and this experimental evidence appeared after the need for such a binding energy had already been deduced[7] from an examination of the spectrum of elementary particle resonances.

Two other recent experimental results which confirm the magnitude of quark binding energies are the "charmed quark" decays $D^{*+} \rightarrow D^0 + \pi^+$ and $D^{*0} \rightarrow D^0 + \pi^0$, which have binding energies of 5.7 ± 0.5 MeV and 7.7 ± 1.7 MeV, respectively.[9]

The quark models of Fermi and Yang and of Gell-Mann and Zweig are both based on sets of spinning quarks; neither model includes a spinless mass quantum M ≃ 70 MeV. Since the assumption of weak quark binding energies mandates the existence of the quantum M, the question of the binding energy of the quark turns into the question of the evidence that can be produced for the existence of the quantum M. This is the subject of the next chapter.

References for Chapter 4.

1. E. Fermi and C. N. Yang, Phys. Rev. 76, 1739 (1949).

2. M. Gell-Mann, Phys. Lett. 8, 214 (1964); G. Zweig, CERN Reports Nos. Th401 and Th412 (1964) (unpublished).

3. B. T. Feld, Models of Elementary Particles, Blaisdell, Waltham, Mass. (1969), Chapter 17, pages 439-440. This book contains an excellent account of the development of elementary particle quark models.

4. M. Gell-Mann, R. J. Oakes, and B. Renner, Phys. Rev. 175, 2195 (1968); S. P. De Alwis, Phys. Lett. 70B, 333 (1977).

5. H. Sazdjian and J. Stern, Nucl. Phys. B94, 163 (1975); M. D. Scadron and H. F. Jones, Phys. Rev. D10, 967 (1974).

6. D. B. Lichtenberg, J. G. Willis, and J. T. Kiehl, Phys. Rev. Lett. 39, 1592 (1977).

7. M. H. Mac Gregor, in Fundamental Interactions at High Energies, proceedings of the 1971 Coral Gables Conference, edited by M. Dal Cin, G. J. Iverson, and A. Perlmutter, Gordon and Breach, New York (1971), Vol. 3, pages 75-154. On page 96 of this article, the following comment is made: "In both the mesons and the baryons, there seem to be two types of resonant structure, one that occurs right on the mass shell (a resonance), and one that has about a 3% binding energy (a bound state)." In this comment, the mass shell binding energy corresponds to the ∿0% binding energy described in the present discussion; and the 3% binding energy corresponds, with a slightly different choice for the masses M^0 and $M^{\pm}$, to the 4% binding energy of the present discussion. The Coral Gables paper was also published in Nuovo Cimento 8A, 235 (1972).

8. L. Gray, P. Hagerty, and T. Kalogeropoulos, Phys. Rev. Lett. 26, 1491 (1971). For an early discussion of this Syracuse experiment, see M. H. Mac Gregor, Lett. Nuovo Cimento 3, 197 (1972).

9. G. Feldman et al., Phys. Rev. Lett. 38, 1313 (1977); I. Peruzzi et al., Phys. Rev. Lett. 39, 130 (1977).

CHAPTER 5. EVIDENCE FOR THE FUNDAMENTAL SPINLESS MASS QUANTUM M = 70 MEV

This is the pivotal chapter in the book, both because of the clarity and comprehensiveness of the experimental results presented here, and because of the theoretical implications of these results. If the existence of the 70 MeV mass quantum M can be established, then the later results in the book follow as a logical consequence; but if the evidence presented here, plus the corollary evidence in Chapters 6 through 9, is not accepted as conclusive, then the later results will not be conclusive either.

5.1 Where Shall we Look for Evidence?

In the present book, the main point we are attempting to establish is that elementary particles are formed from a subset of quark states which are held together with small (less than 5%) binding energies. One unequivocal result which follows from this conclusion is the fact that the mass quantum $M \simeq 70$ MeV must be one of these quark states, so that we can form the 140 MeV pi meson as the quark-antiquark bound state pair $\pi = M\bar{M}$. Unfortunately, the quantum M, like all of the other postulated quark states, has never been observed directly. (As we discussed in the last two chapters, it seems to be impossible to smash an elementary particle apart into its constituent quarks.) However, even though we have no direct evidence for the existence of the mass quantum M, there are two ways in which we can search for indirect evidence that it does exist: (1) we can examine the spectrum of elementary particle masses to see if these occur as levels which are spaced by multiples of 70 MeV; (2) we can determine if the spinless (as we will show) quantum M also occurs in a spinning configuration which can be observed directly. In Chapters 6 and 7 we discuss the second of these two possibilities, and we demonstrate that the spinless 70 MeV quantum M is related by the equations of both special and general relativity to the spinning 105 MeV muon. (The muon interacts only weakly with other particles, but it is observed as a prominent final-state particle in many hadronic interactions.) In the present chapter we discuss the first of these two possibilities, and we demonstrate that the spectrum of narrow-width elementary particle resonances, when arranged in the form of energy level diagrams, abounds with intervals that are accurate multiples of 70 MeV. These intervals can be observed in the low-mass meson resonances, in the baryon resonances, and, interestingly, in the high-mass New Particle resonances.

5.2 The Spinless Meson Resonances

In the task of unraveling the mystery of the elementary particle spectrum, the most useful states to study are: (A) the lowest-mass states; (B) the lowest-spin states; and (C) the most stable (longest-lived) states. The lowest-mass hadronic (strongly-interacting) particles are the $\pi(\sim 140$ MeV), K($\sim$494 MeV), and η(549 MeV) mesons. All three of these particles are spinless, as is the somewhat-higher-mass η'(958 MeV) meson, and probably also its closely-related partner, the M(953 MeV) meson.[†] These particles are quite stable on a hadronic time scale, and their half-lives form an interesting pattern in powers of α, as is discussed in Chapter 10. Thus the π, K, η, and η' mesons fulfill all three of the conditions (A), (B), and (C) described above, and they are in fact the only particles which do. Hence these are undoubtedly the most important resonances to study from the standpoint of deducing the (unseen) basis states from which they are constructed.

Let us now construct an energy level diagram (which in the present book translates into a mass diagram) for these resonances. This diagram is shown in Fig. 5.1, where the π, K, η, M, and η' resonances are arranged in two columns, corresponding to nonstrange (S = 0) and strange (S = ±1) excitations, and where the mass scale that is used is a linear scale plotted in units of 137 MeV.

What can we learn from Fig. 5.1? We can learn a number of things from this figure, which is probably the most important figure in the entire book. First of all, note the following mass relationships: (1) the 137 MeV mass unit is at about the midpoint of the π^0 and $\pi^{\pm}$ masses; (2) the η mass is almost exactly 4 x 137 MeV; and (3) the η' mass is almost exactly 7 x 137 MeV. Since these are the only spinless, low-mass, long-lived nonstrange meson resonances which have been observed, this coincidence in the mass intervals cannot be regarded as accidental. The systematics shown in Fig. 5.1 tell us that the mass intervals for these resonances are linear (to a mass accuracy of better than 0.2%), and that the mass quantum M $\simeq$ 137 MeV is in some sense a basis state for these particles.

For the second point, note that the strange K meson states occur just a little above 3 1/2 x 137 MeV. Since the η' meson appears at 7 x 137 MeV, this suggests that we can form the η' resonance as a $K\bar{K}$ bound state. If we therefore set $\eta' \equiv K\bar{K}$, so that 958 MeV = 494 MeV + 494 MeV - B. E., we obtain a $K\bar{K}$ binding energy B. E. = 30 MeV, which is the $\sim$4% binding energy that we have stated occurs between strange basis state

---

[†] The experimental properties of the particles used in the discussions in this book are summarized in Appendix B. In general, we will not supply references to these particles in the text. Most of the experimental data used here are taken directly from the compilation of the Particle Data Group.

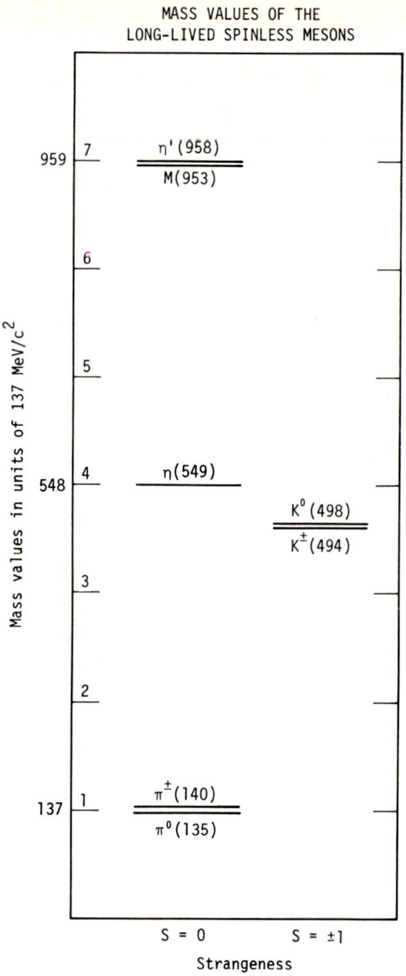

Figure 5.1. An energy level diagram for the metastable spin zero meson and kaon resonances. Since these are the lowest-mass hadronic states, they are the most important states to study from the standpoint of assessing theoretical models. Note the even spacings of the strangeness S = 0 meson masses in units of 137 MeV/c^2. The fact that the S = ±1 kaon masses are just over half the value of the η' and M meson masses suggests that the η' and M mesons are $K\bar{K}$ bound states.

and antistate pairs (see the discussion at the end of Chapter 4). Furthermore, as we show in Chapter 10, the ratio of the η' lifetime to the $K^{\pm}$ lifetime is α^6, where $\alpha = e^2/\hbar c$ is the fine structure constant; and, as we also show in Chapter 10, related resonances are characteristically separated in lifetimes by ratios of α^6. Thus it makes sense from more than one viewpoint to regard the η' resonance as a $K\bar{K}$ bound state.† Conversely, the K meson can be thought of as "one half of an η' meson", and its strangeness value $S = \pm 1$ comes from its internal particle-antiparticle asymmetry.

The third point to be noted in connection with Fig. 5.1 is that although the mass quantum M = 137 MeV appears to be a useful basis state for the nonstrange meson resonances, it clearly does not fit the K mesons, which have mass values that are about 3 1/2 times 137 MeV. Thus, in order to reproduce the kaon masses, we must choose a smaller basis state, namely the mass quantum $M \simeq 70$ MeV. To show that no further subdivision of the basis state masses is necessary, we now demonstrate that a very simple model based on the quantum M = 70 MeV accurately reproduces the mass values of all of the meson states contained in Fig. 5.1. (This calculation, although rather disarming in its simplicity, represents a vast improvement over SU(3)-type mass calculations, and it is one of the most important results to emerge from the present studies; we will extend these results in Chapter 9 to include the fundamental low-mass baryon and hyperon states.) In order to accurately reproduce the $\pi^0(135)$ and $\pi^{\pm}(139.6)$ mass values, we select the basis-state masses to be

$$M^0 = \bar{M}^0 = 70.0 \text{ MeV}$$

and (5.1)

$$M^+ = M^- = 74.6 \text{ MeV},$$

where we assume integral electric charges; and we postulate the following hadronic binding energy:

$$\text{B. E.} \qquad M\bar{M} = -5.0 \text{ MeV}, \qquad (5.2)$$

which is comparable percentagewise to the 4% binding energy that is observed experimentally for the $\bar{p}n$ bound state. We then construct the π mesons as the $M\bar{M}$ basis state

---

†Other examples can be shown where basis states bind together in this manner. For example, if we bind a strangeness S = -1 K^- meson to a nonstrange proton with this same binding energy of 30 MeV (see Chapter 9), we obtain a $\bar{K}p$ bound-state resonance that has a strangeness value S = -1 and a calculated mass of 938 + 494 - 30 = 1402 MeV; this state corresponds precisely to the $\Lambda(1402)$ S-state hyperon resonance.

configurations

$$\pi^0 = M^0\bar{M}^0 = 70.0 + 70.0 - 5.0 = 135.0 \text{ MeV},$$
$$\pi^+ = M^+\bar{M}^0 = 74.6 + 70.0 - 5.0 = 139.6 \text{ MeV}, \qquad (5.3)$$
$$\pi^- = M^0M^- = 70.0 + 74.6 - 5.0 = 139.6 \text{ MeV},$$

which, by this construction, reproduce the exact masses for these particles. Next we form the basis states

$$4^{\pm} \equiv (4M)^{\pm} = 284.6 \text{ MeV}$$

and (5.4)

$$7^{\pm} \equiv (7M)^{\pm} = 494.6 \text{ MeV},$$

where the masses of the subquanta M in these basis states add linearly. Then, constructing the η and η' resonances as the basis state pairs $\eta = 4^+4^-$ and $\eta' = 7^+7^-$, and using the same binding energy as above [Eq. (5.2)], we obtain the mass values

$$\eta = 4^+4^- = 284.6 + 284.6 - 20.0 = 549.2 \text{ MeV} \qquad (5.5)$$

and

$$\eta' = 7^+7^- = 494.6 + 494.6 - 35.0 = 954.2 \text{ MeV}. \qquad (5.6)$$

If we now subtract about 1 MeV from each of these mass values to account for the coulomb attraction between the internal + and - charges (see Chapter 19), we obtain

$$\eta \simeq 548 \text{ MeV} \qquad (5.5')$$

and

$$\eta' \simeq 953 \text{ MeV} \qquad (5.6')$$

as the coulomb-corrected mass values for these particles. The calculated η mass is within 1 MeV of the actual η mass (η = 548.8 ± 0.6 MeV), but the calculated η' mass matches the mass of the M meson (M = 953 ± 2 MeV) rather than the mass of the η' meson (η' = 957.6 ± 0.3 MeV). We will have more to say about the M and η' mesons in point four below. To conclude these mass calculations, we set

$$K^{\pm} = 7^{\pm} = 494.6 \text{ MeV} \quad \text{and} \quad K^0,\bar{K}^0 = 7^{+-} = 499.2 \text{ MeV}, \qquad (5.7)$$

where the K^0 and $\bar{K}^0$ mesons each contain two internal charges, which makes them more massive than the K^+ and K^- mesons. (One of the continuing problems with the quark model of Gell-Mann and Zweig has been the fact that it clearly requires the $K^{\pm}$ mesons to be heavier than the $K^0,\bar{K}^0$ mesons, in contradiction with experiment.) The measured kaon masses are $K^{\pm}$ = 493.7 MeV and $K^0,\bar{K}^0$ = 497.7 MeV, so that the calculated kaon mass values are within 1 or 2 MeV of the actual kaon mass values. We will have more to say about these kaon mass values in point five below.

We can see from these mass calculations that by using the $M^{\pm}$ and M^{o} mass values of Eq. (5.1) together with the binding energy of Eq. (5.2), we not only obtain an exact fit to the π meson masses, but we also obtain very accurate fits to all of the other meson masses in Fig. 5.1. If we can use a 70 MeV building block and thereby reproduce the mass values of Fig. 5.1 to an accuracy of one or two MeV, then this building block must have physical significance. The key point to note in these calculations is that the nonstrange π, η, and η' mesons, which involve matching particle and antiparticle substates, have 4% binding energies, whereas the strange K mesons, which each consist of one unpaired substate, have zero binding energies; this may seem to be a small point, but it is the essential ingredient that makes these calculations successful. The numerical results that we obtain for the masses of these resonances are better than an order-of-magnitude improvement over any mass calculations that have emerged from SU(3)-type theories. Moreover, there are further ramifications which emerge from these calculations, as we now discuss.

The fourth point to be noted in connection with Fig. 5.1 has to do with binding energies. In Eq. (5.2) we defined the binding energy of an $M\bar{M}$ pair to be -5 MeV. This is almost a 4% binding energy, and it is in general agreement with the observed 4% binding energy of the $\bar{p}n(1795)$ bound state that was discussed at the end of the last chapter. Do we have any other evidence for this 5 MeV $M\bar{M}$ binding energy? Yes, we do; and it comes from the 5 MeV spacing of the M(953) and $\eta'(958)$ levels in Fig. 5.1. As we can see in Fig. 5.1, the π^{o} and $\pi^{\pm}$ levels and the $K^{\pm}$ and K^{o} levels are also separated by about 5 MeV, but these latter separations can be attributed to coulomb effects — that is, to the coulomb self-energy differences between the masses M^{o} = 70.0 MeV and $M^{\pm}$ = 74.6 MeV, as shown in the mass calculations of Eqs. (5.3) and (5.7). But the M and η' masses are both neutral states, and thus do not have this coulomb energy difference. How then can we account for their 5 MeV separation in mass values? The clue to the answer to this question is provided by a study of the production and decay modes of the M and η' mesons. The M meson is seen in reactions of the type

$$A + B \rightarrow A + B + M , \tag{5.8}$$

in which the initial-state particles A and B also appear in the final state. The M meson thus represents the straight conversion of kinetic energy into hadronic matter, so that the M necessarily has perfect internal particle-antiparticle symmetry. The η' meson, on the other hand, is produced in reactions of the type

$$K^{-} + p \rightarrow \Lambda + \eta' , \tag{5.9}$$

in which the initial-state particles merge and then separate to form two different final-state particles. There is nothing in this latter reaction which requires the η' to have perfect internal symmetry. This line of reasoning carries over to the decay modes, which in fact form the main experimental basis for separating the M meson from the η' meson. The M meson has the symmetric one-step decay mode

$$M \rightarrow \pi\pi\gamma \quad (\pi\pi \text{ not in } \rho \text{ band}) , \tag{5.10a}$$

whereas the η' has the asymmetric two-step decay mode

$$\eta' \rightarrow \rho\gamma , \quad \rho \rightarrow \pi\pi . \tag{5.10b}$$

(The rho meson, ρ, is discussed in Chapter 13.) Since the M and η' mesons are each formed from 14 subquanta M, and since each π meson is formed from 2 subquanta M, the decay mode shown in (5.10a) represents the annihilation of 10 subquanta M, whereas the decay mode shown in (5.10b) represents the annihilation of 5 subquanta M followed by the subsequent annihilation of 5 more subquanta M [see Eq. (14.22) in Chapter 14]. Now, in the mass calculations of Eqs. (5.6) and (5.6') we used a binding energy of -35 MeV, which assumes perfect internal particle-antiparticle symmetry, and we ended up with a mass value of 953 MeV, which is the experimental mass of the M meson. This mass value, when taken together with the above systematics of the production and decay modes, suggests that the M meson, with perfect internal symmetry, has a binding energy of -35 MeV; whereas the η' meson, with a somewhat scrambled internal symmetry, has one unmatched internal MM pair, and hence a binding energy of -30 MeV. Thus the experimental difference of 5 MeV between the masses of the M and η' mesons becomes a measure of the binding energy of an $M\bar{M}$ pair.

The <u>fifth</u> point we can make in connection with Fig. 5.1 has to do with the $K^0 - K^\pm$ mass difference. The calculation of Eq. (5.7) gave the absolute K^0 and $K^\pm$ masses quite accurately, but it led to a $K^0 - K^\pm$ mass difference of 4.6 MeV, rather than to the experimental value of 4.0 MeV. However, the K^0 and $\bar{K}^0$ mesons in Eq. (5.7) each contain two internal charges, + and - , and we have not yet included the coulomb attraction between these charges, which will lower the calculated masses of the K^0 and $\bar{K}^0$. In order to match the experimental $K^0 - K^\pm$ mass difference of 4.0 MeV, the + and - charges in the K^0 must have an attractive coulomb interactive energy of 0.6 MeV, which means that they must be separated from one another by a distance of a little over 2 fermis (2×10^{-13} cm). This large charge separation distance indicates that the kaon is a linear structure, with the seven subquanta M in the kaon in the form of a linear array. We go into this subject in more detail in Chapter 18, where the geometry of the elementary particle is discussed.

The <u>sixth</u> and final point we wish to make in connection with Fig. 5.1 has to do with the quantum numbers of the mass quantum M = 70 MeV. These can be deduced as follows: (1) Since the spinless kaon contains seven subquanta M [Eq. (5.7)], the quantum M must be spinless. (2) The charge states shown in Eq. (5.1) indicate that the quantum M has isotopic spin I = 1/2. (3) The fact that the kaon, with an odd number of subquanta M, is strange, whereas the π, η, and η' mesons, with an even number of subquanta M, are nonstrange, suggests that a single quantum M is strange; this is a result that seems to be borne out in some of the higher-mass New Particle resonances. (4) An argument similar to that of point (3) suggests that the quantum M may have negative intrinsic parity; furthermore, the $K_L^o \to K_S^o$ regeneration process, which involves a geometrical rearrangement of the basis states, and which changes the parity of the kaon, indicates that the parity of the quantum M may depend on its spatial orientation with respect to the other subquanta in the resonance. We can summarize these points by asserting that the quantum M = 70 MeV appears to have the same basic quantum numbers as the K meson.

Precision mass calculations for all of the hadron resonances are summarized in Chapter 19. But we can already see here that the calculations of Eqs. (5.1)-(5.7), when combined with the ramifications discussed above, are sufficient to give absolute mass values (not just mass ratios) for the particles of Fig. 5.1 that are accurate to within about 1 MeV. Furthermore, as we have tried to show by the above discussions, this approach affords a good deal of physical insight into other properties of these particles. The particle states of Fig. 5.1 contain a whole world of physics information.

One point that has not been emphasized in the above discussion is the manner in which Fig. 5.1 functions as an energy level diagram. This point is important when we go on to the systematics of the other particles. The energy level aspects of Fig. 5.1 are brought out more clearly in Fig. 5.2, which shows the level spacings and excitation quanta for these particles. If we think of the π meson,

$$\pi = 1\bar{1} \equiv M\bar{M} , \qquad (5.11)$$

as constituting a "ground state" resonance, then the η meson,

$$\eta = 4\bar{4} = 1\bar{1}3\bar{3} , \qquad (5.12)$$

can be thought of as a π meson plus a $3\bar{3}$ excitation quantum [see Eq. (4.7)]; and the η' meson,

$$\eta' = 7\bar{7} = 1\bar{1}3\bar{3}3\bar{3} , \qquad (5.13)$$

represents an additional $3\bar{3}$ excitation. The K meson can similarly be thought of, at least in the first approximation, as an unseen M = 1 "ground state" plus an unpaired

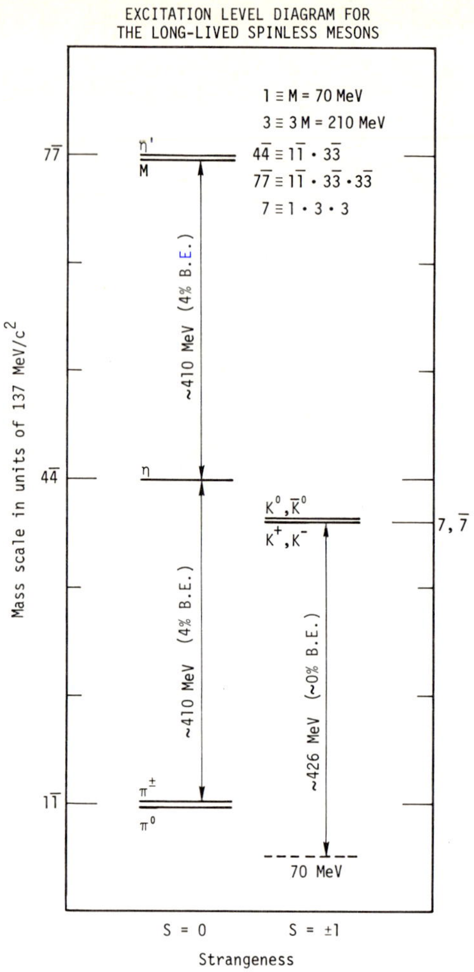

Figure 5.2. The energy intervals and excitation systematics for the meson and kaon levels of Fig. 5.1. The mass values of these levels are accurately reproduced by the light-quark basis states $1 \equiv M = 70$ MeV, $4 \equiv 4M$, and $7 \equiv 7M$, as shown along the figure ordinates [see Eqs. (5.1)-(5.7)]. The meson basis-state pairs $\pi = 1\bar{1}$, $\eta = 4\bar{4}$, and $\eta' = 7\bar{7}$ have 4% binding energies, whereas the unpaired kaon basis states $K = 7$ and $\bar{K} = \bar{7}$ have zero binding energy.

33 excitation:

$$K = 7 = 133 . \tag{5.14}$$

The 33 excitation interval of $\sim$426 MeV in the kaon (see Fig. 5.2) represents $\sim$0% binding energy, whereas the $3\bar{3}$ excitation intervals of $\sim$410 MeV in the η and η' represent 4% binding energy. The usefulness of this energy diagram approach will be more apparent in the succeeding sections, where spinless quanta M are combined with "ground state" resonances that contain spinors S. The spinless basis states that were deduced in the present section apply with equal accuracy to these more complicated resonances.

One of the strongest arguments we can advance for the reality of the quantum M = 70 MeV is the calculational agreement that we have demonstrated here for the meson resonances of Fig. 5.1. In this connection, it is important to contrast the present results with those obtained by using the Gell-Mann and Zweig quark model. The π meson, as the lowest-lying hadronic state, should be the state that is the most simply and accurately reproduced by any proposed model of hadronic particles. As we have seen above, the pion appears here as the basis-state configuration $\pi = M\bar{M}$, which is certainly the simplest configuration we can devise; and the pion mass values are precisely reproduced [Eq. (5.3)] by the same basis-state parameters [Eqs.(5.1)-(5.2)] which give precision fits to the other low-lying states [Eqs. (5.5)-(5.7)]. In the G-Z quark model, on the other hand, the π mesons have more complicated quark configurations than do the ρ mesons, which seems contrary to our intuitive expectation; furthermore, the assumption of weak quark binding energies that now seems to be emerging from the systematics of the high-mass New Particle states (see Fig. 5.5, for example) leads to G-Z quark binding energies in the pion (B.E. $\sim$ 80%) that are completely out of line with the G-Z quark binding energies in the other meson and baryon resonances (which are on the order of 10% or less). If the Gell-Mann and Zweig quark model contains an Achilles heel, it resides in the manner in which the pion, the most fundamental hadron resonance of all, is handled.

5.3 The Pion as a Fundamental Meson Building Block

In the preceding section we saw that the low-mass spinless meson and kaon resonances can be accurately reproduced as the basis-state combinations

$$\pi \equiv M\bar{M} , \tag{5.11}$$

$$\eta = 4\pi , \tag{5.12'}$$

$$\eta' = 7\pi \equiv K\bar{K} . \tag{5.13'}$$

Now the η meson is clearly more than just a combination of four pions; it is a resonance in its own right, with its own set of quantum numbers and interaction characteristics. Similar remarks also apply to the η' and K mesons. However, as we have seen in Fig. 5.1 and in Eqs. (5.1)-(5.7), the mass values of these resonances indicate that they are accurately formed as combinations of pion substates. Furthermore, we note the following principal decay modes:

$$\eta \equiv \pi\pi\pi\pi \rightarrow [\pi]\pi\pi\pi \rightarrow \pi + \pi + \pi , \qquad (5.15)$$

$$\eta' \equiv \eta\pi\pi\pi \rightarrow [\pi]\eta\pi\pi \rightarrow \eta + \pi + \pi , \qquad (5.16)$$

$$K \equiv M\pi\pi\pi \rightarrow [M]\pi\pi\pi \rightarrow \pi + \pi + \pi , \qquad (5.17)$$

where $[\pi]$ and $[M]$ denote the annihilation of the subquanta $\pi \simeq 140$ MeV and $M \simeq 70$ MeV, respectively. Thus the η, η', and K mesons behave in both their production modes (mass values) and decay modes as if they are constructed from linear combinations of $\pi \simeq 140$ MeV substates. We now demonstrate that many other meson resonances, including the high-mass New Particle states, also contain π mesons as substates. The reason for this is that meson resonances have balanced particle-antiparticle symmetry, so that meson excited states tend to contain equal numbers of M and $\bar{M}$ subquanta; thus the meson excitation intervals are characteristically in units or multiples of the excitation quantum $\pi = M\bar{M} \simeq 140$ MeV. Baryon resonances, on the other hand, do not have balanced particle-antiparticle symmetry; and, as we will see in the next section, baryon excited states appear with characteristic excitation intervals that are in multiples of the excitation quanta $\dot{3} = 3M \simeq 210$ MeV and $\dot{4} = 4M \simeq 280$ MeV.

The low-mass π, η, η', K, and $\bar{K}$ meson and kaon resonances discussed above are composed entirely of spinless quanta $M \simeq 70$ MeV and $\pi \simeq 140$ MeV. Hence we refer to them as "spinless" resonances. However, in addition to these spinless substates, we also require a spinning basis state, the spinor $S \simeq 330$ MeV, in order to be able to reproduce the nucleon, $N \equiv S\bar{S}S$. The properties of the spinor S were described briefly at the end of Chapter 4, and they are discussed in detail in Chapter 8. S occurs in baryon resonances in the form of spinor triplets $S\bar{S}S$, with $\sim$4% binding energies; and it occurs in meson resonances in the form of $\ddot{S}\bar{S}$ spinor pairs, with $\sim$0% binding energies. Narrow-width spin 1 (or spin 0) meson and kaon resonances can be formed by combining spinless excitation quanta with an $\ddot{S}\bar{S}$ spinor pair, so that we look among these resonances for additional evidence that the π meson is a fundamental basis state. (As we discuss in Part II of the book, broad-width resonances correspond to rotational excitations, and they follow a quite different mass quantization pattern.)

The lowest-mass narrow-width spin 1 meson that is readily-identifiable is the $\omega(783)\Gamma10$. Figure 5.3 shows this meson state, together with other narrow-width meson

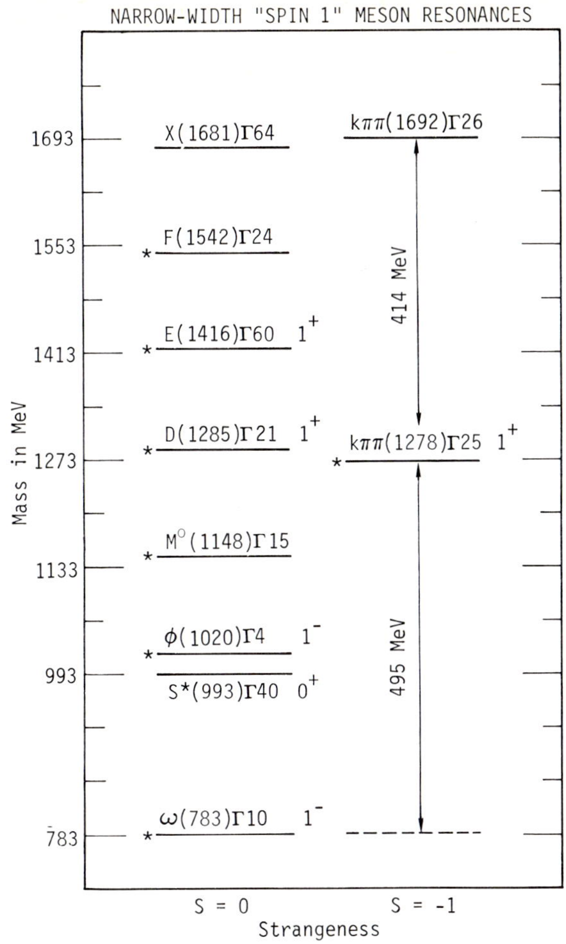

Figure 5.3. An energy level diagram for the narrow-width "spin 1" meson and kaon resonances, arranged as excited states above the ω(783) meson "ground state". Note the quantization of the levels in multiples of 70 and 140 MeV. The Kππ(1278) kaon level is just the mass of a K meson above the ω(783), and the Kππ(1692) - Kππ(1278) mass interval matches the excitation intervals shown in Fig. 5.2. The levels labeled with asterisks serve as bandheads for overlying rotational bands, as described in Chapter 16.

states that have been reported.† As can be seen in Fig. 5.3, the S*(993), ϕ(1020) pair of meson states occurs about 210 MeV above the ω, and then a series of five meson levels appear at 140 MeV intervals above the (S*,ϕ) pair, with at least some of these levels being identified as spin 1 mesons. Also, two narrow-width K$\pi\pi$ kaon resonances have been observed, and at mass values which match two of the meson levels: the K$\pi\pi$(1278) appears 495 MeV (the mass of a kaon K) above the ω(783); and the K$\pi\pi$-(1692) appears 414 MeV (compare this to the mass intervals in Fig. 5.2) above the K$\pi\pi$(1278). Thus we see the same kind of evidence for $\pi \simeq$ 140 MeV building blocks in these narrow-width "spin 1" mesons as we saw in the spinless π, η, η', and K mesons. The systematics of the meson and kaon states shown in Fig. 5.3 is discussed in detail in Chapters 13, 14, and 16. Also, we demonstrate in Sec. 8.3 of Chapter 8 that the ω meson basis-state configuration is $\omega \equiv \dot{S}\dot{S}\pi$, where $\dot{S}\dot{S}$ = 655 MeV is the spinor pair that (in a rotational mode) corresponds to the ρ meson; the decay modes for these particles are

$$\rho \equiv \dot{S}\dot{S} \rightarrow \pi\pi \qquad \text{and} \qquad \omega \equiv \dot{S}\dot{S}\pi \rightarrow \pi\pi\pi \,. \tag{5.18}$$

Thus the ω meson itself contains a $\pi \simeq$ 140 MeV substate, and the spinor pair $\dot{S}\dot{S}$ ultimately emerges as the fundamental "ground state" for the meson and kaon resonances of Fig. 5.3.

One difficulty in evaluating the significance of a figure such as Fig. 5.3 is that a few of the levels shown there have poorly-determined masses or widths, and even the existence of some of the levels is open to question (although we offer additional evidence for the reality of these levels in Chapter 16). However, when we move on to the high-mass, narrow-width meson and kaon resonances — the so-called New Particle states — , we discover an unmistakable and sharply-delineated confirmation of the reality of the $\pi \simeq$ 140 MeV meson building block. This confirmation comes both from the New Particle production modes (the resonance mass values), and from the New Particle decay modes (the final-state mass values), as we now demonstrate.

A striking example of 140 MeV New Particle mass intervals occurs right at the $\bar{p}n$(1795) threshold for New Particle production. Figure 5.4 shows the nonstrange $\bar{N}N$ resonances $\bar{p}n(1795)\Gamma\leq 8$ and $\bar{p}p(1935)\Gamma\leq 4$, plotted together with the strange "charmed"

---

†The mass values of the resonances shown in Fig. 5.3 are quite well determined, but measurements of the widths in many cases show a considerable spread in values. We have, where available, used the average widths as compiled by the Particle Data Group, but some of these resonances probably have narrower widths than the quoted averages indicate. We expect the width of a rotationless resonance to be in the range $\Gamma \simeq$ 10-40 MeV, with the observed width increasing somewhat as the mass of the resonance increases. (The parameters of the ϕ meson are discussed in Chap. 10, Ref. 7)

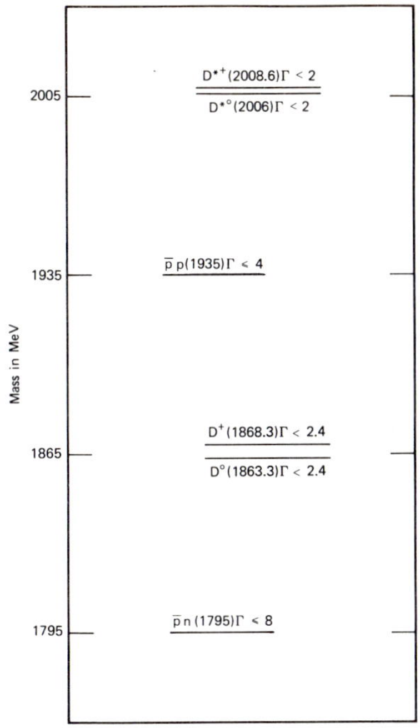

Figure 5.4. An energy level diagram for two narrow-width $\bar{N}N$ meson resonances and the "charmed" D and D* kaon resonances, arranged as excited states above the $\bar{p}n$(1795) "ground state". Note the accurate quantization of the levels in units of 70 MeV. The nonstrange $\bar{p}p$ and $\bar{p}n$ levels are separated by 140 MeV, and so are the strange D* and D levels. An extension of this figure to include higher-mass quantized levels is shown in Fig. 5.6.

resonances $D^{*+}(2008.6)\Gamma<2$, $D^{*o}(2006)\Gamma<2$, $D^{+}(1868.3)\Gamma<2.4$, and $D^{o}(1863.3)\Gamma<2.4$. As can be seen in Fig. 5.4, the $\bar{p}p$ level is just 140 MeV above the $\bar{p}n$ level, and the D* levels are just 140 MeV above the D levels. Also, the D and D* levels are offset with respect to the $\bar{p}n$ and $\bar{p}p$ levels by exactly 70 MeV. All of these resonances have narrow widths and accurately-known mass values, so that there is no question about the observed 70 MeV and 140 MeV quantizations of the levels. The fact that the D and D* resonances are strange whereas the $\bar{p}n$ and $\bar{p}p$ resonances are nonstrange is in agreement with the conclusion reached in point six of Sec. 5.2 that a single mass quantum M = 70 MeV is strange. Although the $\bar{p}p$(1935) level shown in Fig. 5.4, which is also denoted as the S(1935) meson, has only recently been confirmed as a bona fide resonance, its existence has long been indicated by the systematics of the present light-quark model.[1,2] Also, the $\bar{p}n$(1795) level, which is the $\bar{N}N$ "ground state" for the New Particles, was singled out as an important higher-mass meson state well before the discovery of the New Particles.[3]

An equally remarkable example of a meson quantization in units of 140 MeV is shown in the radiative decays of the New Particles. These are illustrated by the levels shown in Fig. 5.5, which include the D and D* "charmed" levels of Fig. 5.4 and the χ(3414), χ(3503), and χ(3551) "charmonium" levels that occur in the radiative decays of the ψ'(3684) New Particle resonance (see Chapter 17). As Fig. 5.5 shows,

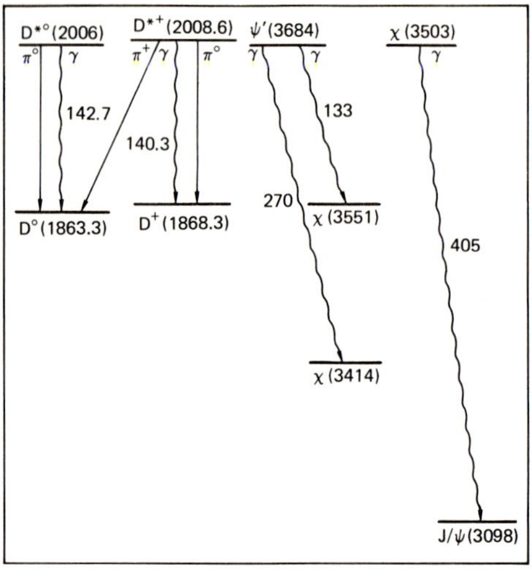

Figure 5.5. Pionic and radiative decays of the New Particles. The decays $D^* \rightarrow D + \pi$ shown in Fig. 5.5, which have very small Q-values, indicate clearly that the pion π is a substate of the D*. The decays $D^* \rightarrow D + \gamma$ reveal that the π substate in the D* can annihilate, with the emission of a γ ray. The decays $\psi' \rightarrow \chi + \gamma$ and the decay $\chi \rightarrow J/\psi + \gamma$ in Fig. 5.5, which are very accurately quantized in multiples of $\pi^0 = 135$ MeV [see Eq. (5.20)], show that internal pion annihilations also take place in the decays of these ψ' and χ New Particle states.

the D* kaons not only are situated just 140 MeV above the D kaons, but they actually decay into D kaons by emitting pions. These D* decays, which have measured Q-values of only a few MeV, furnish perhaps the most convincing evidence of all for the reality of the pion as a basic meson and kaon building block. In these decays, the pion is ejected as an intact unit, and the remainder of the quark structure is left as another intact unit. That is, we have the decay mode

$$D^* \equiv \bar{N}N \cdot M \cdot \pi \rightarrow \bar{N}N \cdot M + \pi \equiv D + \pi , \qquad (5.19)$$

where $\bar{N}N$ is the 1795 MeV $\bar{p}n$ bound state, M is the "strange" mass quantum M = 70 MeV, π = 140 MeV is the pion, and the masses are additive. We can see in Fig. 5.5 that the decay mode $D^* \rightarrow D + \gamma$ also exists, so that the pion substate can be annihilated instead of being ejected. This is a very important result, because it enables us to establish a direct link between the D* decays and the $\psi'(3684)$ radiative decays. When we compare the $D^* \to D + \gamma$ decays in Fig. 5.5 with the $\psi' \to \chi + \gamma$ and $\chi \to J/\psi + \gamma$ decays, we note the following numerical relationships:

$$
\begin{array}{llllll}
D^{*+}(2008.6) & \to [\pi] & + D^{+}(1868.3) + \gamma , & \Delta M = 140.3 \text{ MeV} ; & \\
D^{*0}(2006) & \to [\pi] & + D^{0}(1863.3) + \gamma , & \Delta M = 142.7 \text{ MeV} ; & \\
\psi'(3684) & \to [\pi] & + \chi(3551) \quad + \gamma , & \Delta M = 133 \text{ MeV} ; & (5.20) \\
\psi'(3684) & \to [\pi\pi] & + \chi(3414) \quad + \gamma , & \Delta M = 270 \text{ MeV} ; & \\
\chi(3503) & \to [\pi\pi\pi] & + J/\psi(3098) \quad + \gamma , & \Delta M = 405 \text{ MeV} ; &
\end{array}
$$

where $[\pi]$, $[\pi\pi]$, and $[\pi\pi\pi]$ denote one-pion, two-pion, and three-pion annihilations, respectively. From the numerical systematics of Eq. (5.20), it seems clear that the $\psi' \to \chi(3551)$ and $\psi' \to \chi(3414)$ decays correspond to the annihilation of a one-pion substate and a two-pion substate, respectively, in the ψ'; and the $\chi(3503) \to J/\psi$ decay corresponds to the annihilation of a three-pion substate ($\pi\pi\pi \equiv 3\bar{3}$) in the $\chi(3503)$. These three decay modes, when considered together, form part of an orderly pattern, as we describe in Chapter 17 [also see Eqs. (14.28)-(14.32) in Chapter 14]. But the important point to note here is that — independent of any theoretical interpretation — the $\chi(3414)$, $\chi(3503)$, and $\chi(3551)$ levels are clearly quantized in multiples of the mass quantum $\pi^0 = 135$ MeV, either with respect to the mass of the ψ' or else with respect to the mass of the J/ψ. Indeed, we could hardly ask for more accurate numerical values than are shown in Eq. (5.20). The currectly-favored explanation for these three χ levels is that they are the $^3P_0$, $^3P_1$, and $^3P_2$ positronium-type levels of a pair of "charmed" quarks,[4] and the spin values of these states (as far as they are known) seem to be in remarkable agreement with this hypothesis. However, the observed $\chi(3551)$ - $\chi(3503)$ and $\chi(3503)$ - $\chi(3414)$ mass splittings have continued to defy a convincing "charmonium" explanation. In the opinion of the present author, any valid explanation of these three χ states must involve a recognition of the accurate 135 MeV mass quantizations shown in Eq. (5.20).

There are other examples of 70 MeV mass quantizations in the New Particle resonances. Perhaps the most remarkable of these is the $\psi'(3684)$ resonance itself, which occurs exactly 3 x 3 x 3 x 70 MeV above the mass of the $\bar{p}n(1794.5)$ "ground state" (for a reason that is explained in Chapter 17). Also, another observed χ state, $\chi(3455)$, appears 357 MeV (5 x 70 MeV) above the J/ψ; since this χ state does <u>not</u> fit in with the expected "charmonium" levels,[5] the fact that it does fit in with the present light-quark mass quantization is of more than casual interest. And an important New Particle peak, $\psi(4028)$, appears 344 MeV (5 x 70 MeV) above the ψ'.

The complete spectrum of New Particle resonances is studied in Chapter 17.

We close the present discussion of the New Particles by showing one more figure, Fig. 5.6, which extends the mass systematics of Fig. 5.4 to higher mass values.

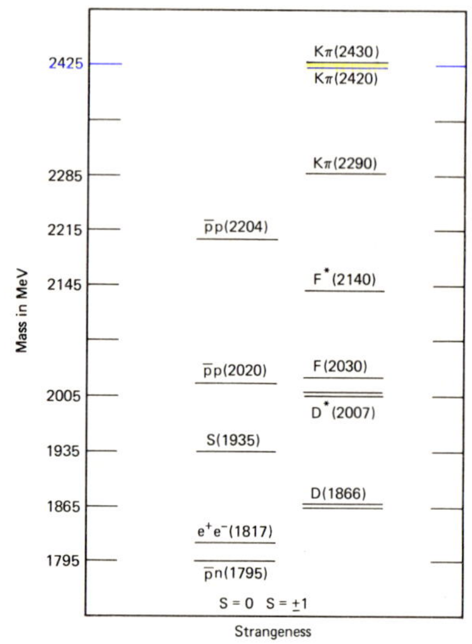

Figure 5.6. A series of narrow-width meson and kaon levels that extend the systematics of Fig. 5.4 to higher mass values. The meson levels appear at intervals of 140 MeV and 210 MeV above the $\bar{p}n(1795)$ meson "ground state", and the kaon levels appear at 140 MeV intervals above the D(1866) kaon "ground state" (compare this last result to Fig. 5.3). The narrow e^+e^- peak seen near the $\bar{p}n$ state is of interest because it strengthens the identification of the $\bar{p}n$ level as the threshold for the New Particles (which are characteristically observed in e^+e^- annihilation experiments).

Figure 5.6 contains the $\bar{p}n$, $\bar{p}p$ or S, D, and D* levels of Fig. 5.4, together with a narrow-width e^+e^- level observed near the mass of the $\bar{p}n$, two narrow-width higher-mass $\bar{p}p$ peaks, a possible sighting of the F and F* resonances that are expected in the "charm" model, and some higher-mass Kπ peaks (all of these states are documented in Appendix B). The important points to note in Fig. 5.6 are that the strangeness S = 0 peaks $\bar{p}p(2020)$ and $\bar{p}p(2204)$ appear approximately 210 MeV and 420 MeV above the nonstrange $\bar{p}n(1795)$ meson "ground state", and the strangeness S = ±1 kaon resonances appear at 140 MeV intervals above the strange D(1866) kaon "ground state". By comparing Fig. 5.6 with Fig. 5.3, we see that meson and kaon resonances separated by mass intervals of 140 MeV (or 70 or 210 MeV) are a universal phenomenon for the narrow-width states — in both the Old Particle and New Particle resonances.

5.4 Baryon Mass Intervals as Multiples of 70 MeV

When we examined the narrow-width meson and kaon resonances, as described above, we found that 70 MeV and 140 MeV mass intervals occur as an essentially universal phenomenon. When we now examine the corresponding baryon resonances, we find this same situation. The baryon states that we select for this purpose are the narrow-width states plus the identified S-states, since these are the resonances that we identify as rotationless excitations. The methods used to sort the baryon resonances into rotationless ($\ell = 0$) and rotational ($\ell > 0$) categories are described in detail in Part II (see Chapter 12). The essential point here is that the mass quantization of the rotationless excitations is completely different from that of the rotational excitations; only the former are expected to exhibit 70 MeV mass intervals.

Figure 5.7 shows the narrow-width and S-state baryon excitations, plotted as energy levels above the fundamental N(939), Λ(1116), Σ(1193), and Ξ(1321) baryon and hyperon "ground states". The 210 MeV "energy gap" that can be observed above the ground states in Fig. 5.7 matches the 210 MeV energy gap seen in the case of the meson resonances in Fig. 5.3. But whereas the narrow-width meson excitations are characteristically separated by excitation intervals of π = 140 MeV, the narrow-width and S-state baryon excitations are characteristically separated by excitation intervals of $\dot{3} \equiv 3M = 210$ MeV and $\dot{4} \equiv 4M = 280$ MeV, as we have indicated at the left in Fig. 5.7. The identified levels in Fig. 5.7 that correspond to the excitations $\dot{4}$ and $\dot{4}\dot{4}$ are prominent S-state resonances, whereas the identified levels that correspond to the excitations $\dot{3}$ and $\dot{3}\dot{3}$ are mostly very-narrow-width P-states and D-states. Thus the $\dot{3}$-type excitations have different characteristics from the $\dot{4}$-type excitations, which is a result that is described in detail in Appendix E. This result lends credence to the systematics displayed in Fig. 5.7, since the levels shown there were originally selected solely on the basis of mass values. The strangeness S = -3 Ω^-(1672) hyperon, which is not shown in Fig. 5.7, has a mass that is 351 MeV above the mass of the strangeness S = -2 Ξ^-(1321) hyperon shown in Fig. 5.7, thus giving still another example of a mass quantization in multiples of 70 MeV.

The main point we are attempting to establish in Fig. 5.7 is that the S-state and narrow-width $\ell = 0$ baryon resonances are all spaced by intervals which are multiples of 70 MeV. This is a consequence of the facts that (1) each of these resonances consists of a "ground state" plus a number of spinless 70 MeV "excitation quanta", and (2) the excitation quanta are nonstrange, so that $\sim$0% binding energies are involved. There are local variations of 10-20 MeV in the spacings of some of the levels in Fig. 5.7, but the overall pattern seems to emerge quite clearly. The mass quantization

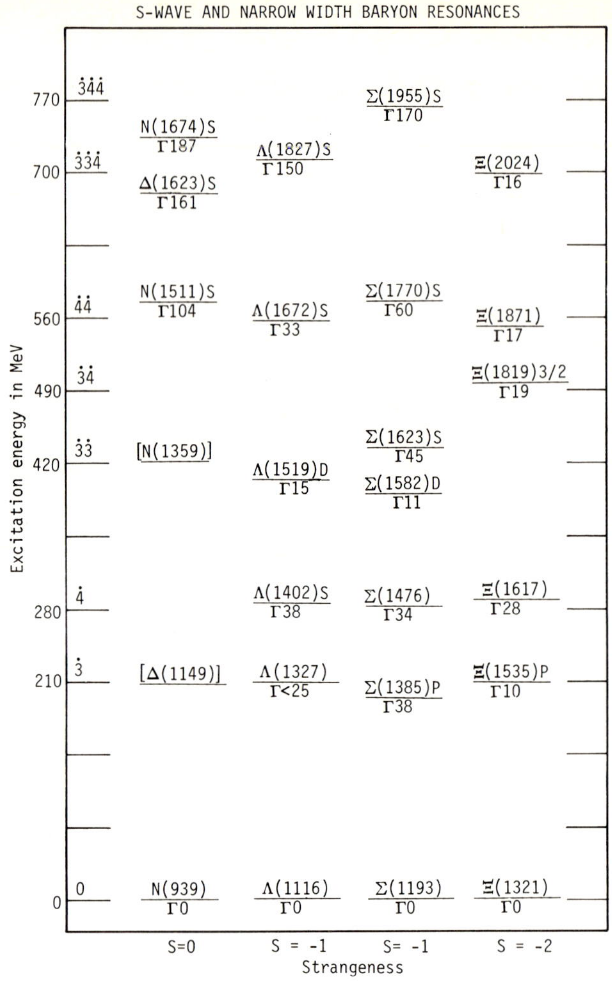

Figure 5.7. An energy level diagram for the S-state and narrow-width baryon and hyperon resonances, arranged as excited states above the fundamental N(939), Λ(1116), Σ(1193), and Ξ(1321) "ground states". Note the quantization of the levels in units of $\dot{3} \equiv 3M = 210$ MeV and $\dot{4} \equiv 4M = 280$ MeV. As is described in Appendix E, the $\dot{3}$-type excitations have characteristics which are quite different from those of the $\dot{4}$-type excitations. The N and Δ levels enclosed in brackets in Fig. 5.7 are unobserved excitations which serve as bandheads for observed overlying rotational bands, as described in Chapter 15.

that is observed for the low-mass <u>baryons</u> in Fig. 5.7 reinforces the mass quantization that is observed for the low-mass <u>meson</u> resonances in Figs. 5.1 and 5.2; after the binding energy effects are properly taken into account, the excitation systematics in these figures are seen to be quite similar to one another.

The systematics of the baryon excitations is discussed at some length in Chapter 12, and the complete mapping of the baryon and hyperon excited states is given in Chapter 15. Most of the S-states shown in Fig. 5.7 have overlying rotational bands, and it turns out that by identifying the overlying rotational levels we can deduce the existence of the unseen S-states $\Delta\dot{3}$ and $N\dot{3}\dot{3}$ that are enclosed in brackets in Fig. 5.7; these states fill out the systematics of Fig. 5.7 in a rather complete manner. The mass quantization of the long-lived N, Λ, Σ, and Ξ "ground states" of Fig. 5.7 is discussed in Chapter 9.

As a final result in connection with Fig. 5.7, we note that there is a close correspondence between the Λ and Σ levels, except that the Σ resonances are shifted upwards in mass by 70 MeV. This suggests that the Σ hyperon is essentially a Λ hyperon to which has been added a quantum M = 70 MeV. We will illustrate this result in Chapter 9. There are a couple of interesting historical remarks that can be made in connection with these Λ and Σ resonances. The 70 MeV-mass-shifted correspondence between the narrow-width and S-state Λ and Σ resonances that is shown in Fig. 5.7 was first noted and published by the author several years ago.[6] At that time the Σ level $\Sigma(1582)D\Gamma 11$, which so strikingly matches the Λ level $\Lambda(1519)D\Gamma 15$, did not exist. Thus its subsequent appearance considerably strengthens the systematics of Fig. 5.7. On the other hand, the $\Lambda(1327)$ $\Gamma<25$ resonance shown in Fig. 5.7 has not been seen in several recent experiments (which involve, it should be noted, different channels than those of the original experiments), and hence has been officially designated by the members of the Particle Data Group as a "dead" resonance; however, the original experiments which indicated the existence of the $\Lambda(1327)$ have not been repeated. The $\Lambda(1327)$ is a $\Lambda\dot{3}$ excitation, and is thus expected to have some rather distinctive properties (such as being difficult to observe directly), as we discuss in Appendix E. Hence it will be interesting to see if future experiments reconfirm its existence.

5.5 <u>Summary</u>

The experimental results that we have displayed in Figs. 5.1-5.7 furnish the strongest possible empirical evidence for the reality of the mass quantum M = 70 MeV as a basic building block for both the meson and baryon resonances. As we have seen, there is ample evidence for 70 MeV mass intervals in all of the <u>narrow-width</u> states — including both meson and baryon families, both low-mass and high-mass resonances, and both strange and nonstrange excitations. It may seem that we have been somewhat overzealous in examining all of these resonances in such detail. However, it must be kept in mind that the conventional quark models of Gell-Mann and Zweig, of Fermi and Yang, and others are all based on the use of spinning quarks only, and quarks with masses which are unknown but presumed to be large. Thus if we attempt to introduce a light spinless quantum M, it is concomitant upon us to supply all of the documentation for its existence that we can obtain.

The one set of narrow-width states that we have not considered in this chapter — and one of the most important sets to consider — is the group of metastable baryon resonances N, Λ, Σ, Ξ, and Ω. We have deferred the discussion of these resonances until Chapter 9, because we first want to examine the properties of the spinor S, which we do in Chapter 8. However, we will find in Chapter 9 that the light-quark configurations $N \equiv S\bar{S}S$, $\Lambda \equiv S\bar{S}S\bar{3}$, $\Sigma \equiv S\bar{S}S\bar{4}$, $\Xi \equiv S\bar{S}S\bar{3}\bar{3}$, and $\Omega \equiv S\bar{S}S\bar{3}\bar{4}\bar{4}$ accurately reproduce both the masses and the strangeness quantum numbers of these states, using the same general rules for binding energies that we discussed in Chapter 4 and applied in Sec. 5.2 of the present chapter.

In Chapter 5 we have summarized evidence based on <u>experimental data</u> which points to the existence of the spinless quantum M = 70 MeV. In Chapters 6 and 7 we produce evidence based on <u>theory</u> — the theory of relativity — which points to the existence of this same mass quantum.

<u>References for Chapter 5.</u>

Since much of the present book consists of an examination of the experimental data on elementary particles, the data base that we use is of utmost importance. We have chosen to assemble and evaluate these data in Appendix B, where the references to the data are contained. The data values quoted in the text are taken directly from Appendix B, and references to the data are included in the individual chapters only when the particular data points are not adequately discussed in Appendix B, or when the use of references facilitates the discussion in a chapter. Wherever possible, we have used the data directly as given by the Particle Data Group in the 1976 compilation of the Review of Particle Properties (RPP76), which was published in <u>Rev. Mod. Phys.</u> <u>48</u>, No. 2, Part II, April 1976, and in the Supplement to the 1976 Edition of Review of Particle Properties (RPP76S), which was published in <u>Phys. Lett.</u> <u>68B</u>, 1 (1977). Updating additions to these data from the research literature were made through March 1978, and include some values taken from RPP78.

1. M. H. Mac Gregor, Phys. Rev. D9, 1259 (1974), Table XXIII.

2. See the discussions in Refs. 42 and 43 of reference 1.

3. M. H. Mac Gregor, Lett. Nuovo Cimento 3, 197 (1972) and 5, 135 (1972). Also see Tables XXII and XXIII in reference 1.

4. See, for example, W. K. H. Panofsky, "The experimental basis for the new quark spectroscopy", SLAC-PUB-2025, October (1977); G. J. Feldman, "Charmed particle spectroscopy", SLAC-PUB-2068, December (1977); K. Gottfried, "The spectroscopy of the new particles", CLNS-376, November (1977).

5. See pages 8-12 in the paper by K. Gottfried in reference 4.

6. See M. H. Mac Gregor, Lett. Nuovo Cimento 4, 1043 (1970), Table II.

CHAPTER 6. THE RELATIVISTIC MASS INCREASE FOR SPINNING SYSTEMS

"In modern physics the laws of conservation are of fundamental importance. Essentially, there are three of them: the principle of inertia, which states the conservation of linear momentum; the energy principle, which asserts the conservation of energy; and the law of the conservation of the quantity of electricity. There are two more conservation laws, dealing with the conservation of the angular momentum and of the inert mass. The first of these two, however, is a necessary consequence of the law of conservation of linear momentum, and the latter, as far as we still acknowledge it to be correct, has become identical with the law of the conservation of energy. Finally the laws of conservation of linear momentum and of energy fuse into one for modern relativistic considerations."

Max von Laue

"The electron spin carries no energy in itself ..."

Leonard Schiff

6.1 The Dilemma Posed by the Electron

One of the most perplexing problems with respect to the elementary particle has been the manner in which the angular momentum of a particle should be regarded. The origin of this conceptual difficulty can be traced back to the discovery of the spin of the electron and its associated magnetic moment. The magnitude of the electron spin, $S = 1/2\ \hbar$, when taken together with the mass of the electron, $m \simeq 0.5$ MeV, indicates that from a classical viewpoint the electron must have a radius $R \sim 5 \times 10^{-11}$ cm. Furthermore, the magnitude of the electron magnetic moment, $\mu = e\hbar/2mc$, suggests that the magnetic moment must arise from a current loop of about this same radius. However, the actual radius of the electron is assumed to be much smaller than this: it is believed to have a value which is comparable to the "classical electron radius", $R_e = e^2/mc^2 = 2.8 \times 10^{-13}$ cm. If the radius of the electron is really this small, then the electron spin and the electron magnetic moment are quantities which cannot be understood in terms of more-or-less classical rotating masses and rotating charges, and we are somewhat at a loss to know how to regard them. The difficulties which have resulted from this conceptual impasse are illustrated by the two quotations given at the beginning of this chapter. As summarized in the quotation by Von Laue, physicists have now tied together the concepts of total energy, inertial mass, linear momentum, and angular momentum in a unified manner. And yet Schiff, stating the operational credo of the present-day elementary particle physicists, asserts that no energy is to be associated with the spin of the electron, so that the spin does not contribute explicitly to the inertial mass of the electron.

The concept of the spin angular momentum of the electron has been carried over and applied, essentially intact, to the spin angular momentum of the quark. In the quark models of Gell-Mann and Zweig and of Fermi and Yang, the strongly-interacting hadrons — mesons and baryons — are assumed to be formed as combinations of spinning quarks, with each quark having the same spin, $S = 1/2\ \hbar$, as the spin of the electron. In these hadron quark models, the spin of the quark is treated on essentially the same footing as the spin of the electron — namely, as a quantity which is postulated rather than calculated, and a quantity which does not contribute in any explicit manner to the total energy of the system. In fact, since these models include only spinning quarks, it is never necessary to ask the question as to what fraction of the quark mass arises from the spin energy of the quark.

The nucleon is the only hadron, and in fact the only elementary particle, whose size has been directly measured, and it is important to note that there is an essential difference between the cases of the nucleon and the electron. As we discuss in the next chapter, the size, mass, spin angular momentum, and magnetic moment of the nucleon are all consistent with the idea that these quantities admit an essentially classical representation, whereas the corresponding quantities for the small electron, which has $R_e \simeq e^2/mc^2$, do not admit such a representation. Hence the conceptual situation with respect to the nucleon is in marked contrast to the conceptual situation that we have described above with respect to the electron: the nucleon can be described classically, whereas the point electron cannot.

In Part II of the present book, we discuss the separation of the total angular momentum of a hadron resonance into its spin and orbital components. In the present chapter we are concerned with a more specific problem — namely, to determine the energy and hence the mass that should be associated with the spin of an elementary particle. As we will see, the answer to this problem has some far-reaching ramifications, both from a conceptual and from a phenomenological point of view.

6.2 Special Relativity, General Relativity, and Rotating Systems

It is generally agreed that rotating systems involve general-relativistic effects, since the accelerations that are produced in a rotating system can be expressed in terms of pseudo-gravitational potentials. However, it is not so generally recognized that special-relativistic effects also logically apply to rotational motion. Moreover, we will demonstrate that if we carry out the relativistic calculations in the (fixed) laboratory frame of reference – the inertial frame – , general relativity and special relativity both lead to the same set of equations, so that it is not even necessary to specify which theory it is that we are using. In the present section, we first list some arguments which indicate that special relativity is applicable to rotating systems; then we discuss some Mössbauer experiments which have been carried out to investigate relativistic effects in these rotating systems. In Section 6.3 we discuss the special properties of the inertial frame of reference, which is the logical frame to use for carrying out relativistic calculations; then, as an example, we derive the equation for the special-relativistic mass increase of a linearly-moving particle in this inertial frame of reference, using a method that circumvents the customary use of the Lorentz transformation equations. In Section 6.4 we work out the problem of a relativistically-spinning sphere, and we obtain answers which appear to be crucial for the spectroscopy of the elementary particle. Quite amazingly, this simplest of all elementary particle models does not seem to have been considered by any of the previous workers in the field; the results that we present here are original. We now turn to the topic of special relativity, and of its application to objects that are in rotational motion.

We ordinarily think of special relativity as applying to objects that are in linear motion. However, as van Laue emphasized in the quotation at the beginning of this chapter, angular momentum is closely related to linear momentum. Thus, while a rotating system will involve effects that are of a general-relativistic nature, it can at the same time also involve special-relativistic effects. The topic that is of particular concern here is the relativistic mass increase that is observed for a particle when it is accelerated. Back in the days when Newtonian physics was first formulated, the mass of a particle was considered to be an invariant quantity. However, subsequent investigations in the areas of electromagnetism and thermodynamics led to the idea that if energy is added to a system, it results in an increase in the mass of the system. The culmination of these ideas was reached in 1905, when Albert Einstein postulated[1] that the entire mass of a system is derived from its energy content, as expressed in the famous equation

$$E = mc^2 \quad . \tag{6.1}$$

If an object, initially at rest, is given a translational velocity, the total energy of the object is thereby increased, and hence its mass[†] must also increase. Correspondingly, if we impart a rotational velocity to a system, we are adding rotational energy to the system, and thus its mass must similarly increase.

Many of the workers in relativity theory, including Einstein, have assumed that special relativity applies to rotational motion as well as to translational motion. For example, in the introduction to his 1916 article on the general theory of relativity, Einstein discussed the case of a rotating disc. With respect to the special-relativistic contraction of length, Einstein pointed out that the ratio of the circumference C of the disc to the radius R is no longer $C = 2\pi R$, and he commented[2]: "This is readily understood if we envisage the whole process of measuring from the 'stationary' system K, and take into consideration that the measuring-rod applied to the periphery undergoes a Lorentzian contraction, while the one applied along the radius does not." Einstein then went on to point out that the special-relativistic dilation of time also applies to the rotating disc[2]: "By a familiar result of the special theory of relativity, the clock at the circumference – judged from K – goes more slowly than the other because the former is in motion and the latter at rest." It should be noted that in both of these comments, Einstein emphasized the use of the inertial frame of reference K for interpreting these phenomena in a special-relativistic context. We will have more to say about this in Section 6.3 below.

The subject of the rotating disc has actually been the center of much controversy over the years, and topics such as the non-Euclidean nature of the geometry experienced by a rotating observer have been discussed at length in the literature. A good review of this subject is given by Berenda[3], who considers both the special-relativistic and the general-relativistic aspects of the problem. A more recent discussion is contained in the book Relativistic Kinematics by H. Arzeliès[4], which has a chapter on The Rotating Disc.

One point that emerges clearly from the theoretical discussions of the rotating disc is the desirability of carrying out experiments which test these concepts. In relativity theory, the measurement process itself becomes such an important part of the problem that only the use of actual experimental results can insure that a proper analysis of the problem has been carried out. In setting up a laboratory experiment that can test relativistic effects, the rotating disc has the advantage that rotational motion is cyclic and is in many respects much easier to deal with than linear motion, both from the standpoint of devising the actual experiment and from the

---

[†]From the Eötvös experiments, which established the equivalence between inertial mass and gravitational mass, we can use the word "mass" to denote both of these quantities.

standpoint of setting up observations which are conceptually straightforward. The problem with the rotating disc is that rotational velocities are limited by the strength of the rotor, so that velocities $v = \omega r \sim 10^{-6}$ c (where c is the velocity of light) are the largest that are typically attained. Since relativistic effects appear in order v^2/c^2, this means that relativistic effects will be in the range of 1 part in 10^{12}. However, Mössbauer experiments can be set up which have accuracies of 1 part in 10^{13}. Thus the combination of a rotating disc with a Mössbauer source and absorber is sufficient to investigate relativistic effects. Experiments of this type have been carried out, for example, by Hay and coworkers[5] and by Champeney and coworkers[6]. In these experiments, when the source and the absorber were placed at the center and the edge of the rotating disc, respectively, a frequency shift of the expected magnitude was observed. However, when the source and the absorber were placed at the opposite ends of a diameter of the disc, no frequency shift was observed. It was pointed out in these papers that the results could be explained either in terms of special relativity or else in terms of general relativity.

A somewhat different Mössbauer relativity experiment was carried out by Pound and Rebka.[7] They measured the frequency shift that is produced when the source and the absorber are at different temperatures. The frequency shift is caused by the time dilation of the nuclei which are undergoing thermal vibrations in the crystal lattice. This experiment, which was discussed by Sherwin,[8] is of interest because the observed frequency shift is proportional to the <u>velocities</u> of the nuclei in the crystals, and not to their <u>accelerations</u>. Sherwin estimates that any specific acceleration-dependent effects must be at least an order of magnitude smaller than the time dilation effects. In the Pound and Rebka temperature difference experiment, accelerations are on the order of 10^{16} g, whereas in the rotating disc experiments the radial accelerations are only about 10^5 g. This suggests that the observed frequency shifts in the rotating disc experiments are not due to specific acceleration-dependent distortions.

Two striking special-relativistic effects that have been verified experimentally for the elementary particle are the increase in mass and the increase in lifetime of the particle when it is accelerated. From the viewpoint of the inertial frame of reference, the mass of a particle increases by the factor $1/(1 - v^2/c^2)^{1/2}$ when it is given a velocity v, and the lifetime of the particle also increases by this same factor. The increase in the mass and lifetime occurs both for linear motion (e.g., a cosmic ray muon) and for circular motion (e.g., a muon moving in a magnetic field), which of course is in agreement with von Laue's comment that angular motion is a special case of linear motion. We can use this empirical relationship between the change in mass and the change in lifetime (frequency) to conclude that the frequency changes which are observed in the rotating disc Mössbauer experiments are also accompanied by mass changes; and Møller, in his book The Theory of Relativity[9], gives the equation for the relativistic mass increase of a rotating ring of matter. Thus, if we extend these ideas to the case of a spinning particle, we expect that the particle has a mass increase which is associated with its rotational motion. In Section 6.4 we calculate this mass increase for the simplest model of a spinning particle that we can devise — a relativistically-spinning sphere of matter. But first, in Section 6.3 we discuss the unique position that is occupied by the inertial frame of reference, and we use this inertial frame of reference in deriving the special-relativistic mass increase of a particle under translational motion, without making any use of relativistic coordinate transformations. Of particular interest from the present standpoint is the fact that in the inertial frame of reference, special-relativistic and general-relativistic explanations of rotational effects result in the same lowest-order mathematical equations, so that we are not forced to distinguish between these two approaches to the problem.

6.3 The Relativistic Mass Increase in the Inertial Frame of Reference

The inertial frame of reference occupies a unique position in the theory of relativity. Although some relativists such as Dingle[10] have maintained that all frames of reference must be equivalent, this does not in fact seem to be the case, either theoretically or experimentally:

Theoretically, we have the unique result that when (and only when) a rotating system is viewed in the inertial frame of reference, the lowest-order relativistic effects can be interpreted as coming either from general relativity or from special relativity. For example, in the Mössbauer rotating disc experiment of Hay *et al.*,[5] the observed frequency shift can be attributed either to the effective gravitational potential that is associated with the acceleration of the moving absorber, or else to the time dilation of the moving absorber; both effects lead to the same equation for the frequency shift. Similarly, the mass increase for a rotating ring of matter can be calculated from general relativity, as was done by Møller,[9] and the answer that is obtained, $m = m_o/(1 - \omega^2r^2/c^2)^{1/2}$, is the same as the answer which is obtained from special relativity if we make the replacement $v \to \omega r$. The fact that in the inertial frame of reference both the general-relativistic and the special-relativistic formulations give the same answer, at least in lowest order, suggests that the inertial frame is uniquely appropriate for considering these relativistic effects. By way of contrast, the Mössbauer rotating disc experiment can be considered in the frame of reference that is fixed in the rotating disc, but then the analysis of the problem becomes very involved,[8] and it requires the inclusion of general relativistic effects in an essential manner. Thus the inertial frame is unique theoretically in that it reduces general relativity and special relativity to the same theory, at least to order v^2/c^2 in the relativistic equation, when we apply these theories to rotating systems.

Experimentally, the uniqueness of the inertial frame of reference is operationally apparent. If we interpret the Mössbauer rotating disc experiment in terms of clocks, one of which is fixed at the center and the other which rotates at the periphery of the disc, the moving clock clearly runs more slowly than the stationary clock.[8] As was pointed out by Ives,[11] this result can be generalized to the case where there are a number of clocks moving with various relative motions; the clock that is running the fastest will be the clock that has the lowest rms velocity with respect to the inertial frame of reference. Thus the inertial frame of reference can be singled out by other means than simply using accelerometers.

As the above discussion indicates, the inertial frame of reference is particularly appropriate for the treatment of relativistic effects. In this context, we now demonstrate that, working entirely in the inertial frame of reference, we can derive the special relativistic mass increase of a linearly-moving particle. The usual derivation of this result involves the use of the Lorentz equations for transforming from one coordinate system to another.[12] However, by staying entirely within the inertial frame of reference, we obtain a clearer insight into the physical significance of this relativistic mass increase.

The Einstein equation

$$E = mc^2 \tag{6.1}$$

expresses the proportionality between the total energy of a system and the mass of the system. If the energy is increased, then the mass is increased accordingly. In the inertial frame of reference, consider a particle that is moving with velocity v. Both the energy E and the mass m are functions of v, and the Einstein equation tells us that this functional dependence is essentially the same for both quantities. Our task is to determine this functional dependence; that is, we must determine m(v) for the particle. This can be done by applying a linear impulse Fdx in the direction of motion of the particle; this impulse changes the energy, the observed mass, and the velocity of the particle. The energy change of the particle is

$$dE = F dx \quad ,$$

which, from the relationship

$$F = d/dt(mv)$$

can be written as

$$dE = d/dt(mv)dx \quad .$$

Since $v = dx/dt \rightarrow d/dt = vd/dx$, we have

$$dE = vd/dx(mv)dx$$

$$= v^2dm + mvdv \quad .$$

Setting

$$dE = c^2dm \quad ,$$

we obtain

$$(c^2 - v^2)dm = mvdv \quad .$$

Integrating this equation gives

$$\ln m = -1/2 \ln (c^2 - v^2) + \text{constant} \quad .$$

Applying the limits

$$m(v) = m \text{ and } m(0) = m_o \quad ,$$

we obtain

$$\ln (m/m_o) = \ln (1 - v^2/c^2)^{-1/2} \quad ,$$

which by cxponentiation gives the final result,

$$m = m_o/(1 - v^2/c^2)^{1/2} \quad . \qquad (6.2)$$

Thus we have derived the equation for the special-relativistic mass increase of a particle in the inertial frame of reference, without making any use of the Lorentz transformation equations. Equation (6.2) happens to be the equation for m(v) whose form is preserved under Lorentz transformations, but the physical significance of this equation is best understood by working entirely within the inertial frame of reference. From this derivation, the increase in mass is simply a consequence of the increase in energy when the particle is given an impulse. In Appendix F we give a slightly different derivation of this same result — a derivation which is more pedestrian but which permits an even clearer interpretation of the physical significance of this relativistic increase in mass.

The relativistic calculation that we have just carried out was for linear motion. However, as we discussed in Section 6.2, relativistic effects can also be studied for the case of rotating systems. In the next section we calculate the mass increase for a relativistically-spinning sphere. Rather incredibly, this calculation appears never to have been carried out by any of the workers in the field; and equally incredibly, this model reproduces the first-order properties of the spinning muon.

6.4 The Relativistically-Spinning Sphere

In the inertial frame of reference, consider a ring of matter that has mass M_0 and radius R, as shown in Fig. 6.1.

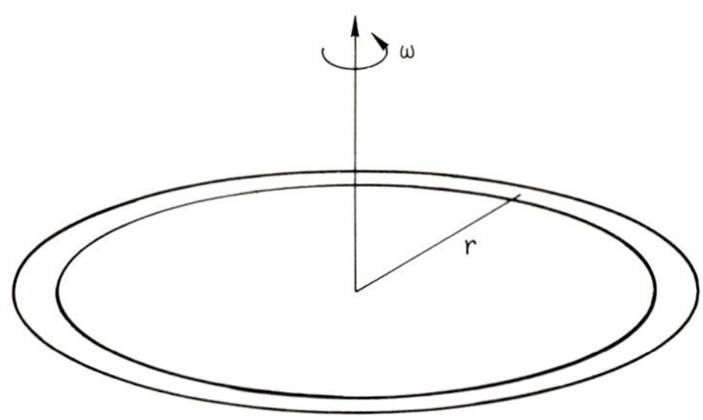

Figure 6.1. A rotating ring of matter. The mass of the ring increases relativistically when the ring is set into motion [Eq. (6.3)], but the coordinate r remains invariant since the motion is at right angles to r. Thus rotating rings can be linearly superimposed to form a rotating disc, or cylinder, or sphere.

The measured mass M_0 of the ring derives from the fact that it exists in a gravitational potential. Now impart an angular velocity ω to the ring. Each of the mass points in the rotating ring then experiences an acceleration, and hence a force, which in the general theory of relativity can be considered as arising from an effective gravitational potential. Thus the rotational motion has resulted in a change in the overall gravitational potential of the ring, which leads in turn to a change in the measured mass of the ring. The general-relativistic equations for the gravitational potential of a rotating ring have been written down, for example, by Møller,[9] and they lead to the mass equation

$$M = M_0/(1 - \omega^2 r^2/c^2)^{1/2} \quad . \tag{6.3}$$

Equation (6.3) is similar to the special-relativistic mass equation, Eq. (6.2), that we derived above for linear motion, except that the linear velocity v of the system in linear motion has been replaced by the "instantaneous linear velocity" ωr of the system in rotational motion. This result is in agreement with our above discussions, wherein we showed that for a rotating system viewed in the inertial frame of reference, both the general theory of relativity and the special theory of relativity lead to the same set of lowest-order relativistic equations.

There is one other property of the rotating ring of matter that is of crucial importance here. When a measuring rod is placed along the circumference of the ring, it undergoes a special-relativistic contraction in length as viewed in the inertial frame; this is in line with the Mössbauer rotating disc experiment that we discussed in Section 6.2. However, as pointed out by Einstein,[2] Berenda,[3] and others,[4] a measuring rod placed along a radius of the ring undergoes no change in length, because the rotational velocities are at right angles to the rod. Thus the circumference C of the rotating ring is no longer equal to $2\pi R$, and the geometry of the ring is non-Euclidean. The crucial point for our purposes, however, lies in the invariance of the length of the radius R under rotations. Consider the rotating ring to be in the x,y plane. If the geometry of the ring is specified by a radius R and by a cross section dR in one dimension or dRdz in two dimensions, then R, dR, and dRdz are all invariant quantities that do not depend on the rotational velocity ω. This means that the external dimensions of the ring are not relativistically distorted by the rotational motion, so that we can linearly combine rotating rings to form a rotating flat disc, or a rotating cylinder, or a rotating sphere. Now the rotating flat disc and the rotating cylinder have been discussed at length in the literature,[2-4,13] and the rotating disc has been used for relativistic experiments,[5-7] but, as far as the present author has been able to determine, and this matter has been discussed with many experts in the field, no one else has considered the case of the rotating sphere,[14] which is, after all, just a superposition of rotating discs or rotating cylinders. And it is at the level of the relativistically-rotating sphere that some very remarkable equations suddenly appear. We now proceed to the derivation of these equations.

We start with a uniform sphere of matter that has radius R and rest mass M_o in the inertial frame of reference. We then impart to this sphere an angular velocity ω, as shown in Fig. 6.2. In particular, we take the relativistic limit for the rotational velocity,

$$\omega = c/R \quad , \tag{6.4}$$

so that the equator of the spinning sphere is moving at the velocity of light c (or infinitesimally below that value). If we visualize the sphere as composed of rings of matter, then each ring has a relativistic mass increase that is given by Eq. (6.3),

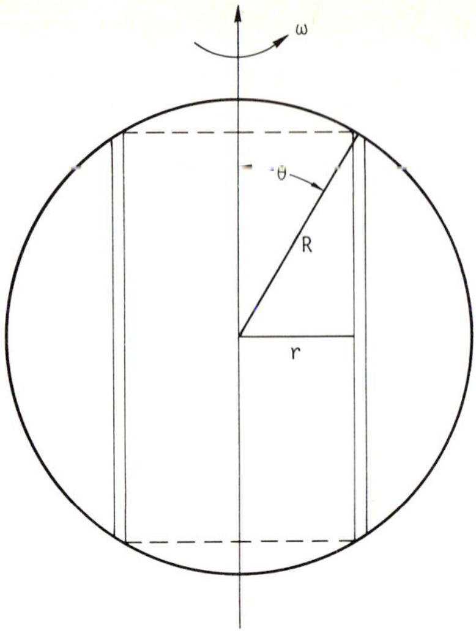

Figure 6.2. A relativistically-spinning sphere of matter. This sphere can be formed by superimposing spinning rings, as described in Fig. 6.1. When the sphere is spinning at the full relativistic limit (with its equator moving at or infinitesimally below the velocity of light), the spinning mass is half again as large as it was at rest. Thus a spinless 70 MeV mass quantum (see Chapter 5) turns into a spinning 105 MeV quantum, and the identification with the spinning 105 MeV muon seems unmistakable. Furthermore, if the sphere is assigned the Compton radius $R = \hbar/mc$, then the muon spin angular momentum $J = 1/2\ \hbar$ emerges as a <u>directly-calculated</u> quantity [see Eqs. (6.9) - (6.13)]; and if the sphere is given an equatorial unit charge distribution, the muon magnetic moment $\mu = e\hbar/2mc$ also emerges as a <u>calculated</u> quantity [see Eqs. (7.1) - (7.3)]. Thus this relativistically-spinning sphere model correctly reproduces the mass, spin angular momentum, and magnetic moment of the muon. Of further interest is the fact that this simplest of all elementary particle models does not appear to have been considered anywhere in the published literature on elementary particles.

and the total mass increase of the sphere is obtained by integrating over the volume of the sphere. In order to carry out this integration, it is useful to use the cylindrical volume elements that are shown in Fig. 6.2, where each cylindrical element is characterized by its radius r with respect to the axis of rotation. Instead of integrating over the radial coordinate r, it is more convenient to integrate over the related angular coordinate θ, where

$$r = R \sin\theta \quad , \qquad \text{and} \qquad dr = R \cos\theta \, d\theta \quad . \tag{6.5}$$

The volume of the sphere is

$$\begin{aligned} V &= \int_0^R 2\cdot(R^2 - r^2)^{1/2} \cdot 2\pi r dr \\ &= 4\pi R^3 \int_0^{\pi/2} \cos^2\theta \sin\theta \, d\theta \\ &= 4\pi R^3/3 \quad , \end{aligned} \tag{6.6}$$

where we are implicitly using the fact, as discussed above, that the coordinate r is relativistically unaffected by the rotation ω. The relativistic mass increase of the rotating sphere is

$$\begin{aligned} M &= \int_0^R 2\cdot(R^2 - r^2)^{1/2} \cdot \rho/(1 - \omega^2 r^2/c^2)^{1/2} \cdot 2\pi r dr \\ &\equiv \int_0^R M(r) dr \quad , \end{aligned} \tag{6.7}$$

where ρ is the uniform density of matter in the sphere when it is at rest. From the fact [see Eq. (6.4)] that

$$(1 - \omega^2 r^2/c^2)^{1/2} = (1 - r^2/R^2)^{1/2} \quad ,$$

equation (6.7) becomes

$$\begin{aligned} M &= 4\pi\rho R^3 \int_0^{\pi/2} \sin\theta \cos\theta \, d\theta \\ &= 2\pi\rho R^3 \quad ; \end{aligned}$$

and since

$$\rho = M_o/(4\pi R^3/3) = 3M_o/4\pi R^3 \quad ,$$

we have finally that

$$M = 3/2\, M_o \quad : \tag{6.8}$$

the relativistically-spinning sphere is half again as massive as it was at rest. Thus a 70 MeV spinless mass becomes a 105 MeV spinning mass — the muon!

We can carry these relativistic mass calculations one step farther by calculating the relativistic moment of inertia I of the spinning sphere about the axis of rotation, working always in the inertial frame of reference. I is given by the equation [see Eq. (6.7)]

$$\begin{aligned} I &= \int_0^R M(r)\, r^2\, dr \\ &= 4\pi\rho R^5 \int_0^{\pi/2} \sin^3\theta \cos\theta\, d\theta \\ &= \pi\rho R^5 \\ &= 3/4\, M_o R^2 \quad , \end{aligned} \tag{6.9}$$

so that

$$I = 1/2\, MR^2 \quad , \tag{6.10}$$

where M is the observed mass of the relativistically-spinning sphere. Thus we have the value $I = 1/2\, MR^2$ rather than the nonrelativistic value $I = 2/5\, M_o R^2$ as the moment of inertia of a relativistically-spinning sphere about its axis of rotation in the laboratory frame of reference, where the rotation is at the full relativistic limit given by Eq.(6.4).

We can now obtain the angular momentum J of the spinning sphere by combining Eqs. (6.4) and (6.10). This gives

$$\begin{aligned} J &= I\omega \\ &= 1/2\, MR^2 \cdot c/R \\ &= 1/2\, MRc \quad . \end{aligned} \tag{6.11}$$

If we set the radius R of the sphere equal to its Compton wavelength,

$$R = \hbar/Mc \quad , \tag{6.12}$$

we thereby obtain

$$J = 1/2\ \hbar \tag{6.13}$$

as the calculated spin angular momentum of the sphere! This is the only model of an elementary particle that reproduces the spin of the particle as a directly calculated quantity.

As we have described at some length in Chapter 5, there is ample evidence for the existence of a spinless mass quantum M = 70 MeV. From the results of the present chapter, we see that if this quantum M has a spherical shape, and if it is set into rotation at the relativistic limit (with its equator at or near the velocity of light), it will have an observed mass of 105 MeV in the inertial (laboratory) frame of reference. Furthermore, if we take the radius of the spinning sphere to be its Compton wavelength, which seems reasonable, then the spin angular momentum is $J = 1/2\ \hbar$. Thus this model of a relativistically spinning sphere uniquely reproduces both the mass and the angular momentum of the muon as directly calculated quantities. In Chapter 7 we discuss the properties of the muon in more detail; and in Chapter 14 we consider some aspects of the topic of spin quantization.

We conclude the present chapter by extending these calculations to nonspherical geometries, and by briefly considering some ramifications of the calculations. As we have discussed above, the model of a relativistically-spinning sphere accurately reproduces the properties of the muon. However, when we extend these results to include the nucleon spinor $S \simeq 330$ MeV (see Chapters 8 and 18), there are a number of arguments which suggest that the geometry of the quantum M = 70 MeV in hadronic matter is that of an oblate spheroid. Thus we should extend the relativistic mass equation, Eq. (6.7), and the relativistic moment of inertia equation, Eq. (6.9), to include the case of nonspherical but axially-symmetric geometries. This can be done by parametrizing the radius R of the axially-symmetric spheroid as

$$\begin{aligned} R_A &\equiv \text{axial radius} \quad , \\ R &= R_A\,(1 + \varepsilon \sin\theta) \quad , \\ R_E &\equiv \text{equatorial radius} \\ &= R_A(1 + \varepsilon) \quad , \end{aligned} \tag{6.14}$$

where

$$\varepsilon = (R_E - R_A)/R_A \tag{6.15}$$

is a deformation parameter, and where θ is the angle shown in Fig. 6.2. In terms of

Eqs. (6.14) and (6.15), Eqs. (6.7) and (6.9) become

$$M = 3M_o \int_o^{\pi/2} d\theta \, \frac{(1 + \varepsilon \sin\theta)^2 (1 + 2\varepsilon \sin\theta) \sin\theta \cos^2\theta}{[1 - \sin^2\theta (1 + \varepsilon \sin\theta)^2/(1 + \varepsilon)^2]^{1/2}} \qquad (6.16)$$

and

$$I = 3M_o R_A^{\,2} \int_o^{\pi/2} d\theta \, \frac{(1 + \varepsilon \sin\theta)^4 (1 + 2\varepsilon \sin\theta) \sin^3\theta \cos^2\theta}{[1 - \sin^2\theta (1 + \varepsilon \sin\theta)^2/(1 + \varepsilon)^2]^{1/2}} \quad . \qquad (6.17)$$

Empirically, the relativistic mass M in Eq. (6.16) is a slowly-varying function of the deformation parameter ε; and the moment-of-inertia I in Eq. (6.17) is also a slowly-varying function of ε if the radius R_E in Eq. (6.14) is held constant as ε is varied. Thus Eqs. (6.8) and (6.10) are reasonable approximations to M and I for all spheroidal geometries.

In the derivation of the above equations for the relativistic mass and moment of inertia of a spinning sphere of matter, two questions come immediately to mind: (1) Suppose the equator of the spinning sphere does not actually reach the velocity $v = \omega R = c$, but instead remains infinitesimally below that value (as it must). (2) What about problems such as rigidity and stresses? The answer to (1) is that if the velocity $v = \omega R$ in Eq. (6.7) is lowered infinitesimally, the change in the integral is also infinitesimal; the relativistic equations that are given here do not have divergence difficulties. Question (2), the problem of rigidity and stresses, is one that has been debated in the literature,[15] and about which no consensus has been reached. It is important to note, as stated by Arzeliès,[16] that no one really knows the environment experienced under rotational motion by a mass point embedded in a matrix of mass points. For example, the non-Euclidean geometry experienced by this mass point may minimize centrifugal effects.[17] By neglecting stresses, we obtain phenomenologically useful results, which is an _a posteriori_ indication that, at least for the spherical geometries considered here, stresses are not of crucial importance. In this regard, it should be pointed out that from the standpoint of calculating relativistic effects, the rotating sphere represents a particularly advantageous geometry: all of the calculational difficulties occur right at the equator, and (fortunately) the contribution of this equatorial region to the integrals in Eqs. (6.7) and (6.9) is minimal.

In the next chapter, Chapter 7, we extend the present results to include electromagnetic effects, which enables us to reproduce the gyromagnetic ratio of the muon. And in Chapter 8 we apply these results to the case of the nucleon spinor $S \simeq 330$ MeV.

We close this chapter by noting that the equality of inertial and gravitational masses, which is contained in Einstein's Principle of Equivalence,[18] and which was established experimentally by Eötvös, is also related to the result (see Section 6.3) that in the inertial frame of reference both the special and general theories of relativity lead to the same equations; the special theory has to do with inertial masses and the general theory with gravitational masses, so that the equality of the equations has as a consequence the equality of the masses.

References for Chapter 6.

The quotation by Max von Laue at the beginning of the chapter is from the article "Inertia and Energy," which was published in the book Albert Einstein: Philosopher-Scientist, Tudor Publishing Company, New York (1949).

The quotation by Leonard Schiff at the beginning of the chapter is from his classic textbook Quantum Mechanics, McGraw-Hill, New York, Second Edition (1955), page 331. Other textbooks are not in agreement with this viewpoint; for example, in the book Introduction to Modern Physics, McGraw-Hill, New York, Fifth Edition (1955), by F. K. Richtmyer, E. H. Kennard, and T. Lauritsen, a discussion is given on page 252 in which a rotating Lorentz electron has a relativistic distortion of the charge distribution which produces a change in the energy of the electron. The reason we included this quotation by Schiff is that it serves to illustrate the conceptual difficulties imposed by the point electron. According to present-day relativity theory, when a relativistic transformation is made that involves spin 1/2 particles, this transformation can cause a change in the spatial orientation of the spin vectors, but it is not assumed to cause any change in the spin energy of the particle.

1. See, for example, The Principle of Relativity, Dover Publications, Inc., New York, page 69.
2. See Ref. 1, page 116.
3. C. W. Berenda, Phys. Rev. 62, 280 (1942).
4. Henri Arzeliès, Relativistic Kinematics, Pergamon Press, Oxford (1966), Chapter IX.
5. H. J. Hay, J. P. Schiffer, T. E. Cranshaw, and P. A. Egelstaff, Phys. Rev. Lett. 4, 165 (1960).
6. D. C. Champeney and P. B. Moon, Proc. Phys. Soc. (London) A77, 350 (1960); D. C. Champeney, G. R. Isaak, and A. M. Khan, ibid. A85, 583 (1965).

7. R. V. Pound and G. A. Rebka, Jr., Phys. Rev. Lett. 4, 274 (1960).

8. C. W. Sherwin, Phys. Rev. 120, 17 (1960).

9. C. Møller, The Theory of Relativity, Oxford Univ. Press, London (1952); see page 318, Eq. (42) and the accompanying discussion.

10. H. Dingle, Nature 144, 888 (1939) and 179, 866 (1957) and 180, 1275 (1957).

11. H. E. Ives, J. Opt. Soc. Am. 27, 305 (1937).

12. For example, see Max Born, Einstein's Theory of Relativity, Dover, New York (1962), pp. 269-273. Also see H. Goldstein, Classical Mechanics, Addison-Wesley, Cambridge, Mass. (1950), Chapter 6.

13. E. L. Hill, Phys. Rev. 69, 488 (1946); N. Rosen ibid. 70, 93 (1946); 71, 54 (1947); B. Kursunoglu, Proc. Camb. Phil. Soc. 47, 177 (1951); H. Takeno, Prog. Theor. Phys. 7, 367 (1952); G. H. F. Gardner, Nature 170, 243 (1952); J. L. Synge, ibid. 170, 243 (1952); T. E. Phipps, ibid. 195, 67 (1962); J. G. Fletcher, ibid. 199, 994 (1963); J. L. Anderson, Principles of Relativity Physics, Academic Press, New York (1967), p. 183; E. L. Hill, Phys. Rev. 71, 318 (1947).

14. M. H. Mac Gregor, Lett. Nuovo Cimento 4, 211 (1970); Phys. Rev. D9, 1259 (1974), Appendix B.

15. G. Salzman and A. H. Taub, Phys. Rev. 95, 1969 (1954); J. E. Hogarth and W. H. McCrea, Proc. Camb. Phil. Soc. 48, 616 (1952); W. H. McCrea, Proc. Phil. Soc. A206, 562 (1951); G. C. Mc Vittie, ibid. A211, 295 (1952); also Ref. 4.

16. See Ref. 4, Sec. IX, [120], p. 237.

17. Arzeliès, Ref. 16, mentions that the special-relativistic stretching strains may not be real. We can extend this result by noting that the non-Euclidean geometry near the equator that is experienced by the moving observer may minimize the stresses that originate as centrifugal effects.

18. See Ref. 9, pages 220-221.

CHAPTER 7. THE MUON: LONELY ORPHAN OR MISSING LINK?

"I often say that when you can measure what you are speaking about and express it in numbers you know something about it; but when you cannot measure it, when you cannot express it in numbers, your knowledge is of a meagre and unsatisfactory kind."

Lord Kelvin

"The muon — who ordered that?"

Wolfgang Pauli

7.1 The Dilemma Posed by the Muon

It is a general rule in physics that when there is an ensemble of related states, the lowest-mass states in the ensemble are usually the most basic. Perhaps the clearest example of this rule occurs in the area of nuclear physics, where more than 1100 different atomic nuclei have been identified. The lowest-mass nuclear state is the nucleon (the proton and the neutron), and nucleons constitute the building blocks for all of the 1100-odd nuclei. Furthermore, there is evidence that the low-mass nuclei D, T, He^3, and He^4 also serve as basic constituents of the more massive nuclei (see Chapter 20).

Now let us apply this rule about the importance of the low-mass states to the domain of elementary particle physics. We will focus our attention on the elementary particles which are more massive than the electron; for convenience, we denote these as "heavy particles". As is shown in Fig. 7.1, the two lowest-mass heavy particles are the half integral spin muon and the integral spin pion. Hence, in analogy to the situation that occurs in nuclear physics, we might expect the muon and pion to serve in some manner as building blocks for half integral spin and integral spin particles, respectively. And, indeed, we have already seen in Fig. 5.1 that the η meson is, from the standpoint of its mass, essentially a cluster of four pions; and the η' meson is essentially a cluster of seven pions. From the same standpoint, namely with a linear addition of the masses, Fig. 7.1 shows that the nucleon appears as a cluster of nine muons. The only difficulty with this formulation is that no one told the muon about it. The π, η, and η' mesons are all strongly-interacting (hadronic)

LOW-MASS "HEAVY PARTICLES"

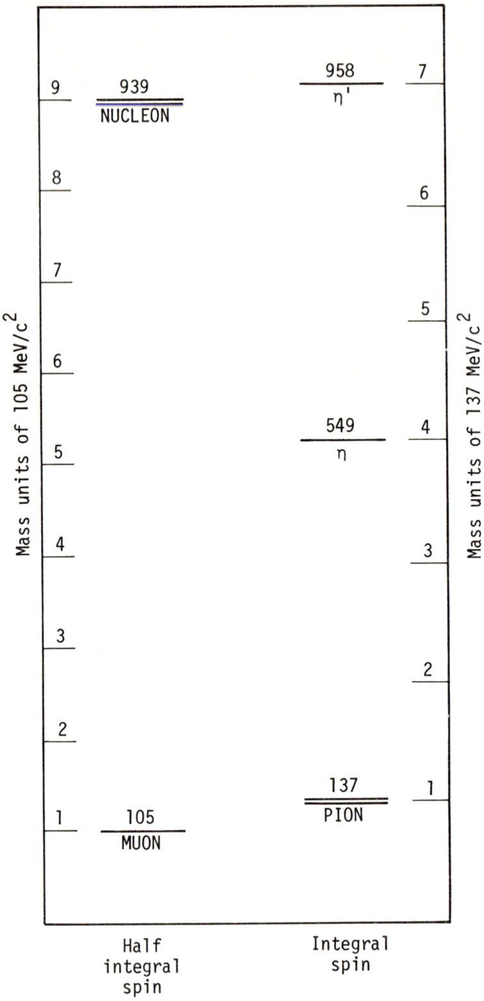

Figure 7.1. The spectrum of low-lying "heavy particles" (particles that are more massive than the electron). The lowest-mass spin 1/2 heavy particle is the muon; and the next-lowest-mass spin 1/2 particle is the nucleon, which appears accurately at nine times the mass of the muon. The lowest-mass spin 0 heavy particle is the pion; and the next-lowest-mass spin 0 mesons are the η and η' mesons, which appear accurately at four and seven times the mass of the pion.

particles, so that it makes a certain amount of sense to group them together as we did in Chapter 5. However, although the nucleon is a hadronic particle, the muon is not — it interacts only electromagnetically (weakly). In view of this deficiency, the muon has been relegated to the role of an outcast. Present-day hadronic elementary particle theories such as SU(3) have no place in which to include the muon. From the standpoint of these theories, the muon is considered to be nothing more than a heavy electron, and in fact there appears to be no rational reason for its existence at all, as Pauli reminds us in the quotation at the beginning of the chapter.

Nevertheless, whether it interacts strongly or not, the muon does exist, and it is the lightest of the "heavy particles". Unless Nature has uncharacteristically taken leave of her senses, the muon must somehow fit logically into the overall scheme of things. Furthermore, from the viewpoint of the light-quark model, where all binding energies are small, we need a spin 1/2 basis state that is related to the muon mass of 105 MeV in order to be able to reproduce the nucleon. In fact, it was this last consideration that led to the muon model which we are about to discuss in this chapter. Suppose we think of the nucleon as being in some manner composed of nine muons. The difficulty with this picture, apart from the fact that the muon is nonhadronic, is that the nucleon is known to be an extended object, with an rms radius of about 0.8 fermi, whereas the muon is commonly regarded as a point particle. How can we combine nine point particles together in a sensible manner so as to reproduce the non-point nucleon? In order to do this, it seems that we are first obliged to construct some kind of a model for the muon.[1]

The simplest model that can be devised for the muon is a spinning sphere. As we have already seen in the last chapter, this is the model which, by means of the relativistic mass equations, establishes a link between the mass of the muon and the masses of the hadrons. In the present chapter we extend these results to include the electromagnetic properties of the muon. However, before entering into this discussion, we first digress briefly and consider some basic properties of the nucleon.

7.2 Properties of the Nucleon

In Chapter 2 we asked a number of questions about elementary particles. In particular, we asked if the angular momentum and magnetic moment of a particle can be attributed to rotating mass and charge distributions in the particle. In order to answer this question, we must know the size of the particle. Since the nucleon is the only particle whose size has been determined experimentally, it is the only particle we can study from this viewpoint with any authority.

If the nucleon were a single Dirac particle, it would logically have a radius

$R = \hbar/mc \simeq 0.2$ fermi, where $m = 939$ MeV. Since the measured rms charge and magnetic moment radii of the nucleon are both about 0.8 fermi, it is apparent that the nucleon is not a single Dirac particle, a fact that is also demonstrated by the anomalous magnetic moments of the proton and neutron; the nucleon must be a composite structure. Once we have reached this conclusion, there are a number of phenomenological arguments which suggest that the nucleon is in fact composed of three subunits of approximately equal mass. As a simple and very general model for a nucleon subunit, we can take a spherical quantum with a mass that is roughly equal to 1/3 the nucleon mass, or about 330 MeV (this mass value is "derived" in Chapter 8). If we set the radius of each subunit equal to its Compton wavelength ($R_s = \hbar/m_s c$, with $m_s = 330$ MeV), and if we spin these subunits at their relativistic limits (with their equators moving at the velocity of light), then each subunit will have an angular momentum $J = 1/2\ \hbar$, as we have shown in the last chapter. This is the angular momentum that is conventionally attributed to a nucleon quark. Furthermore, since the radius of a subunit is $R_s \simeq 0.6$ fermi for $m_s = 330$ MeV, a cluster of three of these subunits will have an rms radius that is about equal to the measured value of 0.8 fermi for the nucleon.

Now consider electromagnetic effects. If a nucleon subunit is spinning at the full relativistic limit (with angular velocity $\omega = c/R$), and if a unit electrical charge e is placed on the subunit and allowed to move freely, it will be forced to the equator of the spinning sphere by the action of the electromagnetic forces. An equatorial current loop formed by the charge e moving at the velocity of light gives rise to a magnetic moment $\mu = \pi R^2 \cdot i = \pi R^2 \cdot e/c \cdot \omega/2\pi = eR/2$, in Gaussian units. Since we have assigned the radius $R = \hbar/mc$ to each subunit, where m is the mass of the subunit, the calculated magnetic moment is $\mu = e\hbar/2mc \simeq 2.8$ nuclear magnetons for $m = 330$ MeV. This is the same magnetic moment that is conventionally assumed for a singly-charged nucleon quark.[2] Thus this simple model for a nucleon subunit has the angular momentum and magnetic moment properties that are customarily attributed to an integrally-charged quark, and three of these subunits form a nucleon that has the proper mass and also the proper rms charge and magnetic moment radii (as can be shown by a simple calculation).

In addition to reproducing both the magnetic moment and the rms charge and magnetic moment radii of a nucleon subunit, an equatorial current distribution has other interesting properties: (1) the charge and magnetic moment radii are equal to one another; (2) low-energy electron scattering off a current loop, where the wavelength is large as compared to the dimensions of the loop, appears as a dipole interaction; (3) high-energy electron scattering off a current loop, where the wavelength is small as compared to the dimensions of the loop, appears as a pointlike (parton) interaction. All three of these properties have been verified experimentally for the nucleon.

We will consider the properties of the nucleon in more detail in Chapter 18, where the geometry of the elementary particle is discussed at some length. The point we are seeking to establish here is that in the case of the nucleon, which is the only elementary particle whose size is known experimentally, the mass, spin, and electromagnetic properties are quantitatively reproduced by a model in which spin angular momenta derive from the rotation of extended mass distributions, and magnetic moments derive from the rotation of extended charge distributions. The fact that the "size" of the nucleon is consistent with these results would not be regarded as surprising by elementary particle theorists. But the point at which our preconceptions become strained is when we move on to the case of the muon, whose size has not been measured, but is commonly believed to be small. Hence, when studying the muon model that is described in the next section, it is important to keep in mind the example of the nucleon — which in fact is the only reliable elementary particle example that we have.

7.3 A Model for the Muon

In this section we develop a mathematical model for the muon. The model is that of a relativistically-spinning mechanical mass with an equatorial charge distribution. The rationale for this model was developed from the results in Chapter 5, where we produced evidence for the existence of a spinless mass quantum M = 70 MeV; and from the results in Chapter 6, where we demonstrated that a relativistically-spinning form M_S of the quantum M has an observed mass M_S = 105 MeV, which is the same as the mass of the muon. We also showed in Chapter 6 that if the spinning quantum M_S is assigned the Compton radius $R_S = \hbar/M_S c$, then it has a calculated spin angular momentum $J = 1/2\ \hbar$. In the present chapter we extend these results and show that this model also reproduces the electromagnetic properties of the muon.

It is the use of the Compton radius R_S for the muon that raises difficulties with our preconceptions about the size of the muon. The Compton radius is a large radius, and it does not square with the notion that the muon is a "point particle". However, as we will demonstrate, the electromagnetic "size" effects of this large muon cancel out in a rather spectacular manner at low energies, and the Compton radius of the muon is what is actually observed at high energies if we use the relativistic mass to define this radius. The principal virtue of this model lies in the fact that it gives the main properties of the muon — its mass, spin angular momentum, and magnetic moment — as directly calculated quantities. However, the model does more than that: it relates the vanishing of the electric quadrupole moment to the quantum mechanical vector relationship for the spin; it suggests an energy conservation mechanism (see Chapter 14) that may account for the spin quantization of the muon; it also suggests a possible reason for the fact that the muon, in contrast to the spinless quantum M, does not interact hadronically; and, as we discuss in Section 7.4, this model seems to serve as a bridge between classical and quantum mechanical concepts with respect to the de Broglie wavelength of the particle.

The muon model that we describe here is highly unorthodox, but it does give correct numerical answers. And, as Lord Kelvin points out in the quotation at the beginning of the chapter, a model that is numerically correct may also possess physical significance.

As the starting point for this model of a muon, we take a solid sphere of mechanical matter that has a rest mass M_o = 70 MeV. This is the mass quantum M whose existence was delineated in Chapter 5. Working always in the laboratory (inertial) frame of reference, for reasons that were described in Chapter 6, we now rotate this sphere until it is spinning at the full relativistic limit, with its equator moving at the velocity of light (or infinitesimally below that value). From Eq.(6.8), the observed mass M_s of the relativistically-spinning sphere is half again as large as the rest mass M_o, so that M_s = 105 MeV, which is the observed mass of the muon.

Our next step is to ascribe to this spinning sphere the Compton radius

$$R_s = \hbar/M_s c = 1.9 \times 10^{-13}\ \text{cm} \quad . \tag{7.1}$$

Then, from Eqs. (6.10)-(6.13), the calculated angular momentum of the spinning sphere is $J = 1/2\ \hbar$, which is the observed spin of the muon.

Finally, we place an electric charge e on the spinning sphere and allow it to move freely. A rotating electric charge constitutes a current loop, and the mutual repulsion of the magnetic field lines that pass through this current loop makes it expand. Thus the freely-moving electric charge e is forced to the equator of the spinning sphere, where it appears as an electric current that is moving at essentially the velocity of light. The magnetic moment produced by this current loop is

$$\mu = \pi R_s^2 \cdot i = \pi R_s^2 \cdot e/c \cdot \omega/2\pi = eR_s/2 \quad , \tag{7.2}$$

where we use Gaussian units for the current i. Inserting the Compton radius, Eq.(7.1), into this equation, we obtain

$$\mu = e\hbar/2M_s c \tag{7.3}$$

as the calculated magnetic moment of the spinning sphere, which is the observed magnetic moment of the muon.

This calculational model has accurately accounted for the first-order properties of the muon without using any arbitrarily-adjusted parameters, which is a considerable accomplishment. We must now examine some of the ramifications of this model, which involve mainly its electromagnetic aspects.

One electromagnetic problem about which we can say very little has to do with the question of radiation: a point charge that is moving in a circle should radiate. However, if the charge on the spinning sphere is actually in the form of a ring rather than a discrete point (which gives rise to the same magnetic moment), it will not radiate. As an alternate explanation, the point charge e, due to the non-Euclidean geometry it experiences on the spinning sphere, may think it is going in a straight line, and hence will not radiate. In any event, electrons move in circular orbits in atoms without radiating, so that the problem posed here by the rotating charge e is probably not a serious one.

There is another electromagnetic problem with this model of the muon. But this problem has a straightforward, and in fact intriguing, solution. The solution is intriguing because it leads directly to a quantum mechanical result that was formulated on quite different grounds. The problem we are referring to here is the fact that the equatorial current loop on the muon, Eq. (7.2), gives rise to electric quadrupole, octupole, ... moments.[3] There is a general theorem[4] which states that a spin 1/2 particle cannot have an observable quadrupole moment. Thus, if we are attempting to produce a realistic model for the muon, we should construct a model in which the electric quadrupole moment vanishes with respect to the z axis of quantization (this is the only observable axis for a spin 1/2 particle, since the particle aligns itself parallel or antiparallel to this axis). The way that we can make the electric quadrupole moment of the muon vanish is to tilt the equator of the muon with respect to the z axis, as is shown is Figure 7.2. The projection of the electric quadrupole moment onto the z axis is proportional to $P_2(\cos\theta)$, where P_2 is a Legendre function and θ is the angle between the z axis and the axis of the current loop. Now

$$P_2(\cos\theta) = 1/2\,(3\cos^2\theta - 1) \quad , \tag{7.4}$$

so if we set $\cos\theta = 1/\sqrt{3}$, then $P_2(\cos\theta)$ vanishes identically along the z axis. Thus, by tilting the rotating muon in this manner, we have almost solved our problem. In order to completely solve this problem, we must now increase the radius of the spinning sphere by a factor of $\sqrt{3}$, so that

$$R_s = \sqrt{3}\hbar/M_s c \quad . \tag{7.5}$$

When we do this, the spin angular momentum, Eq. (6.13), and the magnetic moment,

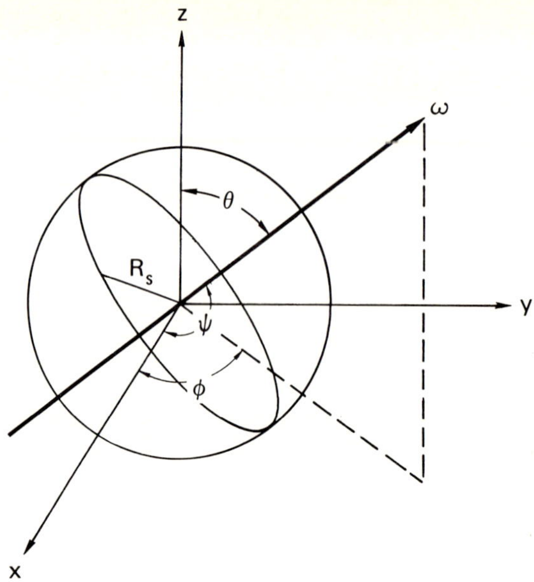

Figure 7.2. The muon as a relativistically-spinning sphere with an equatorial charge distribution. This model correctly reproduces the mass, spin angular momentum, and magnetic moment of the muon, using no adjustable parameters. The radius R_S of the spinning sphere is proportional to its Compton radius [see Eqs. (7.1) and (7.5)]. When the axis of rotation of the muon is tipped by the angle $\theta = \cos^{-1} 1/\sqrt{3}$ with respect to the z-axis of quantization, as is specified by the rules of quantum mechanics, the electric quadrupole moment of the equatorial charge distribution vanishes identically along the z axis, and it vanishes along the x and y axes when averaged over a cycle of spin precession ϕ. Thus this large muon appears in its electromagnetic interactions as a much smaller object (by one order in $\alpha = e^2/\hbar c$).

Eq. (7.3), are each increased by a factor of $\sqrt{3}$, and their projections along the z axis of quantization assume the proper values:

$$J_z = 1/2\ \hbar \qquad (6.13')$$

and

$$\mu_z = e\hbar/2mc\ . \qquad (7.3')$$

Now let us study what we have just accomplished. In order to achieve a vanishing electric quadrupole moment along the z axis, and at the same time maintain the proper z axis projections, we tilted the spinning sphere by the angle $\cos\theta = 1/\sqrt{3}$ and increased its radius by a factor of $\sqrt{3}$. The spin angular momentum of this tilted and expanded sphere is $J = \sqrt{3}/2\ \hbar$, and its projection along the z axis is $J_z = 1/2\ \hbar$. But this classical procedure has just led us to a well-known quantum mechanical result: according to the rules of quantum mechanics, a spin $J = 1/2$ particle is characterized by a spin vector $\vec{J}$ which has a total length $|\vec{J}| = \sqrt{J(J+1)}\ \hbar = \sqrt{3}/2\ \hbar$, and a projection along the z axis $J_z = 1/2\ \hbar$, so that the vector $\vec{J}$ is inclined at an angle $\cos\theta = 1/\sqrt{3}$ with respect to the spin axis; hence we can see that the quantum mechanical spin vector $\vec{J}$ is identical to the spin vector of the spinning sphere shown in Fig. 7.2. Thus the requirement that a spin 1/2 particle should obey the vector rules of quantum mechanics is identical in the present model to the requirement that the particle should have a vanishing electric quadrupole moment along the z axis of quantization. In requiring this model of the muon to correspond to physical reality, we are thereby requiring it to follow a rule of quantum mechanics which was conceived quite independently of the present model.

There are further ramifications of this result. The muon, as we know, interacts electromagnetically with other particles. In the present model, it is the equatorial current loop of the muon which is responsible for this interaction. When we make a multipole expansion† for the potential of the current loop,[3] the lowest term is the coulomb potential, which appears as a point-like interaction. The second term is the electric quadrupole moment, and it is this term which gives a finite "size" to the muon. Now we have already seen that the electric quadrupole moment of the tilted spinning sphere (Fig. 7.2) vanishes identically along the z axis. Thus the muon behaves like a point particle with respect to the z axis. But this same condition also holds, on the average, for the x and y axes as well. In Fig. 7.2, the angle ψ is the angle between the spin axis and the x axis, so that the projection of the electric quadrupole moment along the x axis is proportional to $P_2(\cos\psi)$. But $\cos\psi$

---

†This multipole expansion is in powers of α^2, just as in Q.E.D.

$= \sin\theta \cos\phi$, where ϕ is the angle of precession of the spin axis of the sphere as it precesses slowly around the z axis of quantization. Since $\cos\theta = 1/\sqrt{3}$ means that $\sin\theta = \sqrt{2/3}$, we have

$$P_2(\cos\psi) = 1/2(2\cos^2\phi - 1) \quad . \tag{7.6}$$

If we average this equation over a cycle of precession, the fact that the average value of $\cos^2\phi$ is 1/2 leads to the vanishing of the average value of $P_2(\cos\psi)$. Hence the electric "size" of the muon, which cancels out identically along the z axis, cancels out on the average along the x and y axes as well. Thus this large model of the muon, which we require in order to reproduce the angular momentum and the magnetic moment of the muon, appears observationally in its electromagnetic interactions as a much smaller object. In particular, the vanishing of the electric quadrupole moment decreases the observed electric "size" of the muon by one power of $\alpha = e^2/\hbar c$, so that, from an observational point of view, the muon Compton radius, $R_s = \hbar/M_s c$, goes over into the "classical muon radius", $R_{cl} = R_s \times \alpha = e^2/M_s c^2$.

The results that we have been discussing thus far apply mainly to low energy interactions of the muon. At high energies, it has been well established experimentally that the muon behaves as a point particle — that is, as a particle whose interaction range is consistent with its de Broglie wavelength, $\lambda\!\!\!^{-} = \hbar/mv$. But at high energies $v \simeq c$, so that $\lambda\!\!\!^{-}$ is essentially just the Compton radius which corresponds to the relativistic mass m. Now, the spin angular momentum of the relativistically-spinning sphere is $J_z = 1/2\ mRc$ [Eq. (6.11)]. When the sphere is given a translational velocity v, its mass m increases relativistically, and the radius R, in order to conserve angular momentum, must correspondingly decrease, as we discuss in Section 7.4. Hence the sphere must obey Eq. (7.5) for all values of the mass m, which is in agreement with the high energy experimental observations.

There is another ramification of this muon model that is of interest here. In this model we have posited the existence of a mechanical mass that is principally responsible for the inertial properties of the muon; the electromagnetic characteristics of the muon are supplied by the equatorial current loop. Since the interaction of the muon with other particles is known to be accurately electromagnetic, this means that the mechanical mass itself must have only a very small interaction with these other particles. Thus if we could in some manner remove the electric charge e from the equator of the spinning muon, we would be left with a neutral spinning mass that has angular momentum $J = 1/2\ \hbar$ and a very weak interaction with hadronic matter. Is there experimental evidence for an object of this nature? Yes, there is; these are precisely the properties of the neutrino. Hence, in introducing the weakly-interacting mechanical mass of the muon, we are not going beyond the bounds of observation. Of course, the masses of the known neutrinos are much smaller than the mechanical mass that is used in this muon model, and neutrinos are in fact customarily assumed to be massless particles. But the masslessness of the neutrino has not been established experimentally, and arguments appear in the literature from time to time which suggest that neutrinos may possess rest masses[5]; the present results suggest this same conclusion. The importance of this discussion lies in the fact that if the muon model we have been studying does have physical content, then we expect the ramifications of the model to lead to testable consequences in other areas of particle physics, and this seems to be the case.

In this model of the muon, spin quantization is associated with the fact that the muon is rotating at the relativistic limit, with its equator moving at the velocity of light. The question of spin quantization in general is discussed in Chapter 14. We note here that since a fully-relativistic spinning sphere is half again as massive as its nonspinning counterpart [Eq. (6.8)], the transformation of <u>three</u> nonspinning quanta into <u>two</u> spinning quanta (with the annihilation of one of the three original quanta) is isoergic if and only if the two spinning quanta are rotating at the full relativistic limit. From this viewpoint, spin quantization appears as a direct consequence of energy conservation. Spin quantization seems to be associated with the transition between spinless and spinning mass quanta, and in the meson resonances this transformation can be traced in both directions — from spinless quanta to spinor pairs and back again (Chapter 14).

7.4 The de Broglie Wavelength of the Muon

In Section 7.3 we demonstrated that a relativistically-spinning sphere with an equatorial charge distribution accurately reproduces the main "particle" properties of the muon — namely, its mass, spin angular momentum, and magnetic moment. We also demonstrated that the electric quadrupole moment of this muon model vanishes if, and only if, we require the spin vector $\vec{J}$ of the muon to obey the usual quantum mechanical vector relationship for a spin 1/2 particle: $|\vec{J}| = \sqrt{1/2(1/2+1)}\,\hbar$ and $J_z = 1/2\,\hbar$. The significance of this last result is that, starting from a strictly classical viewpoint, we have ended up with an important quantum mechanical rule. Thus we have established a bridge between classical physics and quantum physics — that is, between the "particle" aspects and the "wave" aspects of the muon. In Section 7.4 we extend these results by demonstrating that this "particle" model of the muon also accounts for its most important "wave" property — its de Broglie wavelength.

We start by sketching the general derivation of the de Broglie wavelength. The de Broglie wavelength of a muon (or electron) is given by the equation

$$\lambda\!\!\!^{-} = \hbar/p \quad . \tag{7.7}$$

This hybrid equation, which Neils Bohr called an "irrationality",[6] combines a patently particle property, the linear momentum $p = mv$, with an equally-patent wave property, the de Broglie wavelength $\lambdabar$. To understand the significance of the wave length $\lambdabar$, we first consider a muon at rest, with energy E_o, mass m_o, and frequency ω_o, as related by the equations†

$$E_o = m_o c^2 = \hbar\omega_o \quad . \tag{7.8}$$

The frequency ω_o is attributed to an unspecified "periodic internal phenomenon" of the muon.[7] Associated with this muon is a vibration

$$\psi = \sin \omega_o t_o \quad . \tag{7.9}$$

If we now view the muon from a frame of reference in which it is moving along the x axis with velocity $v \equiv \beta c$, the special-relativistic transformation of the time coordinate, $t_o \to (t - vx/c^2)/(1 - \beta^2)^{1/2}$, leads to the equation

$$\psi = \sin \omega(t - vx/c^2) \quad , \tag{7.10}$$

where

$$\omega = \omega_o/(1 - \beta^2)^{1/2} \tag{7.11}$$

is the transformed frequency of the vibration. Thus ψ is now a traveling wave that moves with phase velocity

$$V = c^2/v \quad . \tag{7.12}$$

The de Broglie wavelength associated with this traveling ψ-wave is

$$\lambdabar = V/\omega = \hbar/mv \quad , \tag{7.13}$$

where

$$m = m_o/(1 - \beta^2)^{1/2} \quad . \tag{7.14}$$

We also have

$$E = mc^2 = \hbar\omega \quad , \tag{7.15}$$

---

†The $\sqrt{3}$ projection factor that enters into these equations [see Eq. (7.5)] is discussed briefly at the end of this section.

so that the equation $E = \hbar\omega$, which relates the de Broglie frequency ω to the energy E of the muon, is the same in all frames of reference. Finally, we replace the single wave ψ by a wave packet $\psi(\omega)$ centered about the frequency ω. The group velocity of this wave packet is

$$V_g = d\omega/d(1/\lambda) = v \quad , \tag{7.16}$$

so that the wave packet $\psi(\omega)$ moves at the same velocity as the muon.

We must now relate these de Broglie wave equations to the muon particle equations that we developed in Sec. 7.3 and in Chapter 6. To do this, we briefly recapitulate the relevant equations, using a slightly different notation. We start with a sphere that has a rest mass M_{oo} in the inertial frame of reference. We then spin it at the full relativistic limit,

$$\omega_o = c/R_o \quad , \tag{6.4}$$

where R_o is the radius of the spinning sphere, so that it acquires a spinning mass

$$m_o = 3/2\ m_{oo} \quad . \tag{6.8}$$

The calculated spin angular momentum is

$$J = I_o\omega_o = 1/2\ m_oR_oc = 1/2\ \hbar \tag{6.11}$$

where we take

$$R_o = \hbar/m_oc \quad . \tag{6.12}$$

Placing a unit charge e at the equator gives rise to a magnetic moment

$$\mu_o = \pi R_o^2 \cdot i = e\hbar/2m_oc \quad . \tag{7.3}$$

The contribution of this equatorial ring of charge to the total energy of the spinning muon is small; it is of order $e^2/R_o = e^2/\hbar c \cdot m_oc^2 = \alpha m_oc^2$, which is the order of magnitude of the fine-structure correction to the Dirac magnetic moment of the muon. Thus the total energy of the spinning muon is essentially

$$E_o = m_oc^2 = \hbar\omega_o \quad , \tag{7.17}$$

which follows from Eqs. (6.4) and (6.12). Finally, we give the spinning muon a translational velocity $v \equiv \beta c$, where, as emphasized above, we work always in the laboratory frame of reference. The muon experiences a mass increase

$$m = m_o/(1 - \beta^2)^{1/2} \quad . \tag{7.18}$$

In order to conserve spin angular momentum, the radius of the muon decreases, so that

$$R = R_o(1 - \beta^2)^{1/2} \quad ; \tag{7.19}$$

and the angular velocity $\omega = c/R$ of the spinning sphere correspondingly increases, giving

$$\omega = \omega_o/(1 - \beta^2)^{1/2} \quad . \tag{7.20}$$

This is the important result that we require in order to account for the systematic behavior of the de Broglie wave equation, as embodied in Eqs. (7.11)-(7.13).

From the standpoint of special relativity, we customarily think of a moving object as being contracted in size only in the direction of motion — the argument being that the other two coordinates are orthogonal to the motion, and hence are unaffected relativistically. However, in the case of a spatially-extended spinning object, as we have here, the effect of the spin angular momentum is to couple all three of the spatial axes together dynamically, so that they are all affected by the translational motion. It should be noted that this shrinking of a relativistically-spinning muon or electron† under translational motion is not just a mathematical artifact; it has observable consequences which can be, and in fact have been, measured experimentally. From Eqs. (7.2), (7.19), and (7.20), we see that the magnetic moment of a moving electron is

$$\mu = \mu_o(1 - \beta^2)^{1/2} \quad , \tag{7.21}$$

so that μ is a function of the electron velocity. Since, by our construction, the spin angular momentum $J = 1/2\ \hbar$ of the electron [Eq. (6.11)] is an invariant, the ratio of the spin to the magnetic moment — that is, the gyromagnetic ratio g of the electron — is also a function of the electron velocity: $g = g(v)$. The gyromagnetic ratio $g(v)$ has been measured in the Larmor rotation experiment of Louisell, Pidd, and Crane,[8] and it exhibits just the dependence on v that we expect from Eqs. (6.11) and (7.21). In analyzing the results of the LPC experiment,[8] it is significant, as has been commented upon elsewhere,[9] and as we have emphasized above, to view these results

---

†We have been applying these equations to the muon, but the same equations apply (with a scaling in mass) to the electron.

in the laboratory frame of reference.

From Eqs. (7.18) and (7.20), we see that this model of a spinning muon (or electron) gives the relations

$$E = \hbar\omega = mc^2 \tag{7.22}$$

in all frames of reference. Furthermore, by comparing Eqs. (7.20) and (7.22) with Eqs. (7.11) to (7.16), we can ascertain that the properties of the muon coincide with the properties of the moving de Broglie wave packet $\psi(\omega)$. In particular, we now have a physical muon frequency that corresponds to the de Broglie frequency, since the frequency of rotation ω of the relativistically-spinning sphere [Eq. (7.20)] is equal to the de Broglie frequency that appears in Eq. (7.13), where everything is measured in the laboratory frame of reference. In order to understand the force of this remark, it is important to note that the derivation of the de Broglie wavelength given in Eqs. (7.8)-(7.16), while a standard one,[10] is not the picture that was originally envisioned by de Broglie. In his derivation, de Broglie started[7] by assuming that an electron (or any particle) has a natural vibrational frequency

$$\omega_o = E_o/\hbar = m_o c^2/\hbar \quad , \tag{7.23}$$

as observed in the rest frame of the particle, where the frequency ω_o is ascribed to a periodic internal phenomenon of unknown origin. When the electron is set in motion with velocity $v = \beta c$, a stationary observer measures a decreased vibrational frequency

$$\omega_1 = \omega_o(1 - \beta^2)^{1/2} < \omega_o \quad , \tag{7.24}$$

due to the relativistic dilation of time. However, since the mass of the moving electron — and hence its energy — is observed to increase as

$$E = m_o c^2/(1 - \beta^2)^{1/2} \quad , \tag{7.25}$$

de Broglie ascribed "a fictitious wave" ("une onde fictive"[7]) with a fictitious angular frequency

$$\omega = E/\hbar = \omega_o/(1 - \beta^2)^{1/2} > \omega_o \tag{7.26}$$

to the electron. He then demonstrated[7] that if the vibrational frequency of the electron is initially in phase with the vibrational frequency of this "fictitious wave", then the phase relationship is maintained; and he also demonstrated that a wave packet formed from these fictitious waves moves with the group velocity v.

However, the de Broglie frequency ω that is connected with the electron, from this de Broglie viewpoint, is not a measurable property of the electron even in principle: the "observable" frequency in this viewpoint is $\omega_1 < \omega_0$, and not $\omega > \omega_0$. Also, this approach gives $E = \hbar\omega$, whereas we would logically expect to make the identification $E = \hbar\omega_1$ since ω_1 is the "observable" frequency. The virtue of the present relativistically-spinning model of the electron or muon is that it leads to Eq. (7.26) rather than to Eq. (7.24).

In the above discussion, we have not inserted the $\sqrt{3}$ projection factor that we introduced in Sec. 7.3 [see Eq. (7.5)]. If we now use this projection factor, we increase the radius of the muon by $\sqrt{3}$ and correspondingly decrease its angular velocity by the same factor, which in turn increases the de Broglie wavelength by $\sqrt{3}$ [Eq. (7.13)]. However, if the de Broglie waves are propagated in a direction which is parallel to the spin axis of the electron, then the projection of the de Broglie wave vector along the axis of spin precession is the same as it was before we introduced this factor.

The model of the muon as a relativistically-spinning sphere has the interesting property that it correctly reproduces the muon mass, spin angular momentum, and magnetic moment without using adjustable parameters. Moreover, by identifying the rotational frequency of the spinning sphere with the de Broglie frequency, and by leading us to the correct quantum mechanical relationship for the spin vector, it seems to be combining together the separate domains of classical physics and quantum physics. And, by the properties that we have to ascribe to the mechanical mass of the muon, it suggests a relationship to the very-weakly-interacting spin 1/2 neutrinos. As a final result, the equatorial current loop of the muon, which is responsible for its interactive properties, appears mathematically as an expansion in powers of the fine structure constant $\alpha = e^2/\hbar c$,[†] and hence is suggestive of the formalism of quantum electrodynamics, which is known to apply to the muon. Thus the ramifications of this mathematical model for the muon extend well beyond the original purpose for which it was devised. It seems almost inexplicable that a model which ties together so many different facets of elementary particle physics has apparently never been discussed, except by the present author, anywhere in the scientific literature.

[†]We have not demonstrated this result, but it is a simple one to ascertain by making a multipole expansion of the current loop, using the muon Compton radius as the radius of the loop.

7.5 Comparison of the Muon with the Hadronic Quantum M = 70 MeV

As the final result in this chapter, we compare the geometry of the spinning muon with the geometry of the hadronic quantum M = 70 MeV. It must be apparent by this time that there is a discrepancy here. We stated in Chapter 5 that the spinless quantum M = 70 MeV is a fundamental building block for hadrons. We then showed in Chapter 6 that a spinning form M_s of the quantum M will have an observed mass of 105 MeV, which we related to the mass of the muon. And we went on in the present chapter to construct a model for the muon based on this mass quantum. However, the muon model that we ended up with (Fig. 7.2) has a radius of about 3 fermis, whereas the nucleon, which contains nine spinning quanta M_s, has an rms radius of only 0.8 fermi. Thus the hadronic 70 MeV quantum M must have a much smaller size than does the muonic 70 MeV rest mass (we assume that the hadronic quanta M = 70 MeV and $M_s \simeq 110$ MeV have the same "sizes"). The nucleon spinor S contains three subquanta M_s in a linear array, and the geometry of the spinor S is discussed in Chapter 18; one principal result which emerges from this discussion is that, just as the muon has a radius R_μ that reflects its Compton wavelength $\lambda\!\!\!^{-} = \hbar/m_\mu c$, so also the nucleon spinor S has an equatorial radius R_s = 0.6 fermi that reflects its Compton wavelength $\lambda\!\!\!^{-} = \hbar/m_s c$, where m_s = 330 MeV. Thus, since the spinor S is three times as massive as the muon, its radius is three times as small. Furthermore, the spinor S does not appear to have the factor of $\sqrt{3}$ that we have added [Eq. (7.5)] to the muon radius (otherwise the nucleon would be too large), so that the radius of the spinor S is actually a factor of $3\sqrt{3}$ smaller than that of the muon. The total volume of the nucleon, as obtained from the measured volumes of large atomic nuclei, indicates that the nucleon spinor S has a height of 2 fermis. Hence the three spinning quanta M_s in the spinor S must be in the form of flattened oblate spheroids. (One consequence of this last result is that the spinning mass of the spinor S is slightly larger than 3/2 times its rest mass [Eq. (6.16)], as is evidenced by the fact that the spinor mass of 330 MeV (see Appendix D) is slightly more than three times the mass of the muon, m_μ = 105.7 MeV.)

The comparative sizes of the muon and the spinor S (which has three subquanta M_s) are shown in Fig. 7.3. As can be seen, the 70 MeV mass of the muon is much more diffuse than the 70 MeV mass of the quantum M. At first thought, it is somewhat disconcerting to find two mass quanta which both have rest masses of 70 MeV, but which have radically different sizes. However, this is phenomenologically just what we require. The compact quantum M interacts hadronically, whereas the expanded muon does not, and the difference in their geometries furnishes a logical explanation for this difference in their interactions. (As we will see in the ensuing chapters, the geometry shown in Fig. 7.3 for the hadronic quantum M seems to apply to all of the meson and baryon substates — both spinning and nonspinning.†) In the dominant decay

†However, see the comments in Sec. 8.7 of Chapter 8 and in Sec. 14.6 of Chapter 14.

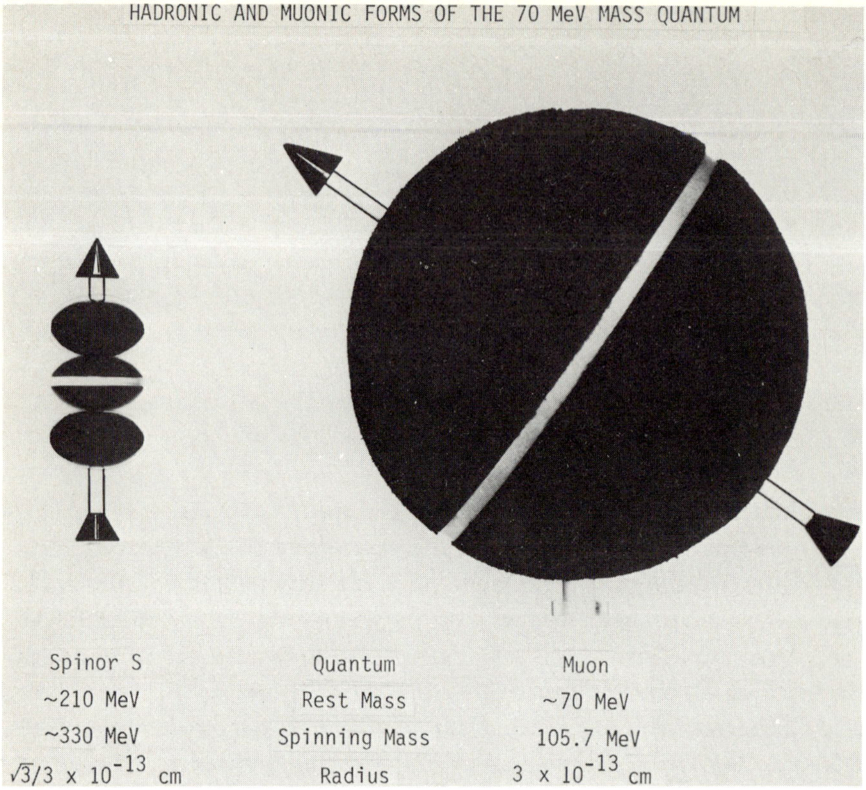

Figure 7.3. A comparison between the leptonic mass quantum μ = 70 MeV as it appears in the muon, and the hadronic mass quantum M = 70 MeV as it appears in the spinor S. The muon contains one quantum μ in a relativistically-spinning configuration, μ_S; and the spinor S contains three quanta M in a relativistically-spinning configuration, $M_S M_S M_S$. As can be seen in Fig. 7.3, the non-hadronic mass quantum μ is an order-of-magnitude larger than the hadronic mass quantum M. This may explain why the quantum M interacts hadronically whereas the quantum μ does not; the muon interacts only via its equatorial charge distribution, so that the spin 1/2 mechanical mass μ_S of the muon has a very small neutrino-like interaction cross section.

process

$$\pi^+ \rightarrow \mu^+ + \nu \tag{7.27}$$

which we can write as

$$\pi^+ \equiv M^+\overline{M}^o \rightarrow M^+[M^o] \rightarrow M^+_{spinning} + \nu \equiv \mu^+ + \nu \quad , \tag{7.28}$$

the neutral hadronic quantum $\overline{M}^o$ is annihilated, and the charged hadronic quantum M^+ is transformed into a spinning muon μ^+. Thus, in this decay process we have the direct transformation of a hadronic 70 MeV quantum M into a muonic 70 MeV quantum. The hadronic quantum M can expand to form the muon; however, from its lack of hadronic interactions, the muon does not seem to be able to reverse this process.

If we combine the observations of Bertrand Russell in Chapter 2 and Lord Kelvin in the present chapter, then, in view of the fact that we can't actually "see" a muon, the best thing we can do is to construct a mathematical model which reproduces its properties. The muon model that we have presented here does reproduce the principal spectroscopic properties of the muon; and, as we have tried to illustrate, it has ramifications which are pertinent to other areas of elementary particle physics. In particular, the muon is at least distantly related to its hadronic counterpart, the spinor S = 330 MeV, whose existence is required in order to reproduce the nucleon. The properties of the spinor S are discussed in the next chapter.

References for Chapter 7.

1. As an historical note, it was the necessity in the light-quark model for reproducing the nucleon as a multiple of the muon mass which led to the search for a model of the muon. And the investigation of the relativistically-spinning sphere of matter was motivated by the observation that if the moment-of-inertia I of a spinning sphere is $I = 1/2\ MR^2$ rather than the classical value $I = 2/5\ MR^2$, then the gyromagnetic ratio is given correctly for an equatorial current loop. Thus it was phenomenological necessity that led to the considerations described in Chapters 6 and 7, which may account for the reason that the relativistically-spinning sphere has not previously been investigated. As the first worker in the field to investigate the light-quark model in detail, the present author was the first elementary particle physicist to be forcibly confronted with these problems.

2. See B. T. Feld, Models of Elementary Particles, Blaisdell, Waltham (1969), page 339, Eq. (18a'). However, it should be noted that if we form the nucleon from three spinors that have equatorial charge distributions, then the proton and neutron magnetic moments are each too large by a factor of three (see Chapter 18). Since an equatorial charge distribution represents the largest possible magnetic moment, this is in principle an easy difficulty to correct. But it is desirable to base this correction on some definable aspect of the phenomenology rather than on just an arbitrary alteration of the charge distribution; and, to date, no convincing phenomenological basis for making this correction has suggested itself.

3. See, for example, W. R. Smythe, Static and Dynamic Electricity, McGraw-Hill, New York (1939), page 137.

4. N. F. Ramsey, Experimental Nuclear Physics, edited by E. Segre, Wiley, New York (1953), Vol. I, page 365.

5. For example, see J. N. Bachall, N. Cabibbo, and A. Yahil, Phys. Rev. Lett. 28, 316 (1972). The current experimental status with respect to neutrino masses is summarized in a Brookhaven National Laboratory report by G. R. Kalbfleisch: BNL-20227, "The Velocity of the Neutrino", May (1975).

6. M. Born, Atomic Physics, Hafner, New York (1959), page 89.

7. See M. Jammer, The Conceptual Development of Quantum Mechanics, McGraw-Hill, New York (1966), page 243.

8. W. H. Louisell, R. W. Pidd, and H. R. Crane, Phys. Rev. 94, 7 (1954).

9. See Comment I on page 296 of B. Cagnac and J. C. Pebay-Peyroula, Modern Atomic Physics, Halsted Press, New York (1975).

10. See A. P. French, Principles of Modern Physics, Wiley, New York (1958), pages 175-177.

The quotation by Lord Kelvin at the beginning of Chapter 7 is taken from W. Thomson, Popular Lectures and Addresses, Macmillan, London, ed. 2 (1891), Vol. 1, page 80.

W. Pauli's quotation about the muon at the beginning of Chapter 7 is a somewhat apocryphal but popular statement that has been attributed to a number of different physicists.

CHAPTER 8. EVIDENCE FOR THE SPIN 1/2 SPINOR S = 330 MEV

"In our framework, the masses of the quarks have a realistic meaning .... Their masses may be estimated from observed magnetic moments .... We obtain $m_p = p/\mu(p) = 336$ MeV."

De Rújula, Georgi, and Glashow, Physical Review D12, 147 (1975).

8.1 The Analogy Between the Spinor S and the Spinless State 3

In Chapter 5 we summarized the experimental evidence which points to the existence of the spinless hadronic mass quantum M = 70 MeV. The quantum M characteristically appears in meson and baryon resonances in the form of substates that are composed of 1, 3, 4, or 7 quanta M, which we denote for simplicity as the spinless basis states 1, 3, 4, and 7. Since the masses in these states are additive, they have the mass values, apart from coulomb effects, of 70, 210, 280, and 490 MeV, respectively. The π meson, $\pi \equiv 1\bar{1} = 140$ MeV, also serves as a spinless basis state in the meson resonances. The main properties of these spinless basis states were sketched in Chapter 5, and they are developed in more detail in Chapters 15 - 19 and in Appendices C, D, and E.

The delineation of these spinless hadronic basis states is perhaps the most unique result that has emerged from the present light-quark model, since the more conventional quark models of Fermi and Yang and of Gell-Mann and Zweig are based entirely on the use of spin 1/2 basis states. But, in addition to spinless basis states, the light-quark model must of course also include at least one spin 1/2 basis state in order to be able to reproduce the spin 1/2 baryon resonances. As we have already seen in Chapters 6 and 7, a spin 1/2 nonhadronic basis state occurs in the form of the spinning muon. The question to which we must now address ourselves concerns the number of spin 1/2 hadronic basis states that are required in order to reproduce the observed spectrum of baryon and meson resonances. The answer to this question turns out to be remarkably simple: only one spin 1/2 hadronic basis state is required — the spinor S ≃ 330 MeV (however, see Sec. 8.7). The reason for this simplicity lies in the fact that spinors S occur in combination with spinless quanta, and it is the spinless quanta which are responsible for properties such as the strangeness quantum number.

Our aim in the present chapter is to analyze the experimental evidence which points to the existence of the spinor S. At the end of Chapter 4 we briefly discussed the properties of the spinor S, and we related them to the properties of its spinless counterpart, the spinless basis state 3 ≡ 3M = 210 MeV. Since this information is essential for a proper examination of the experimental data, we recapitulate these results here. A complete description of these basis states is given in Appendix C.

The spinor S ≃ 330 MeV has an observed mass that is slightly more than half again as large as the mass of the spinless basis state 3 = 210 MeV. Thus, from the discussions in Chapters 6 and 7 [see Eqs. (6.8) and (6.16)], it is clear that $S \equiv (3)_{\text{spinning}}$. In fact, it is possible to identify transformations between the states S and 3, as we describe in detail in Chapters 14 and 17. Now, the fundamental basis state M = 70 MeV occurs in both particle (M) and antiparticle ($\bar{M}$) forms, corresponding to its isotopic spin I = 1/2. Thus we have several ways in which we can construct the basis states 3 ≡ 3M and $S \equiv (3M)_{\text{spinning}}$. The spinless quantum 3 appears in both unmixed (all subquanta M or else all subquanta $\bar{M}$) and mixed (subquanta M and $\bar{M}$) representations, which we denote as 3 ≡ 3M and $\bar{3} \equiv 3\bar{M}$ (unmixed), and $\dot{3} \equiv M\bar{M}M$ or $\bar{M}M\bar{M}$ (mixed). As might be expected, the unmixed states have much stronger binding energies than do the mixed states. In fact, with the mass values assumed here [Eq. (5.1)], the $3\bar{3}$ binding energy is ∿4% [see Eq. (5.2)], whereas the $\dot{3}$ binding energy (e.g., $\dot{3}\dot{3}$ or $\dot{3}3$ or $\dot{3}\bar{3}$) is ∿0% (that is, right on the mass shell). The pairs 33 and $\bar{3}\bar{3}$ do not bind hadronically. In direct analogy to these spinless basis states, the spinor S appears in both unmixed (S and $\bar{S}$) and mixed ($\dot{S}$) forms, with ∿4% and ∿0% binding energies, respectively, and the spinor pairs SS and $\bar{S}\bar{S}$ do not bind hadronically. (Although the mixed states $\dot{3}$, $\dot{4}$, and $\dot{S}$ have ∿0% binding energies, they still bind to each other and to the unmixed states hadronically.) The fundamental (unbound) mass $S^{\pm} \simeq 330$ MeV is the same in both mixed ($\dot{S}$) and unmixed (S and $\bar{S}$) forms, just as the basis states $\dot{3}$, 3, and $\bar{3}$ all have the same (unbound) mass. Mixed and unmixed forms of the spinor S appear in hadron resonances in a very characteristic manner: <u>baryon</u> resonances each contain a Fermi-Yang spinor triplet — $S\bar{S}S$ — which is composed of <u>unmixed</u> spinors, and hence has a 4% binding energy; <u>meson</u> resonances, on the other hand, contain pairs of <u>mixed</u> spinors — $\dot{S}\dot{S}$ or $\dot{S}\dot{S}\dot{S}\dot{S}$ —, which have ∿0% binding energies. Both the baryon resonances and the meson resonances usually also contain spinless excitation quanta which are combined with the spinors S.

Unmixed spinor triplets $S\bar{S}S$ and mixed spinor pairs $\dot{S}\dot{S}$ differ drastically in their decay properties. Whereas the mixed spinor pairs $\dot{S}\dot{S}$ and $\dot{S}\dot{S}\dot{S}\dot{S}$ in the meson resonances usually decay quickly ($\tau < 10^{-21}$ sec), the unmixed spinor triplets $S\bar{S}S$ in the baryon resonances do not decay at all. The $S\bar{S}S$ spinor triplets are stable, and the conservation of the baryon quantum number is precisely equivalent to the conservation of these $S\bar{S}S$ triplets.

With this preamble, let us now turn to the experimental and theoretical evidence which indicates the existence of the spinor S ≃ 330 MeV.

8.2 The Nucleon N and the Bound State $\bar{N}N$

The most direct evidence for the properties of the spinor S comes from its stablest configurations, the nucleon N and antinucleon $\bar{N}$, which are the spinor clusters

$$N \equiv S\bar{S}S \qquad \text{and} \qquad \bar{N} \equiv \bar{S}S\bar{S} \ . \tag{8.1}$$

When we include coulomb effects (see Chapters 18 and 19), the proton and neutron particle and antiparticle states appear with the following charge and spin configurations (where we assume integral electric charges):

$$\begin{aligned} p &\equiv S^{+}_{\uparrow}S^{-}_{\downarrow}S^{+}_{\uparrow} \ , & n &\equiv S^{+}_{\uparrow}S^{--}_{\downarrow}S^{+}_{\downarrow} \ , \\ \bar{p} &\equiv S^{-}_{\uparrow}S^{+}_{\downarrow}S^{-}_{\uparrow} \ , & \bar{n} &\equiv S^{-}_{\uparrow}S^{++}_{\downarrow}S^{-}_{\downarrow} \ . \end{aligned} \tag{8.2}$$

If the three spinors in the nucleon had zero binding energies, they would each, apart from coulomb differences, have a mass that is 1/3 the nucleon mass, or 313 MeV. However, if they have appreciable binding energies, then the (unbound) spinor masses must be correspondingly increased. By using the fact that the nucleon we have constructed, $N = S\bar{S}S$, is a Fermi-Yang[1] nucleon, we can make an estimate of the spinor binding energies and masses in the following manner.

In the Fermi-Yang model of the nucleon[1], the nucleon N serves as the prototype for the spinor S, and both N and S are assumed to have the same basic quantum numbers. We can extend this model by assuming that N and S also have the same binding energies. The experimental nucleon and antinucleon binding energies are

$$\text{B.E.} \quad pp = nn = 0, \quad pn \simeq 0.1\%, \quad \bar{p}n = 4.4\%^{2}, \tag{8.3}$$

which we can generalize as

$$\text{B.E.} \quad NN = \bar{N}\bar{N} \simeq 0, \quad \bar{N}N \simeq 4\%. \tag{8.4}$$

If we now, in the spirit of the Fermi-Yang quark model, attribute these same binding energies also to the spinors S and $\bar{S}$, we have

$$\text{B.E.} \quad SS = \bar{S}\bar{S} = 0, \quad \bar{S}S = 4\%. \tag{8.5}$$

The proton, $p \equiv S\bar{S}S = S^+_\uparrow S^-_\downarrow S^+_\uparrow$, contains three singly-charged spinors $S^\pm$, and it has two hadronic $\bar{S}S$ bonds. Thus if we denote the unknown mass of the spinor $S^\pm$ as $M^\pm_S$, we have the following equation for $M^\pm_S$:

$$M_p = 938 \text{ MeV} = 3M^\pm_S - 2(4\% \times 2M^\pm_S) = 2.84\, M^\pm_S, \tag{8.6}$$

which gives $M^\pm_S = 330.3$ MeV as the calculated mass of the unbound spinor $S^\pm$.

Now let us calculate the problem in the other direction. We take

$$S = 330 \text{ MeV} \tag{8.7}$$

as the mass of a singly-charged spinor S. For simplicity, we ignore Coulomb effects, which are quite small (these are discussed in Chapters 18 and 19). The binding energy of Eq. (8.5) becomes

$$\text{B.E.} \quad SS = \bar{S}\bar{S} = 0, \quad \bar{S}S = 4\% = 26.4 \text{ MeV}. \tag{8.5'}$$

As we discussed briefly in Sec. 7.5 of the last chapter, the spinor S has what is essentially a cylindrical geometry (see Fig. 7.3), with a radius of 0.6 fermi and an overall height of 2 fermis. Thus, as will be made amply clear in the ensuing chapters, spinors S bind together in a side-by-side manner. The proton p appears as the spinor cluster

$$p = S\bar{S}S \quad \text{or} \quad \bar{S}^{S}_{S}\,, \tag{8.8}$$

with a calculated mass of 3 x 330 - 2 x 26.4 = 937.2 MeV; the experimental proton mass is 938.3 MeV. The $\bar{p}n$ bound state[2] appears as the spinor cluster

$$\bar{p}n = \begin{matrix} \bar{S} & S & \bar{S} \\ S & \bar{S} & S \end{matrix} \quad \text{or} \quad S^{\bar{S}}_{\bar{S}}{}^{S}_{S}\bar{S}\,, \tag{8.9}$$

with a calculated mass† of 6 x 330 - 7 x 26.4 = 1795.2 Mev; the experimental $\bar{p}n$ bound state mass[2] is 1794.5 MeV. Thus this model accurately reproduces the masses of both the nucleon N and the nucleon-antinucleon bound state $\bar{N}N$.

The calculation that we have just carried out may seem to be a simple result, but it is the <u>only</u> quantitative explanation that has been advanced to account for

---

†In the mass calculation of Eq. (8.9), we assume that the hadronic $\bar{S}S$ binding energy is very short-ranged, and thus operates only between adjacent $\bar{S}S$ pairs; this assumption is consistent with the large binding energy value (4%), which is more than four times as large as nuclear binding energies and hence has a much shorter range.

the observed mass of the $\bar{p}n$ bound state (the potential model approach to this problem runs into difficulties in reproducing the very narrow width, $\Gamma \leqslant 8$ MeV, of the $\bar{p}n$ resonance). The essential point in the present calculation is that, with side-by-side binding of the nucleon N and antinucleon $\bar{N}$, all possible configurations of the nucleon spinors S and $\bar{S}$ involve <u>three</u> external $\bar{S}S$ bonds plus <u>four</u> internal $\bar{S}S$ bonds, for a total of <u>seven</u> $\bar{S}S$ bonds in all. In Chapter 9 we show that a direct extension of this same model leads to accurate masses for the Λ, Σ, Ξ, and Ω hyperon resonances. The tendency for nucleons to bind in a side-by-side manner can also be traced to the case of atomic nuclei, where, as is shown in Chapter 20, this side-by-side binding accounts quantitatively for the asymmetric fission of the uranium nucleus.

As one final result in this section, we note that the spinless basis state $3^{\pm}$ has a mass of 214.6 MeV. Thus if its spinning counterpart $S^{\pm}$ is half again as heavy [Eq. (6.8)], it will have an observed mass of 321.9 MeV; but if the subquanta M in the spinor $S \equiv (3M)_{\text{spinning}}$ are flattened into oblate spheroids (see Fig. 7.3), the spinning mass will be a few MeV larger than 321.9 MeV [see Eq. (6.16); and see Fig. 18.2 in Chapter 18]. Hence the $S^{\pm}$ mass of about 330 MeV that we quoted in Eq. (8.7) is consistent with the interpretation of the spinor S as a spinning form of the spinless state $3 \equiv 3M$.

By using the measured masses of the nucleon N and the $\bar{N}N$ bound state, we have been able to pinpoint the mass of the nucleon spinor S at 330 MeV (apart from small coulomb effects, which are considered in Chapter 19). In the excited states of the nucleon, the spinor triplet $S\bar{S}S$ occurs in combination with spinless basis states, and in a manner that involves 4% binding energies and/or broad resonance widths, as we discuss in detail in Chapters 9 and 15. Thus the nucleon excited states give no further direct information about the spinor S. Hence we must now turn to the meson resonances for further experimental evidence; these are discussed in Sections 8.3-8.5 below. In Sec. 8.6 we briefly describe some theoretical ideas which are in agreement with these experimental results. And in Sec. 8.7 we mention a required phenomenological extension of the systematics of the spinor S.

8.3 <u>The Low-Mass Spin 1 Meson Resonances</u>

The properties of the spinor S were summarized in Sec. 8.1, where it was pointed out that the spinor S occurs in unmixed (S and $\bar{S}$) and mixed ($\dot{S}$) forms, with the same spinor mass in each form, and with ∿4% and ∿0% binding energies for the unmixed and mixed forms, respectively. Unmixed spinors occur in baryon resonances in the form of spinor triplets $S\bar{S}S$, and mixed spinors occur in meson resonances in the form of pairs of spinors $\dot{S}\dot{S}$. In Sec. 8.2 we used the N and $\bar{N}N$ resonances to delineate the 4% binding energy effects of the spinors S and $\bar{S}$ in detail, and we arrived at the unbound spinor mass $S \simeq 330$ MeV. In the meson resonances of the present section and the two following sections, we will be dealing with mixed spinor pairs $\dot{S}\dot{S}$. Since these, according to our above assertion, have ∿0% binding energies, we expect to find evidence for $\dot{S}\dot{S}$ spinor pairs with masses of about 660 MeV, and $\dot{S}\dot{S}\dot{S}\dot{S}$ spinor quartets with masses of about 1320 MeV (small binding energy and coulomb effects can lower these values slightly). It turns out, as we will demonstrate, that an experimental mass of 1310 MeV is indicated for the spinor quartet $\dot{S}\dot{S}\dot{S}\dot{S}$, and a mass of 650-660 MeV for the spinor pair $\dot{S}\dot{S}$, in close agreement with this estimate. Thus, after we have taken the difference in binding energies into account, the same basic spinor S does seem to occur in both the baryon and meson resonances.

Let us first consider the narrow-width spin 1 meson and kaon resonances that were plotted in Fig. 5.3. In analogy to the spin 1/2 <u>baryon</u> resonances, which are formed as excited states of the nucleon ground state $S\bar{S}S$, it seems logical that these spin 1 <u>meson</u> resonances are formed as excited states of an $\dot{S}\dot{S}$ spinor pair. Now, the lowest-mass resonance shown in Fig. 5.3 is the $\omega(783)$ meson. Does the ω serve as the spin 1 "ground state" for all of these resonances? If we think of the ω as being constructed from a pair of spinors, so that $\omega = \dot{S}\dot{S}$, then each spinor $\dot{S}$ in the ω would have a mass of at least 391 MeV, which is in disagreement with our remarks above. However, if we instead think of the ω as the light-quark cluster

$$\omega = \dot{S}^{+}\dot{S}^{-}\pi^{0}, \tag{8.10}$$

then the $\dot{S}^{+}\dot{S}^{-}$ spinor pair has a calculated mass of 783 - 135 = 648 MeV, which is close to the estimated value of 660 MeV. Furthermore, there is another argument which suggests that Eq. (8.10) is the correct basis-state assignment for the ω. The ρ meson, as we discuss in considerable detail in Chapter 13, is the spinor pair $\dot{S}\dot{S}$ in an $\ell = 1$ rotational mode. If we compare the principal decay modes of the ρ^{0} and ω mesons, we have

$$\rho^{0} \;=\; \dot{S}^{+}\dot{S}^{-} \quad\rightarrow\quad \pi^{+}\pi^{-} \tag{8.11}$$

and

$$\omega \;=\; \dot{S}^{+}\dot{S}^{-}\pi^{0} \quad\rightarrow\quad \pi^{+}\pi^{-}\pi^{0}\;, \tag{8.10'}$$

which reinforces this basis-state assignment for the ω.

The above line of reasoning can be carried over to the other spin 1 resonances of Fig. 5.3. Suppose we assume that these resonances are each composed of an $\dot{S}\dot{S}$ spinor pair combined with spinless excitation quanta. As can be seen in Fig. 5.3, the ϕ meson appears roughly $\dot{3}$ = 210 MeV† above the $\omega = \dot{S}\dot{S}\pi$ meson, so that it represents the light-quark configuration $\phi = \dot{3}\dot{S}\dot{S}\pi$. The other nonstrange levels shown in Fig. 5.3 appear at intervals of 140 MeV above the ϕ, so that they have additional excitation quanta $\pi = M\bar{M}$ (compare these nonstrange 140 MeV meson intervals with the strange kaon intervals shown in Fig. 5.6). The detailed mapping of these meson resonances is discussed in Chapter 16.

The two strange kaon levels shown in Fig. 5.3, the $K\pi\pi(1278)\Gamma 25$ and $K\pi\pi(1692)$ $\Gamma 26$ resonances, can be constructed as the light-quark configurations $K\dot{S}\dot{S}\pi^{o}$ and $KSS\eta^{o}$, respectively, with the strangeness quantum number S = ±1 being carried in each case by the spinless substate K. If we assume ∿0% binding energies for these light-quark substates, and if we use the substate mass values $K^{\pm}$ = 494 MeV, π^{o} = 135 MeV, and η^{o} = 549 MeV (which are the experimental values for these states), we obtain $\dot{S}\dot{S}$ = 649 MeV as the calculated mass for the $\dot{S}\dot{S}$ spinor pair in each of these $K\pi\pi$ resonances. This agrees closely with the value of 648 MeV that we obtained above from the $\omega = \dot{S}\dot{S}\pi^{o}$ configuration, so that the same spinor pair $\dot{S}\dot{S}$ seems to be appearing in all of these meson and kaon resonances. Hence we can properly regard the spinor pair $\dot{S}\dot{S}$ as the basic "ground state" for the narrow-width spin 1 resonances of Fig. 5.3.

The mass value $\dot{S}\dot{S} \simeq$ 650 MeV that we obtain here indicates that the mixed spinor mass is $\dot{S} \simeq$ 325 MeV. This is close to the (unbound) mass value S = 330 MeV that we obtained in the previous section for the unmixed spinor S. (Some experiments show $\dot{S}\dot{S} \simeq$ 660 MeV, as is summarized in Appendix D.) Thus we seem to have the same basic spin 1/2 spinor appearing in both the meson and baryon resonances.

The meson and kaon resonances that we studied in Fig. 5.3 all contain spinor pairs $\dot{S}\dot{S}$ which are combined with spinless excitation quanta. In the next section we examine evidence for spinor pairs $\dot{S}\dot{S}$ and spinor quartets $\dot{S}\dot{S}\dot{S}\dot{S}$ that occur <u>without</u> any accompanying spinless excitation quanta. This evidence is both meager and controversial, but it does exist.

8.4 Interference Effects

The existence of a meson or baryon resonance is usually revealed by a cross section peak which occurs at the mass of the resonance. However, the existence of the resonance can sometimes be revealed by an interference effect that it produces in another nearby resonance which has similar quantum numbers. As a practical matter in the hadron resonances, this interference effect is readily analyzed only

---

†As we discuss in Ref. 7 of Chapter 10, there is a kinematic constraint that may serve to give the ϕ an anomalously large mass and narrow width.

for the case where the easily-observed resonance has a broad width and the interfering resonance has a much narrower width, so that the interfering resonance produces a localized effect on the shape of the broad resonance.

The most clear-cut example where the properties of a resonance are revealed by its interference effects occurs in the case of the ρ and ω mesons. Both of these resonances have spin and parity $J^P = 1^-$, so that they can interfere with one another. Also, their masses, $\rho(773)$ and $\omega(783)$, are close together, so that the interference occurs near the peak of the broad ($\Gamma = 152$ MeV) ρ resonance, as produced by the narrow ($\Gamma = 10$ MeV) ω resonance. The shape of this interference effect can vary, depending on the reaction channel that is involved. Goldhaber et al.[3] studied the reaction $\pi^+p \to \pi^+\pi^-\Delta^{++}$, in which the final-state $\pi^+\pi^-$ spectrum is dominated by the ρ^o, and observed a dip in the $\pi^+\pi^-$ spectrum which mirrors both the position and the width of the ω; this destructive $\rho - \omega$ interference dip is shown in Fig. 8.1a. Hagopian et al.[4] studied the reaction $\pi^-p \to \pi^+\pi^-n$ and found a constructive $\rho - \omega$ peak in the $\pi^+\pi^-$ spectrum. And Biggs et al.[5], in the reaction $\gamma + C \to \pi^+\pi^-C$, obtained a distorted ρ spectrum that could be analyzed to reveal the shape of the interfering resonance.

In the energy level diagram of Fig. 5.3 we displayed a number of spin 1 meson resonances which contain spinor pairs $\ddot{S}\dot{S}$ that are combined with spinless excitation quanta. However, the spinor pair $\ddot{S}\dot{S}$ by itself, which should have an observed mass of 650 - 660 MeV, produces only the barest indication of a resonance peak[6]. Similarly, the spinor quartet $\dot{S}\dot{S}\dot{S}\dot{S}$, which should have an observed mass of 1300 - 1320 MeV, gives almost no evidence of a peak[7]. Thus we must turn to interference effects in order to delineate the properties of these two resonances.

The only experimental interference effect which indicates the presence of the resonance $\dot{S}\dot{S} \simeq 660$ MeV comes from an analysis of low-energy $\pi\pi$ scattering amplitudes that was carried out by Gutay et al.[8] These workers projected out the rotation-invariant components of the $\pi\pi$ amplitudes, and they found a statistically-significant dip that appears at a mass of about 658 MeV and has a width of about 15 MeV. This dip, which occurs only in the isotopic spin $I = 0$ amplitudes and which was observed in two different reaction channels, is shown here in Fig. 8.1b.

The most interesting, and most controversial, interference effect arises in the case of the $\dot{S}\dot{S}\dot{S}\dot{S}$ spinor quartet. The $\dot{S}\dot{S}\dot{S}\dot{S}$ quartet has spin and parity $J^P = 2^+$, and it is expected to appear as a narrow-width ($\Gamma \simeq 10$ MeV) resonance at a mass of 1300 - 1320 MeV. Now, the broad-width ($\Gamma = 102$ MeV) A_2 meson has spin and parity $J^P = 2^+$, the same as that of the $\dot{S}\dot{S}\dot{S}\dot{S}$ quartet, and it has a mass of 1310 MeV. Thus the $\dot{S}\dot{S}\dot{S}\dot{S}$ spinor quartet should produce a small interference effect that occurs exactly at the peak of the A_2. An early CERN missing-mass experiment[9], which showed

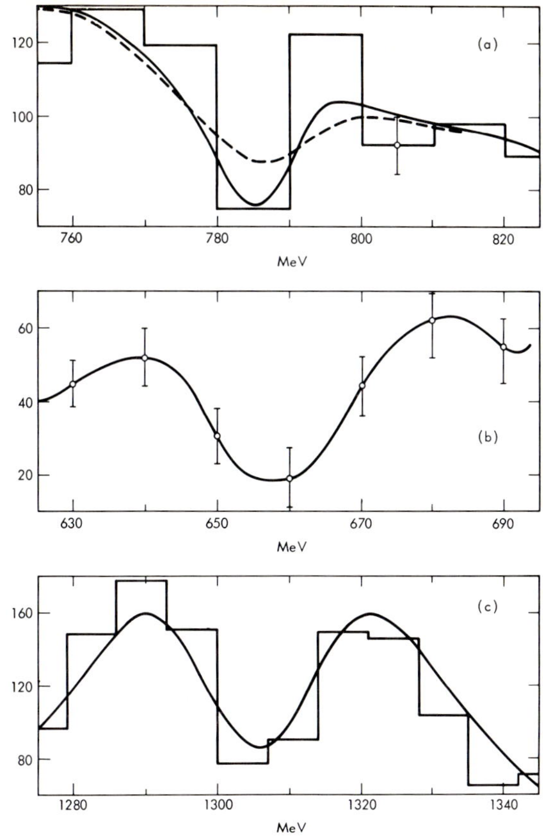

Figure 8.1. Meson interference dips. Figure 8.1a shows a well-known spin 1 ρ-ω interference dip (Ref. 3), which is given here for comparison purposes. Figure 8.1b shows a (spin 1?) ππ interference dip (Ref. 8) that was observed at about 658 MeV, and Fig. 8.1c shows a spin 2 interference dip (Ref. 10) that was observed at just twice this energy — 1310 MeV — , and with the same width. The interference dips shown in Figs. 8.1b and 8.1c are logically associated with the production of a spin 1 $\dot{S}\dot{S}$ spinor pair and a spin 2 $\dot{S}\dot{S}\dot{S}\dot{S}$ spinor quartet, respectively.

a deeply-split A_2 meson, caused a great deal of excitement and led to a number of follow-up experiments. Some of these experiments showed a split A_2 shape (one of these[10] is illustrated in Fig. 8.1c), but other experiments did not. A subsequent NE/SUNY missing-mass experiment[11], which essentially repeated the original CERN experiment, revealed that the A_2 meson is not deeply split; and a number of other recent A_2 experiments[12] also show no evidence for a split A_2. However, a reanalysis of the original CERN data by W. Kienzle[13] indicated that the high-momentum-transfer (high $|t'|$) data still show evidence of a dip, and this dip occurs at 1310 ± 2 MeV. Furthermore, three different bubble chamber experiments[14-16] have also shown this dip, which appears in every case right at 1310 MeV. In fact, the authors of the third bubble chamber experiment[16] made the following comment: "The appearance of some structure in this and in two other experiments [14,15] could clearly be due to statistical fluctuations, although it seems surprising that the effect occurs in the same few bins." Finally, a recent high statistics measurement of the A_2 decay process $A_2^- \to K^0K^-$ by Margulies <u>et al.</u>[17], which was believed by these authors to show "no substructure in any region," does in fact show a small dip in the high $|t'|$ data[18], and this dip again appears at exactly 1310 MeV. This small dip in the $A_2^- \to K^0K^-$ data is shown here in Fig. 8.2. Thus, while there is no supported

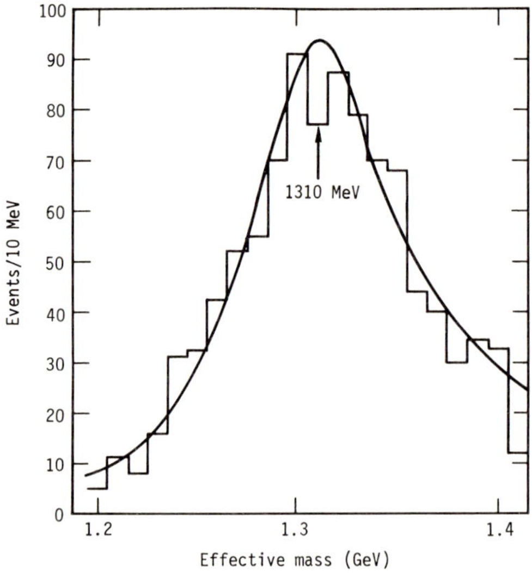

Figure 8.2. A small dip that appears right at 1310 MeV in a recent A_2 experiment[17,18]. It seems clear that a <u>deeply-split</u> A_2 meson shape (the dipole resonance) is ruled out experimentally, but a <u>small interference dip</u> continues to persist in a number of experiments[13-16] and always right at 1310 MeV (see the data summary in Appendix D).

evidence for a deeply-split A_2 meson, there is considerable evidence for a small interference dip which appears right at 1310 MeV. The data on this spin 2 interference dip are summarized in Appendix D.

The importance of the narrow spin 2 interference dip in the A_2 meson is twofold: (1) the position of this dip can be used to pinpoint the mass of the spinor $\dot{S}$ (see Appendix D), just as we used the N and $\bar{N}N$ masses to pinpoint the mass values of the spinors S and $\bar{S}$; (2) the $\dot{S}\dot{S}\dot{S}\dot{S}$ spinor quartet does not play an important role in the low-mass meson resonances, but it appears to be a dominant excitation in the high-mass New Particle resonances, where, for example, we have $J/\psi(3098) \equiv \bar{N}N \cdot \dot{S}\dot{S}\dot{S}\dot{S}$. We now turn to a discussion of these New Particle resonances.

8.5 The New Particle Resonances

The New Particle resonances were discussed briefly in Sec. 5.3 of Chapter 5, and they are treated in detail in Chapter 17. These narrow-width high-mass resonances appear to be associated with the opening up of the $\bar{N}N$ channel at 1795 MeV. In Fig. 5.4, several New Particle resonances were shown that appear at accurate intervals of 70 MeV above the $\bar{N}N(1795)$ ground state, and in Fig. 5.6 these results were extended upward in energy to show several other resonances that also appear at accurate multiples of 70 MeV above the $\bar{N}N$ ground state. However, the $J/\psi(3098)$, the first New Particle state to be discovered, follows a different excitation pattern: by extending the on-shell (zero binding energy) mass systematics of the narrow-width low-mass meson resonances to higher mass values, we are led to the conclusion that the J/ψ resonance has the light-quark structure

$$J/\psi(3098) \equiv \bar{N}N \cdot \dot{S}\dot{S}\dot{S}\dot{S} \quad , \tag{8.12}$$

with a calculated mass of 1795 + 1310 = 3105 MeV. Similarly, the $\psi(4414)$ resonance has the light-quark structure

$$\psi(4414) \equiv \dot{S}\dot{S}\dot{S}\dot{S} \cdot \bar{N}N \cdot \dot{S}\dot{S}\dot{S}\dot{S} \quad , \tag{8.13}$$

with a calculated mass of 1795 + 1310 + 1310 = 4415 MeV. These two resonances reflect a light-quark mass quantization that is in excitation intervals of 1310 MeV, where $\dot{S}\dot{S}\dot{S}\dot{S}$ = 1310 MeV is the spinor quartet that interferes with the spin 2 A_2 meson (as we discussed in the preceding section). These resonances are shown in Fig. 8.3, where it can be seen that the average of the level spacings is just about 1310 MeV.

The light-quark basis state assignment $J/\psi \equiv \bar{N}N \cdot \dot{S}\dot{S}\dot{S}\dot{S}$ follows from the meson excitation series shown in Table 13.1 of Chapter 13, and it accurately reproduces

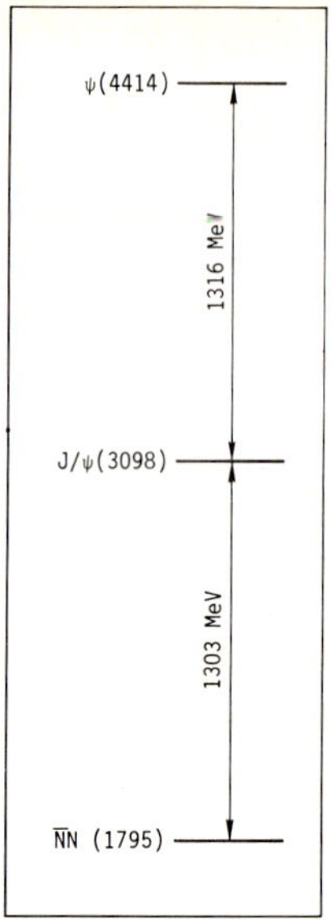

Figure 8.3. The equal spacings of the prominent J/ψ(3098) and ψ(4414) New Particle peaks in intervals of about 1310 MeV above the $\bar{p}n(1795)$ "ground state". These intervals reflect the production of $\dot{S}\dot{S}\dot{S}\dot{S}$ and $(\dot{S}\dot{S}\dot{S}\dot{S})^2$ meson excitation quanta, respectively. Independent evidence for the existence of the excitation quantum $\dot{S}\dot{S}\dot{S}\dot{S}$ = 1310 MeV is shown in Figs. 8.1c and 8.2.

the mass of the J/ψ. Furthermore, it enables us to account for the mass values of the $\psi' \to X + \gamma$ radiative decays, as we show in Eqs. (14.28)-(14.32) of Chapter 14. In attempting to account for the very large J/ψ cross section, we note that the nucleon state N is the dominant spin 1/2 baryon excitation, and the meson state $\dot{S}\dot{S}$ [which corresponds to the ρ meson, as shown in Eq. (8.11)] is the dominant meson excitation. Furthermore, as we have seen in Sec. 5.4 of Chapter 5, spin 1/2 baryon resonances are typically formed by combining the nucleon ground state N with what are essentially meson excitation quanta. Thus the combination $N\cdot\dot{S}\dot{S}$ logically corresponds to a dominant hadron excitation. This particular spin 1/2 baryon excitation does not occur in the observed spectrum of baryon resonances in any obvious manner, but it evidently is readily produced in a pairwise manner in the annihilation of two spin 1/2 particles, e^+e^-, where we have the reaction

$$\underset{J=1/2}{e^+} + \underset{J=1/2}{e^-} \to \underset{J=1/2}{(\bar{N}\cdot\dot{S}\dot{S})}\cdot\underset{J=1/2}{(N\cdot\dot{S}\dot{S})} \to \underset{J=1}{\bar{N}N\cdot\dot{S}\dot{S}\dot{S}\dot{S}} \equiv J/\psi \quad . \tag{8.14}$$

From this light-quark viewpoint, the unusual feature that the New Particle resonances introduce is not new types of quarks ("charmed" quarks), but rather new combinations of old quarks. In particular, these new combinations involve the nucleon-antinucleon pair $N\bar{N}$ combined with excitation quanta $\dot{3}\dot{3}\dot{3}$ or $\dot{S}\dot{S}\dot{S}\dot{S}$, as is shown in Table 13.1.

There are two other New Particle states that may also involve the $\dot{S}\dot{S}\dot{S}\dot{S}$ spinor quartet. These are the "charmed baryon" excitations

$$\Lambda_c^+ (2260) = \bar{N}\cdot\dot{S}\dot{S}\dot{S}\dot{S} = 939 + 1310 = 2249 \text{ MeV} \tag{8.15}$$

$$\Sigma_c (2430) = \Lambda\cdot\dot{S}\dot{S}\dot{S}\dot{S} = 1116 + 1310 = 2426 \text{ MeV} \tag{8.16}$$

which are shown in Table B10 of Appendix B. The Λ_c^+ and Σ_c are the only two "charmed baryon" states to be reported, and their mass values are accurately reproduced by the basis state configurations shown in Eqs. (8.15) and (8.16). In the case of the Σ_c, the strangeness quantum number S = -1 is carried by the Λ substate; however, in the case of the Λ_c^+, the manner in which the strangeness quantum number is accounted for is not apparent from the light-quark basis state configuration shown in Eq. (8.15). The baryon state $\Sigma_c(2430)$ (where we have used the notation of RPP76S in Table B10) is identified experimentally with the resonance X(2426) that is produced in the reaction[20]

$$\begin{aligned} \nu + p &\rightarrow \mu^- + X^{++}(2426) \quad , \\ X^{++}(2426) &\rightarrow \Lambda\pi^+\pi^+\pi^+\pi^- \quad . \end{aligned} \tag{8.17}$$

In Sec. 8.3 we substantiated the light quark assignment $\omega = \dot{S}\dot{S}\pi$ by comparing the decay modes of the ρ and ω mesons [Eqs. (8.11) and (8.10')]. We can similarly substantiate the light quark assignment $X(2426) = \Lambda\cdot\dot{S}\dot{S}\dot{S}\dot{S}$ [Eq. (8.16)] by comparing the decay modes of the ρ and the X, which are

$$\rho = \dot{S}\dot{S} \rightarrow \pi\pi$$

and

$$X = \Lambda\cdot\dot{S}\dot{S}\dot{S}\dot{S} \rightarrow \Lambda\pi\pi\pi\pi \quad . \tag{8.18}$$

Although the $\dot{S}\dot{S}\dot{S}\dot{S}$ spinor quartet is too massive to be an important factor in the low-mass resonances, it appears to be a fundamental excitation unit for the high-mass New Particle resonances.

8.6 Evidence from Theory

In the preceding sections we used the mass systematics of the baryon and meson resonances to establish that the mass of the nucleon spinor is S ≃ 330 MeV. In the present section we briefly summarize some theoretical arguments which also lead to this same mass value for the spinor S.

In Chapter 15 of his book on elementary particles, B. Feld gives tabulated values[21] that enable us to relate the intrinsic magnetic moment of the nucleon quark to both the range of the quark interaction and the mass of the quark; if we use these tabulated values, and if we set the intrinsic magnetic moment of the quark equal to that of the proton, we obtain a range of 0.6 fermi (which agrees with our value for the radius of a quark, as shown in Chapter 18), and we obtain a quark mass of 322 MeV. Using a slightly different argument based on the Dirac form for the magnetic moment[22], Feld obtained an estimate of 336 MeV for the mass of the quark. These two estimates bracket our value of 330 MeV for the mass of the nucleon quark. J. Franklin[23] similarly used Dirac magnetic moments to estimate the masses of quarks, and he obtained values of 339 MeV for the p quark and 323 MeV for the n quark, which again bracket the value 330 MeV; he also made the following comment: "These might be considered reasonable quark masses, except that quarks of this mass have never been seen." More recently, De Rújula, Georgi, and Glashow have expressed somewhat similar ideas, as shown by the quotation at the beginning of the present chapter.

In other theoretical approaches, H. Kleinert[24] used quark gluon theory with very small bare quark masses to deduce an average nucleon quark mass of 312 MeV; M. Pavkovic[25] used parton theory to obtain an estimate for the parton mass of 320 MeV; and D. Robson[26] used an SU_{12} classification scheme to deduce a mass value for the nonstrange quark of 338 MeV. Also, Drell and Johnson[27] studied a quark model that featured a quark mass of about 300 MeV. And a number of recent workers[28], guided by the linear systematics of the New Particles, have assumed a nonstrange quark mass that is one-half of the rho meson mass – or roughly 390 MeV. Finally, F. Gilman[29], again in a New Particle context, has considered a nonstrange quark mass of about 350 MeV, although he cautions that this is a "constituent mass" that "cannot have the usual meaning since we do not see the quarks as free particles."

The conclusion we are attempting to establish here is that, from many different viewpoints, both experimental and theoretical, a mass value of about 330 MeV seems reasonable for the spin 1/2 basis state which appears in the baryon and meson resonances. This result is tied up closely with another conclusion that we have already reached – namely, that quark binding energies are weak (a few percent) in all of these resonances.

8.7 Excited States of the Spinor S

There is one phenomenologically-required result that we should discuss here, even though it is a topic about which we cannot draw firm conclusions on the basis of the present studies. We saw from the π, K, η, and η' mesons of Fig. 5.1 that spinless mass quanta M = 70 MeV exist as separate entities [see Eqs. (5.1)-(5.7)]. We also saw from the nucleon states of Eqs. (8.1) and (8.2) that the spinor S = 330 MeV exists as a separate and geometrically-well-defined entity (this result will be brought out clearly in Chapter 18). And we saw from the baryon and meson excited states of Figs. 5.3 – 5.7 that spinors S occur in combination with spinless excitations M. Now, in the case of the fundamental hyperon states Λ, Σ, Ξ, and Ω, in which 4% binding energies are involved, we can draw some conclusions as to the geometries of the light-quark clusters in these states (see Chapters 9 and 18). However, in the case of the mixed excitations which involve ∿0% binding energies, we have no evidence based on binding energies which tells us anything about the actual geometries of these states. Nevertheless, there is one way to deduce the overall geometries of these states, and it comes from the rotational systematics of the baryon and meson resonances.

We saw in Fig. 5.7 that the rotationless (ℓ = 0) S-state baryon resonances are formed by adding spinless excitation quanta to the fundamental ground states N, Λ, Σ, and Ξ; and we saw in Fig. 5.3 that the narrow-width (ℓ = 0) spin 1 meson resonances are similarly formed. Furthermore, we have asserted (although we have not yet demonstrated it) that these rotationless baryon and meson states serve as the rotors (bandheads) for overlying rotational bands. If this is true, and if we assume that the rotational energies arise from rotations of the entire resonance structure in a "rigid-body" manner,† then we can use the rotational energies to deduce the moments of inertia of the rotors, and hence their geometries. When we carry out these calculations, which we do in Chapters 15 and 16, we discover that as excitation quanta are added to the basic "ground states" to produce successively higher-mass rotors, the "size" of the rotor does not increase, but it stays quite constant and actually seems to decrease. This indicates that, for these resonances, the spinless excitation energy that is added is not in the form of spinless mass quanta which are "tacked onto the outside" of the resonance structure, but is rather in the form of quantized excitations which are internal to the spinors S. These excitations are analogous to the vibrational excitations that occur in atomic nuclei. Thus we can refer to these as "vibrational excitations," although we have no model to explain the manner in which these vibrations occur.

---

†This assumption seems logical from the present light-quark viewpoint, although it has not been demonstrated to be correct.

The relevance of this topic to the present chapter lies in the fact that if the spinors S absorb this "vibrational energy," then they are transformed into higher mass spinor states, so that we have more than just the basic spinors S, $\bar{S}$, and $\dot{S}$ appearing in these resonances. However, from a phenomenological viewpoint, we obtain the correct answers by using a single 330 MeV spinor and treating the 70 MeV-quantized excitation energies as separate entities. Thus, we will implicitly assume this procedure throughout the present book. As we discuss in Chapter 15 (see the section on "doorway excitations"), even the rotational energies exhibit a quantized behavior that is somewhat puzzling. Hence it is clear that the present studies do not represent the last work on this subject. But they hopefully serve to further our knowledge about this difficult and interesting topic.

In Chapter 5 we displayed the experimental evidence which points to the existence of the spinless basis states 1, 3, 4, and 7; and in the present chapter we displayed the evidence which points to the existence of the spinor S. In Chapter 9 we combine these spinless and spinning basis states so as to reproduce the fundamental baryon and hyperon states N, Λ, Σ, Ξ, and Ω. As we will see in Chapter 9, the light-quark clusters $N \equiv S\bar{S}S$, $\Lambda \equiv S\bar{S}S\bar{3}$, $\Sigma \equiv S\bar{S}S\bar{4}$, $\Xi \equiv S\bar{S}S\bar{3}\bar{3}$, and $\Omega \equiv S\bar{S}S\bar{3}\bar{4}\bar{4}$ accurately reproduce both the mass values and the strangeness quantum numbers of these fundamental baryon excitations.

References for Chapter 8.

1. E. Fermi and C. N. Yang, Phys. Rev. 76, 1739 (1949).
2. L. Gray, P. Hagerty, and T. Kalogeropoulos, Phys. Rev. Lett. 26, 1491 (1971).
3. G. Goldhaber et al., Phys. Rev. Lett. 23, 1351 (1960).
4. S. Hagopian et al., Phys. Rev. Lett. 25, 1050 (1970).
5. P. J. Biggs et al., Phys. Rev. Lett. 24, 1201 (1970).
6. See Figs. 13.8 and 13.9 in Chapter 13; also, M. H. Mac Gregor, Nuovo Cimento 18A, 169 (1973), Figs. 8 and 9.
7. An A_2 peak was reported by D. J. Crennell et al., Phys. Rev. Lett. 20, 1318 (1968).
8. L. J. Gutay et al., Nucl. Phys. B27, 486 (1971).
9. G. Chikovani et al., Phys. Lett. 25B, 44 (1967).
10. Bologna-CERN-Strasbourg Collaboration, Lett. Nuovo Cimento 4, 838 (1970).
11. D. Bowen et al., Phys. Rev. Lett. 26, 1663 (1971).

12. A compilation of A_2 experiments is given in the 1976 report of the Particle Data Group, Rev. Mod. Phys. 48, No. 2, Part II, April (1976).

13. W. Kienzle, "MMS, 1967" (unpublished CERN report, 1972).

14. Bonn-Durham-Nijmegen-Paris (E.P.)-Torino Collaboration, Nucl. Phys. B16, 221 (1969).

15. K. W. J. Barnham et al., Phys. Rev. Lett. 26, 1494 (1971).

16. I. J. Bloodworth et al., Nucl. Phys. B37, 203 (1972).

17. M. Margulies et al., Phys. Rev. D14, 667 (1976).

18. M. H. Mac Gregor, UCRL-78601, Aug. 20, 1976 (unpublished).

19. See M. H. Mac Gregor, UCRL-76423, Jan. 17 (1975); Phys. Rev. D12, 1492 (1975).

20. E. G. Cazzoli, Phys. Rev. Lett. 34, 1125 (1975).

21. B. T. Feld, Models of Elementary Particles, Blaisdell, Waltham (1969), Table 15.8 on page 377.

22. Ref. 21, Eq. 18.c' on page 340.

23. J. Franklin, Phys. Rev. 172, 1807 (1969); see Eq. 9 and the accompanying discussion.

24. H. Kleinert, Phys. Lett. 62B, 77 (1976).

25. M. I. Pavković, Annals of Physics 85, 465 (1974).

26. D. Robson, Phys. Rev. D13, 2004 (1976).

27. S. D. Drell and K. Johnson, Phys. Rev. D6, 3248 (1972).

28. For example, see D. Lichtenberg et al., Phys. Rev. Lett. 39, 1592 (1977); and see W. K. H. Panofsky, SLAC-PUB-2025, October (1977).

29. F. J. Gilman, SLAC-PUB-2054, November (1977).

CHAPTER 9. THE FUNDAMENTAL BARYON STATES N, Λ, Σ, Ξ, AND Ω

In our search for a set of hadronic basis states, we have been examining the properties of the narrow-width (long-lived) particles, because these are the elementary particle states which most clearly reveal the underlying mass quantization of their quark substates. In carrying out this search, we have studied the low-mass spinless mesons, the low-mass spin 1 mesons, the high-mass New Particle mesons, and the S-state and narrow-width baryons. The one group of very-narrow-width particle states that we have not yet considered is the set of long-lived hyperons Λ, Σ, Ξ, and Ω. We have put off the study of these states until the present chapter, because it was first necessary to examine the properties of the spinor S and the nucleon, which we did in the preceding chapter. In Chapter 9, the last chapter in Part I, we reproduce the masses and strangeness quantum numbers of the basic baryon states N, Λ, Σ, Ξ, and Ω, and we also consider a few ramifications of this light-quark approach to elementary particle structure. Then we move on, in Part II of the book, to a general consideration of the broad-width (short-lived) elementary particle states.

9.1 The 4% Hadronic Binding Energy

As we have discussed in the preceding chapters, hadronic binding energies occur with two different strengths, $\sim$4% and $\sim$0%, which correspond to the binding of unmixed and mixed basis states, respectively. It is important to note that both the 4% and the 0% binding energies are hadronic, which means empirically that the hadronically-bound resonances cannot be broken apart into their individual quark substates. Now, the baryon excitations that were illustrated in Fig. 5.7 of Chapter 5 involve 0% binding energies, and some aspects of these 0% (mass shell) excitations were mentioned in Sec. 8.7 of Chapter 8; the baryon ground states of Fig. 5.7, on the other hand, involve 4% binding energies, and it is these 4% binding energies that we wish to consider in more detail in the present chapter. In the present section we discuss the general systematics of the 4% binding energies; then in the next section we demonstrate the accurate manner in which they account for the masses of (1) the spinless mesons, (2) the nucleon and the $\bar{N}N$ bound state, and (3) the Λ, Σ, Ξ, and Ω hyperons. Some of these results have already been described in preceding chapters, but we very briefly recapitulate them here in order to present a comprehensive picture of the workings of the light-quark model as applied to the most fundamental meson and baryon particle states.

The 4% hadronic binding energy that we use here was deduced on the basis of the present light-quark systematics before it was reported experimentally in the form of the $\bar{p}n$ bound state (see Ref. 7 in Chapter 4). Since nuclear binding energies are less than 1%, this 4% hadronic binding energy is much stronger than nuclear binding energies, and it therefore has a much shorter range. Hence these 4% hadronic binding energies operate only between adjacent substates in a light-quark cluster. Also, they operate only between an unmixed substate and a matching antistate (e.g., $3\bar{3}$), so that basis state pairings such as 33 and $\bar{3}\bar{3}$ do not bind together hadronically.

When we formed the spinless meson states of Sec. 5.2 in Chapter 5, we used the ∿4% light-quark binding energies

$$\begin{aligned} 1\bar{1} &= -5 \text{ MeV}, \\ 4\bar{4} &= -20 \text{ MeV}, \\ 7\bar{7} &= -35 \text{ MeV}. \end{aligned} \tag{9.1}$$

That is, we used a binding energy that is proportional to the mass of the resonance. The simplest way to produce this result is to consider the basic states $1 \equiv M$, $4 \equiv 4M$, and $7 \equiv 7M$ as linear arrays of subquanta M = 70 MeV. This assumption is substantiated by the circumstance (which was discussed in point five of Sec. 5.2) that the plus and minus charges in the neutral kaon ($K^o \equiv 7^{\pm}$) seem to be separated from one another by a distance of a little over two fermis. We denote these linear arrays of subquanta M as cabers. (A caber is a linear mass that is used in Scottish sporting events.) We will see in a moment that the linear structure of these cabers leads to another important result with respect to hadronic binding energies.

In Chapter 8, where we reproduced the mass of the nucleon, we used the ∿4% spinor binding energies

$$S\bar{S} = -26.4 \text{ MeV}, \qquad SS = \bar{S}\bar{S} = 0. \tag{9.2}$$

(As we will see in Chapter 19, the introduction of electromagnetic corrections increases the $S\bar{S}$ spinor binding energy by about one MeV.) The only 4% binding energy that we have not as yet considered is the hadronic binding energy that occurs between the spinor S and the spinless cabers $\bar{1},\bar{3},\bar{4},\bar{7}$. In establishing a binding energy relationship between the spinless cabers and the spinor S, we can make use of the fact that $S \equiv (3)_{\text{spinning}}$, which was deduced in Chapters 7 and 8 on the basis of mass values. It turns out, as we will see, that from its binding energy relationship to the cabers $\bar{3}$, $\bar{4}$, and $\bar{7}$, the spinor S also appears to be nothing more or less than a spinning caber. Now, since we have already determined the binding energies $3\bar{3}$ = -15 MeV and $S\bar{S}$ = - 26.4 MeV, it seems logical that the $S\bar{3} = \bar{S}3$ binding energy should fall in the range of - 15 to -26.4 MeV, and it turns out empirically that $S\bar{3} = -15$ MeV. Moreover, the really interesting result that we obtain here comes (as we have hinted above) when we determine the binding energies $S\bar{4}$ and $S\bar{7}$. We deduced above that the basis states $\bar{3}$, $\bar{4}$, and $\bar{7}$ are in the form of linear arrays (cabers). When we combine this fact with the fact that $S \equiv (3)_{\text{spinning}}$, we are led to the conclusion that

$$S\bar{3} = -15 \text{ MeV},$$
$$S\bar{4} = -15 \text{ MeV}, \tag{9.3}$$
$$S\bar{7} = -15 \text{ MeV}.$$

As we will see in the calculations of the next section, these are precisely the binding energies that are required in order to reproduce the mass values of the basic hyperon states.

In reproducing the spinless mesons in Chapter 5, we used spinless caber-anticaber pairs. In reproducing the nucleon N in Chapter 8, we used a cluster of three spinors; and in reproducing the $\bar{N}N$ bound state, we used a cluster of six spinors. In reproducing the fundamental hyperon states in the present chapter, we will use clusters that include three spinors plus one or more additional spinless cabers. Although our main goal in the next section is to reproduce the fundamental Λ, Σ, Ξ, and Ω hyperon resonances, we will also briefly repeat the calculations of the mass values for the π, K, η, M, η', N, and $\bar{N}N$ resonances, since it is important to have all of these calculations together in one place for an overall examination of the results. Although the light-quark model that we present here is a very simple model, it yields mass values for all of these particles which represent a major improvement over the values obtained from other approaches to this problem. Furthermore, this model reproduces the strangeness quantum numbers correctly, as we demonstrate in the next section; and it reproduces the electromagnetic properties of these particles, including the n-p mass difference, as we demonstrate in Chapter 19.

9.2 The Fundamental Long-Lived Hadrons

The long-lived hadrons include the low-mass spinless mesons (π,K,η,η'), the nucleon (N), and the low-mass hyperons (Λ,Σ,Ξ,Ω). For completeness in discussing these states, we also include the M meson, the $\bar{N}N$ bound state, and the $\bar{K}N$ bound state, which have somewhat shorter lifetimes. The lifetimes of these states are discussed in Chapter 10, where they are shown to exhibit interesting regularities with respect to the fine structure constant $\alpha = e^2/\hbar c$. We note in passing that the stablest hadrons are formed entirely from <u>unmixed</u> basis states (a partial exception to this rule comes from the narrow-width New Particle states, which are formed here as combinations of an $\bar{N}N$ "ground state" plus mixed excitation quanta).

The spinless mesons were reconstructed in Sec. 5.2 of Chapter 5. These are the states $\pi = 1\bar{1}$, $\eta = 4\bar{4}$, $(M,\eta') = 7\bar{7}$, $K = 7$, and $\bar{K} = \bar{7}$. If we (1) use the mass values $M^o = \bar{M}^o = 70.0$ MeV and $M^{\pm} = 74.6$ MeV, (2) assume a linear addition of substate masses in the cabers 4 and 7, and (3) neglect coulomb effects (which typically lower the masses by about 1 MeV), then the binding energies of Eq. (9.1) give the following calculated mass values for these resonances:

$$\begin{aligned}
\pi^o(135.0) &\equiv 1^o\bar{1}^o = 70 + 70 - 5 = 135.0 \text{ MeV};\\
\pi^{\pm}(139.6) &\equiv 1^{\pm}1^o = 74.6 + 70 - 5 = 139.6 \text{ MeV};\\
\eta(549) &\equiv 4^+4^- = 284.6 + 284.6 - 20 = 549 \text{ MeV};\\
M(953) &\equiv 7^+7^- = 494.6 + 494.6 - 35 = 954 \text{ MeV};\\
\eta'(958) &\equiv 7^+7^- = 494.6 + 494.6 - 30^{\dagger} = 959 \text{ MeV};\\
K^{\pm}(493.7) &\equiv 7^{\pm} = 494.6 \text{ MeV};\\
K^o(497.7) &\equiv 7^{+-} = 499.2 \text{ MeV}^{\dagger\dagger}.
\end{aligned} \tag{9.4}$$

All of these calculated mass values are within about 1 MeV of the experimental mass values. The strangeness values S = ±1 for the kaon K follow from the fact that they are composed of unpaired cabers (recall that the single cabers, $1,\bar{1},3,\bar{3},4,\bar{4},7,\bar{7}$ are all strange; only the $\dot{3}$ and $\dot{4}$ are nonstrange).

The nucleon, $N \equiv S\bar{S}S$, was reconstructed in Sec. 8.2 of Chapter 8. Since spinor pairs $S\bar{S}$ bind together hadronically, whereas spinor pairs SS do not, the nucleon N is a hinged structure, and it can assume either a linear or a clustered shape:

$$N = S\bar{S}S \qquad \text{or} \qquad N = \bar{S}^{S}_{S} . \tag{9.5}$$

When electromagnetic effects are included (Chapter 19), it can be seen that the free proton p has a linear shape, whereas the free neutron n has a clustered shape; this fact is of importance in reproducing the small (1.3 MeV) n-p mass difference. For either of the nucleon configurations shown in (9.5), the calculated mass value, apart from electromagnetic effects, is

---

†See the discussion of this binding energy in the fourth point of Sec. 5.2 in Ch. 5.

††See the discussion of this mass value in the fifth point of Sec. 5.2 in Chapter 5.

$$\mathrm{N}(939) = 3 \times 330 - 2 \times 26.4 - 937 \text{ MeV}, \tag{9.6}$$

where 330 MeV is the mass of the spinor S, and 26.4 MeV is the S$\bar{\mathrm{S}}$ spinor binding energy of Eq. (9.2). Thus this calculated absolute mass value for the nucleon is within 2 MeV of the experimental mass value. The $\bar{\mathrm{N}}$N bound state can be formed either from linear or from clustered nucleons:

$$\bar{\mathrm{N}}\mathrm{N} = \begin{matrix} \mathrm{S}\ \bar{\mathrm{S}}\ \mathrm{S}\phantom{\ \bar{\mathrm{S}}} \\ \phantom{\mathrm{S}\ }\bar{\mathrm{S}}\ \mathrm{S}\ \bar{\mathrm{S}} \end{matrix} \qquad \text{or} \qquad \begin{matrix} & \mathrm{S} & \\ \bar{\mathrm{S}} & & \bar{\mathrm{S}} \\ \mathrm{S} & & \mathrm{S} \\ & \bar{\mathrm{S}} & \end{matrix} \quad . \tag{9.7}$$

The calculated mass value in either case (again neglecting electromagnetic effects) is

$$\bar{\mathrm{N}}\mathrm{N}(1795) = 6 \times 330 - 7 \times 26.4 = 1795 \text{ MeV}. \tag{9.8}$$

Thus the calculated $\bar{\mathrm{N}}$N mass agrees with the experimental mass.

The above calculations for the π, K, η, M, and η' meson states and the N and $\bar{\mathrm{N}}$N nucleon states have been shown before. However, we now extend these results to include the Λ, Σ, Ξ, and Ω hyperons, using the same general systematics that we have already developed. We will again neglect electromagnetic effects, as we did above; electromagnetic corrections are discussed in Chapter 19, and they lead to more accurate mass values than we show here.

The strangeness S = -1 lambda hyperon is the light-quark configuration

$$\Lambda = \begin{matrix} & \mathrm{S} & \\ \bar{\mathrm{S}} & & \bar{3} \\ & \mathrm{S} & \end{matrix} \quad . \tag{9.9}$$

In this configuration, the spinless caber $\bar{3}$, which carries the strangeness quantum number S = -1, binds to the two spinors S with a binding energy of -15 MeV each [Eq. (9.3)]. Thus the calculated Λ mass is

$$\Lambda(1116) = 939 + 210 - 30 = 1119 \text{ MeV}. \tag{9.10}$$

The strangeness S = -1 sigma hyperon is the light-quark configuration

$$\Sigma = \begin{matrix} & \mathrm{S} & \\ \bar{\mathrm{S}} & & \bar{4} \\ & \mathrm{S} & \end{matrix} \quad . \tag{9.11}$$

In this configuration, the spinless caber $\bar{4}$, which carries the strangeness quantum number S = -1, also binds to the two spinors S with a binding energy of -15 MeV each [Eq. (9.3)]. Thus the calculated Σ mass is

$$\Sigma(1193) = 939 + 280 - 30 = 1189 \text{ MeV}. \tag{9.12}$$

The strangeness S = -2 cascade hyperon is the light-quark cluster

$$\Xi = \bar{S}{}^{S\bar{3}}_{S_{\bar{3}}} \quad . \tag{9.13}$$

In this configuration, each of the spinless cabers $\bar{3}$ carries strangeness S = -1. The first $\bar{3}$ has a binding energy of -30 MeV, as in the case of the Λ hyperon above; however, the second $\bar{3}$ has a binding energy of only -15 MeV, since it can bind to only one spinor S. Thus the calculated Ξ mass is

$$\Xi(1318) = 939 + 420 - 45 = 1314 \text{ MeV}. \tag{9.14}$$

All of the hyperon absolute mass values that we have calculated thus far have been within a few MeV of the experimental values. But the omega hyperon presents us with a minor puzzle. From its strangeness S = -3, we would logically construct the Ω as the cluster

$$\Omega = \bar{S}{}^{S^{\bar{4}}_{\bar{3}}}_{S_{\bar{4}}} \quad , \tag{9.15}$$

in which each spinless cable $\bar{3}$ and $\bar{4}$ carries strangeness S = -1. However, the mass value Ω^- = 1672 MeV is 351 MeV ≃ 5 x 70 MeV above the corresponding mass value Ξ^- = 1321 MeV, which indicates that the third spinless caber in the Ω must bind with essentially zero binding energy. The calculated mass value for the configuration of Eq. (9.15) is

$$\Omega(1672) = 939 + 770 - 60 = 1649 \text{ MeV}, \tag{9.16}$$

which is 23 MeV (1.4%) too low. If we use either of the configurations

$$\Omega = \bar{S}{}^{S\bar{3}}_{S_{\bar{4}}\bar{4}} \qquad \text{or} \qquad \bar{S}{}^{S\bar{4}}_{S_{\bar{7}}} \quad , \tag{9.17}$$

then we get about the correct mass value; but the first of these two configurations presents a problem with respect to the binding of the outermost caber $\bar{4}$, and the second presents a problem with respect to the strangeness quantum number. We note that recent experimental results[1] seem to indicate two different lifetimes for the Ω^-, so that it may actually be produced in more than one form. We leave the light-quark configuration of the Ω^- as a partially-unsolved problem.

The final hyperon resonance that we discuss here is the $\bar{K}N$ bound state. As we saw above, the Λ hyperon is the configuration $\Lambda = S\bar{S}S\bar{3}$, and the Σ hyperon is the configuration $\Sigma = S\bar{S}S\bar{4}$. We can similarly form the light-quark configuration

$$K^{-}p = \bar{S}_{S}^{S}\bar{7} \quad , \tag{9.18}$$

in which the spinless caber $\bar{7}$, which carries strangeness S = -1, binds to the nucleon $S\bar{S}S$ with a binding energy of -30 MeV [Eq. (9.3)]. If we consider this configuration to be a $K^{-}p$ bound state, then its calculated mass is

$$p(938) + K^{-}(494) - 30 = 1402 \text{ MeV}, \tag{9.19}$$

which is just the mass of the $\Lambda(1402.4)$ resonance.

In Chapter 19 we will redo these calculations and include all of the electromagnetic effects. However, these effects are quite small (a few MeV). The important point to note here is that by using the present light-quark model in a very straightforward manner, we have obtained absolute mass values which are accurate to a few tenths of a percent, and we have correctly reproduced the strangeness quantum numbers of these states. When the electromagnetic effects are added, we obtain mass values for these resonances which have an average accuracy of about ±0.1%.

It seems apparent, at least to the author, that mass calculations which are as simple, as comprehensive, and as accurate as the ones we have just presented in this section must have physical content.

We conclude this section with the following friendly challenge to the reader: try to change any of the essential features of this light-quark model and see if the mass accuracy that we have achieved here can still be maintained. At an accuracy level of a few tenths of a percent, the meson and baryon resonances that we have just discussed constitute a very restrictive set of data.

9.3 Comparison to SU(3)

In assessing the significance of the mass values that were calculated in the preceding section, it is of some importance to compare them to the mass values that are obtained from SU(3), since SU(3) symmetries were originally invoked[2] to account for these particular baryon and meson resonances. In Eqs. (9.6), (9.10), (9.12), and (9.14), the absolute N, Λ, Σ, and Ξ baryon and hyperon masses were calculated to an average accuracy of ±3 MeV, or about ±0.3%. Now, the original SU(3) group-theoretic classification of these baryon octet states implied that they all have the same mass value. Since this is obviously incorrect, a broken SU(3) symmetry was soon devised[2] which gives the ratios (not absolute values) of these masses to an accuracy of

1%. However, when this same broken SU(3) symmetry is applied to the π, K, and η pseudoscalar mesons, it gives mass ratios that are only accurate to 13% (linear mass ratios) or 6% (quadratic mass ratios). This compares to the light-quark absolute mass accuracies of ±1 MeV (±0.3%) shown for these resonances in Eq. (9.4) of the preceding section. Furthermore, the SU(3) model clearly predicts[3] that the $K^{\pm}(494)$ kaons should be heavier than the $K^{o},\bar{K}^{o}(498)$ kaons, in contradiction to experiment. Also, SU(3) has nothing to say about the $\eta'(958)$ meson [the E(1416) meson is often used instead of the $\eta'(958)$ meson as the ninth member of this SU(3) nonet]; and the $\bar{N}N(1795)$ meson state is outside of the scope of the established SU(3) groups.

It has sometimes been asserted in conference summary talks that since broken SU(3) reproduces the baryon octet mass ratios to an accuracy of 1%, it must have physical relevance to these baryon states. However, this is a two-edged argument, because if we accept a 1% mass accuracy as a criterion of validity for a theoretical model, then the fact that this same broken SU(3) only gives 13% or 6% accuracy for the pseudoscalar mesons implies that it is not relevant to these meson states. And the SU(3) mass accuracies for the other meson and baryon SU(3) groups are very much worse than for the pseudoscalar mesons. SU(3) symmetries have been quite successful in correlating strangeness quantum numbers and isotopic spins, but in general they have been rather unsuccessful in correlating hadronic mass values.

9.4 The Strangeness of the Spinor S and the Nucleon N

At the end of Chapter 4, where we discussed the properties of the spinless cabers 3 and 4, we found that they occur in two distinct forms — unmixed and mixed [see Eqs. (4.7) and (4.8)]. The unmixed cabers 3, $\bar{3}$ and 4,$\bar{4}$ are strange, and they have 4% binding energies; the mixed cabers $\dot{3}$ and $\dot{4}$ are nonstrange, and they have 0% binding energies. In Chapter 8 we showed that the spinor $S \equiv (3)_{\text{spinning}}$ also occurs in both unmixed (S,$\bar{S}$) and mixed ($\dot{S}$) forms, with 4% and 0% binding energies, respectively. What does this analogy between the spinor S and the caber 3 imply about the "strangeness" of the unmixed spinors S and $\bar{S}$, and hence also about the "strangeness" of the nucleon, $N \equiv S\bar{S}S$?

Technically speaking, we should regard the spinors $S \equiv (3M)_{\text{spinning}}$ and $S \equiv (3\bar{M})_{\text{spinning}}$ as "strange" basic states, since they have 4% binding energies. However, the strangeness value that is carried by a spinor triplet $S\bar{S}S$ in the nucleon or in a baryon or hyperon resonance is an invariant, because the triplet $S\bar{S}S$ is invariant in baryon production and decay processes. And strangeness, per se, is not an observable phenomenon; only changes in strangeness lead to observable effects (such as long lifetimes). Thus the invariant strangeness that is carried by a spinor triplet $S\bar{S}S$ will not affect the bookkeeping of Associated Production (see Appendices

A and I), since the spinor triplet $S\bar{S}S$ appears in both the initial and final states with the same strangeness value. Spinless excitation quanta, on the other hand, which are created and then annihilated in the process of Associated Production and decay, can change their strangeness values during the decay process, and they can also break apart asymmetrically prior to the decay process. This is why we asserted that the spinless cabers $\bar{3}$ and $\bar{4}$ in the hyperon resonances Λ - $S\bar{S}S\bar{3}$, Σ = $S\bar{S}S\bar{4}$, Ξ = $S\bar{S}S\bar{3}\bar{3}$, and Ω = $S\bar{S}S\bar{3}\bar{4}\bar{4}$ are the carriers of the usual strangeness quantum numbers.

On the basis of the above remarks we can formally assign any strangeness value that we like to the spinor S and hence also to the nucleon N; to be in conformance with the conventional strangeness assignments used in SU(3) we should assign to S and N the strangeness value zero. However, this procedure is somewhat misleading, because the particle-antiparticle asymmetry that is inherent in the spinor triplet $S\bar{S}S$ does lead to observable consequences: the triplet N = $S\bar{S}S$ combines with the anti-cabers $\bar{3}$ and $\bar{4}$ to form strangeness S = -1 hyperons (the so-called Y* resonances[4]), but it does not combine with the cabers 3 and 4 to form strangeness S = +1 hyperons (the so-called Z* resonances[4]). Hence we can formally assign the strangeness value S = 0 to the nucleon in order to conform with standard usage (both in SU(3) and elsewhere), but we should keep in mind the fact that the nucleon actually has an unbalanced internal particle-antiparticle asymmetry, in contrast to the strangeness S = 0 mesons; this internal asymmetry in the nucleon is revealed both by the systematics of the Y* and Z* resonances, and also by the fact that nucleons bind to antinucleons with a 4% binding energy. A somewhat more formal treatment of the strangeness of the spinor S is given in Appendix C [Eq. (12a)].

9.5 Looking Back and Looking Ahead

We have now come to the end of the search for the basis states. In Chapter 2 we posed nine spectroscopic questions about the elementary particle? What answers have we obtained to these questions? From the discussions in Chapters 3 - 9, we concluded that the elementary particle is a composite structure formed from a set of quark substates (Question 1); and that quark binding energies occur with two values, ~4% and ~0% (Question 2), where these values are correlated with strangeness. We also found that the muon is mathematically related to the hadronic basis states (Question 3); and that relativistic equations for rotating systems (Question 4) constitute the link between the muon and the hadrons. In the cases of the nucleon and the muon, relativistic models exist which give the spin angular momentum and magnetic moment as directly-calculated quantities derived from the rotation of extended mass and charge distributions (Questions 5 and 6). The baryon quantum number B emerged in the form of the invariant spinor triplet $S\bar{S}S$ (Question 8); and the strangeness quantum number S appeared as an internal particle-antiparticle asymmetry in the light-quark basis states (Question 9).

The one question that we have not dealt with as yet is Question 7 — the concept of orbital angular momentum as applied to elementary particles. Without an accurate delineation of the basis states, it is a difficult problem to decompose the total angular momentum J of a resonance into its orbital ℓ and spin S components. However, with the aid of the light-quark basis states 1,3,4,7,S that have evolved from the studies in Part I, we are now in a position to tackle this problem. It turns out, as we have already suggested above, that the spinless resonances ($\vec{J} = \vec{\ell} = \vec{S} = 0$), the rotationless resonances ($\vec{J} = \vec{S}$, $\vec{\ell} = 0$), and the rotational resonances ($\vec{J} = \vec{\ell} + \vec{S}$, $\vec{\ell} > 0$) each have their own characteristic widths — and hence their own characteristic angular momentum uncertainties. The behavior of rotational excitations in elementary particles shows certain similarities to rotational excitations in atomic nuclei, as we would expect; but it also has important differences, which arise mainly from the highly nonadiabatic nature of elementary particle rotations. The development of these ideas forms Part II of the book, where we move from the problem of mass quantization to the problem of angular momentum quantization.

References for Chapter 9

1. See the discussion of the Ω^- lifetime in Refs. 13 and 14 of Chapter 10.

2. A good review of SU(3) systematics is given in Part III of B. T. Feld's book Models of Elementary Particles, Blaisdell, Waltham (1969).

3. See page 338 in Ref. 2.

4. The experimental situation with respect to Y* and Z* resonances is discussed in Review of Particle Properties, Rev. Mod. Phys. 48, No. 2, Part II, April (1976) (see pages S188 - S197).

PART II. ANGULAR MOMENTUM QUANTIZATION: SPIN AND ORBITAL ANGULAR MOMENTA

INTRODUCTION AND SUMMARY

In Part II of the present book we are faced with a different problem than we were in Part I. Our task in Part I was to search for evidence of mass quantization in the light-quark basis states — a mass quantization whose existence had to be demonstrated. Our task in Part II is to study the angular momentum quantization of elementary particles — an angular momentum quantization that is well known to exist. Thus the problem in Part II is not to establish that elementary particle angular momenta are quantized, but rather to examine the various factors that enter into this quantization. Molecules, atoms, and atomic nuclei have angular momentum characteristics that are similar to one another. However, when we come to the domain of the elementary particle, we are at the limit of very small masses and very short lifetimes. These factors modify the angular momentum characteristics of the elementary particle and cause these characteristics to differ in essential respects from the angular momentum characteristics of the more massive and more stable systems. We find, interestingly, that the mass quantization studies of Part I are also important for the angular momentum quantization studies of Part II. Mass quantization effects enter into angular momentum systematics in two different ways: (1) Rotationless ($\ell = 0$) hadron resonances exhibit a 70 MeV mass quantization of the observed levels, whereas rotational ($\ell > 0$) hadron resonances do not. (2) The width of a resonance turns out to be its most revealing characteristic when we attempt to classify hadron resonances on the basis of their angular momentum properties. The width is of course just the uncertainty in the mass quantization, so that the sharpness of the mass quantization evidently depends on the angular momentum properties of the resonance. This relationship between mass values and angular momentum values is a direct consequence of the results derived in Chapters 6 and 7, wherein it was demonstrated that the rotational energy contributes in an important manner to the total energy (mass) of an elementary particle. Thus, uncertainties in rotational energies are reflected as uncertainties in mass values. The studies that are discussed in Part II represent an extension of the ideas of Part I to include not only the mass quantization of the elementary particle levels, but also the uncertainty in the mass quantization; that is, we now study not only the positions of levels, but also their widths.

The empirical importance of resonance widths lies in the fact that when we take the total angular momentum J of a resonance (which is the only measurable angular momentum) and decompose it into its orbital ℓ and spin S components, where

$\vec{J} = \vec{\ell} + \vec{S}$, we discover that each of these two components has its own characteristic width. In particular, rotational ($\ell > 0$) resonances typically have widths $\Gamma > 50$ MeV that reflect the uncertainty in the quantization of the orbital angular momentum ℓ, whereas rotationless ($\ell = 0$) resonances typically have widths $\Gamma << 50$ MeV that reflect the uncertainty in the quantization of the spin angular momentum S. As a result of this characteristic difference in the quantization uncertainties of the orbital and spin angular momenta, we can use the observed width of a resonance as a guide in deducing the angular momentum components of the resonance.

We should state at the outset of these studies in Part II that the interpretation we are giving to the orbital rotation of the elementary particle is not the usual one. We find that the orbital rotation of a particle is analogous to the orbital rotation of an atomic nucleus – namely, a rotation of the entire structure in which the observed moment of inertia is roughly comparable (within a factor of two) to the rigid-body moment of inertia. In the more customary way of picturing the elementary particle, the orbital motion, if it is ascribed any literal significance at all, is taken to be a rotation of individual quark pairs within the particle; and since the quark masses and quark "sizes" in these conventional quark models are regarded as essentially unknown, the orbital rotations of these internal quark pairs do not in general lead to calculable results. One advantage of our present nuclear physics viewpoint is, as we will see, that we can make direct estimates of elementary particle moments of inertia and rotational energies, and these turn out to be in reasonable agreement with experiment.

Our main task in Part II is to learn how to decompose the total angular momentum $\vec{J}$ of a resonance into its $\vec{\ell}$ and $\vec{S}$ components. There are three interrelated factors which combine to make this a difficult task: (1) Elementary particles are very light, and consequently have small moments of inertia, so that their rotational energies are very large. (2) The broad widths of rotational ($\ell > 0$) excitations make it difficult to pin down their mass values experimentally. (3) Hadron resonances seem to occur in rotational bands, just as in nuclear and molecular physics, but the large rotational energies cause an overlapping of levels from different rotational bands, and a consequent difficulty in sorting out the bandhead masses. Also, hadron rotational bands generally lack the completeness of nuclear-physics-type rotational bands, and the number of levels in a band is limited by the requirement that the periphery of the rotor cannot exceed the velocity of light. We now sketch the organization of the chapters in Part II, where we come to grips with these difficulties.

Chapter 10 is a phenomenological analysis of the systematics of elementary particle lifetimes. This analysis reveals that the lifetimes fall into two separate and distinct domains – the domain of long-lived "particles," whose lifetimes are all spaced by powers of $\alpha \simeq 1/137$; and the domain of short-lived "resonances," whose lifetimes occur with essentially a continuum of values. The "resonance" lifetimes appear to be strongly influenced by the dictates of final-state phase space, whereas the "particle" lifetimes are essentially independent of phase space effects (phase space corrections are discussed in Appendix G). In addition to an overall scaling in powers of α, "particle" lifetimes also show a fine structure scaling in powers of two, with supposedly-independent particles grouped together in lifetime multiplets. This factor-of-two fine structure in the lifetimes furnishes a vital clue about the nature of "particle" decay mechanisms, and it indicates the reality of the two different lifetimes that have recently been reported for the Ω^-. As was the case for the relativistic relationships developed in Chapters 6 and 7, the lifetime relationships shown in Chapter 10 represent original results. No one other than the present author appears to have undertaken any comprehensive analysis of this complete spectrum of elementary particle lifetimes.†

In Chapter 11 we study the relationship between elementary particle lifetimes and elementary particle angular momentum values. The important result which emerges from this study is that particles with higher and higher angular momentum values have systematically shorter and shorter lifetimes. This is just the opposite of the situation that occurs in nuclear physics (which is a fact that does not seem to be generally recognized), and it indicates that the angular momentum barriers which dominate nuclear rotations are not dominant factors in elementary particle rotations. The reason for the difference in the two cases lies both in the highly non-adiabatic nature of elementary particle rotations as compared to the relatively gentle nuclear rotations, and in the explosive nature of elementary particle decays. Since many elementary particle theorists have implicitly assumed that nuclear angular momentum barrier factors do apply to elementary particle rotations, and consequently have applied these barrier factors in calculating (for example) SU(3) decay rates, the present empirical results offer a challenge to the validity of these calculations.

In Chapter 12 we make a formal study of the systematics of the baryon resonances. It turns out that we can isolate rotationless ($\ell = 0$) baryon excitations by a procedure which is independent of theoretical assumptions. This procedure consists very simply of dividing the baryons and hyperons into two groups: (1) the J = 1/2 and J = 3/2 resonances; (2) the J = 5/2, 7/2, ... resonances. Since baryons and hyperons each contain three spin 1/2 quarks (three spinors S), the possible

---

†As evidence for the relevance of these lifetime studies, it should be noted here that when an early preprint on this work — UCRL-73961 — was sent to the CERN international high energy physics research center in Geneva, Switzerland, the CERN preprint librarian wrote back and reported that 65 physicists had requested permanent copies of the paper.

values for the intrinsic spins of these resonances are $S = 1/2$ and $S = 3/2$. Thus the $J \geq 5/2$ resonances are necessarily $\ell > 0$ excitations, whereas the $J = 1/2$ and $J = 3/2$ resonances include both $\ell = 0$ and $\ell > 0$ excitations. Hence, if $\ell = 0$ resonances have unique properties, we expect to find a unique class of resonances which occur in the $J \leq 3/2$ group and not in the $J \geq 5/2$ group. This unique class is there, and it consists of a group of narrow-width resonances. Thus $\ell = 0$ resonances have narrower widths than do $\ell > 0$ resonances. This result is in agreement with the results of Chapter 11, wherein it was demonstrated that low-spin resonances have longer lifetimes (narrower widths) than do high-spin resonances. When we use this result to sort the baryon and hyperon excitations into $\ell = 0$ and $\ell > 0$ groups, we find the additional confirmatory result that the mass values of the $\ell = 0$ resonances are quantized in levels which are separated by multiples of 70 MeV, whereas the mass values of $\ell > 0$ resonances are not quantized.

In Chapter 13 we carry the study of rotational and rotationless excitations over to the meson resonances, and we analyze the ρ and ω mesons as typical examples. The spin 1 $\rho(773)\Gamma152$ and spin 1 $\omega(783)\Gamma10$ mesons have comparable masses (773 MeV and 783 MeV), but radically different widths (152 MeV and 10 MeV). By using the light-quark basis states, we show that the ρ is a rotational ($\ell = 1$) excitation, whereas the ω is a rotationless ($\ell = 0$) state. The distribution of rotational energies that characterizes the broad width of the ρ is associated with a corresponding distribution of angular momentum values, so that the ρ is a spin 1 meson only at the 773 MeV peak of the resonance; for mass values that are greater than (less than) 773 MeV, the ρ has angular momentum values that are greater than (less than) $J = 1$. This distribution of angular momentum values is reflected in the experimental $\pi\pi$ phase shifts which have been determined over the energy region that is dominated by the ρ meson.

In Chapter 14, the final chapter in Part II, we study a special topic — the transition between spinless and spinning light-quark basis states. This is the transition $\dot{3}\dot{3}\dot{3} \leftrightarrow \dot{S}\dot{S}$, which is almost (but not quite) isoergic, and which can be followed as it occurs in both directions in meson resonances. This transition is of special importance in the New Particle states, where it occurs in the form $\dot{3}\dot{3}\dot{3}\cdot\dot{3}\dot{3}\dot{3} \leftrightarrow \dot{S}\dot{S}\dot{S}\dot{S}$, which serves as a bridge between the $J/\psi \equiv \bar{N}N\cdot\dot{S}\dot{S}\dot{S}\dot{S}$ and $\psi' \equiv \bar{N}N\cdot(\dot{3}\dot{3}\dot{3})^3$ resonances. In particular, this transition is responsible for an upward energy shift noted in a radiative decay level of the ψ'. The transition $\dot{3}\dot{3}\dot{3} \rightarrow \dot{S}\dot{S}$ is also of interest from a conceptual point of view in that it establishes a relationship between spin quantization and energy conservation.

The angular momentum studies of Part II together with the mass quantization studies of Part I complete the specification of the mechanical attributes of the elementary particle. On the basis of these studies, we move on in Part III to the mapping of the observed spectrum of elementary particle resonances. Then in Part IV we return again to the spectroscopic aspects of these particles, and we discuss their geometries and electromagnetic properties.

We now begin Part II with the relatively-unexplored topic of elementary particle lifetimes. In a sense this is a separate subject, but it leads to some important empirical results which independently suggest, in line with our conclusion from Part I, that elementary particles have a common (light-quark) substructure.

CHAPTER 10. ELEMENTARY PARTICLE LIFETIMES: AN UNTAPPED GOLDMINE OF INFORMATION

10.1 Introduction and Summary

The lifetimes of the elementary particles, apart from the stable particles, range from that of the neutron, $\tau \sim 10^3$ sec, down to the lifetimes of the broadest meson and baryon resonances, $\tau \sim 10^{-24}$ sec. This is a span of 27 orders of magnitude, and it is so great a range that physicists have not usually attempted to group these lifetimes together into one comprehensive picture. However, quite surprisingly, this can be done, and it reveals some very interesting facts about the lifetimes of these particles. Furthermore, some of the results which emerge from this study are in disagreement with generally-accepted notions about the lifetimes of these particles. The systematics of elementary particle lifetimes is largely an unexplored subject, and, as we have indicated in the title to this chapter, it is a subject that deserves to be actively pursued.

Elementary particle lifetimes, when properly arrayed, reveal interesting patterns. In the present chapter we systematically lay out these patterns, which in a phenomenological sense are quite clear-cut. In Sec. 10.2 we search for a proper scaling factor that can tie together the widely separated lifetimes of the long-lived particles. This scaling factor turns out to be the fine structure constant $\alpha = e^2/\hbar c \simeq 1/137$, which applies even in regions that were thought to be dominated by the so-called "weak interaction"; hence the "weak interaction", which is usually regarded as one of the fundamental (and thus independent) interactions, must in some deep sense be electromagnetic. In Sec. 10.3 we bring the short-lived states into the picture, and we demonstrate that they have a completely different type of lifetime systematics than do the long-lived states. For convenience, we denote the long-lived states as "particles" and the short-lived states as "resonances".

In Sections 10.4 - 10.6 we probe for some of the finer details of these lifetimes. We show in Sec. 10.4 that the lifetimes of half-integral-spin fermion "particles" follow a different pattern in powers of α than do the lifetimes of integral-spin boson "particles". In Sec. 10.5 we investigate some SU(3) groupings of particles, and we discover that the lifetimes of these groups have interesting regularities. In Sec. 10.6 we demonstrate that the long-lived "particles" have a factor-of-two fine structure that is superimposed on the overall scaling in powers of α; these fine-structure lifetime multiplets include groupings of supposedly-independent particles, which means that these particles must have a common (light-quark) substructure.

In Sec. 10.7 we briefly comment on the question of phase space corrections as applied to elementary particle lifetimes. Although it is commonly believed that phase space factors are important in determining the lifetime of an elementary particle, this belief does not appear to be correct in the case of the long-lived "particles", although it is probably correct for the short-lived "resonances". Phase space corrections are discussed in more detail in Appendix G.

The lifetime of an elementary particle is an indication of the stability of the particle. Thus the systematics of elementary particle lifetimes furnishes a clue as to the nature of elementary particle decay mechanisms. It is clear from the appearance of the fine structure constant α that the decays of the long-lived "particles" are ultimately dominated by electromagnetic effects, and yet present-day theories of these decays do not include α as a factor in the transition matrix elements. The aspect of the lifetime systematics which is most important for our present purposes lies in the implications that can be drawn with respect to the underlying substates that form these particles. As we will see, the lifetimes of the long-lived "particles" form what is in effect a "lifetime democracy", with baryons, hyperons, mesons, kaons, and leptons (the muon) all appearing together in one comprehensive over-all pattern. Hence it seems clear that all of these particles must be intimately related — that is, they must all have a common set of quark substates. This, of course, is the conclusion that we have drawn from our studies in Part I, which were based on totally different considerations.

10.2 Particle Lifetimes as a Scaling in Powers of $\alpha = e^2/\hbar c$

There are 18 elementary particles that have measured lifetimes τ which are longer than 10^{-21} sec. These lifetimes are summarized in Table 10.1.[1-3] Ratios of these lifetimes are also shown in Table 10.1, with the lifetime of the longest-lived particle — the neutron — set equal to unity. As can be seen in Table 10.1, these lifetimes span 24 orders of magnitude. This span of lifetimes is so large that physicists do not customarily group them together in one category. Instead, the lifetimes $\tau > 10^{-16}$ sec are assigned to the domain of "weak" decays, the lifetimes $\tau < 10^{-16}$ sec are assigned to the domain of "electromagnetic" and "strong" decays, and these domains are treated separately[4]. Is it possible to combine these separate domains in one unified picture? If we are to accomplish this, we must somehow find a meaningful framework that relates these two domains to one another. Since our interest here is in relationships among lifetimes rather than absolute values of lifetimes, it is more convenient to work with the lifetime ratios R in Table 10.1 rather than directly with the lifetimes themselves. The ratio R is defined as

$$R_i = \tau_i/\tau_{neutron}, \qquad (10.1)$$

Table 10.1. The eighteen elementary particles that have measured lifetimes $\tau > 10^{-21}$ sec. The experimental lifetimes are listed, together with the ratio R of each lifetime with respect to the lifetime of the neutron, and also the logarithm X of the ratio R to the base $\alpha = e^2/\hbar c$.

Particle	Experimental lifetime τ (in seconds)	Lifetime ratio R	Logarithm X of R to the base α	Ref.
neutron	918 ± 14	1.0	0.000 ± (0.003)	a
$\mu^\pm$	2.1971 ± 0.0001 x 10^{-6}	2.393 x 10^{-9}	4.034 ± 0.00001	
K^o_L	5.181 ± 0.040 x 10^{-8}	5.644 x 10^{-11}	4.796 ± 0.002	
$\pi^\pm$	2.6030 ± 0.0023 x 10^{-8}	2.836 x 10^{-11}	4.936 ± 0.0002	
$K^\pm$	1.2371 ± 0.0026 x 10^{-8}	1.348 x 10^{-11}	5.087 ± 0.0004	
Ξ^o	2.89 ± 0.10 x 10^{-10}	3.148 x 10^{-13}	5.851 ± 0.007	2
Λ	2.635 ± 0.012 x 10^{-10}	2.870 x 10^{-13}	5.869 ± 0.001	2
Ξ^-	1.652 ± 0.023 x 10^{-10}	1.800 x 10^{-13}	5.964 ± 0.003	
Σ^-	1.482 ± 0.017 x 10^{-10}	1.614 x 10^{-13}	5.986 ± 0.002	
Ω^-	1.3 +0.3, -0.2 x 10^{-10}	1.416 x 10^{-13}	6.013-0.042,+0.034	b
K^o_S	0.8930 ± 0.0023 x 10^{-10}	9.728 x 10^{-14}	6.089 ± 0.001	
Σ^+	0.800 ± 0.006 x 10^{-10}	8.715 x 10^{-14}	6.112 ± 0.002	
π^o	8.28 ± 0.057 x 10^{-17}	9.020 x 10^{-20}	8.913 ± 0.002	
η	7.74 ± 1.09 x 10^{-19}	8.435 x 10^{-22}	9.862-0.027,+0.031	
Σ^o	0.58 ± 0.13 x 10^{-19}	6.318 x 10^{-23}	10.389-0.041,+0.052	2
ψ(3098)	9.54 ± 2.07 x 10^{-21}	1.039 x 10^{-23}	10.756-0.040,+0.050	
ψ'(3684)	2.89 ± 0.73 x 10^{-21}	3.144 x 10^{-24}	10.999-0.045,+0.058	
η'	2.1 +1.8, -0.7 x 10^{-21}	2.241 x 10^{-24}	11.068-0.078,+0.129	3

[a]Zero neutron error was assumed in calculating errors for the logarithms X.

[b]See the two Ω^- lifetimes listed in Refs. 13 and 14.

where τ_i is the lifetime of the ith particle and $\tau_{neutron}$ is the reference lifetime. The lifetime ratio R_i can in turn be expressed in terms of a parameter X_i, which is the logarithm of the ratio R_i to some suitably chosen base B:

$$R_i = B^{X_i}, \tag{10.2}$$

where B < 1. The task of finding a suitable framework for studying elementary particle lifetimes thus becomes a matter of selecting an appropriate base B.

As a first guess for the logarithmic base B, we can simply use powers of 10 and select the value $B = 10^{-1}$. Figure 10.1 shows a plot of the lifetime logarithms X

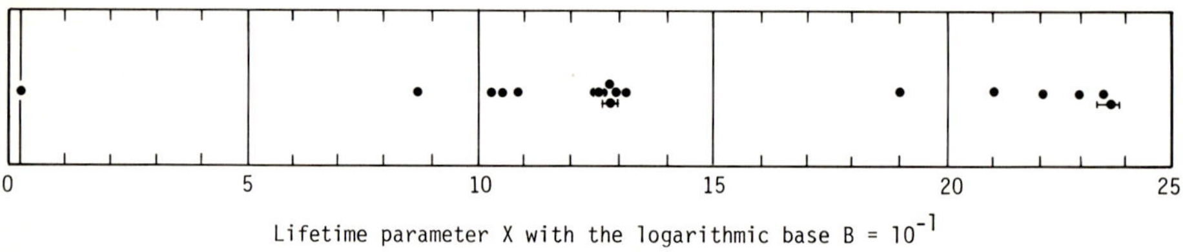

Figure 10.1. The experimental lifetimes of Table 10.1 plotted along the abscissa as lifetime logarithms X, using Eq. (10.2) with the logarithmic base $B = 10^{-1}$. As can be seen, this base has no phenomenological significance.

for the 18 elementary particle lifetimes of Table 10.1, with this choice for the base B in Eq. (10.2). As can be seen in the figure, the base $B = 10^{-1}$ has no phenomenological significance. However, several lifetime groupings can be observed in Fig. 10.1, and the centroids of these groupings are separated by factors of roughly 10^2, or 100. This suggests that the base $\alpha \simeq 1/137$ might be a suitable one to use for representing these lifetimes. In support of this conjecture, we note that the scaling of elementary particle total or partial decay widths in powers of α is well-recognized for certain electromagnetic decays[4]. These decays involve a number n of photon interactions, with each interaction contributing a power of α, so that the factor α^n appears in the decay rate[†]. However, this scaling in powers of α has not

---

[†]As an example, the lifetimes of the electromagnetically-inhibited η meson and the strongly-decaying ω meson occur in the ratio $\tau_\eta \simeq \tau_\omega \times \alpha^{-2}$.

been extended into the domain that encompasses the "weak" decays — namely, the domain of lifetimes $\tau > 10^{-16}$ sec. We now make this extension.

Figure 10.2 shows a plot of the lifetime logarithms X for the 18 elementary particle lifetimes of Table 10.1, with the base B = α used in Eq. (10.2); that is, we use the equation

$$\tau_i/\tau_{neutron} \equiv R_i = \alpha^{X_i}. \qquad (10.3)$$

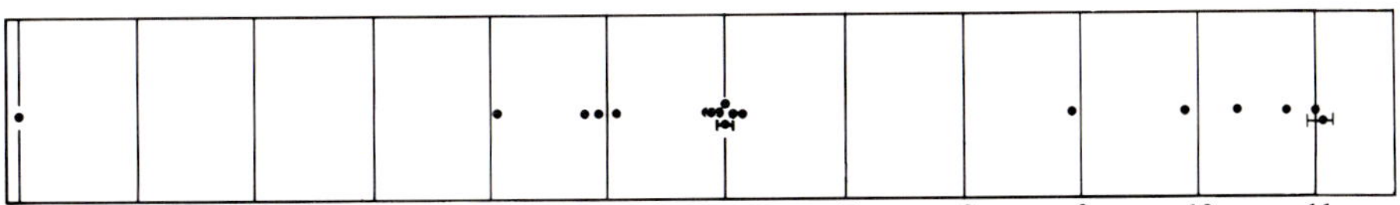

Figure 10.2. This is the same plot as Fig. 10.1, but with the base B = α. When plotted in this manner, the lifetime logarithms X exhibit a clear-cut integer periodicity, which indicates that α is a relevant scaling factor for these lifetimes. For logarithms X > 10, the periodicity starts to break down, and for X $\simeq$ 12 it breaks down completely (see Fig. 10.5).

In contrast to the X-plot of Fig. 10.1, the X-values in Fig. 10.2 exhibit a clear-cut integer periodicity, which indicates that these lifetimes scale in powers of α. This scaling extends accurately over the range of lifetimes from X = 0 to X = 10 ($\tau \simeq 1000$ sec to $\tau \simeq 10^{-19}$ sec), and it probably also includes the lifetimes grouped near X = 11. As can be seen in Fig. 10.2, there is some breakdown in the scaling pattern for X > 10; and as we will see in the next section, there is a total breakdown near X = 12, where a continuum of lifetimes suddenly appears.

Can the observed periodicity of lifetimes in powers of α that is shown over the range X = 0 to X = 10 in Fig. 10.2 be accidental? Intuitively, it seems that it cannot. Figure 10.2 includes all 14 particles whose measured lifetimes fall into this range, and there are no counterexamples: all 14 of these lifetimes appear with approximately integer values for X (the numerical values for X are shown in Table 10.1, together with the experimental errors on these values). However, in providing an answer to this question, we do not have to rely solely on intuition. There are three arguments we can use to demonstrate that the lifetime periodicity shown over the range X = 0 to X = 10 in Fig. 10.2 is not accidental:

(1) Statistical probability. In Eqs. (10.2) and (10.3), the deviations ΔX of the lifetime logarithms X from integer values can range from $\Delta X = -0.5$ to $\Delta X = +0.5$. Figure 10.3 shows plots of the deviations ΔX for the 14 long-lived particles of Table 10.1, where a whole series of logarithmic bases B, ranging from B = 1/2 down to B = 1/1000, are used in Equation (10.2). As can be seen in Fig. 10.3, a significant bunching of the deviations ΔX occurs only for values of B near B = 1/137; for other values of B, the deviations ΔX are spaced rather uniformly over the entire "deviation space", thus indicating random correlations. The deviations ΔX for $B \simeq 1/137$ are concentrated in a sector that is just under 1/3 of the available deviation space, thus indicating a significant correlation; from Fig. 10.2 we see that these 14 lifetimes occur in 6 separate groups, so that the accidental probability for a bunching of this compactness is given by a conservative estimate† as $(1/3)^5 = 0.3\%$, where the base B is treated as a given (fixed) parameter, and where one group (the neutron) is used as a reference group. The smallness of this accidental probability stems of course from the fact that out of the 14 long-lived particles in Table 10.1, there are no counter-examples to the observed scaling of lifetimes in powers of α; all 14 of these particles fit into this scaling pattern.

(2) Demonstrated predictive power. The scaling of elementary particle lifetimes in powers of α was first recognized[5] on the basis of the 13 particles in Table 10.1 which have lifetimes $\tau > 10^{17}$ sec. Subsequent to the publication[5] of this discovery, the $\psi(3098)$ and $\psi'(3684)$ lifetimes appeared[1], the $\eta(549)$ lifetime was revised upward by a factor of three[1], and the $\eta'(958)$ lifetime was dractically revised upward[3]. All four of these lifetimes — ψ, ψ', η, η' — now fit this previously-established[5] scaling in powers of α, and they extend the scaling by two powers of α. Only the recently-determined Σ^0 lifetime[2] appears with a markedly non-integer value for X, and even here we note the lifetime ratios $\pi^0/\pi^\pm \simeq \alpha^4$ and $\Sigma^0/\Sigma^\pm \simeq \alpha^4$.

(3) Theoretical content. If the elementary particles of Fig. 10.2 are separated into theoretically-suggested groupings (see Sections 10.4 and 10.6), striking lifetime patterns emerge which reveal some of the underlying electromagnetic content of these particle lifetimes. It seems highly improbable that these patterns can be accidental.

When the experimental systematics of Fig. 10.2 are considered in the light of the three arguments outlined above, it seems probable that this scaling of particle

---

†This estimate is based on the assumption that the lifetimes within a group are somehow related to aone another, so that the group as a whole represents only one independent lifetime. If we were to instead assume, as is generally believed, that these 14 lifetimes are in fact all independent of one another, then the accidental probability for a bunching of this compactness becomes $(1/3)^{13} = 3 \times 10^{-7}$.

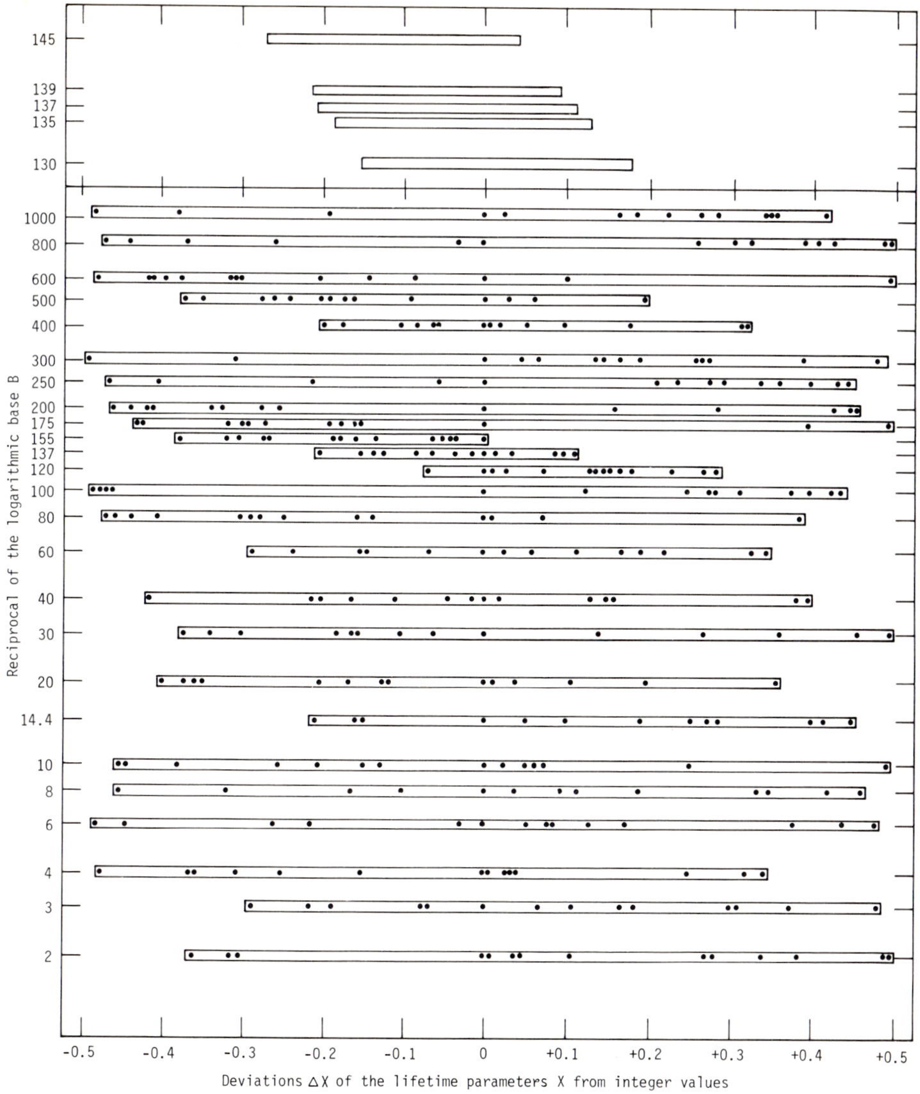

Figure 10.3. An analysis of the logarithmic scaling of lifetimes for the fourteen longest-lived particles of Table 10.1. In this analysis, Eq. (10.2) is used together with a series of values for the logarithmic base B, ranging from B = 1/2 to B = 1/1000. As can be seen in the figure, a significant bunching of the logarithms X occurs only for the base B = $\alpha \simeq 1/137$. A statistical analysis of these results is given in the text.

lifetimes in powers of α — a scaling which extends over 11 powers of α, or 24 orders of magnitude — is not accidental. One of the ramifications of this lifetime scaling is shown in Fig. 10.4, which is an expanded plot of the X-values in Fig. 10.2, and which illustrates the types of particles that are grouped together in the lifetime scaling pattern. As can be seen in Fig. 10.4, particle lifetimes form a real "lifetime democracy", with all types of particle — hadrons and leptons, baryons and

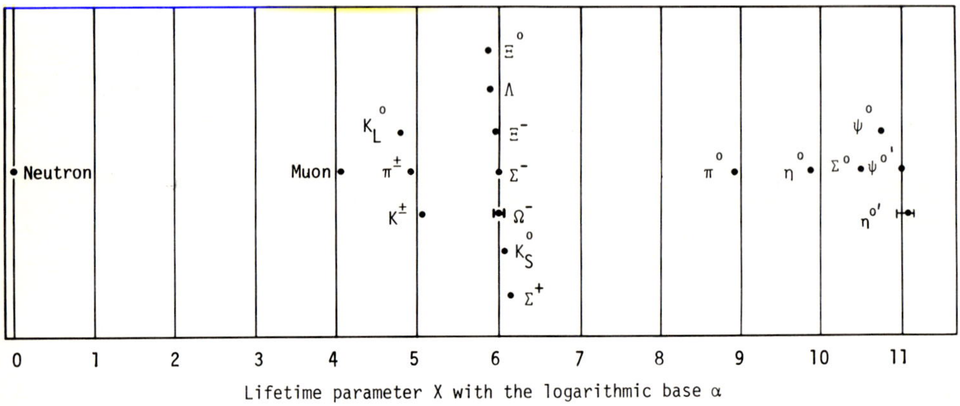

Figure 10.4. This is the same plot as Fig. 10.2, but with the dots displaced vertically and labeled. As this figure reveals, we have a "lifetime democracy", with baryons, mesons, and leptons all combining together in one overall lifetime pattern that is a scaling in powers of α.

mesons, strange particles and nonstrange particles — combining together in one overall lifetime pattern. Phenomenologically speaking, this is a very powerful result. For example, it shows us, without the need for any recourse to theory, that the muon is closely related to the hadronic states of Fig. 10.4. (Note the lifetime ratios $\mu/n \simeq \alpha^4$, $\pi^0/\pi^\pm \simeq \alpha^4$, and $\Sigma^0/\Sigma^\pm \simeq \alpha^4$.) This is a result that we were led to in Chapters 6 and 7 for quite different reasons, and it indicates that the quark model states which are devised to account for the hadron resonances should also encompass the muon. Thus the lifetime systematics of Fig. 10.4 pose a real challenge to quark models such as that of Gell-Mann and Zweig, which are unable to accommodate the muon except as a totally independent entity.

Before delving any farther into the theoretical sorting out of the lifetimes of Fig. 10.4, we first extend these results to include the lifetimes of the short-lived baryons and mesons.

10.3 The Lifetime Dichotomy of "Particles" and "Resonances"

Figure 10.2 was a plot of the lifetime logarithms X for the long-lived particles of Table 10.1. Figure 10.5 is the same plot extended in X so as to include the short-lived (broad-width) resonances. These short-lived resonances are the approximately 138 resonances listed in the Review of Particle Properties[1] which have

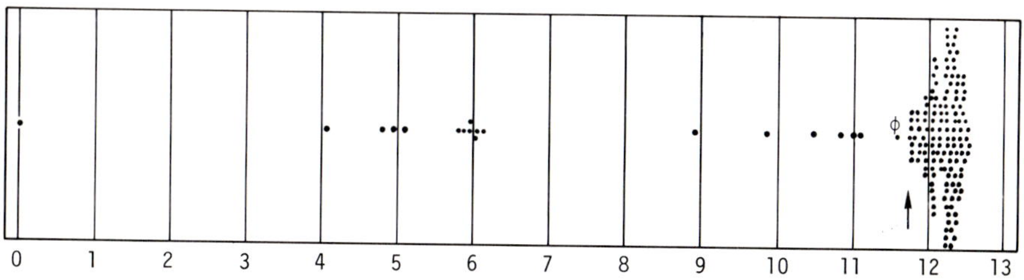

Figure 10.5. This is Fig. 10.2 extended in X so as to include the ∿138 resonances listed in RPP76 that have widths $\Gamma > 1$ MeV (lifetimes $\tau < 10^{-21}$ sec). The ordinate is an evenly-spaced distribution of lifetime parameters X after they have been sorted into $\Delta X = 0.1$ bins. As can be seen in the figure, a continuum of resonance lifetimes commences abruptly at the position of the arrow. The observed lifetime cutoff near $X = 12.6$ is a natural cutoff imposed by transit-time limitations.

measured widths $\Gamma > 1$ MeV[6], where the width Γ and lifetime τ of a particle state are related to one another by means of the equation

$$\Gamma = \hbar/\tau, \tag{10.4}$$

with $\hbar \simeq 6.6 \times 10^{-22}$ MeV sec. In the case of the long-lived particle states the lifetime τ can be measured directly, but in the case of the short-lived states it is the width Γ which must be measured.

The lifetime logarithms X shown in Fig. 10.5 divide empirically into two quite different groups: (1) the long-lived particles which appear in the range X = 0 - 11 in Fig. 10.5, and which have lifetimes that are spaced by powers of α; (2) the short-lived particles which appear to the right of the arrow in Fig. 10.5 ($X \geq 11.7$), and which have essentially a continuum of lifetime values. It seems to be phenomenologically significant to distinguish between these two groups of particles. Hence we denote these two groups as "particles" and "resonances", respectively, with the "particles" in the "particle region" having lifetimes $\tau > 10^{-21}$ sec and the "resonances" in the "resonance region" having lifetimes $\tau < 10^{-22}$ sec (widths $\Gamma > 6$ MeV).

As is shown in Fig. 10.5, the resonance region commences abruptly at the position of the arrow, and there is a gap of roughly one power of α between the particle region and the resonance region. The only resonance which appears in this gap is the ϕ meson[7]. The cutoff on lifetimes at $X \sim 12.6$ ($\Gamma \sim 600$ MeV) that is observed in Fig. 10.5 is due to lifetime limitations: the reasonance must persist for a time that is longer than the transit time of the interaction.

There is one other structural feature that can be observed in the lifetime logarithms X shown in Fig. 10.5. As can be seen in the figure, the X-values in the resonance region form two groups, $X \leq 12.1$ and $X \geq 12.2$, with a gap in the middle. These two groups correspond to narrow-width $\ell = 0$ rotationless resonances and broad-width $\ell > 0$ rotational resonances, respectively (see Chapter 12).

From the many relationships which have been established among resonance widths on the basis of phase-space arguments, it seems clear that phase-space effects are important in the resonance region of Fig. 10.5; phase-space effects can lead to changes by factors of 10 - 100 in the observed decay widths[4], and these phase-space effects are probably responsible for the continuum distribution of lifetimes in the resonance region. However, the accurate spacings in powers of α that are observed for the experimental "particle" lifetimes indicate that phase-space corrections in the particle region must be small; particles evidently have decays which are electromagnetically inhibited (as shown by the observed scaling in powers of α), and this inhibiting factor is essentially independent of the dictates of phase space. Phase-space corrections are discussed in Sec. 10.7 and in Appendix G.

10.4 The Lifetime Dichotomy of Fermions and Bosons

In Fig. 10.5, the lifetime logarithms X of 18 particles and 138 resonances were plotted together. If we now make a separation of these 156 states into half-integral-spin fermions and integral-spin bosons, we obtain 88 fermion states — 9 particles and 79 resonances — and 68 boson states — 9 particles and 59 resonances. Figure 10.6 shows the results of plotting the fermion and boson states separately, using the same representation as in Fig. 10.5. As can be seen in Fig. 10.6, the fermion resonance region and the boson resonance region are essentially identical to one another. However, whereas boson particle lifetimes are separated from one another by single powers of α, fermion particle lifetimes are separated from one another and from the fermion resonance region by powers of α^2. Hence we have phenomenologically related the number of virtual photons in the decay process[4] to the type of spin of the decaying particle (integral or half-integral), which is not a particularly surprising result.

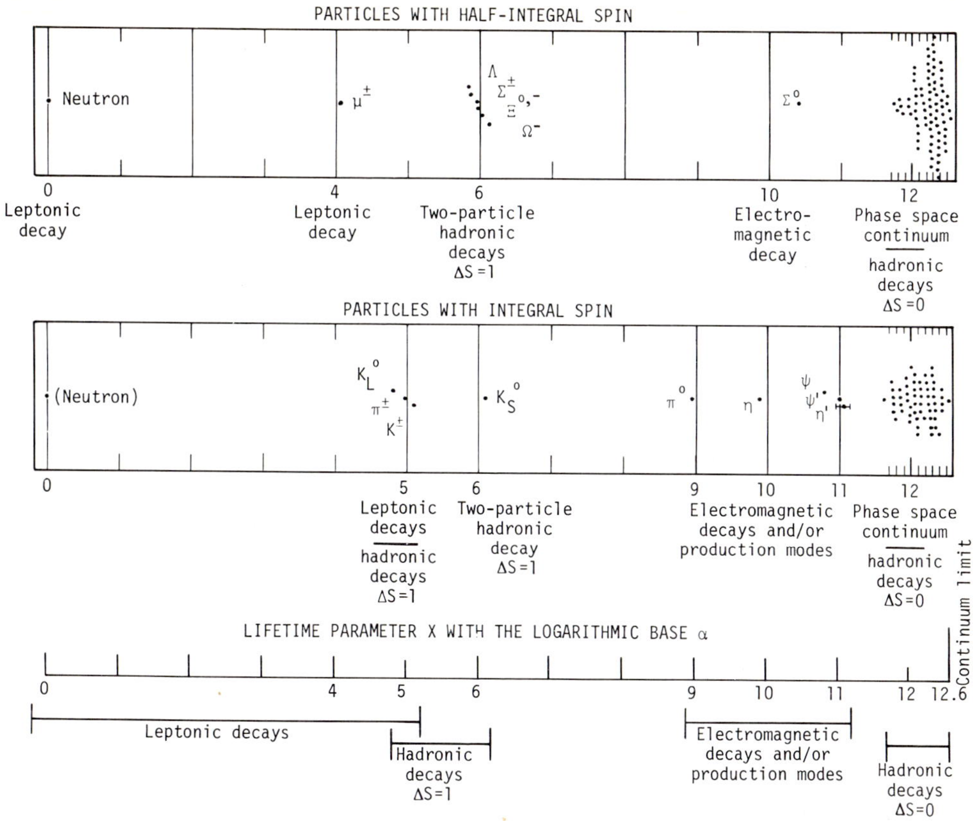

Figure 10.6. A separation of the 156 particles and resonances of Fig. 10.5 into 88 half-integral-spin fermions and 68 integral-spin bosons, using the same logarithmic plot and the same neutron reference lifetime for both fermions and bosons. As can be seen in the figure, the boson particle lifetimes are spaced by powers of α, whereas the fermion particle lifetimes are spaced by powers of α^2.

Although the α-spaced lifetime grid of Figs. 10.2 - 10.6 is common to all types of elementary particles, the positions where "particles" appear on this grid seem to be determined by the type of decay process involved, as is shown by the breakdown of decay modes given at the bottom of Fig. 10.6. This is further confirmation of the conjecture, stated above, that the long lifetimes of particles are due to characteristic electromagnetic inhibiting factors.

The separation of elementary particles into fermion and boson states has revealed some of the electromagnetic structure that is inherent in the decay processes of these particles (compare Figs. 10.5 and 10.6). This suggests that a further separation on the basis of spins and parities might reveal even more of this structure, a suggestion which is borne out by the results shown in the next section.

10.5 SU(3) Groupings of Lifetimes

SU(3) classification schemes[8] have demonstrated that when elementary particles are sorted out according to baryon number, spin, and parity, some very interesting isotopic spin regularities appear. This raises the question as to whether these same SU(3) selection rules also lead to interesting lifetime regularities. If we make an SU(3) breakdown of the 18 particle lifetimes in Table 10.1, we obtain 7 baryon octet lifetimes (the proton is missing), one baryon decimet lifetime (the Ω^-), 7 pseudoscalar-meson lifetimes (the pseudoscalar nonet), two vector-meson lifetimes (the ψ and ψ'), and one lepton lifetime (the $\mu^{\pm}$). Figure 10.7 shows plots of the baryon lifetimes, including $J^P = 1/2^+$ and $3/2^+$ SU(3) groupings, and Fig. 10.8 shows plots of the meson lifetimes, including $J^P = 0^-$ and 1^- SU(3) groupings, where the short-lived resonances[6] are added to the long-lived particles of Table 10.1. A plot of lepton lifetimes is also included in Fig. 10.8. The lifetimes in Figs. 10.7 and 10.8 are plotted on the same universal lifetime grid that was used for Figs. 10.2 - 10.6. We now discuss the plots of Figs. 10.7 and 10.8 in turn.

A. The baryons. Figure 10.7(a) is a plot of the lifetime logarithms X for the 87 measured[6] baryon and hyperon states. This plot is very striking, both in its scope and in its simplicity. If we start with the neutron as a basic reference lifetime, a jump by a factor of α^6 (13 orders of magnitude) gives the lifetimes of the long-lived hyperon states, and a second jump by another factor of α^6 gives the lifetimes of the short-lived excited states in the baryon and hyperon resonance region.

If we now project relevant SU(3) groups out of the states in Fig. 10.7(a), we obtain the $J^P = 1/2^+$ baryon octet lifetimes shown in Fig. 7(b) and the $J^P = 3/2^+$ baryon decimet lifetimes shown in Fig. 7(c). Comparing Fig. 7(b) with Fig. 7(c), we see that the lifetimes of the octet and decimet groups are similar in that each group

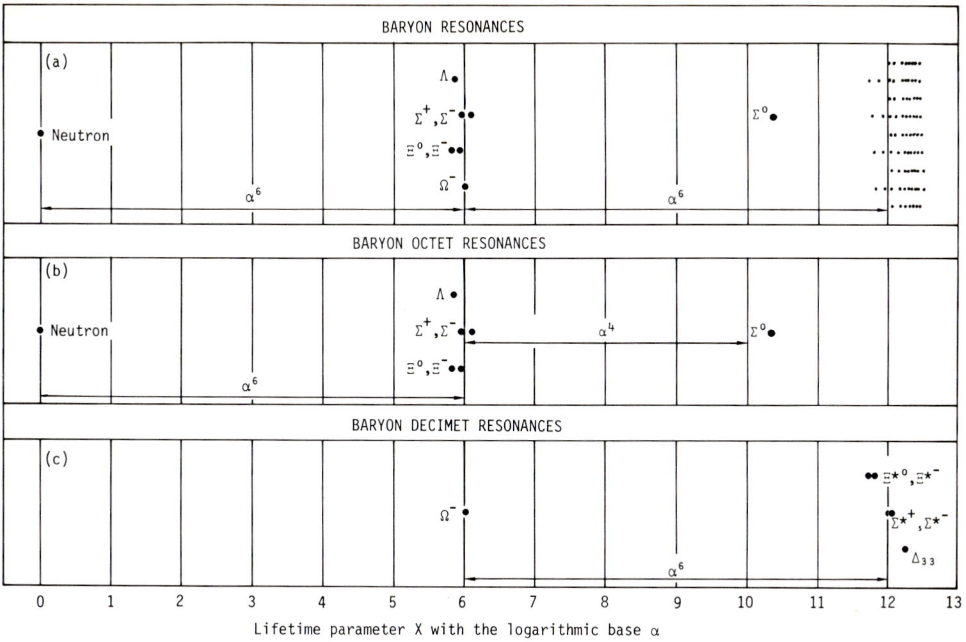

Figure 10.7. Figure 10.7(a) is a plot of the 87 hadron fermion lifetimes of Fig. 10.6. As can be seen, these divide into three main groups, which are separated from one another by factors of α^6. Figure 10.7(b) shows the baryon octet resonances from Fig. 10.7(a), and Fig. 10.7(c) shows the baryon decimet resonances from Fig. 10.7(a). The Σ^0 hyperon has a single-γ-ray electromagnetic decay mode which is not available to the other hyperons, and which evidently shortens its lifetime by a factor of α^4 (a similar situation occurs in the case of the π^0 meson).

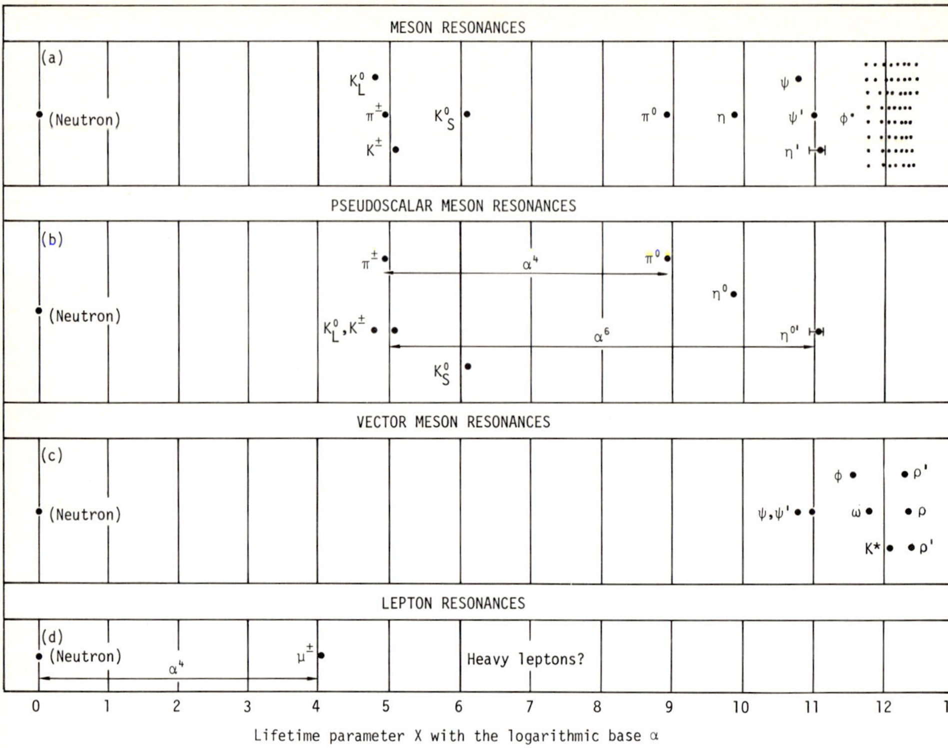

Figure 10.8. Figure 10.8(a) is a plot of the 68 boson lifetimes of Fig. 10.6. Figure 10.8(b) shows the spin-parity $J^P = 0^-$ pseudoscalar meson states from Fig. 10.8(a), and Fig. 10.8(c) shows the $J^P = 1^-$ vector meson states from Fig. 10.8(a). In Fig. 10.8(b), the π^0, η^0, and $\eta^{0\prime}$ mesons appear as a regularly-spaced lifetime sequence in powers of α, and the kaons form part of a similar sequence. The $(\pi^\pm - \pi^0)$ interval of α^4 echoes the $(\Sigma^\pm - \Sigma^0)$ interval of α^4 shown in Fig. 10.7(b). In Fig. 10.8(c), the ψ,ψ' lifetime group is displaced from the vector meson resonance region by about one power of α, which suggests that the key to their unique properties is electromagnetic. Figure 10.8(d) is a lepton lifetime plot that is placed here for convenience; as can be seen, the $(n - \mu^\pm)$ lifetime interval is just α^4.

has one long-lived member which is separated from the shorter-lived members of the group by a factor of α^6; also, the octet and decimet lifetime groups are separated from one another by a factor of α^6. The only exceptional cases are: (1) the Σ^0 hyperon in Fig. 7(b), which is separated from the Σ^+ and Σ^- hyperons by a factor of α^4 (compare this to the $\pi^0 - \pi^\pm$ meson separation of α^4 shown in Fig. 10.8(b)); and (2) the proton, which is stable.

There is one ramification of the baryon lifetime systematics of Fig. 10.7 which is worth mentioning here. The two related jumps of α^6 which are shown in Fig. 10.7 constitute a total lifetime span of α^{12}, or 25 orders of magnitude. This is a much larger span of values than physicists are ordinarily accustomed to dealing with. Now the magnitude of the span between the electromagnetic and gravitational domains, which has always seemed to be so prohibitively large, is in fact just α^{18}. In view of the systematics of Fig. 10.7, this span of α^{18} may not be as totally impossible to bridge as we usually assume it to be.[9]

B. The pseudoscalar mesons. Figure 10.8(a) is the meson counterpart of the baryon plot of Fig. 10.7(a), and it shows the lifetime logarithms X for 68 measured[6] meson and kaon states. As can be seen in Fig. 10.8(a), the meson and kaon lifetimes are separated by single powers of α, which is in marked contrast to the α^2 and α^6 lifetime intervals observed in Fig. 10.7.

Seven of the lifetimes shown in Fig. 10.8(a) — $\pi^\pm$, π^0, η, η', $K^\pm$, K_L^0, and K_S^0 — correspond to $J^P = 0^-$ spin-parity states[1]. These states, which are used to form the SU(3) pseudoscalar nonet[1], are plotted together in Fig. 10.8(b). As can be seen in Fig. 10.8(b), the neutral meson states $\pi^0(135)$, $\eta^0(549)$, and $\eta^0{}'(958)$, which have successively increasing masses, also have lifetimes that are successively shorter by single powers of α. Similarly, the K_S^0 state is separated from the K_L^0 and $K^\pm$ kaon states by one power of α.

We saw in Fig. 10.7(b) that the Σ^0 hyperon, which has an electromagnetic decay, has a lifetime that is shorter than the Σ^+ and Σ^- lifetimes by a factor of α^4. Correspondingly, we see in Fig. 10.8(b) that the π^0 meson, which also has an electromagnetic decay, has a lifetime that is shorter than the $\pi^\pm$ lifetime by this same factor of α^4.

It was demonstrated in Fig. 10.7 that related groups of strange and nonstrange states are characteristically separated by factors of α^6. Hence the observed separation of α^6 between the strange kaons and the nonstrange η' meson in Fig. 10.8(b) suggests that these states may be related. If we form the η' meson as a $K\bar{K}$ bound state, then the $\eta' = K\bar{K}$ system has a calculated binding energy of 4% (see the discussion of this $\eta' = K\bar{K}$ bound state in point two of Sec. 5.2 in Chapter 5).

C. The vector mesons. Figure 10.8(c) is a plot of the $J^P = 1^-$ spin-parity states from Fig. 10.8(a). Of these states, the ρ, ω, φ, and K* constitute the SU(3) vector meson nonet[1], the ρ'(1250) and ρ'(1600) are other recently-identified vector mesons[1], and the ψ and ψ' are New Particles[1]. The lifetimes of the ρ, ω, K*, ρ'(1250), and ρ'(1600) mesons (and possibly also the φ meson; see Ref. 7) lie in the meson resonance region, whereas the lifetimes of the ψ and ψ' mesons are shifted by one power of α. Hence the ψ and ψ' are analogous to the η' meson in Fig. 10.8(b), whose lifetime is also shifted by one power of α from the meson resonance region. While this does not constitute an "explanation" for the existence of the ψ and ψ' New Particles, it indicates that their unique properties are electromagnetic in origin (which is hardly a surprise), and that the same underlying factor which is responsible for the single-power-of-α intervals observed in Fig. 10.8(b) is probably also responsible for the single-power-of-α interval observed in Fig. 10.8(c). Thus a theoretical explanation for the uniqueness of the ψ and ψ' particles should in some manner incorporate the fine-structure constant $\alpha = e^2/\hbar c$ (which none of the presently-fashionable theories do).

D. The leptons. The lepton lifetime plot, Fig. 10.8(d), is shown for convenience at the bottom of the meson lifetime plots. The $\mu^\pm$ is the only lepton with a measured lifetime[1], and its lifetime falls rather surprisingly into the same lifetime grid that fits the baryons and mesons. The neutron-to-muon lifetime spacing of α^4 shown in Fig. 10.8(d) is a part of the general fermion lifetime systematics of Fig. 10.6. (The ratio of the neutron and muon lifetimes has been calculated using current algebra techniques and certain assumptions about matrix elements,[10] but the present results suggest that the current algebra calculation, which does not involve α, may not be physically significant.) There are indications that other heavy leptons may exist[11], with lifetimes possibly in the general range shown in Fig. 10.8(d). If these particles are clearly identified, it will be interesting to see where their lifetimes fall on the plot of Fig. 10.8(d).

We see from the systematics of Figs. 10.7 and 10.8 that SU(3)-suggested groups of particles exhibit interesting lifetime patterns when these lifetimes are plotted on the universal α-logarithmic scale of Eq. (10.3). It seems clear from these results that the scaling of lifetimes in powers of α must be a real physical occurrence, and not just a statistical accident. Hence electromagnetic effects play a dominant role in all of the "particle" decays — that is, in the decays of the elementary particle states which persist for longer than 10^{-21} seconds. This tells us, among other things, that strangeness, which is traditionally assumed to be responsible for the long lifetimes of the lowest-mass hyperons, is in some way related to electromagnetic effects; and it tells us that SU(3), to the extent that it is a valid symmetry for these particle states, is also related to electromagnetic phenomena. We now go on to demonstrate a totally different kind of lifetime regularity — a regularity which reveals still another facet of "particle" decay modes.

10.6 A Factor-of-Two Fine Structure in "Particle" Lifetimes

Figure 10.9 is taken from Fig. 10.4; it shows the three $X \simeq 5$ lifetimes and the seven $X \simeq 6$ lifetimes plotted as relative lifetimes on an expanded logarithmic scale, with the seven $X \simeq 6$ lifetimes sorted out according to isotopic spin, and with the shortest lifetime in each group set equal to unity. As can be seen at the top of Fig. 10.9, the $K^{\pm}$, $\pi^{\pm}$, and K^0_L lifetimes occur in accurate ratios of 1:2:4. Now these are supposedly-independent particles, whose lifetimes should all be more-or-less unrelated. Thus, is the factor-of-two correlation in these three lifetimes just an accident? As a way of answering this question, we can examine the lifetimes shown at the bottom of Fig. 10.9. These seven lifetimes, which are one power of α shorter than the three lifetimes at the top of Fig. 10.9, also occur with factor-of-two ratios in the lifetimes. Hence, since all ten particles in Fig. 10.9 — which is to say, all of the particles in this lifetime range — exhibit lifetime factor-of-two ratios, it seems clear that this is a physically-significant result, and not just an accidental happenstance.[12]

In addition to the results shown in Fig. 10.9, there is another very important conclusion that can be obtained from these factor-of-two lifetime intervals. In Fig. 10.9 we used the world-average lifetime for the Ω^-, which is the lifetime shown in Table 10.1. However, the number of Ω^- events has been greatly increased in the past few years, and these events now reveal two distinct lifetimes for the Ω^-, $\tau = 0.75 \times 10^{-10}$ sec for low-momentum experiments[13] (4.2 GeV/c), and $\tau = 1.41 \times 10^{-10}$ sec for high-momentum experiments[14] (10-16 GeV/c). These two lifetimes differ by a factor of two, and they fit in with the systematics of Fig. 10.9. Denoting these two Ω^- states as Ω^-_S and Ω^-_L, in analogy with K^0_S and K^0_L, we demonstrate in Table 10.2 the manner in which all of these lifetimes fit together. The importance of the regularities shown in Table 10.2 is the strong implication they give that the states Ω^-_S and Ω^-_L both exist, and are not simply due to a statistical fluctuation in the data. As we saw in Chapter 9, the Ω^- has a rather complicated light-quark substructure, so it is not surprising to find that it can occur in two different forms. This situation seems to closely parallel that of the kaon. The spin of the Ω^-_L hyperon has now been measured[14], and it is $J = 3/2$ (with a statistical probability of 70%).

Most of the factor-of-two lifetime ratios shown in Fig. 10.9 and Table 10.2 are between dissimilar particles. But in three cases — namely, in the Ω^-_S, Ω^-_L and Ξ^-, Ξ^0 and Σ^+, Σ^- pairings — the particle states are closely related. This brings up the question of final-state phase-space effects. Can these factor-of-two ratios be

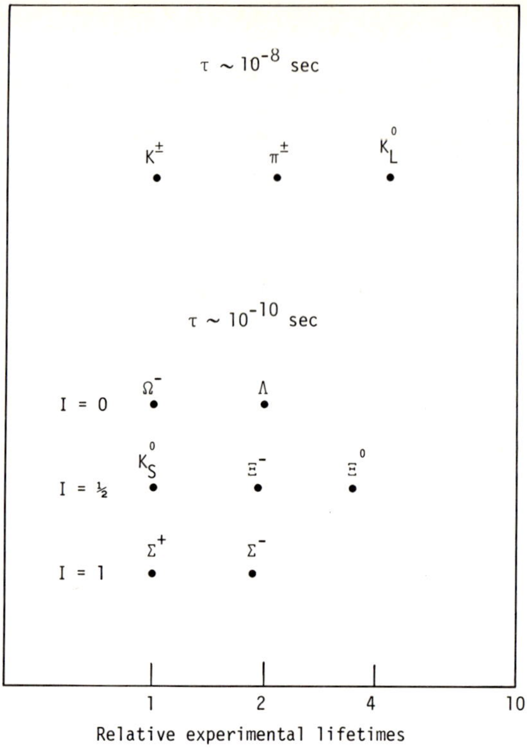

Figure 10.9. A logarithmic plot of the relative experimental lifetimes for the three particles observed near X = 5 ($\tau \sim 10^{-8}$ sec) and for the seven particles observed near X = 6 ($\tau \sim 10^{-10}$ sec) in Fig. 10.4. The three $\tau \sim 10^{-8}$ lifetimes occur in the ratio 1:2:4, which is surprising for these supposedly independent particles. The seven $\tau \sim 10^{-10}$ lifetimes, when sorted out according to isotopic spin, also exhibit factor-of-two lifetime ratios, which indicates that these results are not accidental. If phase-space corrections are applied [see Eq. (10.5)], the (Ξ^-, Ξ^0) and (Σ^+, Σ^-) pairs shift slightly to even more accurate 1:2 ratios. This result indicates two things: (1) the factors of two shown in this figure do not arise from phase-space considerations; and (2) phase-space corrections, if properly applied, are probably small for all of the particles shown in Fig. 10.9. These results are extended even farther in Table 10.2, where it is demonstrated that the two lifetimes which have recently been reported for the Ω^- also fit into these factor-of-two lifetime ratios. (The world-average lifetime for the Ω^- was used in Fig. 10.9; see Table 10.1.)

Table 10.2. The lifetime systematics of the particles with lifetimes in the range $\tau \sim 10^{-8}$ to 10^{-10} sec. The three particles with lifetimes $\tau \sim 10^{-8}$ sec have lifetime ratios of 1:2:4. The eight particles with lifetimes $\tau \sim 10^{-10}$ sec also have lifetime ratios of 1:2 and 1:2:4 when they are grouped according to isotopic spin I. Phase space corrections are discussed in the text. The results shown in Table 10.2 duplicate those displayed in Fig. 10.9, except that in Table 10.2 we have replaced the world-average lifetime of the Ω^- (Table 10.1) by the two different Ω^- lifetimes that have recently been reported: the low-momentum (4-6 GeV/c) lifetime Ω_S^- (Ref. 13); and the high-momentum (10-16 GeV/c) lifetime Ω_L^- (Ref. 14). The lifetime regularities displayed in Table 10.2 suggest that the two different lifetimes Ω_S^- and Ω_L^- probably really do exist, which indicates that the light-quark substructure of the Ω^- is dependent on the momentum-transfer conditions under which the Ω^- is produced.

Particle	$K^{\pm}$	$\pi^{\pm}$	K_L^0
Exper. lifetime (sec)	1.2371×10^{-8}	2.6030×10^{-8}	5.181×10^{-8}
Lifetime ratio	1.0	2.1	4.2
Scaled lifetime ($\tau \times \alpha$)	0.903×10^{-10}	1.900×10^{-10}	3.781×10^{-10}
Particle (I = 1/2)	K_S^0	Ξ^-	Ξ^0
Exper. lifetime (sec)	0.893×10^{-10}	1.652×10^{-10}	2.89×10^{-10}
Particle (I = 1)	Σ^+	Σ^-	
Exper. lifetime (sec)	0.800×10^{-10}	1.482×10^{-10}	
Particle (I = 0)	Ω_S^-	Ω_L^-	Λ^0
Exper. lifetime (sec)	0.75×10^{-10}	1.41×10^{-10}	2.635×10^{-10}

attributed to phase-space effects? The importance of the Ω_S^-, Ω_L^- and Ξ^-, Ξ^0 and Σ^+, Σ^- pairs in this regard is that these particles have similar decay modes, so that we can make reliable comparisons of the phase-space corrections to their lifetimes. Phase-space corrections are discussed in the next section, where it is made clear that these factor-of-two lifetime ratios are not due to phase-space effects, and neither are the factor-of-α lifetime ratios discussed above. If this is so, then, just as the factor-of-α lifetime ratios indicate the presence of electromagnetic factors in the decays, so also the factor-of-two lifetime ratios must correspond to some underlying aspect of the decays. But what aspect is involved?

In order to shed some light on these lifetime factors-of-two, we can resort to a statistical argument. We know from the workings of the light-quark model that the decay of an elementary particle state usually involves the annihilation of some of the subquanta M = 70 MeV: for example, it was shown in Chapter 5 that the kaon contains seven quanta M and the pion contains two quanta M, so that the decay modes $K \to \pi\pi\pi$ and $K \to \pi\pi$ involve the annihilation of 1 and 3 subquanta M, respectively; similarly, the decays $\pi^\pm \to \mu^\pm + \nu$ and $K^\pm \to \mu^\pm + \nu$ involve the annihilation of 1 and 6 subquanta M, respectively, with the remaining subquantum M in each case being transformed into a muon-neutrino pair (see Eqs. (7.27) and (7.28) in Chapter 7). We also know from the decays of mixed and unmixed substates that resonances which contain matching internal $M\bar{M}$ pairs decay much faster than do particles which contain only unmatched internal subquanta M or $\bar{M}$. Thus it seems apparent that the decays of the long-lived particles are triggered by the spontaneous annihilation of an unmatched subquantum M or $\bar{M}$. Now, the appearance of lifetimes which occur with ratios of 1:2:4 suggests to a statistician that the particles involved have 4:2:1 independent triggers (subquanta M) which can initiate the decay process. This explanation is illustrated in Figure 10.10, which indicates possible assignments for the decay triggers of the $K^\pm$, $\pi^\pm$, K_L^0, and K_S^0 mesons. This is of course an a posteriori explanation for these lifetime ratios, but it is the only explanation of any kind that has been advanced to account for the striking 1:2:4 ratios in the lifetimes of these supposedly-independent particles. As we are trying to stress throughout this book, the experimental regularities that we discover in these groups of elementary particles are not accidental, and they must be explained. A theoretical model that has no explanations for these phenomena is manifestly incomplete.

The importance of the lifetime explanation we have illustrated in Fig. 10.10 lies in the fact that it ties together a number of different facets of the present light-quark phenomenology: (1) the π and K mesons share a common set of 70 MeV light-quark substates; (2) the decays of these particles involve the annihilation of some of the constituent subquanta M; (3) the factor-of-two scaling in elementary particle lifetimes is superimposed on an over-all factor-of-α scaling in these lifetimes; (4) the decay rates of these particles are essentially independent of phase

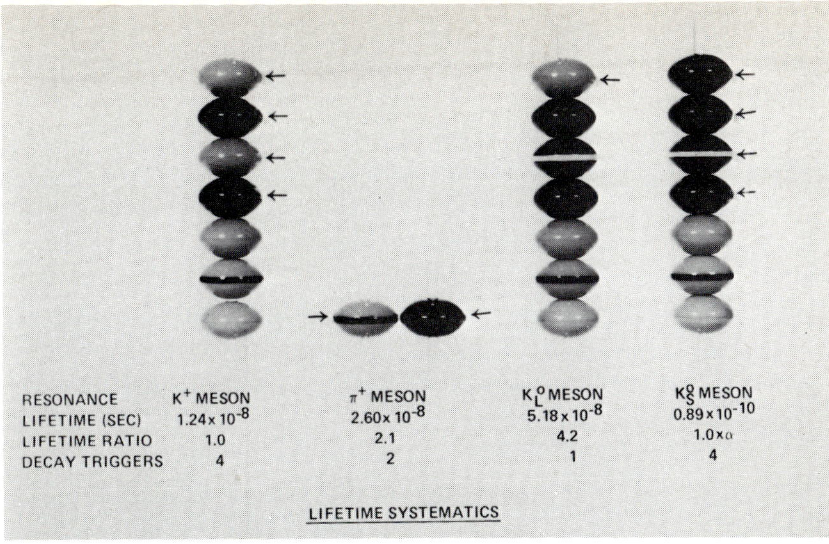

Figure 10.10. Lifetime systematics of the K^+, π^+, K_L^o, and K_S^o mesons. As shown in this figure, the $K^\pm$, $\pi^\pm$, and K_L^o lifetimes form an accurate 1:2:4 lifetime triplet, and the K_S^o is shifted from the $K^\pm$ by one power of $\alpha = e^2/\hbar c$. Statistically, these accurate 1:2:4 lifetime ratios can be explained by assuming that 4:2:1 subquanta M are available to independently annihilate and trigger the decays, as is indicated by the arrows. The author is aware of no other explanation for these striking lifetime ratios.

space effects; (5) the particle structures that are shown in Fig. 10.10, in addition to accounting for these lifetime effects, also accurately reproduce the masses, spins, strangeness quantum numbers, and baryon quantum numbers of these resonances (see Chapters 5, 9, 18, and 19).

We now turn to the one question that hangs over all of the lifetime results presented here — namely, the question of final-state phase-space effects.

10.7 Phase Space Corrections to Elementary Particle Lifetimes

It is often assumed that the lifetimes of elementary particles are strongly influenced by final-state phase-space effects. In this connection, there are two comments which should be made here: (1) The scaling of particle lifetimes in powers of α that is observed in Figs. 10.2 and 10.4 is for the experimental lifetimes, and not for phase-space-corrected lifetimes; hence, if this scaling in α is physically significant, phase-space corrections to these lifetimes must necessarily be small. (2) Phase-space corrections involve an unknown normalization constant whose dimensionality is proportional to the number of particles (number of degrees of freedom) in the final state; this fact makes it impossible to apply phase-space corrections in a uniform manner to systems of particles that have different degrees of freedom in their final states, as do the particles of Table 10.1. The first of these two comments tells us that, from an empirical point of view, the lifetimes of the long-lived "particles" of Table 10.1 are not strongly influenced by final-state phase-space effects; and the second comment tells us that we are fortunate to have the first comment, because phase-space corrections to these lifetimes are, as a practical matter, impossible to apply with any reliability.

Instead of discussing phase-space corrections at any length here, we have placed this discussion in Appendix G, and we merely summarize the conclusions of Appendix G that are important for our present purposes. In the "resonance region" of Fig. 10.5, considerable success has been achieved in correlating widths on the basis of phase-space arguments; hence these short-lived "resonances", which have "strong" decays, seem to be responsive to phase-space considerations. However, in the "particle region" of Fig. 10.5, where the decays are "weak", it seems clear that these long-lived "particles" have an (electromagnetic) inhibiting factor which determines their lifetimes, and which is essentially independent of phase-space effects. This conclusion is reinforced by the systematics of Fig. 10.4, wherein it is shown that many types of "particles", with many different decay modes, all share a common lifetime framework; hence these "particle" lifetimes are not dependent on the decay modes, and therefore are not dependent on the final-state phase-space freedom of the decay modes. In Appendix G we apply, or attempt to apply, phase-space corrections to these "particle" lifetimes, and we demonstrate that in general these corrections cannot be applied in a physically-significant manner.

There is one case, however, where phase-space corrections can be applied to "particle" lifetimes with some degree of confidence. This is the case, cited above, of the Ω_S^-, Ω_L^- and Ξ^0, Ξ^- and Σ^-, Σ^+ pairs, in which both of the particles in each pair have very similar two-body final states. As was shown in Fig. 10.9 and Table 10.2, the experimental lifetimes for these pairs occur approximately in the ratio of 2:1. The Ω_S^- and Ω_L^- particles presumably have identical phase space corrections, but the particles in the Ξ^0, Ξ^- and Σ^-, Σ^+ pairs do not. If we apply phase-space corrections to these lifetimes, as is described in detail in Appendix G, we obtain the following result:

Lifetime ratio	Experimental lifetimes	Phase-space-corrected lifetimes	
Σ^-/Σ^+	1.85	1.90	(10.5)
Ξ^0/Ξ^-	1.75	2.00	

This result tells us that the factor-of-two lifetime ratios shown in Fig. 10.9 are not due to phase-space factors: phase-space corrections, instead of removing the factor-of-two differences in these lifetimes, actually make them more precise. The result shown in Eq. (10.5) is discussed at more length in Appendix G.

It seems apparent from the studies of the present chapter and of Appendix G that for the long-lived "particles", the experimental lifetimes are the most relevant quantities to work with from the standpoint of determining the over-all factors which govern their decays.

In the present chapter we have demonstrated a number of interesting regularities in the lifetimes of the elementary particles, and we have shown how these regularities give some clues as to the nature of the substructures which form these particles. We have not attempted to construct a theoretical model to account for these decays, although this is a task which is obviously worthwhile to do. One point which emerges from this study of particle lifetimes is that electromagnetic effects are important throughout the entire region of long lifetimes ($\tau > 10^{-21}$ sec). Another point we have tried to emphasize here is that the decay systematics of these particles are consistent with the light-quark viewpoint of elementary particle structure that was developed in Part I of this book. In the next chapter we investigate the correlation between particle lifetimes and particle angular momentum values. This correlation has some surprising features, and it furnishes us with another basic clue in our quest to unravel the secrets of the elementary particle.

References for Chapter 10.

1. Particle Data Group, "Review of Particle Properties", Rev. Mod. Phys. 48, No. 2, Part II, April (1976).

2. The Σ^0 lifetime is from F. Dyak et al., Nucl. Phys. B118, 1 (1977). The Ξ^0 and Λ lifetimes are from G. Zech et al., Nucl. Phys. B124, 413 (1977).

3. A Duane et al., Phys. Rev. Lett. 32, 425 (1974). The η' lifetime shown in Table 10.1 was obtained by fitting parabolas to the two sets of data in Table I of

Duane et al. and averaging the results; the error limit quoted here is an average of the $\Delta\chi^2 = +1$ values obtained from these two parabolas.

4. See B. T. Feld, Models of Elementary Particles, Blaisdell, Waltham (1969), Sec. 7.4.

5. M. H. Mac Gregor, in Fundamental Interactions at High Energies, proceedings of the 1971 Coral Gables Conference, edited by M. Dal Cin, G. J. Iverson, and A. Perlmutter (Gordon and Breach, New York, 1971), Vol. 3, p. 75; Nuovo Cimento 8A, 235 (1972) and 20A, 471 (1974); Phys. Rev. D9, 1259 (1974), Sec. XII.

6. The short-lived resonances include all resonances in RPP76, Ref. 1, except those of Table 10.1 in the present paper, which have quoted lifetime or width values. In cases where a spread of values is given, the RPP76 quoted average was used where available. In some cases where different charge states of a resonance have separately quoted lifetimes, both of these lifetimes were included. In Figs. 10.5 and 10.6, the abscissa corresponds to the lifetime logarithms X as obtained from Eq. 10.3, and the ordinate gives an equally spaced distribution of the logarithms X after they have been sorted out into $\Delta X = 0.1$ bins.

7. We ordinarily expect a resonance to exhibit the same mass and the same observed total width in all of its decay modes. However, both the observed mass and the observed width for a particular decay mode can be shifted if there is a severe phase-space constraint which occurs for just this specific decay mode. The ϕ meson is the only example known to the author in which this condition occurs. The $\phi \to K\bar{K}$ decay mode, which has been used for most experimental determinations of the ϕ-meson parameters, is highly unusual in that it is a spin-1 decay with a very small Q value. From the Q value of 30 MeV, each final-state kaon has a kinetic energy of only 15 MeV, or a linear momentum of 123 MeV/c. Thus, in order to carry away the spin-1 angular momentum of the ϕ, the two kaons must have a noncollinear separation distance of 2.3 fermis. If we assume that the 1020-MeV ϕ meson has about the same "size" as the measured size of the 939-MeV nucleon, this large separation distance required for the final-state kaons constitutes a severe kinematic constraint, and it limits the available phase space for the decay by cutting off the low-momentum portion, thus increasing the observed mass of the ϕ in this decay channel and decreasing its observed width. The ϕ in its $\phi \to \pi\pi\pi$ decay mode does not have this severe kinematic constraint, and it appears in some experiments at a lower mass and with an apparently broader width [see M. Aguilar-Benitez et al., Phys. Rev. D9, 29 (1971), Fig. 61 and the accompanying discussion; J.-E. Augustin et al., Phys. Lett. 28B, 517 (1969); G. R. Kalbfleisch et al., Phys. Rev. D13, 22 (1976)], which would place it closer to the "resonance region" of Fig. 10.5 in the present paper. However, in Orsay Storage Ring experiments [G. Cosme et al., Phys. Lett. 48B, 155 (1974); G. Parrour et al., Orsay Report No. LAL 1280 (1975), (unpublished)], the $e^+e^- \to \phi \to 3\pi$ and $e^+e^- \to \phi \to K_L^0 K_S^0$ measurements of the ϕ mass and width give very similar results.

8. A review of SU(3) classification schemes is given in Ref. 1. Also see Chapter 14 in Ref. 4.

9. See M. H. Mac Gregor, Lett. Nuovo Cimento 1, 759 (1971). The factor α^{18} is approximately equal to the square of the ratio of the proton mass to the Planck mass ($m^* = \sqrt{\hbar c/G}$); see J. A. Wheeler, Analytical Methods in Mathematical Physics, edited by R. P. Gilbert and R. G. Newton, Gordon and Breach, New York (1970), p. 335.

10. J. Bernstein, Elementary Particles and Their Currents, Freeman, San Francisco (1968), p. 128.

11. M. L. Perl et al., Phys. Rev. Lett. 35,1489 (1975); also see a theoretical paper by A. De Rújula et al., Phys. Rev. Lett. 35, 628 (1975).

12. The ratio of the lifetimes of the J/ψ and ψ' particles (see Table 10.1) is currently given in Ref. 1 as 3.3 ± 1.1, which may or may not be in agreement with the factor-of-two lifetime ratios shown in Fig. 10.9.

13. Amsterdam — CERN — Nijmegen — Oxford Collaboration, CERN/EP/PHYS 76-19, June (1976); a total of 32 Ω^- events at 4.2 GeV/c gave a lifetime $\tau = 0.75 \pm 0.15 \times 10^{-10}$ sec. Also, in a recent compilation of Ω^- events [R. J. Hemingway, CERN/EP/PHYS 76-50, Aug. 10 (1976) (unpublished)], 50 low-momentum events (4 - 6 GeV/c) gave a lifetime $\tau_{\Omega^-} = 0.79 \pm 0.11 \times 10^{-10}$ sec.

14. M. Deutschmann et al., Phys. Lett. 73B, 96 (1978). A total of 101 high-momentum Ω^- events at 10-16 GeV/c gave a lifetime $\tau = 1.41 + 0.15, -0.24 \times 10^{-10}$ sec.

CHAPTER 11. DO ORBITAL ANGULAR MOMENTUM BARRIERS APPLY TO ELEMENTARY PARTICLES?

"The main obstacles to progress are not the things we don't know, but rather the things we do know that aren't right."

In the present chapter we discuss a simple but important experimental result. This result is important for two reasons: (1) it furnishes us with a useful method for classifying hadron resonances; (2) it is a result that differs from comparable results in the area of nuclear physics, and that, as a consequence, also differs from the preconceptions of many elementary particle theorists. This experimental result has to do with the relationship between elementary particle lifetimes (or widths) and elementary particle angular momentum values.

In the preceding chapter we studied the general systematics of elementary particle lifetimes. By plotting these lifetimes in one overall picture as logarithms to the base $\alpha = e^2/\hbar c$, we were able to identify two distinct lifetime regions — a "resonance region" that contains short-lived (broad-width) "resonances", and a "particle region" that contains long-lived (narrow-width) "particles". When the lifetimes of the half-integral-spin fermions and the integral-spin bosons were plotted separately, characteristic differences appeared between the fermions and bosons in their power-of-α spacings, but they each still exhibited both a "particle region" and a "resonance region" (see Fig. 10.6). In the present chapter we investigate the differences between the narrow-width states which populate the "particle region" and the broad-width states which populate the "resonance region". The key factor which emerges from this investigation is that the width of a resonance is dominated by the angular momentum value of the resonance.

Many of the ideas that are used in elementary particle physics have been taken over directly from nuclear physics. As an example, the characteristic Breit-Wigner resonance shapes that were deduced by fitting to nuclear resonances are now also used in fitting to elementary particle resonances.[1] As another example, and one that is pertinent to the present discussion, the angular momentum barrier concepts of nuclear physics are customarily assumed to apply to elementary particle decays, and centrifugal barrier factors are used in calculations involving elementary particle rotational modes.[2] It is the assumption in this last example that seems to be at variance with the experimental systematics of elementary particle decays.

In <u>nuclear physics</u> there is a well-established relationship between the orbital angular momentum ℓ of a resonance and the width or lifetime of the resonance: high-ℓ resonances, which have large centrifugal bariers, have narrow widths and long lifetimes; low-ℓ resonances, which have small centrifugal barriers, have broad widths and short lifetimes. In fact, this relationship is accurate enough that the observed width of a nuclear resonance is often used as a means of determining the spin of the resonance. Experimentally, many high-ℓ nuclear resonances have such narrow widths that it is difficult to observe them.

Does this same kind of angular momentum versus width or lifetime relationship carry over to <u>particle physics</u>? The usual assumption is that it <u>does</u> carry over,[2,3] and explicit centrifugal-barrier factors are inserted when elementary particle widths are related to decay coupling constants in SU(3) calculations.[2] However, there is an important difference between orbital rotations in nuclear physics and orbital rotations in elementary particle physics. In nuclear physics, the nucleus is assumed to have an overall potential V_o which is essentially independent of the orbital quantum number ℓ. When a resonant state is formed with a certain ℓ-value, a centrifugal-barrier term $V_{cb}(\ell)$ is added to V_o (see Ref. 4). Since V_o is a constant (i.e., independent of ℓ) for a given nucleus, and since $V_{cb}(\ell)$ has a known[2,3] functional dependence on ℓ, it is a straightforward matter to intercompare a series of nuclear resonances and relate their widths to their ℓ-values. The essential point here is that nuclear rotations are adiabatic (slow), so that the rotational motion does not distort the shape of the nucleus, which would lead to changes in the potential V_o. As a typical example, a nucleus with a mass number A $\sim$ 100 has a total mass of about 10^5 MeV/c^2, and the rotational energies for such a nucleus are $\sim$ 1 MeV, so that the (adiabatic) rotational energy is only $\sim$ 0.001% of the total energy of the nucleus.

In the case of elementary particles, on the other hand, the orbital angular momentum values are the same as they are for atomic nuclei (that is, small multiples of $\hbar$), but the masses and consequently the moments of inertia are very small, which means that orbital rotations are highly nonadiabatic, and orbital rotational energies are comparable (as we will show in subsequent chapters) to the rest mass of the rotating system. In this situation, it seems clear that substantial rotational deformations can occur; if these rotational deformations lead to changes in the potential V_o which are larger than the centrifugal-barrier effects V_{cb}, these barrier effects will be completely overshadowed.

There is another aspect of elementary particle decays which also suggests that the centrifugal-barrier considerations of nuclear physics may not apply to elementary particles. The effectiveness of the nuclear centrifugal barrier in inhibiting nuclear decay rates is essentially limited to resonance energies which are below the height of the barrier — a few MeV. However, in elementary particle decays the energy release is enormous: the elementary particle literally explodes when it decays. A striking example is provided by the high-mass $\psi' \rightarrow J/\psi + \pi\pi$ decay, in which the two pions are blown off in opposite directions while the J/ψ remains essentially at rest (see Chapter 17). In the decay $\rho(773) \rightarrow \pi\pi$, which is a typical low-mass example, more than 60% of the total energy of the resonance appears as kinetic energy of the two emitted pions (this compares to 0.001% as the typical kinetic energy in the decay of a nuclear resonance). Thus, by the standards of nuclear physics, these very energetic elementary particle decays result in kinetic energies which are well above the height of any reasonable elementary particle centrifugal barrier.

From the considerations we have just discussed, it is not-at-all obvious that the decays of rotating elementary particle systems bear much relationship to the decays of rotating nuclear systems. Since we do not know the form of the overall potential V_o inside an elementary particle, or even if it makes sense to define such a potential, we have no reliable theoretical method for inserting centrifugal-barrier effects, and thus relating elementary particle spins and widths. However, we can determine this relationship empirically by examining the available experimental data. There are 74 hadrons whose spins and lifetimes (or widths) have been measured most reliably. These hadron states are listed in Table 11.1. If we group these hadrons into baryon number B and strangeness S families, and then divide these (B,S) families into spin states J, we obtain the lifetime plot shown in Fig. 11.1, where the same universal lifetime plot in powers of α (Eq. 10.3) is used as for the lifetime studies of Chapter 10 (but note the expanded abscissa in Fig. 11.1 for X-values $X > 12$). As is shown in Fig. 11.1, the maximum observed lifetime in each (B,S) family decreases monotonically as the spin J increases, which is just the opposite of the situation that occurs in nuclear physics. This elementary particle result holds uniformly for all combinations of the baryon number B and the strangeness S. In particular, 14 mesons and baryons which have lifetimes $\tau > 10^{-20}$ sec all occur in the lowest-spin states, $J = 0$ and $J = 1/2$; and 12 baryons which have the slightly higher spin value $J = 3/2$ all have lifetimes $\tau < 10^{-22}$ sec. The only exceptional case is the $J = 3/2$ Ω^- hyperon; as the lowest-mass strangeness $S = -3$ state, it has a long lifetime in spite of its higher spin value. This decrease in the lifetimes (and increase in the widths) with increasing spin values is brought out more clearly in Fig. 11.2, which is the same plot as that of Fig. 11.1, but with the lifetimes arranged in the order of the J values, and without the expanded abscissa for $X > 12$.

From the experimental results shown in Figs. 11.1 and 11.2, we see that short-lived (broad-width)resonances occur for all values of the angular momentum J, but long-lived (narrow-width) resonances occur only for low values of J; as J increases, the maximum observed lifetime decreases monotonically. Thus, instead of having a centrifugal-barrier effect which inhibits high-angular-momentum decays, as in nuclear physics, it appears from these experimental results that rapidly-rotating elementary particles experience a centrifugal-force effect which actually enhances

Table 11.1. A listing of the 74 hadrons whose spins and lifetimes (or widths) have been measured most reliably.

Mesons:	(J = 0)	π^0, $\pi^\pm$, η, η', S*
(22)	(J = 1)	ρ, ω, ϕ, A_1, B, ρ', D, E, ρ', J/ψ, ψ'
	(J = 2)	f, A_2, f', A_3
	(J = 3)	g
	(J = 4)	h
Kaons:	(J = 0)	$K^\pm$, K_L^0, K_S^0
(8)	(J = 1)	K*, Q, Q
	(J = 2)	K_N, L
Baryons:	(J = 1/2)	p, n, N(1413,1511,1674,1720), Δ(1623,1900)
(20)	(J = 3/2)	N(1517,1722), Δ(1232,1676)
	(J = 5/2)	N(1674,1680), Δ(1869)
	(J = 7/2)	N(2132), Δ(1928)
	(J = 9/2)	N(2183)
	(J = 11/2)	N(2650), Δ(2406)
Hyperons:	(S = -1)	(J = 1/2) Λ, Λ(1402,1672), Σ^+, Σ^0, Σ^-, Σ(1770)
(24)		(J = 3/2) Λ(1519,1690), Σ(1383,1387,1670,1920)
		(J = 5/2) Λ(1822,1825), Σ(1774,1920)
		(J = 7/2) Λ(2110), Σ(2040)
	(S = -2)	(J = 1/2) Ξ^0, Ξ^-
		(J = 3/2) Ξ^0(1532), Ξ^-(1535)
	(S = -3)	(J = 3/2) Ω^-

the decay rate. Thus, as we suspected from the theoretical considerations discussed above, elementary particle rotations bear little resemblance to nuclear rotations.

The analysis of elementary particle spins and lifetimes that we have given here is incomplete in one respect. In discussing nuclear angular momentum barriers, the orbital angular momentum ℓ is the pertinent quantity to consider. However, the elementary particle lifetime-versus-spin plots of Figs. 11.1 and 11.2 are in terms of the total angular momentum J rather than the orbital angular momentum ℓ. We should, properly speaking, decompose the total angular momentum J of an elementary particle state into its ℓ and S components in order to make a comparison with nuclear systematics (the decomposition of J into its ℓ and S components is discussed for baryons in Chapter 12 and for mesons in Chapter 13). However, the experimental relationship

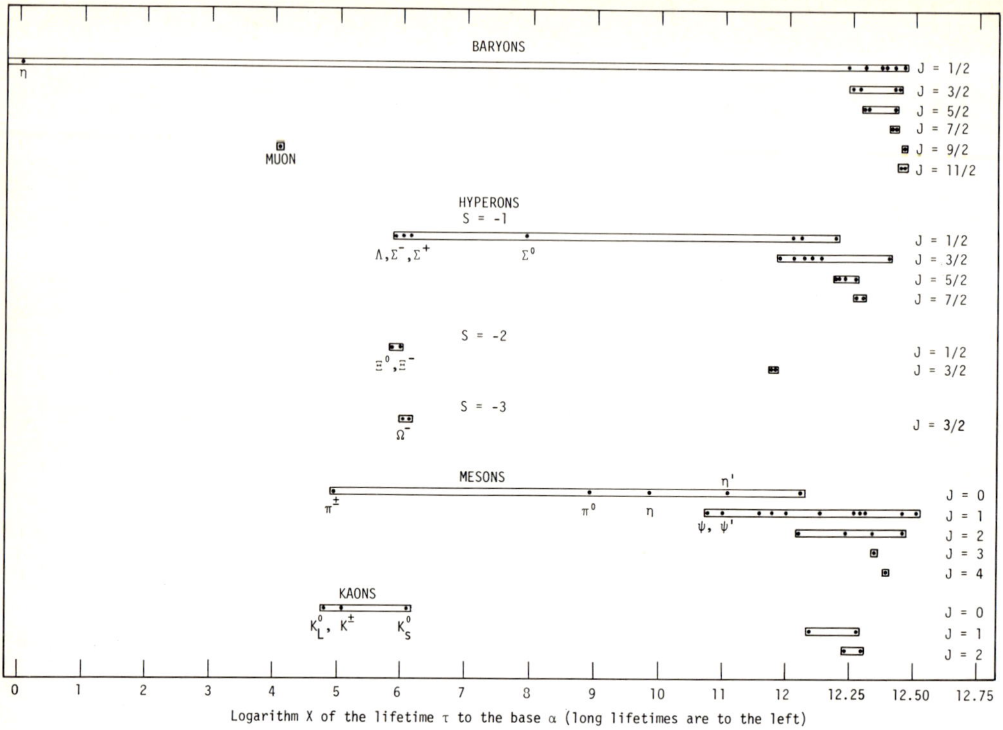

Figure 11.1. The lifetime logarithms X for the particles in the hadron families of Table 11.1, which are grouped according to baryon number B and strangeness S. The dots denote the logarithms X, and the envelopes surrounding the dots indicate the range of experimental lifetimes observed in each spin state J (note the expanded scale for $X > 12$). As can be seen in the figure, there is a well-defined correlation in each (B,S) family between the spin J and the maximum observed lifetime (the minimum value of X): the higher the spin value, the shorter the maximum observed lifetime. This result holds uniformly for baryons and mesons, and for all values of the strangeness quantum number S.

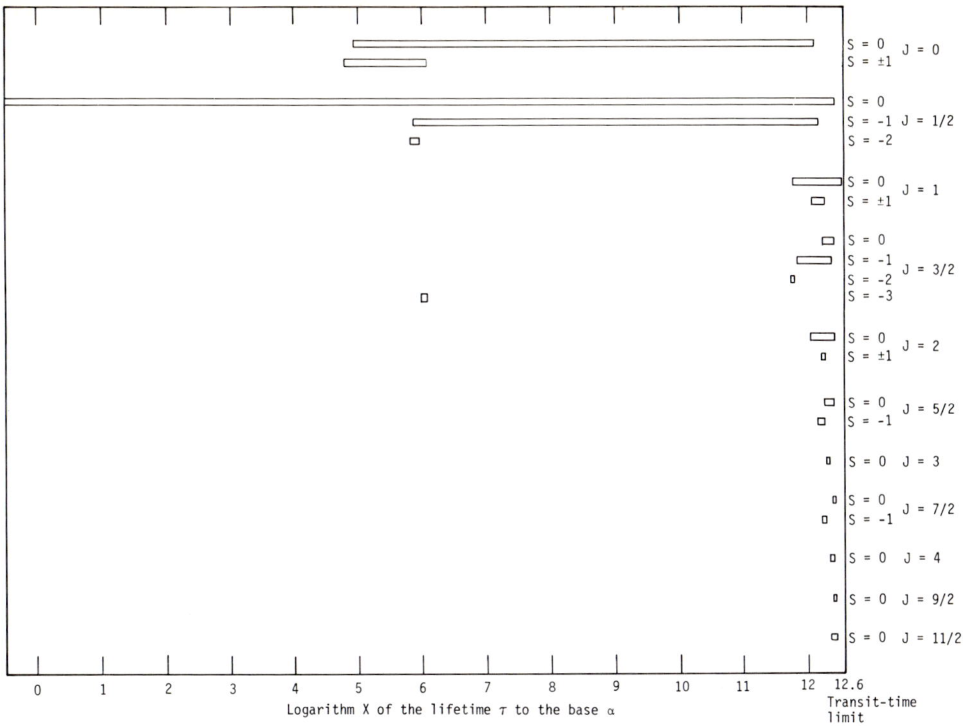

Figure 11.2. The lifetime logarithms X of Fig. 11.1 regrouped according to the spin value J, and without the expanded abscissa for X > 12. As can be seen in the figure, the long lifetimes occur in the lowest spin states, J = 0 and J = 1/2. This is just the opposite of the situation that occurs in nuclear physics, and it indicates that the centrifugal barriers which inhibit nuclear decays are not effective in elementary particle decays. Theoretically, as we discuss in the text, there is no reason to expect centrifugal barriers to function in the same manner in the violent elementary particle decays as they do in the relatively gentle nuclear decays.

between elementary particle lifetimes and angular momentum values that is shown in Figs. 11.1 and 11.2 is valid for both J and ℓ: the large J-values shown in these figures necessarily correspond to large ℓ-values as well. In the baryon resonances, for example, the intrinsic spin S is usually S = 1/2 (in a few cases it is S = 3/2), so that in general $J = \ell \pm 1/2$, and J and ℓ are approximately equal to one another.

Let us recapitulate the results that have been discussed in the present chapter. In the area of nuclear physics, there is a well-established relationship between the spin of a nuclear state and the width or lifetime of that state: high-spin nuclear states have narrow widths and long lifetimes. In the area of elementary particle physics, there is also a relationship between the spin of an elementary particle state and the width or lifetime of that state: high-spin elementary particle states have broad widths and short lifetimes. Thus the situation which occurs experimentally for rotating elementary particles is just the opposite of the situation which occurs experimentally for rotating atomic nuclei. Theoretically, this difference between the two situations comes as no surprise. Nuclear rotations are slow (adiabatic), and the decay of a rotating nuclear state involves a transition from one well-ordered internal nucleon configuration to another; in these internal nuclear transitions, high spin values serve as an impediment to the transformation, so that high-spin nuclear states persist for a long time before they decay. Elementary particle rotational states, on the other hand, which have the same orbital angular momentum values (small multiples of $\hbar$) as do atomic nuclei, but which have much smaller masses and therefore much smaller moments of inertia, are violently nonadiabatic, with the periphery of the rotating system approaching the velocity of light. Also, the decay of an elementary particle is an explosive process in which some of the internal 70 MeV quanta annihilate and blow out the remaining quanta as a shower of final-state particles (for example, in Chapter 17 it is shown that the five principal high-mass decay modes of the ψ' involve the annihilation of 1, 2, 3, 4, and 5 subquanta M, respectively). Hence there is no reason to expect elementary particle decays to behave similarly to nuclear decays, and experimentally they do not. However, many present-day elementary particle theorists still adhere to the concepts of nuclear physics, so that they find it difficult to accept the differences between these two situations. We emphasize this point because, as we will see, the width of an elementary particle resonance is one of its most revealing characteristics: rotational ($\ell > 0$) excitations have widths $\Gamma \geq 50$ MeV, whereas most rotationless ($\ell = 0$) excitations have widths $\Gamma \ll 50$ MeV (except for some of the higher-mass excitations). To convince workers in the field of this fact, we not only have the problem of properly displaying the experimental data (which are not all that clean-cut), but we also have the problem of overcoming the prejudices which have been carried over from the field of nuclear physics.

We now go on, in the next chapter, to the problem of decomposing the total angular momentum J of a baryon or hyperon resonance into its orbital ℓ and spin S components.

References for Chapter 11.

1. Particle Data Group, "Review of Particle Properties", Rev. Mod. Phys. 48, No. 2 Part II, April (1976), Sec. IV.
2. Ref. 1, page S240.
3. D. E. Plane et al., Nucl. Phys. B22, 93 (1970) (see page 102).
4. J. M. Blatt and V. F. Weisskopf, Theoretical Nuclear Physics, Wiley, New York (1952), pages 360 - 361.

CHAPTER 12. SPIN AND ORBITAL ANGULAR MOMENTUM COMPONENTS OF BARYONS

12.1 Observable Properties of the Quantum Number ℓ

In our present-day picture of the elementary particle, the concept of orbital angular momentum has emerged as a rather obscure quantity. If we assume, as seems to be correct, that each baryon and hyperon state contains 3 spin 1/2 quark substates, then the intrinsic spin S of a state can be either S = 1/2 or S = 3/2. However, baryons and hyperons are known experimentally to have total angular momentum values J = 1/2, 3/2, 5/2, 7/2, 9/2, ..., so it is clear that some kind of orbital motion ℓ is present which adds to S to produce $\vec{J} = \vec{\ell} + \vec{S} > 3/2$ for the higher-spin states. In the quark model of Gell-Mann and Zweig, orbital motion is attributed to the individual quark pairs[1]. From the standpoint of the present light-quark model, it seems more logical to interpret this orbital motion as a rotation of the entire particle, in a manner that is similar to the orbital rotations of deformed heavy muclei. But in either case, an angular momentum degree of freedom ℓ must exist, and if we assume that it has characteristically different properties from the spin degree of freedom S, we can separate the ℓ and S components of the total angular momentum J without having to specify the exact nature of the rotational motion.

In addition to its angular momentum properties, which are reflected in the spin J of a particle or resonance, the ℓ quantum number also contributes to two observable mass effects: it contributes to the total mass of the resonance via its rotational spin energy; and it contributes to the width of the resonance via the uncertainty in the quantization of the orbital motion. The essential point here is that since elementary particles are so light, they have very large rotational energies which contribute prominently to the total mass of the resonance; this leads in turn to the result that a distribution of rotational energies arising from a distribution of angular momentum values contributes prominently to the total width of the resonance. We can illustrate this point with the example cited in the preceding chapter. An atomic nucleus with a mass number A $\sim$ 100 has a total mass of about 10^5 MeV/c^2, and it has rotational energies that are on the order of 1 MeV. Since elementary particles have masses that are two orders of magnitude smaller, their rotational energies are approximately two orders of magnitude larger, which puts them in the range of 100 MeV or more. Uncertainties in the rotational motion can thus lead to uncertainties in the mass distribution (as expressed by the width of the resonance) that amount to 100 MeV or more. If the rotational motion of a resonance contributes to both the mass and the width of a resonance, as we have just argued, then the rotationless ($\ell = 0$) baryon and hyperon states will in general be states that have small masses and narrow widths. We now make a search of the experimental baryon and hyperon data to see if we can identify these states.

12.2 The Unique Property of $\ell = 0$ Baryon Resonances: Narrow Widths

Let us consider the mass of an elementary particle resonance as arising from two components: (1) a rotationless ($\ell = 0$) component, which contains contributions from both spinless and spinning basis states, and which includes binding energy effects; (2) a rotational ($\ell > 0$) component, which reflects the moment-of-inertia I of the $\ell = 0$ rotor. In terms of these two components, we can denote the mass and width of the resonance schematically as

$$M = M(\ell{=}0) + M(\ell{>}0) \equiv M_o + M(\ell) \tag{12.1}$$

and

$$\Delta M = \Delta M(\ell{=}0) + \Delta M(\ell{>}0) \equiv \Delta M_o + \Delta M(\ell). \tag{12.2}$$

Theoretically, we have no way of separating M_o and ΔM_o from $M(\ell)$ and $\Delta M(\ell)$. However, this separation can be accomplished phenomenologically by dividing the experimental baryon and hyperon resonances into two groups:

$$J = 1/2 \text{ and } 3/2 \text{ resonances:} \quad (S = 1/2 \text{ or } 3/2;\ \ell = 0,1,2,3); \tag{12.3a}$$

$$J > 3/2 \text{ resonances:} \quad (S = 1/2 \text{ or } 3/2;\ \ell > 0). \tag{12.3b}$$

The essential point in this division is that the (12.3a) resonances contain both $\ell = 0$ and $\ell > 0$ states, whereas the (12.3b) resonances only contain $\ell > 0$ states. Thus if the $\ell = 0$ states have different experimental properties than do the $\ell > 0$ states, we can identify the $\ell = 0$ states and also determine their special properties by comparing the (12.3a) and (12.3b) resonances to see if the (12.3a) set contains a unique class of resonances which are not contained in the (12.3b) set.

When we make an experimental comparison of the (12.3a) and (12.3b) resonances, it turns out that the unique property we discover in the (12.3a) set has to do with the widths of the resonances. Figure 12.1 is a plot of the measured widths for the well-established[2] J = 1/2, 3/2, 5/2, and 7/2 N, Δ, Λ, Σ, and Ξ resonances. As can be seen in Fig. 12.1, the J = 5/2 and J = 7/2 resonances all have broad widths ($\Gamma > 80$ MeV), whereas the J = 1/2 and J = 3/2 resonances include states with broad widths, but also a class of states with narrow widths ($\Gamma < 50$ MeV). This suggests that the narrow-width states correspond to $\ell = 0$ excitations, a suggestion which is in

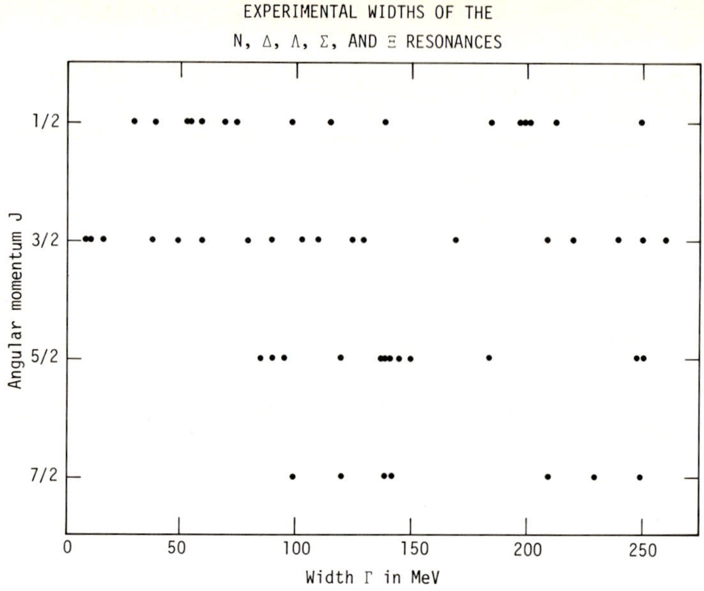

Figure 12.1. A plot of the measured widths for the well-established J = 1/2 to J = 7/2 N, Δ, Λ, Σ, and Ξ resonances. As can be seen, narrow-width resonances (Γ < 80 MeV) occur only in J = 1/2 and J = 3/2 spin states.

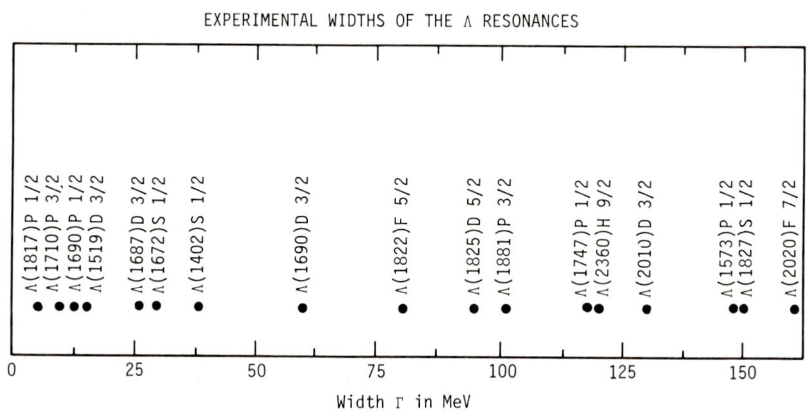

Figure 12.2. The measured widths of the Λ hyperons. All seven resonances with widths Γ < 50 MeV are either J = 1/2 or J = 3/2 states. The narrowest-width J = 5/2 resonance, the Λ(1822)F state, has a width Γ = 81 MeV.

agreement with the lifetime systematics discussed in the preceding chapter (see Figs. 11.1 and 11.2).

Confirmation for the suggestion that the narrow-width resonances correspond to $\ell = 0$ excitations can be obtained by studying the measured widths of the Λ hyperons in more detail; these are shown in Fig. 12.2. Figure 12.2 includes, in addition to the well-established resonances of Fig. 12.1, two additional groups: (1) the less-well-established resonances; (2) several narrow-width P and D peaks which have been identified in various phase shift analyses (some of these coincide with broad-width peaks). The Λ resonances shown in Fig. 12.2 are documented in Appendix B, and they are discussed in detail in Chapter 15, where it is demonstrated that the narrow-width P and D peaks occur at the same mass values which typically correspond to S-state excitations. As can be seen in Fig. 12.2, there are seven Λ enhancements with widths $\Gamma < 50$ MeV: two J = 1/2 S-states (which obviously have $\ell = 0$), two J = 1/2 P-states, one J = 3/2 P-state, and two J = 3/2 D-states. The narrowest-width $J \geq 5/2$ state in Fig. 12.2 is the $\Lambda(1815)$F state, with a width $\Gamma = 81$ MeV; thus we see in Fig. 12.2 that whereas the $J \leq 3/2$ states occur with both broad and narrow widths, the $J \geq 5/2$ states only occur with broad widths

The Λ resonances in Fig. 12.2 that have widths $\Gamma < 50$ MeV are of particular interest here. The J = 1/2 $\Lambda(1402)S\Gamma38$ state is the K^-p bound state that was reproduced in Chapter 9 (see Eqs. 9.18 and 9.19). The J = 3/2 $\Lambda(1519)D\Gamma15$ state, as we will demonstrate below, serves as the $\ell = 0$ bandhead for a prominent Λ hyperon rotational band. The J = 1/2 $\Lambda(1672)S\Gamma33$, J = 3/2 $\Lambda(1687)D\Gamma30$, J = 1/2 $\Lambda(1690)P\Gamma22$, and J = 3/2 $\Lambda(1710)P\Gamma20$ states all appear at about the same mass value and with about the same width as manifestations of the same basic $\ell = 0$ light-quark configuration; the J = 1/2 and J = 3/2 angular momentum values of these states arise from the intrinsic quark spins, and the parities of the states arise in part from the intrinsic properties of the decay modes.†

†The parity of a hadron resonance is another rather obscure concept. Since hadrons decay via the annihilation of some of the internal subquanta M, and since we can have (for example) both [M] and [3M] annihilation processes in the decay modes of the same particle (e.g., $K \equiv 7M \to [M]\pi\pi\pi$ and $K \equiv 7M \to [3M]\pi\pi$), there are states that differ by one (negative-parity) pion, $\pi = M\bar{M}$, and thus have different parities due to their different internal annihilations. The orbital rotation ℓ of a state can contribute to the parity P of the state ($P = (-1)^{\ell}$), but the annihilation process can also contribute, so that for hadron decays the parity P and the orbital angular momentum ℓ are not necessarily uniquely related.

It is a general rule in both the baryon and hyperon resonances that the higher-mass S-states tend to have broader widths than do the lower-mass S-states. In this regard, we note that the recently-identified very-narrow-width J = 1/2 $\Lambda(1817)P\Gamma 12$ state (see the comments on this state in Sec. 15.2 of Ch. 15) corresponds in mass to the broad-width J = 1/2 $\Lambda(1827)S\Gamma 150$ state. This correspondence suggests that the broad widths of the high-mass S-states may be an artifact of the phase shift analyses or of the experimental data rather than a genuine effect.

The lowest-mass nonstrange baryon excitation, the $\Delta(1232)P\Gamma 115$ state, which is the historically-famous Δ_{33} resonance (the first baryon resonance whose quantum numbers were determined), has angular momentum J = 3/2. However, from its combination of a broad width with a low mass value, it cannot be an $\ell = 0$ state; as we will demonstrate in Chapter 15, it is an $N\dot{3}$ excitation in an $\ell = 1$ rotational mode, with a rest mass of about 1150 MeV and a rotational energy of about 80 MeV. In accounting for this result, we note, as is described in Appendix E, that baryon and hyperon excitations which contain a single $\dot{3}$ are characteristically difficult to observe as S-state resonances. A comprehensive study of all of the baryon and hyperon resonances is contained in Chapter 15, where we discuss the problem of mapping these resonances.

12.3 The Difference in the Mass Spectra of $\ell = 0$ and $\ell > 0$ Baryon Resonances

The conclusion we draw from Figs. 12.1 and 12.2 and Eqs. (12.3a) and (12.3b) is that the narrow-width ($\Gamma < 50$ MeV) baryon resonances and the identified S-state baryon resonances correspond to $\ell = 0$ excitations, and the broad-width ($\Gamma > 50$ MeV) P-state, D-state, and higher-spin baryon resonances correspond to $\ell > 0$ excitations. In terms of Eq. (12.2), this means that, at least for the low-mass resonances, $\Delta M_o <$ 50 MeV and $\Delta M(\ell) > 50$ MeV. An independent verification of the physical significance of this dichotomy between $\ell = 0$ and $\ell > 0$ baryon excitations can be obtained by studying the mass systematics of the $\ell = 0$ and $\ell > 0$ baryon resonances, where we use the above classification scheme for assigning ℓ values, and where we use Eq. (12.1) as a guide for the mass values. To do this, we divide the baryon and hyperon resonances into two complementary groups:

Group I: S-state resonances and narrow-width ($\Gamma < 50$ MeV) resonances; (12.4a)

Group II: all other resonances. (12.4b)

If the Group I and Group II resonances in fact correspond to $\ell = 0$ and $\ell > 0$ excitations, as we have suggested above, then we expect them to correspond to different mass equations, as follows:

Group I: $M = M_o$ (12.5a)

Group II: $M = M_o + M(\ell)$, (12.5b)

where we use the schematic notation of Eq. (12.1). It seems plausible that the M_o excitations will follow a different pattern from the $M(\ell)$ excitations. Thus, to test if the Group I and Group II division of resonances is physically significant, we should plot the mass spectra for these two groups. These plots are given in Figs. 12.3 and 12.4. Figure 12.3 shows the mass spectra for the Group I N, Λ, Σ, and Ξ resonances, and Fig. 12.4 shows the mass spectra for the Group II N, Λ, and Σ resonances[3], where the levels in both figures are plotted as excitation energies above the basic "ground states" N, Λ, Σ, and Ξ. As can be seen, the pattern of excitation energies in Fig. 12.3 is completely different from the pattern in Fig. 12.4. The Fig. 12.3 excitation levels are regularly spaced, and they appear to be quantized in units that are multiples of 70 MeV; also, matching excitations appear at the same energy and with the same spin values for states that differ in strangeness and in isotopic spin. The Fig. 12.4 excitation levels, on the other hand, are spaced more or less randomly, and the N, Λ, and Σ excitations in Fig. 12.4 bear essentially no relation to one another. The marked contrast between the mass systematics of Fig. 12.3 and that of Fig. 12.4 indicates in a direct manner that the distinction between $\ell = 0$ and $\ell > 0$ baryon resonances is a meaningful one, and it shows that ℓ is a well-defined quantum number for baryon and hyperon resonances. A similar type of analysis can be carried out for meson and kaon resonances, and it leads to the same general conclusion: broad-width meson resonances are $\ell > 0$ excitations, and narrow-width meson resonances are $\ell = 0$ excitations (see Chapter 13).

12.4 The Nature of Baryon Orbital Rotations

Although we have identified a set of baryon rotational excitations — the Group II excitations shown in Fig. 12.4 — , we have not yet discussed the manner in which the baryon rotational levels occur. It seems logical that the rotational systematics of elementary particles should bear some relationship to the rotational systematics of atomic nuclei. Experimentally, this appears to be the case, but with some important (and understandable) differences[4]. In Chapter 11 we described one difference between rotational excitations in nuclear physics and in elementary physics: centrifugal barriers, which play a dominant role in nuclear rotations, are unimportant for elementary particle rotations. The difference between the two cases is attributed both to the nonadiabatic nature of elementary particle rotations, and also to the explosive nature of elementary particle decays — which puts the kinetic energies of the final-state particles well above the height of any reasonable centrifugal barrier. The nonadiabatic nature of elementary particle rotations also has an important effect

ENERGY LEVELS FOR NARROW-WIDTH AND
S-STATE N, Λ, Σ, and Ξ RESONANCES,
PLOTTED ON A MASS SCALE THAT IS IN UNITS OF 70 MeV

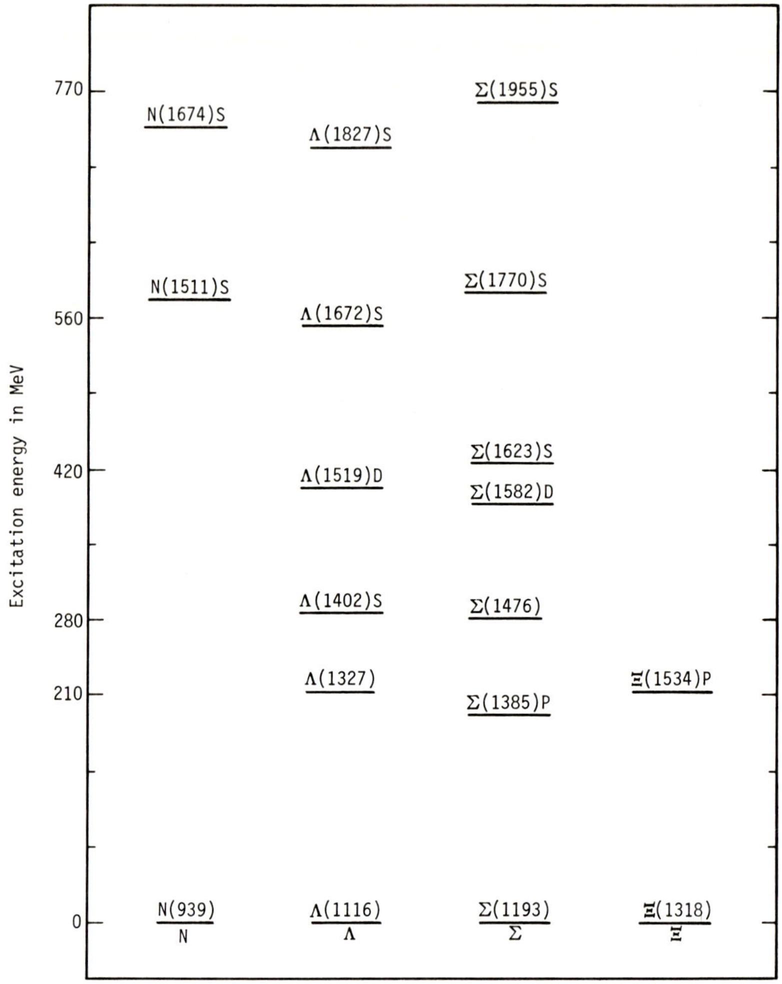

Figure 12.3. The mass spectra for the S-state and narrow-width (Γ < 50 MeV) baryon and hyperon resonances [Eq. (12.4a)]. These levels are well separated, and they appear for all values of strangeness at characteristic quantization intervals of 210, 280, 420, 560, and ∿770 MeV above the corresponding "ground states", thus indicating a quantization that is in multiples of 70 MeV [Eq. (12.5a)]. Compare these results to the mass spectra shown in Fig. 12.4 for the broad-width baryon and hyperon resonances.

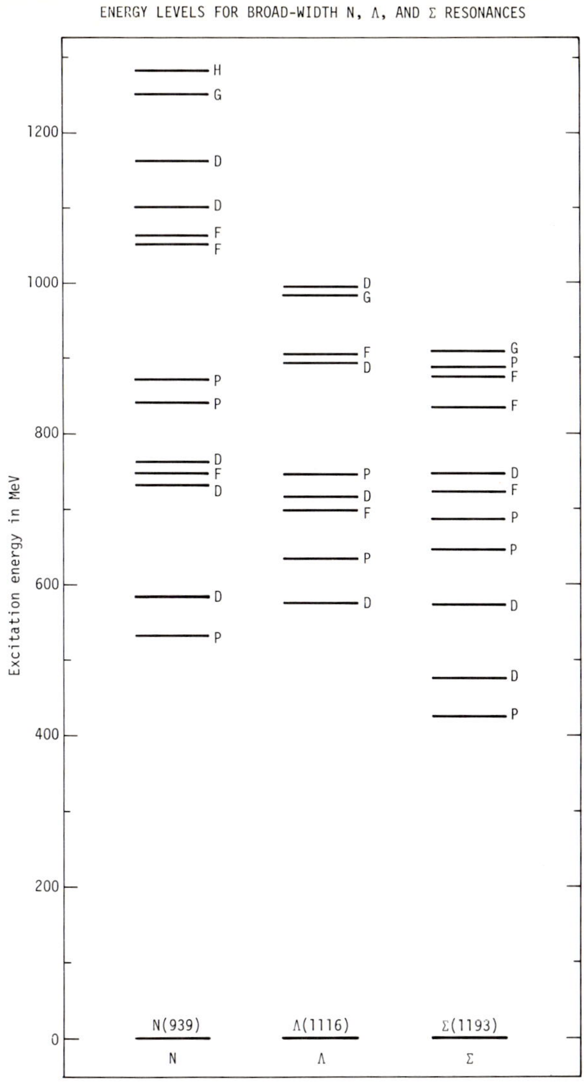

Figure 12.4. The mass spectra for the broad-width baryon and hyperon resonances [Eq. (12.4b)]. In contrast to the mass spectra shown in Fig. 12.3, these levels are randomly spaced, and they are not correlated from one resonance family to the next. They seem to follow a completely different mass quantization equation [Eq. (12.5b)] from that of Fig. 12.3 [Eq. (12.5a)].

on the energies of baryon rotational levels. To see how this comes about, we briefly review the systematics of rotational motion in nuclear physics.

The rotational motion of a quantum-mechanical system is characterized by the Hamiltonian[5]

$$\hbar^2/2I \times \hat{L}^2, \tag{12.6}$$

where $\hat{L}$ is the orbital-angular-momentum operator and I is the effective moment of inertia. In nuclear physics it is customary to write[5-7]

$$\hat{L} = \hat{J} - \hat{S}, \tag{12.7}$$

where $\hat{J}$ is the total angular momentum operator and $\hat{S}$ represents the "intrinsic angular momentum" of the rotor. Thus the rotational energy is given (writing schematically) by the expectation value of the operator

$$\hat{L}^2 = \hat{J}^2 + \hat{S}^2 - 2\,\hat{J}\cdot\hat{S}. \tag{12.8}$$

The usefulness of this representation lies in the fact that the "Coriolis term"

$$\hat{J}\cdot\hat{S} \tag{12.9}$$

does not contribute to the rotational energy if the rotating system satisfies the following conditions:

(A) the system is markedly unspherical, so that rotational axes are clearly delineated;

(B) the rotational velocity is "slow" (the adiabatic approximation), so that the rotation does not appreciably affect the intrinsic spin S; (12.10)

(C) the intrinsic spin S does not have the value 1/2.

Conditions (A) and (B), which are not independent of one another, combine to make it a valid approximation to write the wave function as a product of total angular momentum and intrinsic spin components[8]. Once conditions (A) and (B) are satisfied, a direct calculation using the product wave function shows[9] that the Coriolis coupling $\hat{J}\cdot\hat{S}$ vanishes unless S = 1/2. If the Coriolis term vanishes, then the rotational energies will, from Eqs. (12.6) and (12.8), follow a J(J+1) interval relationship. However, near closed shells, where condition (A) fails, and for nuclei with S = 1/2, where condition (C) fails, the simple J(J+1) interval rule no longer obtains[10].

In the case of even-even nuclei, which usually have S = 0, Eq. (12.8) reduces to Eq. (12.6). Thus the even-even nuclei that satisfy conditions (A) and (B) have ground-state rotational spectra which exhibit an accurate $\ell(\ell+1)$ interval relationship:

$$E_\ell = E_o + \hbar^2/2I \times \ell(\ell+1), \qquad (12.11)$$

where E_ℓ is the energy of the rotational level, E_o is the energy of the rotor, and I is the effective moment of inertia. These nuclei may in addition have vibrational excitation modes, with each vibrational mode serving as the ground state for a superposed rotational band. (The spinless excitation quanta $\dot{3}$ and $\dot{4}$ which combine with the N, Λ, Σ, and Ξ baryon "ground states" to form a series of $\ell = 0$ excited baryon states, as in Figs. 5.7 and 12.3, are roughly analogous to these vibrational excitations of atomic nuclei.) If the nucleus has <u>inversion symmetry</u> (e.g., is not pear-shaped)[11], then the ℓ-values in the rotational band will be all even or all odd, depending on the parity of the vibrational mode[12]. In odd-A nuclei, on the other hand, S has half-integral values, and inversion symmetry does not apply. Thus, for the odd-A nuclei, which follow the J(J+1) interval rule, all successive values of J are observed in the rotational series that is superposed on each vibrational ground state.

Now, how do we apply these results to rotational levels for baryon and hyperon states? Suppose we assume that baryon resonances are physical structures which are similar to atomic nuclei. Will conditions (A), (B), and (C) then be fulfilled, so that a J(J+1) splitting is observed in the rotational levels? The answer to this question seems to be clearly no, mainly because of a violation of condition (B). As we discussed in Chapter 11, elementary particle rotations are highly nonadiabatic, so that condition (B) is strongly violated. Also, condition (C) is violated for most of the baryon rotational bands, and condition (A) may not even be fulfilled by the light-quark clusters of the present model. As a consequence of these effects, the $\hat{J} \cdot \hat{S}$ "Coriolis term" of Eq. (12.9) does not vanish, and we are thrown back to the $\ell(\ell+1)$ interval rule of Eq. (12.6) rather than the J(J+1) interval rule which follows from Eq. (12.9) and conditions (A), (B), and (C). In Chapter 11, and also at the beginning of the present chapter, we illustrated the nonadiabatic nature of hadron rotations by scaling from the rotational energies in nuclear physics to corresponding rotational energies in particle physics. We can demonstrate this result in another way by making a direct estimate of the moment of inertia of a rotating baryon system, where we again use the situation in nuclear physics as an example. As we will see, this estimate again shows that baryon rotational energies are very large.

12.5 Moments of Inertia for Baryon Bandheads

In atomic nuclei the nuclear shapes are well known, and the moment-of-inertia I for a nucleus can be calculated under the assumption (for example) that the nucleus is a rigid body. The quantity

$$E_{rot} = \hbar^2/2I \tag{12.12}$$

then gives the expected rotational energy. When E_{rot} for nuclei is determined experimentally, it turns out that the effective moment-of-inertia I is typically about half of the rigid-body value[13], showing that the nucleus in fact has a collective rotation that is intermediate between the limits of a rigid body and of irrotational flow[13]. However, the pertinent point for the present discussion is the fact that by using the rigid-body assumption in Eq. (12.12), we obtain an estimate for E_{rot} which is accurate to within a factor of two. When we now make a corresponding estimate for E_{rot} in the case of the baryon and hyperon resonances, we can insert the values $R \sim 10^{13}$ cm and $M \sim 1500$ MeV $= 2.7 \times 10^{-24}$ gram into the equation $I \sim 2/5\ MR^2$ to obtain $I \sim 10^{-50}$ gram cm^2, which gives $E_{rot} \sim 30$ MeV. With this value for E_{rot}, rotational energies for intermediate ℓ-values will be a few hundred MeV. A rotational energy of this magnitude begins to approach the rest mass of the rotating system, and it signals a strong violation of the adiabatic approximation (condition B). From the nonadiabatic nature of these baryon rotations, which effectively decouple the orbital and spin angular momenta, it follows that baryon rotational energies will be given by Eq. (12.6), which we can write in the form

$$E_\ell = E_o + E_{rot} \times \ell(\ell+1), \tag{12.13}$$

where $E_{rot} = \hbar^2/2I$. There are two points that should be emphasized here: (1) the applicability of Eq. (12.13) to baryon rotational motion follows directly from the known systematics of rotational motion in nuclear physics; (2) the rotational energies which are actually "observed" in the baryon resonances do in fact correspond to the moment-of-inertia values for E_{rot} that we obtain from straightforward estimates, as we will demonstrate below. Thus it seems logical to investigate the spectrum of baryon and hyperon resonances from the point-of-view of nuclear-physics-type rotational bands (which we do in Chapter 15). This is an interesting point to consider because, whereas present-day elementary particle theorists use nuclear centrifugal barrier factors in cases where they do not apply (see Chapter 11), these same theorists seem reluctant to use nuclear rotational systematics in cases where they probably do apply.

We now give two examples which illustrate the manner in which the experimental baryon and hyperon rotational energies are in agreement with their estimated moments of inertia. The first example is provided by the Δ_{33} resonance, $\Delta(1232)P\Gamma 115$, which we have already discussed above. As we have asserted (and will show in more detail in Chapter 15), the Δ_{33} appears as an $\dot{N3}$ excitation in an $\ell = 1$ rotational mode. The $\dot{N3}$ bandhead† energy is about 939 + 210 = 1149 MeV. If we assume that the Δ_{33} has an effective radius R of 1 fermi (about the same as that of the nucleon), and that its moment of inertia can be approximated by the equation $I = 2/5\ MR^2$, then from these values we have $I = 460$ MeV fm^2, so that the calculated value for E_{rot} is $E_{rot} = \hbar^2/2I = 42.4$ MeV, where $\hbar = 197.3$ MeV fm in the appropriate units. Since the peak energy of the Δ_{33} resonance is 1232 MeV, its rotational energy is 1232 − 1149 = 83 MeV, which corresponds to an experimental value for E_{rot} of 41.5 MeV. In view of the assumptions that were made in the calculation of E_{rot}, the close numerical agreement shown here is fortuitous. But the two essential points to be noted are that the experimental rotational energy of the Δ_{33} is of the expected order of magnitude, and that when other rotational energies are calculated with these same assumptions, good agreement is also obtained. We now illustrate this second point with our next example, which is the dominant rotational band in the Λ resonances.

Although the J = 3/2 resonance $\Lambda(1519)D\Gamma 15$ has the overall quantum numbers of a D-state resonance, it is actually, from the arguments of the present chapter, an S-state ($\ell = 0$) resonance, with its total spin of 3/2 arising from the intrinsic spin of the quark configuration. As we now demonstrate, the $\Lambda(1519)$ serves as the bandhead for an accurately-spaced baryon band that contains P, D, F, G, and H rotational levels, corresponding to the values $\ell = 1, 2, 3, 4$, and 5 in Eq. (12.13). We can calculate the moment-of-inertia I for the $\Lambda(1519)$ bandhead in the same manner as we did above for the Δ_{33} bandhead: assuming a radius of 1 fermi, and using a bandhead mass of 1519 MeV, we obtain a moment-of-inertia $I = 608$ MeV fm^2 which corresponds to the value $E_{rot} = 32$ MeV. The experimental levels for this Λ rotational band are shown in Fig. 12.5, where they are compared to levels calculated from Eq. (12.13) with a value of $E_{rot} = 28$ MeV. As can be seen in Fig. 12.5, the experimental levels closely match the $\ell(\ell+1)$ intervals of the calculated levels, and the average value $E_{rot} \simeq 28$ MeV for these levels is in fair agreement with our theoretical estimate of 32 MeV for this quantity. The ℓ_J values for the rotational levels in this Λ rotational band have been identified experimentally[2] as $P_{1/2}$, $D_{3/2}$, $F_{5/2}$, $G_{7/2}$, and $H_{9/2}$; each of these five levels corresponds naturally to an intrinsic spin S = 1/2 that is directed antiparallel to the orbital angular momentum ℓ. Thus, although the intrinsic spin of the $\Lambda(1519)$ S-state bandhead is S = 3/2, it is possible that the light quark cluster which forms the $\Lambda(1519)$ has the more customary spin value S = 1/2 when it occurs in a rotational mode (although a spin S = 3/2 can of course combine with the observed ℓ-values to produce the observed J-values). In Chapter 15 we study the problem of mapping all of the baryon and hyperon resonances into rotational bands.

---

†The bandhead is the rotor; i.e., the resonance structure in its $\ell = 0$ mode.

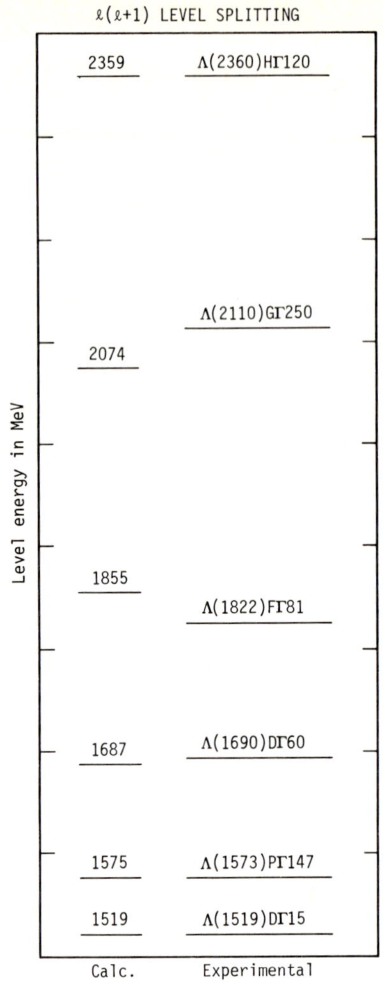

Figure 12.5. Comparison of the experimental energy levels in the Λ(1519) rotational band with the theoretical levels as calculated from Eq. (12.13) with E_{rot} set equal to 28 MeV. The agreement, while not perfect, is consistent with the large uncertainties in the positions of some of these levels. There are two points to note here: (1) these results are consistent with the results for the other baryon rotational bands, as shown in Chapter 15; (2) a J(J+1) interval rule for these resonances would not work — the baryon rotational bands clearly follow an $\ell(\ell+1)$ interval rule rather than a J(J+1) interval rule. This last result is substantiated in Fig. 16.4 of Chapter 16, where the dominant meson rotational family is shown to obey a very accurate $\ell(\ell+1)$ mass interval rule.

12.6 Nonadiabatic Orbital Rotations in Very Light Atomic Nuclei

We started this discussion of rotational systematics by describing the adiabatic rotational bands that are observed in the energy levels of heavy atomic nuclei. Then we considered the nonadiabatic rotations that seem to occur for the very light baryon and hyperon excitations. This brings up the question as to whether very light atomic nuclei, whose masses are in the range of a few GeV/c^2, and hence are comparable to the masses of baryons and hyperons, also have nonadiabatic rotational levels of the same type as we have ascribed here to baryons and hyperons. If this is so, then these light atomic nuclei can serve to bridge the gap between nuclear rotations and elementary particle rotations. Experimentally, this seems to be the case. Table 12.1 shows a series of light nuclei that have identifiable rotational levels, and it contains the experimentally-determined values of E_{rot} for these levels together with a few calculated values[14-17] of E_{rot}. The moments-of-inertia I that correspond to the E_{rot} values of Table 12.1 can be obtained from Eq. (12.12); these nuclear moments of inertia are shown in Fig. 12.6, where they are compared with the experimentally-determined moments of inertia for meson rotational bands (Chapter 16) and baryon rotational bands (Chapter 15). As can be seen in Fig. 12.6, the hadronic moments of inertia extrapolate smoothly into the moments of inertia of the very light nuclei. This indicates that nonadiabatic rotations do occur in very light nuclei, and that the rotational motions of all of these states depend mainly on their overall masses, and not on whether they are stable atomic nuclei or short-lived hadrons.

Table 12.1. Experimental values of $E_{rot} \equiv \hbar^2/2I$ for energy levels in very light nuclei. A few estimates of E_{rot} based on rigid rotors are also shown.

Nucleus	Observed rotational levels	E_{rot} (MeV) experimental	E_{rot} (MeV) estimated	Basis for estimate	Ref.
Si^{28}	0^+ 2^+ 4^+	0.26			
Mg^{24}	0^+ 2^+ 4^+	0.22			
Ne^{20}	0^+ 2^+ 4^+	0.24			
O^{18}	0^+ 2^+ 4^+	0.34			
O^{16}	0^+ 2^+ 4^+ 6^+	0.22	0.27	$I = 2/5\ MR^2$, with	
	1^- 3^- 5^- 7^-	0.22	0.27	R = 3.5 fermis	14
C^{12}	0^+ 2^+ 4^+	0.72	∿0.7	3α at 135°	15
Be^8	0^+ 2^+ 4^+	0.53	0.7	α-α dumbbell	16
Li^5, He^5	1^-	2.5	2.3	α-N dumbbell	16
He^3	1^-	∿11.5	∿11	3N at 135°	17

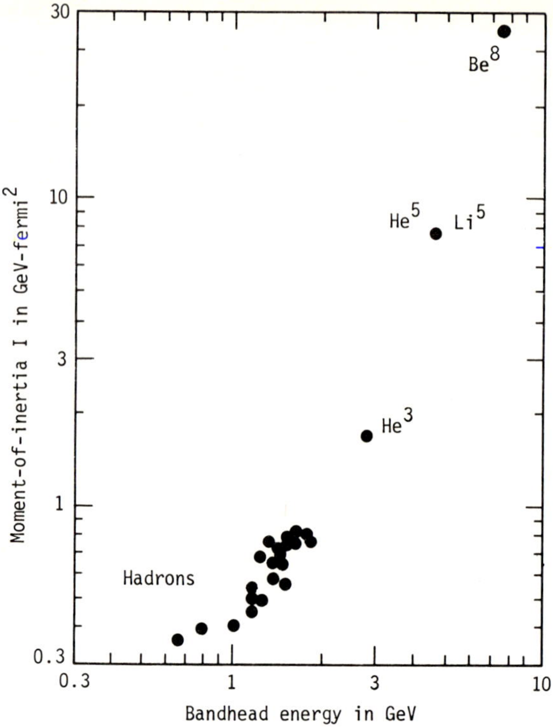

Figure 12.6. Experimental values of the moment of inertia I for rotational bands in mesons (Chapter 16), baryons (Chapter 15), and very light atomic nuclei (Table 12.1). As can be seen in the figure, the moments of inertia for all of these different types of particle exhibit the same unified dependence on the mass of the bandhead (rotor). This indicates that the rotational excitations in these particles are mechanical excitations which depend only on the overall geometries of the particles, and not in any essential manner on their constituent substates.

In summary, it appears that the orbital angular momentum ℓ is a meaningful quantity for baryon and hyperon resonances, and that a separation can be made of the total angular momentum J into its orbital ℓ and spin S components. The orbital motion that is associated with the quantum number ℓ in baryons seems to be a rotation of the entire resonance structure, in more-or-less the same manner as is observed in atomic nuclei. In the next chapter we examine meson resonances from this same point of view. As we have suggested in Fig. 12.6, meson resonances appear to behave similarly to baryon resonances with respect to their angular momentum characteristics.

References for Chapter 12.

1. See B. T. Feld, Models of Elementary Particles, Blaisdell, Waltham (1969), Fig. 16.6 on page 382.
2. Particle Data Group, "Review of Particle Properties", Rev. Mod. Phys. 48, No. 2, Part II, April (1976). The "well-established" states have *** and **** ratings.
3. The broad-width Ξ resonances are not well-enough established to include in Fig. 12.4.
4. For an early discussion of these baryon rotations, see M. H. Mac Gregor, Lett. Nuovo Cimento 1, 427 (1971).
5. S. A. Moszkowski, Handbüch der Physik, Vol. 34, Berlin (1957), p. 413, Eq. 36.3.
6. A. Bohr and B. Mottelson, Kgl. Danske Videnskab Selskab Mat. fys. Medd. 27, No. 16 (1953); J. P. Davidson, Collective Models of the Nucleus, New York (1968).
7. Z. Szymanski, Trieste Lectures, Vienna (1970), p. 67.
8. Ref. 7, p. 94.
9. Ref. 7, p. 101.
10. See, for example, A. K. Kerman, Nuclear Reactions, edited by P. M. Endt and M. Demeur, Amsterdam (1959), Chapter X.
11. Ref. 5, p. 483.
12. See the U^{238} spectrum in Fig. 7.1 of Ref. 7. Note that for vibrational modes with $S \neq 0$, all successive values of ℓ appear.
13. Ref. 5, p. 495.
14. See M. H. Mac Gregor, Phys. Rev. D9, 1259 (1974), Ref. 150.
15. See Ref. 151 in Ref. 14.

16. See Ref. 152 in Ref. 14.

17. See Ref. 153 in Ref. 14.

CHAPTER 13. SPIN AND ORBITAL ANGULAR MOMENTA IN MESONS: THE ω AND ρ MESONS

13.1 Introduction; the Importance of the Lowest-Mass States

We now consider orbital excitations in mesons. Theoretically, the question has been raised as to whether the assignment of orbital angular momentum states to particles is consistent with the requirements of relativistic (Lorentz) invariance[1], but recent studies[2] suggest that these rotational excitations are in fact consistent with the requirements of relativity. The question of orbital excitations for baryons was discussed in Chapter 12, where it was shown that if we make an extrapolation of the theory of rotational motion so as to extend it from the case of adiabatic (slow) rotations in heavy atomic nuclei (where the theory was first developed) to the case of nonadiabatic (rapid) rotations in very light systems, then the baryon resonances seem to be occurring in a manner that is consistent with their identification as nuclear-physics-type rotational bands. We cited only two examples in Chapter 12 — the nonstrange Δ_{33} baryon resonance, which appears as a single rotational level; and the strange Λ(1519) hyperon rotational band, which contains six well-identified levels. (The complete mapping of the baryon and hyperon rotational bands is given in Chapter 15.) We also described some nonadiabatic rotational levels that appear to be occurring in very light atomic nuclei; the rotations of these light nuclei bridge the gap between the rotations of the even lighter baryon and hyperon resonances and the rotations of heavy deformed atomic nuclei. A comparison of the rotational moments of inertia for baryons with the rotational moments of inertia for very light nuclei (Fig. 12.6) indicates that these are rotations of the entire structure (essentially rigid body rotations), and that the rotations depend only on the masses involved — and not on whether the rotating state is a short-lived baryon resonance or a stable atomic nucleus.

If baryons and atomic nuclei both have rotational excitations, then it seems logical that mesons should also have rotational excitations. However, there are two factors which make it difficult to identify meson rotational bands: (1) meson bandheads (rotors) are lighter than baryon bandheads, which means that meson rotations are even more rapid than baryon rotations, so that relativistic constraints will serve to limit the number of observed levels in a rotational band (the periphery of the rotating bandhead cannot exceed the velocity of light); (2) in some cases the meson bandhead does not appear as a rotationless (ℓ = 0) S-state, but only in the form of a rotational level (as was the case for the Δ_{33} level described in Chapter 12). Since meson resonances do not have the stable triplet spinor cores that characterize baryon resonances, a meson literally blows itself apart when it decays; thus the rotating meson bandhead is destroyed in the decay process of a rotational level, and we cannot determine the nature of the rotor from the nature of its decay products. Hence we must develop a systematic method for cataloging meson bandheads. This situation is of course also true for baryons, but baryon bandheads follow a simpler excitation pattern — namely, N, Λ, Σ, and Ξ "ground states" to which are added $\dot{3}$ and $\dot{4}$ excitation quanta (see Figs. 5.7 and 12.3). To offset these difficulties in identifying meson rotational bands, there is one factor which helps us in the classification of the individual meson resonances; this is the information that is obtained from the widths of the resonances. If mesons behave similarly to baryons, then we expect meson rotational states to have broad widths ($\Gamma \geq 50$ MeV) and meson S-states to have narrow widths ($\Gamma \ll 50$ MeV). Empirically, this result seems to be correct, as we discuss below and in Chapter 16.

In attempting to sort out $\ell = 0$ and $\ell > 0$ excitations in meson resonances, we have one important point in our favor: we know where to look for the significant examples. As we discussed in Sec. 5.2 of Chapter 5, the most important states in a physical system are the lowest-mass states of the system. For example, it was the hydrogen atom that led to the unraveling of the mystery of atomic structure, and it is the nucleon that is the decisive element in nuclear structure. Now, the lowest-mass hadron state is the 140 MeV pion; and, as we saw in Chapter 5, the η and η' mesons are essentially combinations of 4 and 7 pions, respectively (see Figs. 5.1 and 5.2). Also, the strange kaon can be thought of as 1/2 of an η' meson (or, to put it more accurately, the η' is a $K\bar{K}$ bound state), so that the pion provides the decisive clue to the structures of all of these particles. Furthermore, the pion and the even-lower-mass muon provide the strongest evidence for the existence of the basic spinless quantum M = 70 MeV. Thus, in our search for meson rotational levels, we should examine the lowest-mass states that have angular momentum J > 0.

Although the mapping of the experimental spectrum of meson and kaon resonances[3] is still incomplete, and the data for a number of reported resonances are contradictory, it appears that the pseudoscalar ($J^P = 0^-$) resonances π, K, η, and η' do not have associated rotational levels. Hence we cannot use these low-mass resonances in our search for evidence of meson rotational bands. Other than the π, K, η, and η' mesons, the two lowest-mass meson states are the 773 MeV spin 1 ρ meson and the 783 MeV spin 1 ω meson. Since the ρ has a very broad width (Γ = 152 MeV) and the ω has a very narrow width (Γ = 10 MeV), it seems natural, on the basis of the above arguments, to identify the ρ as an $\ell = 1$ rotational level and the ω as an $\ell = 0$ rotationless S-state (with an intrinsic spin S = 1). Most of the remainder of the present chapter is devoted to the development of this viewpoint.

In the next section of this chapter we compare the properties of the ω and ρ mesons. Then, in the rest of the chapter, we spend most of our time discussing various aspects of the ρ. The ρ meson is one of the most interesting of all the hadron resonances. The cross section for producing the ρ is very large, and the ρ occurs in a low-mass region where there are few competing resonances. Thus the ρ dominates the mass region from about 650 to 900 MeV, and we can study its properties in a manner that is almost free from background effects. If the ρ is a rotational ($\ell = 1$) resonance, as we have argued above, and if the large width of the ρ is due to a spread in its rotational energies, then this means that the ρ actually has a distribution of ℓ-values: it has $\ell = 1$ at the 773 MeV peak of the resonance, but it has $\ell < 1$ below 773 MeV and $\ell > 1$ above 773 MeV, and this distribution of ℓ-values is reflected in the $\pi\pi$ phase shifts which characterize the ρ.

In the ρ meson, we are seeing the quantization of angular momentum values in its most transparent form, and in the present chapter we show some interesting aspects which follow from the distribution of ℓ-values in the ρ. We also discuss the distribution of ℓ-values in the spin 2 $f(1271) \to \pi\pi$ meson, and we show how the low-energy (spin 1) tail of the f seems to be responsible for a mysterious interference effect that has been observed[3] in another resonance, the spin 2 $\eta_N(1083) \to \pi\pi$ meson. This result, which helps to establish the validity of the η_N resonance, is important from the standpoint of testing the applicability of SU(3) symmetries to elementary particles, because the $\eta_N(1083)$, if it exists, overfills the $J^P = 2^+$ meson nonet, just as the $A_{1.5}(1197)$ meson (see App. B) overfills the $J^P = 1^-$ nonet, and the M(953) meson (see Fig. 5.1) overfills the $J^P = 0^-$ nonet. The accurate SU(3) mapping that is claimed for the $J^P = 0^-$, 1^-, and 2^+ meson nonets (see page S29 of Ref. 3) is obtained only by disregarding some of the experimental data; and, of course, the other predicted SU(3) meson nonets are unfilled. Thus, in spite of its early successes, SU(3) has not led to a complete or accurate mapping of the meson resonances.

We now go on to a comparison of the properties of the spin 1 ω and ρ mesons. These serve as the clearest examples of a rotationless ($\ell = 0$) meson resonance and a rotational ($\ell = 1$) meson resonance, respectively.

13.2 Comparison of the ω and ρ Mesons

The ω and ρ are the lowest-mass spin 1 meson resonances. The ω has isotopic spin $I = 0$, corresponding to its charge state ω^0; and the ρ has $I = 1$, corresponding to its charge states ρ^0 and $\rho^\pm$. The dominant decay modes of the ω and ρ^0 mesons are

$$\omega(783)\Gamma 10 \to \pi^+\pi^-\pi^0 \tag{13.1}$$

and

$$\rho^0(773)\Gamma 152 \rightarrow \pi^+\pi^- . \qquad (13.2)$$

If the decay mechanisms of the ω and ρ are similar [see Eq. (14.20) in Chapter 14], then Eqs. (13.1) and (13.2) suggest that, from a light-quark viewpoint, the ω has a structure which is similar to that of the ρ^0, but which contains an additional π^0. In Chapter 8 we presented evidence for the existence of the spinor S ≃ 330 MeV, which appears in meson resonances in the form of mixed spinor pairs $\dot{S}\dot{S}$ that have a mass of about 655 MeV (see Appendix D). Thus if we add a 135 MeV π^0 to an $\dot{S}\dot{S}$ pair, we obtain a spin 1 resonance structure $\dot{S}\dot{S}\pi^0$ that has a mass of about 790 MeV, which is close to the observed mass of 783 MeV for the ω (we can allow a few MeV for binding energy). Furthermore, the spin S = 1 and the isotopic spin I = 0 of the ω may be related (see Sec. 14.3 in Chapter 14); and the narrow width of the ω follows from its identification as an S-state resonance. Thus the configuration

$$\omega = \dot{S}^+\dot{S}^-\pi^0 \rightarrow \pi^+\pi^-\pi^0 \qquad (13.3)$$

accurately reproduces the main features of the ω. Now if the ρ^0 is in turn constructed as an ω minus a π^0, as the decay modes in Eqs. (13.1) and (13.2) suggest, then

$$\rho^0 = \dot{S}^+\dot{S}^- \rightarrow \pi^+\pi^- . \qquad (13.4)$$

If this basis state assignment is correct, it follows that the ρ has an "intrinsic" mass of 655 MeV (the mass of the $\dot{S}\dot{S}$ pair). Hence, in order to account for its observed mass of 773 MeV, an additional mass of 118 MeV must be attributed to rotational energy, so that the ρ is necessarily a rotational excitation. In support of this idea, we saw from the baryon resonances of Chapter 12 that narrow-width and broad-width states correspond to rotationless (ℓ = 0) and rotational (ℓ > 0) excitations, respectively; thus, from the observed widths Γ_ω = 10 MeV and Γ_ρ = 152 MeV shown in Eqs. (13.1) and (13.2), it seems natural to regard the ω as an ℓ = 0 excitation and the ρ as an ℓ = 1 excitation, in agreement with the requirements of Eqs. (13.3) and (13.4).

The identification of the 773 MeV ρ meson as a 655 MeV $\dot{S}\dot{S}$ spinor pair in an ℓ = 1 rotational mode means that the rotational energy is 118 MeV, and it raises the question as to whether this is a reasonable value to assign as the rotational energy of the ρ. We can provide one answer to this question (two other answers are provided in Sections 13.4 and 13.5) by extrapolating from the case of the baryon resonances. We saw in Chapter 12 that the Λ(1519) rotational band has an ℓ = 1 rotational level with a rotational energy of 54 MeV (see Fig. 12.5), and that the Δ_{33}(1232) resonance, which corresponds to the $N\dot{3}(1149)$ bandhead, has an $\ell = 1$ rotational energy of 83 MeV. The task of scaling these results down to the mass of the ρ involves a knowledge of both the masses and the radii of the bandheads, and whereas the masses are known, the radii are not. However, we can obtain a reasonable estimate by making a linear extrapolation of the rotational energy as a function of the bandhead mass, as follows: when we go from the $\Lambda(1519)$ to the Δ_{33}, the change in bandhead masses is 1519 - 1149 = 370 MeV, and the change in rotational energies is 54 - 83 = -29 MeV; when we then extrapolate from the Δ_{33} to the ρ, the change in bandhead masses is 1149 - 655 = 494 MeV, so that the change in rotational energies, from a linear extrapolation, is -29 x (494/370) = -39 MeV, which gives an estimated rotational energy of 83 + 39 = 122 MeV for the ρ. Thus the rotational energy of 118 MeV that we deduced for the ρ on the basis of the light-quark systematics of Eq. (13.4) is about the value that we expect from an extrapolation of the rotational systematics of the baryon resonances. As mentioned above, in Sections 13.4 and 13.5 we give two more estimates for the rotational energy of the ρ: one estimate that is obtained by calculating the moment of inertia of the ρ, and another estimate that is obtained directly from the behavior of the $\ell = 1$ component of the experimental $\pi\pi$ phase shifts. All of these estimates are in essential agreement. Hence we conclude that

$$\omega = \dot{S}^{+}\dot{S}^{-}\pi^{0} \quad (J = 1,\ \ell = 0,\ S = 1) \tag{13.5}$$

and

$$\rho^{0} = \dot{S}^{+}\dot{S}^{-} \quad (J = 1,\ \ell = 1,\ S = 0). \tag{13.6}$$

As is shown in Eqs. (13.1) and (13.2), the narrow-width ω meson has a dominant $\pi\pi\pi$ decay mode and the broad-width ρ meson has a dominant $\pi\pi$ decay mode. Although we have attributed the difference in the widths of these two resonances to the difference in their ℓ-values, the conventional explanation (see Sec. 13.10) attributes this difference to the difference in the final-state phase-space freedom of a $\pi\pi\pi$ final state as compared to that of a $\pi\pi$ final state. While this phase-space effect is undoubtedly a factor that influences the ω and ρ widths, it is by no means the whole story, as we can tell by the experimental systematics of the hadron resonances. We saw in Chapter 10 that the long-lived "particles" have lifetimes which scale in powers of $\alpha \simeq 1/137$, so that these particular lifetimes are dictated by electromagnetic effects; as an example (and one that is similar to the ω and ρ in its decay modes), the $K_L^0 \to \pi\pi\pi$ and $K_S^0 \to \pi\pi$ lifetimes are separated by one power of α, which indicates that the difference in these lifetimes is <u>not</u> due to phase-space effects. However, phase-space effects should be more important for the "strong" ω and ρ decays than they are for the "weak" K_L^0 and K_S^0 decays (although these effects cannot really be calculated, as we discuss in Appendix G), but factors other than phase space must be important even in the "strong" decays, as we see from the following examples:

(1) The ω has a $\pi\pi$ decay mode in addition to its $\pi\pi\pi$ decay mode, and the observed width for each of these decay modes is the same[4]: $\Gamma_{\omega\to\pi\pi} \simeq 10$ MeV and $\Gamma_{\omega\to\pi\pi\pi} = 10$ MeV; (2) Comparison of the decay widths[3] $\Gamma_{S^*(993)\to\pi\pi} = 40$ MeV (spin 0), $\Gamma_{\rho(773)\to\pi\pi} = 152$ MeV (spin 1), and $\Gamma_{f(1271)\to\pi\pi} = 180$ MeV (spin 2) suggests that the spin of the resonance (J = 0 or J > 0) rather than final-state phase space is mainly responsible for the differences in these decay widths; (3) Comparison of the decay widths[3] $\Gamma_{\eta'(958)\to\eta\pi\pi(68\%)} < 1$ MeV (spin 0) and $\Gamma_{E(1416)\to\eta\pi\pi(\sim 60\%)} \simeq 60$ MeV (spin 1) also indicates that spin effects are more important than phase-space effects; (4) Comparison of the decay widths[3] $\Gamma_{\rho(773)\to\pi\pi} = 152$ MeV (spin 1), $\Gamma_{\omega(783)\to\pi\pi\pi} = 10$ MeV (spin 1), and $\Gamma_{\rho'(1600)\to\pi\pi\pi\pi} \sim 200 - 800$ MeV (spin 1) again suggests that factors other than final-state phase space are determining these particular widths.

In general, the observed decay width for a particle or resonance is determined by its dominant decay mode. Relative phase-space effects for a set of particles can be reliably estimated only in cases where the number of final-state particles is the same (so that the dimensionality of the phase-space volume is the same), which is not the case for the ω and ρ decays that we have been discussing. Thus, from this fact, and from the examples cited above, we conclude that the factor-of-fifteen difference in the ω and ρ widths is not simply a phase-space effect, but is determined primarily by the difference in the ℓ-values of these two resonances: $\ell = 0$ for the ω and $\ell = 1$ for the ρ.

We now discuss the fundamental light-quark excitation that generates the ρ meson. This is a different type of excitation from the one that generates the spinless π, K, η, and η' mesons.

13.3 The Fundamental Meson Excitation Series $(333)^n$

The π meson is the lowest-mass spinless meson excitation, and the ρ meson is the lowest-mass spin 1 meson excitation. Thus it seems logical that these two meson states should each correspond to a fundamental light-quark excitation unit. We have already seen in Chapter 5 that the pion is the configuration $\pi = 1\bar{1}$, and that the other spinless mesons can be constructed as $\eta = 3\bar{3}\pi = 3\bar{3}1\bar{1}$ and $\eta' = 3\bar{3}\pi 3\bar{3} = 3\bar{3}1\bar{1}3\bar{3} \equiv K\bar{K}$ (see Fig. 5.2). Hence we can in a sense think of the π meson as being the generator for these resonances, with the higher-mass states being formed by attaching excitation quanta 3 and $\bar{3}$ to the π "core" of the resonance. What, then, is the fundamental excitation unit that generates the ρ meson? As we now demonstrate, it is the spinless basis state combination 333, which appears in an alternate form as the spinor pair $\dot{S}\dot{S}$.

In Chapter 8 it was shown that the spinor S is a relativistically-spinning form of the spinless state $3 \equiv 3M = 210$ MeV. Now the fundamental <u>baryon</u> basis state configuration is the spinor triplet $S\bar{S}S$, which carries the baryon quantum number B. If we had these three spinors appearing in a spinless instead of spinning form, they would correspond to the light-quark cluster 333 (where we do not distinguish among the states 3, $\bar{3}$, $\dot{3}$, and $\dot{\bar{3}}$; see Appendix I). The cluster 333 is in fact a basic <u>meson</u> excitation unit, and the excitation series $(333)^n$ generates a sequence of observed meson resonances, starting with the ρ meson, and including the J/ψ and ψ' New Particle resonances. However, in order to discuss the $(333)^n$ excitation series, we must first introduce one more facet of this light-quark phenomenology — namely, the transition between spinless and spinning basis states.

We saw from the relativistic equations of Chapter 6 that a relativistically-spinning sphere with its equator moving at the velocity of light is half again as massive as it was at rest. Thus two spinning spheres have the same energy as three nonspinning spheres, and a transition from one situation to the other would be isoergic. We can extend these results to apply to the spinor S and the corresponding spinless state 3: the excitation unit $\dot{3}^{o}\dot{3}^{o}\dot{3}^{o}$ has a mass of about 630 MeV; a bound spinor pair $S\bar{S}$ has a mass of about this same value (see Eq. 8.5'); and a nonstrange spinor pair $\dot{S}^{+}\dot{S}^{-}$ has a mass of about 655 MeV. In the energy regime that commences with the opening up of the $\bar{N}N$ channel at 1800 MeV, basis state transitions $333 \leftrightarrow S\bar{S}$ occur, and they play an important role in the generation of $\bar{N}N$ pairs and the higher-mass New Particle states (see Chapter 17). These transitions are of the form $(333)^3 \equiv 333 \cdot 333 \cdot 333 \leftrightarrow \bar{S}S\bar{S} \cdot S\bar{S}S$. The (almost isoergic) transition $333 \leftrightarrow \dot{S}\dot{S}$ also occurs, and it plays an important role in both the lower-mass and higher-mass meson excitations. These transitions are discussed in detail in Chapter 14.

The fundamental excitation series $(333)^n$ is illustrated in Table 13.1, where the first seven terms are shown (also see Table 17.3 in Chapter 17). As can be seen in Table 13.1, the ρ meson corresponds to the first term in this series, and the dominant J/ψ(3098), ψ'(3684), and ψ(4414) peaks correspond to the fifth, sixth, and seventh terms. (The only candidate to date for the fourth term is the Kπ(2425) kaon peak of Fig. 5.6 in Chapter 5.) The alternating manner in which $333 \rightarrow SS$ transitions occur is illustrated in Table 13.1, and also the very accurate mass values that are obtained from these basis state configurations. The ρ meson, which represents the first term in this series, is the only resonance that is not manifested as an S-state excitation. The fact that the ρ is not observed as an S-state is consistent with the general systematics of low-mass 3-type excitations (see Appendix E), which are characteristically difficult to observe as S-states. The $\dot{S}\dot{S}\dot{S}\dot{S}$ spinor quartet shown in Table 13.1 is also difficult to observe as an S-state, and both the $\dot{S}\dot{S}\dot{S}\dot{S}$ excitation and the $\dot{S}\dot{S}$ bandhead of the ρ have been observed mainly by means of the interference effects that they produce[6].

Table 13.1. The fundamental meson excitation series $(333)^n$. The n = 1 term in this series corresponds to the rho meson, which is the only one of these resonances that does not appear directly as a rotationless excitation. The n = 3 term corresponds to the $\bar{N}N$ resonance that is the "ground state" for the New Particles (Chapter 17), and the n = 5 and n = 6 terms correspond to the J/ψ and ψ' New Particle states. Table 13.1 is discussed in more detail in Chapter 17, where some possible higher terms in the series are also included. It should be noted here that the systematics of the basic $(333)^n$ excitation series was published (Ref. 5) prior to the appearance of the J/ψ and ψ' New Particles.

Excitation	Observed form	Calculated mass (MeV)[a]	Observed resonance	Spin
$(333)^1$	$\dot{S}\dot{S}$ or 333 (ℓ = 1 rotational mode)	∿770[b]	$\rho(773)\Gamma 152$	1
$(333)^2$	$\dot{S}\dot{S} \cdot 333$	1285	$D(12185)\Gamma 21$	1
	$\dot{S}\dot{S}\dot{S}\dot{S}$	1310	$A_2 dip(1310)\Gamma \simeq 10$	2
$(333)^3$	$\bar{S}\bar{S}S \cdot S\bar{S}S$	1795[c]	$\bar{p}n(1795)\Gamma \leq 8$	
$(333)^4$	$\bar{N}N \cdot 333$	2425	$K\pi(\sim 2425)$	
$(333)^5$	$\bar{N}N \cdot \dot{S}\dot{S}\dot{S}\dot{S}$	3105	$J/\psi(3098)\Gamma 0.067$	1
$(333)^6$	$\bar{N}N \cdot (333)^3$	3685	$\psi'(3684)\Gamma 0.228$	1
$(333)^7$	$\bar{N}N \cdot (\dot{S}\dot{S}\dot{S}\dot{S})^2$	4415	$\psi(4414)\Gamma 33$	

[a]The masses are calculated from: 333 = 630 MeV, $\dot{S}\dot{S}\dot{S}\dot{S}$ = 1310 MeV, $\bar{N}N$ = 1795 MeV, binding energy = 0.

[b]See Sections 13.4 and 13.5 in Chapter 13.

[c]See Eq. (8.9) in Chapter 8.

The systematics of Table 13.1 is discussed in more detail in Chapters 14 and 17. The main point we are trying to bring out here is that the series $(333)^n$ represents a dominant meson excitation mechanism, and the ρ meson represents the first term in this series. Hence it has a dominant effect on meson interactions. In the next section we examine the specific properties of the ρ meson in more detail.

13.4 The Moment of Inertia of the ρ from its Basis State Configuration

As we discussed in the previous section, the ρ meson is generated from the basic meson excitation unit 333, and it appears only in a rotational ($\ell = 1$) mode, and not in a rotationless ($\ell = 0$) S-state mode. Now, in view of the basis state transition 333 ↔ $\dot{S}\dot{S}$ (see Chapter 14), we cannot really determine whether the ρ is the configuration 333 ($\ell = 1$), or the configuration $\dot{S}\dot{S}$ ($\ell = 1$), or both. From the considerations discussed in Chapter 18, we can define the overall geometry of the spinor $\dot{S}$ quite accurately, but we cannot similarly delineate the geometry of the spinless state 3. Thus, for calculational purposes in the present studies we will consider the ρ to be formed as an $\dot{S}\dot{S}$ spinor pair, although it may actually include a 333 component. We will also disregard the intrinsic spin S of the $\dot{S}\dot{S}$ spinor pair; this spin can in principle be either S = 1 or S = 0, but there are some suggestions (see Chapter 14) that an isotopic spin I = 1 spinor pair $\dot{S}\dot{S}$ has spin S = 0.

The spinor pair $\dot{S}\dot{S}$ appears in meson resonances with an effective mass of just about 655 MeV (see Appendix D). Also, from its spin angular momentum and magnetic moment, the spinor $\dot{S}$ has a radius of about 0.6 fermi (Chapter 18). Hence we can obtain an estimate for the moment-of-inertia I of an $\dot{S}\dot{S}$ spinor pair from the configuration that is illustrated in Fig. 13.1. Classically, each spinor $\dot{S}$ in Fig. 13.1 has $I \simeq 2/5\ MR^2 + MR^2$, so that the $\dot{S}\dot{S}$ pair has $I \simeq 14/5\ MR^2$, and we obtain I = 330 MeV fm^2 for M = 327.5 MeV and R = 0.6 fermi. From Eq. (12.13), the $\ell = 1$ rotational energy of the ρ corresponding to this value of I is $\hbar^2/I$ = 118 MeV, which is in agreement with the estimates for this quantity given in the previous section.

It should be noted that in carrying out the above calculation, we have neglected the special-relativistic mass increase of the rapidly-rotating system, and we have assumed that the orbital motion is parallel to the spin axis. However, it must be remembered that the ρ is a very short-lived resonance, with a lifetime $\tau_\rho \sim 4 \times 10^{-24}$ sec, so that it makes very few revolutions before it breaks up, and its rotational axis is probably very poorly defined. The quoted mass values for the ρ in Ref. 3 range from about 750 to 780 MeV. What we have done in the present section is to demonstrate that a rotational energy of 118 MeV for the ρ is in reasonable agreement with the known geometrical features of an $\dot{S}\dot{S}$ spinor pair. This result is similar in spirit to the result given in Sec. 13.2, wherein we showed that the required rotational energy for the $\dot{S}\dot{S}$ bandhead which constitutes the ρ is in reasonable agreement with the value for the rotational energy that we obtain by extrapolating from higher-mass baryon rotational excitations. We now go on, in the next section, to obtain still another estimate for the rotational energy of the ρ, an estimate which is deduced from the experimental ππ phase shifts in the energy region of the ρ.

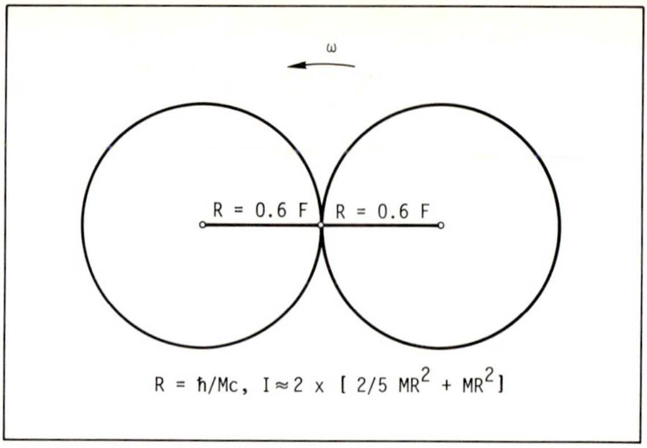

Figure 13.1. A "classical" estimate for the moment of inertia of the rho meson (pictured as an $\ddot{S}\ddot{S}$ spinor pair, with each spinor $\dot{S} \simeq 327.5$ MeV having a radius R equal to its Compton wavelength $R = \hbar/Mc$). The configuration shown here gives the value $I = 330$ MeV fm^2, which is in close agreement with several other determinations of this quantity that are described in the text.

In the present studies, we are not attempting to develop a detailed mathematical model for the ρ in the sense that we developed a model for the muon in Chapter 7. We are merely trying to demonstrate, from as many viewpoints as possible, that it makes sense to regard the ρ meson as an $\ell = 1$ nuclear-physics-type orbital excitation of an $\ddot{S}\ddot{S}$ spinor pair.

13.5 The Moment of Inertia of the ρ from Experimental ππ Phase Shifts

The rho meson is, from the present viewpoint, an $\ell = 1$ rotational excitation. Thus its behavior is characterized by the behavior of the ππ P-wave phase shift. Figure 13.2 illustrates that this is in fact the case. In Fig. 13.2 the ππ mass spectrum (the histogram) from the reaction[7] $\pi^+ p \rightarrow \Delta^{++}\pi^+\pi^-$ is plotted, and also the energy-dependent P-wave phase shifts (the dots) from a Berkeley analysis[8] of these same data. As can be seen in the figure, the δ_1^1 P-wave phase shifts[†] quite

[†]The phase-shift notation is δ_ℓ^I, where I and ℓ are the isotopic spin and the angular momentum, respectively.

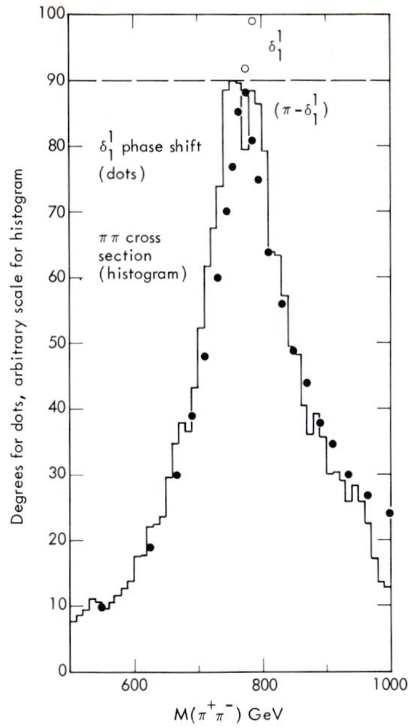

Figure 13.2. Comparison of an experimental rho-meson $\pi\pi$ invariant-mass spectrum (the histogram) with the P-wave $\pi\pi$ phase shifts δ_1^1 (the dots) that were obtained from an analysis (Ref. 8) of these same data (Ref. 7). (Note the P-wave reflection above 90°.) As can be seen, the P-waves closely follow the shape of the experimental $\pi\pi$ mass spectrum. This indicates that the rho is in fact an $\ell = 1$ orbital P-wave excitation.

accurately reproduce both the position and the width (FWHM) of the rho meson (note that the P-wave phase shifts above the peak of the resonance are shown reflected about 90°).

If we represent the ρ meson by the energy-dependent δ_1^1 P-wave phase shifts, as is suggested in Fig. 13.2, then we can use the quantum-mechanical implications of this representation to deduce the moment-of-inertia of the rho. This comes about in the following manner. The peak of the ρ meson occurs at the energy where the P-wave has the value $\delta_1^1 = 90°$. Now, as we mentioned in Sec. 13.1, the observed spread in energies of the ρ should be associated with a corresponding spread in angular momentum values (ℓ-values), in accordance with the equation

$$E_\rho(\ell) = E_o + \hbar^2/2I \times \ell(\ell + 1). \tag{13.7}$$

From the results described in the previous sections of this chapter, we concluded that E_o is equal to 655 MeV (the energy of an $\ddot{S}S$ spinor pair), and we also concluded (from three different calculations) that $\hbar^2/2I \simeq 118$ MeV, which corresponds to the value I = 330 MeV fm^2. In the present section we demonstrate that we can obtain independent determinations of both E_o and I directly from the experimental ππ phase shifts, without making any use of light-quark systematics. We do this by noting that, at the values $\delta_1^1 = 45°$ and $\delta_1^1 = 135°$, the P-wave scattering amplitudes have fallen to 70.7% of their value at $\delta_1^1 = 90°$, and the P-wave probabilities have fallen to 50%. Now for energies below the peak energy, a 50% P-wave probability ($\delta_1^1 = 45°$) can be thought of as half P-wave and half S-wave, or an ℓ-value of 1/2. Correspondingly, for energies above the peak energy, a 50% P-wave probability ($\delta_1^1 = 135°$) can be thought of as half P-wave and half D-wave, or an ℓ-value of 3/2. In the Berkeley ππ phase-shift analysis[8], precise values were obtained for the energies at which the 45°, 90°, and 135° phase shifts occur. These values are shown in Table 13.2. Inserting the 45° and 135° values from Table 13.2 into Eq. (13.7), we obtain

$$E_\rho(3/2) - E_\rho(1/2) = \hbar^2/2I \times [3/2 \times 5/2 - 1/2 \times 3/2] = 163 \text{ MeV}, \qquad (13.8)$$

so that I = 358 MeV fm^2, which is in good agreement with the value I = 330 MeV fm^2 that we had previously deduced. Now I = 358 MeV fm^2 corresponds to $E_{rot} = \hbar^2/2I$ = 54.4 MeV, so that Eq. (13.7), with ℓ = 1, gives

$$E_o = E_\rho(1) - E_{rot} \times 1(1+1) = 772 - 109 = 663 \text{ MeV} \qquad (13.9)$$

as the calculated bandhead mass of the ρ, where $E_\rho(1)$ = 772 MeV is taken from Table 13.2; this calculated value of 663 MeV agrees closely with the value of 655 MeV which follows from the light-quark model as the mass of an $\ddot{S}S$ spinor pair. (The experimental data of Ref. 7 also provide a tantalizing hint of a ππ enhancement at about 660 MeV, as we discuss in Sec. 13.9.)

Table 13.2. The energies at which the ππ P-wave phase shift (Ref. 8) passes through 45°, 90°, and 135°. By identifying these energies with the rho meson ℓ-values ℓ = 1/2, 1, and 3/2, respectively, as described in the text, we can calculate a value for the bandhead mass of the rho without using any theoretical assumptions [see Eqs. (13.8) and (13.9)].

P-wave phase shift	45°		90°		135°
ππ invariant mass	703 MeV		772 MeV		866 MeV
ℓ-value (see text)	1/2		1		3/2
Mass interval spacing		69 MeV		94 MeV	

The significance of the ρ meson calculations shown in Eqs. (13.8) and (13.9) is not in the particular numerical values that are obtained for the bandhead mass E_o and the moment-of-inertia I, but rather in the fact that we are using a completely independent method to obtain these values. The ππ phase shifts[8] are experimentally-derived quantities, and the application of these phase shifts to Eq. (13.7) involves some assumptions about quantum mechanics but no assumptions about the ρ meson (other than that it is an $\ell = 1$ orbital excitation). Thus the close agreement of the present calculations with our previously-deduced values for E_o and I provides an interesting confirmation of the correctness of our ideas about the nature of the rho meson. We now go on to examine the angular momentum components of the rho in more detail.

13.6 Angular Momentum Components of the ρ Meson

As we have discussed above, the broad spread in energy values for the ρ meson is due to a corresponding spread in ℓ-values: the ρ is a spin 1 resonance only at the peak energy (773 MeV) of the resonance. The relationship $E_\rho(\ell)$ between the energy of the ρ and its spin value ℓ is given by Eq. (13.7). Guided by the results of the preceding sections in this chapter, let us choose the constants in Eq. (13.7) so that $E_\rho(0) = 655$ MeV (which is the bandhead mass of the ρ) and $E_\rho(1) = 773$ MeV (which is the energy at the peak of the ρ). Equation (13.7) then uniquely relates the energy and angular momentum of the ρ, so that instead of plotting the δ_1^1 P-wave phase shifts against an energy abscissa, as was done in Fig. 13.2, we can plot them against both an energy abscissa and an angular momentum abscissa. This plot is shown in Fig. 13.3.

From Fig. 13.3, we expect to see D-wave components appearing at energies just above 773 MeV; that is, for $\ell > 1$ on the angular momentum abscissa. Figure 13.4 shows the Berkeley[8] ππ D-wave phase shifts δ_2^0 plotted against the angular momentum abscissa of Fig. 13.3. As can be seen in Fig. 13.4, the D-wave appears at just the place that we would expect it from the present systematics. The occurrence of this low-energy D-wave component has been known for some time, but there has been no clear theoretical justification for its existence. From the present viewpoint, we see it as a consequence of the spread in quantization values of the orbital angular momentum. This low-energy D-wave component has also been observed in the $\pi^+\pi^- \rightarrow \pi^0\pi^0$ cross section.[9] The large D-wave values shown above 1 GeV ($\ell = 2$) in Fig. 13.4 have also presented a puzzle; this puzzle is resolved in Sec. 13.7 below.

We can extend the above results by plotting the Berkeley ππ F-wave phase shifts δ_3^1 against the angular momentum axis of Fig. 13.3. This result is shown in Fig. 13.5. As can be seen, the F-wave starts to rise at the point where $\ell > 2$ on the abscissa.

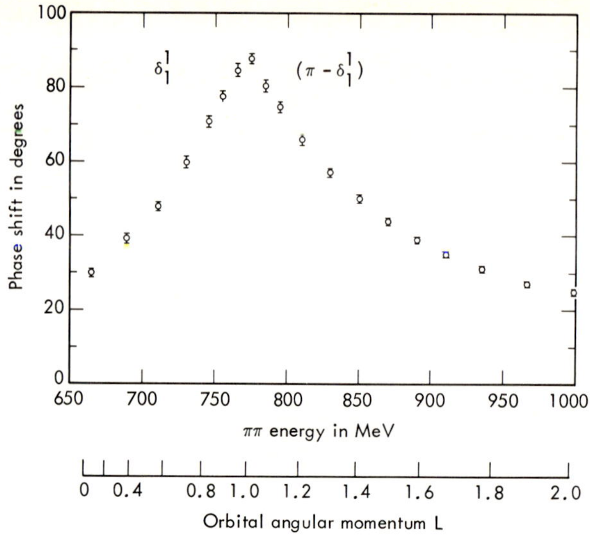

Figure 13.3. The δ_1^1 P-wave phase shifts of the rho meson (Ref. 8; see Fig. 13.2), shown plotted against the conventional energy axis, and also against an orbital angular momentum axis ℓ, where the energy and orbital angular momentum are related to one another by Eq. (13.7) (see the discussion in the text). If the rho is a rotational resonance, as is indicated in Fig. 13.2, then a spread in energies of the rho is associated with a spread in ℓ-values for the rho; thus the rho is a spin 1 particle only at the peak energy of the resonance.

The results presented in this section serve to confirm the interpretation of ℓ-values given in Sec. 13.5. If the ρ has a spin value $\ell < 1$, then this is reflected in the $\pi\pi$ phase shift analysis as a mixture of S-waves and P-waves, and if the ρ has a spin value $\ell > 1$, then this is reflected as a mixture of P-waves and D-waves. When we interpreted the 45° and 135° phase shift values as representing 50% mixtures of these phases ($\ell = 1/2$ and $\ell = 3/2$), we were able to normalize the constants in Eq. (13.7) so as to reproduce both the bandhead mass and the moment of inertia of the ρ. Then, using these normalized values to define the angular momentum abscissa which appears in Figs. 13.3 - 13.5, we saw that the D-wave and F-wave components of the $\pi\pi$ phase shifts do in fact appear at just the expected ℓ-values. Thus the observed distribution of ℓ-values in the ρ is a very real and very understandable phenomenon. It is apparent that this same distribution of ℓ-values must occur for all systems which

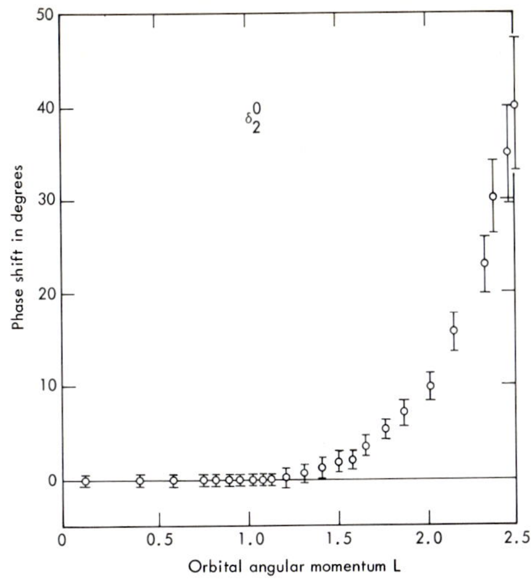

Figure 13.4. The δ_2^0 D-wave rho meson phase shifts of Ref. 8 plotted against the orbital angular momentum axis of Fig. 13.3. As can be seen, the D-wave starts to rise at just the energy where $\ell > 1$. The existence of this D-wave component in low-energy $\pi\pi$ scattering has constituted a long-standing theoretical puzzle.

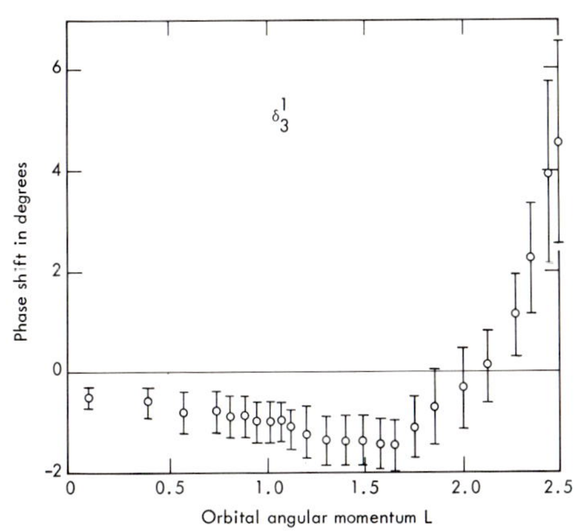

Figure 13.5. The δ_3^1 F-wave rho meson phase shifts of Ref. 8 plotted against the orbital angular momentum axis of Fig. 13.3. As can be seen, the F-wave starts to rise at just the energy where $\ell > 2$.

have orbital excitations. However, for massive systems, the orbital energies are so small as compared to the total energies that the orbital rotational levels have very narrow widths. In the ρ meson, due to its low mass, broad width, and freedom from background effects, we are seeing the distribution of ℓ-values in its most easily-observed form.

We now discuss the distribution of ℓ-values in another rotational resonance, the 1271 MeV spin 2 $f \to \pi\pi$ meson. The f meson is in a higher-mass region that contains other $\pi\pi$ resonances. However, this turns out to be a fortunate circumstance, because it is the interference between the f and another spin 2 $\pi\pi$ meson, the the $\eta_N(1083)$, which makes this such an interesting example to study.

13.7 Angular Momentum Components of the f Meson

In addition to the ρ meson, the other dominant $\pi\pi$ resonance is the spin 2 f(1271) meson, which has a width $\Gamma = 180$ MeV. By analogy to the systematics of the ρ, we identify the f as a rotational excitation, and we attribute the broad range of energies in the f to a corresponding range of angular momentum values. Since extensive $\pi\pi$ phase shift analyses have not yet been carried out in the region of the f, we cannot investigate this conjecture in detail, as we did for the ρ. However, there is one empirical result that appears to support this conclusion. In addition to the $f(1271)\Gamma 180$ $\pi\pi$ resonance, the Review of Particle Properties[3] also lists another broad-width spin 2 $\pi\pi$ resonance, the $\eta_N(1083)\Gamma 98$, whose existence has been the subject of some controversy (see the discussion in Sec. B.1 of Appendix B). The mass and width of the η_N have been reported in half a dozen different experiments[3]; the angular momentum J of the η_N is $J > 1$, and the most probable value is $J = 2$. Oh and coworkers[10] discussed several possible explanations for the η_N, and their favored explanation, based on a number of stated reasons, is that of a D-state resonance riding on a strong P-wave background. However, there has been no explanation for the origin of this P-wave background. This is where the f meson enters the picture. Figure 13.6 shows the f(1271) and $\eta_N(1083)$ resonances plotted schematically on a common energy abscissa, with individual angular momentum abscissae also shown for each resonance[†], in a manner analogous to the ρ meson plot of Fig. 13.3. As can be seen in Fig. 13.6, the strong P-wave background underneath the η_N arises from the low-energy spin 1 tail of the spin 2 f meson. The results shown in Fig. 13.6 not only explain

[†]In order to plot angular momentum abscissae for the f(1271) and $\eta_N(1083)$ mesons in Fig. 12.6, it is necessary to know their bandhead masses. As is shown in Chapter 16, the $\omega(783)$ meson forms the bandhead for the η_N, and the $\phi(1020)$ meson forms the bandhead for the f.

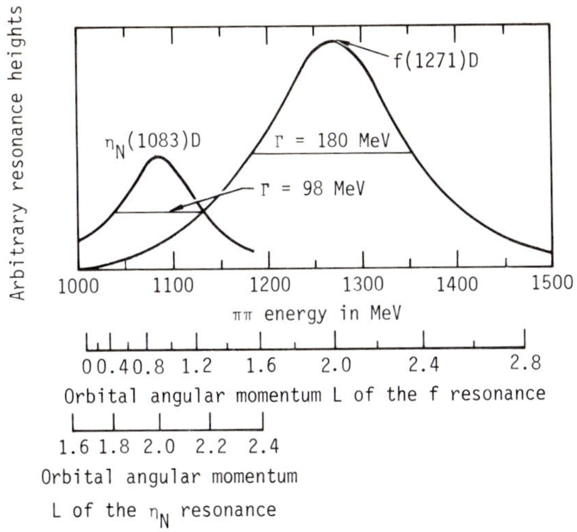

Figure 13.6. The spin 2, isotopic spin I = 0, ππ resonances η_N(1083)DΓ98 and f(1271)DΓ180, shown schematically on a common energy axis. In analogy to Fig. 13.3, the orbital angular momentum abscissae are also shown for these resonances (see the text). As can be seen, the low-energy spin-one tail of the f accounts for the strong P-wave background which occurs under the η_N (Ref. 10). Also, the D-wave component of the η_N accounts for the sharp rise in the D-wave above 1000 MeV that was noted in Fig. 13.4 (1000 MeV equals ℓ = 2; see Fig. 13.3).

the mystery of the P-wave background under the η_N, but they thereby furnish evidence (albeit somewhat indirectly) that the f meson, just like the ρ, has an observable distribution of orbital angular momentum ℓ-values.

Figure 13.6 also helps to clear up another puzzle. In Fig. 13.4 it can be seen that the D-wave phase shift from the Berkeley analysis[8] rises very rapidly above $\ell = 2$ or 1 GeV (see Fig. 13.3). As the Berkeley workers commented[8]: "The I = 0 D-wave phase shift δ_2^0 around 1 GeV is larger than we would expect for the f^o meson alone". This observation is doubly relevant from the present viewpoint, since Fig. 13.6 shows that the f^o meson has no ℓ = 2 component left at an energy as low as 1 GeV. However, the η_N still has a substantial ℓ = 2 component at 1 GeV. Thus the large D-wave phase shifts above 1 GeV that are observed in Fig. 12.4 arise from the η_N resonance, and not from the f. We can turn this argument around and assert that the large D-wave phase shifts above 1 GeV are direct evidence for the existence of the spin 2 η_N resonance.

The establishment of the η_N as a bona fide resonance has, as we mentioned in Sec. 13.1, some theoretical implications. The $J^P = 2^+$ meson nonet already has its full complement of resonances without the η_N. Thus the η_N, if it exists, overfills the 2^+ nonet, which is somewhat of an embarrassment from an SU(3) point-of-view. This fact, as we discuss in Appendix B, has apparently led to some reluctance on the part of the authors of Review of Particle Properties[3] to accept the η_N as an established resonance. By way of contrast, the spin 2 f'(1522)Γ62 resonance, which was needed to fill the 2^+ nonet, was accepted as an established resonance[11] on the basis of experimental data that were less extensive than the data for the η_N. However, as usual, it will be the experiments themselves, and not the preconceptions of theorists, which will ultimately determine the validity of these enhancements.

There is one more interesting point to be noticed with regard to the f(1271) resonance. In a recent experiment[12], the f meson was observed with two different kinematic constraints, and two different mass values were obtained:

$$\begin{aligned} &f(1262\pm3)\Gamma230 \quad |\cos\theta^*| \leq 0.5, \\ &f(1291\pm7)\Gamma220 \quad |\cos\theta^*| \geq 0.5. \end{aligned} \tag{13.10}$$

If broad rotational excitations have a large spread in angular momentum values – that is, a rather loose quantization constraint – , then it seems plausible that the peak energy of the resonance might depend on momentum transfer conditions in the manner that is demonstrated in Eq. (13.10) for the f meson. Somewhat similar results have been observed in the case of the ρ meson[13].

13.8 Symmetrization of the Shape of the ρ Meson

As can be seen in Table 13.2, and also in Fig. 13.2, the shape of the ρ meson is not symmetric: the ρ is broader on the high-energy side. This asymmetry occurs both in the ππ invariant-mass distribution and in the P-wave phase shifts, as is shown in Fig. 13.2. From Fig. 13.3, it seems that the asymmetry in the shape of the ρ matches the stretching of the angular momentum axis. Thus a plot of the ππ P-wave phase shifts against a linear angular momentum axis should yield a symmetric shape. This plot is shown in Fig. 13.7, where it can be seen that the P-wave phase shifts in this angular momentum representation appear as a symmetric bell-shaped distribution.

It should be noted that, for histograms, the representation $N(E)d\ell$, which is essentially the representation used in Fig. 13.7, is a hybrid form. Pure forms would

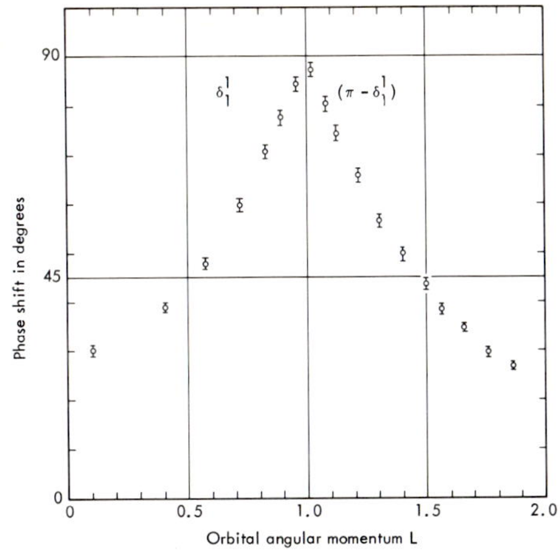

Figure 13.7. The δ_1^1 P-wave phase shifts of Fig. 13.3 shown against a linearized plot of the orbital angular momentum abscissa of Fig. 13.3. In this "angular momentum representation", the P-waves exhibit a symmetric distribution, in contrast to the asymmetric distributions shown in Fig. 13.3 and Table 13.2.

be $N(E)dE$ or $N(\ell)d\ell$. The Jacobian for the transformation $N(E)dE \to N(\ell)d\ell$ is, from Eq. (13.7), $dE/d\ell = \hbar^2/2I \times (2\ell+1)$. Thus from $N(E)dE = N(\ell)d\ell$, we have $N(E) \propto N(\ell)/(2\ell+1)$. Hence $N(E)d\ell$ is in effect a cross section per unit angular momentum: $N(E)d\ell \propto |\alpha_\ell|^2 d\ell$ (where the usual form for the scattering amplitude, $f_\ell(\mu) \propto \alpha_\ell P_\ell(\mu)(2\ell+1)$, gives the ℓ-dependence of these quantities).

13.9 Experimental ππ Cross Section Data

Our discussions in the present chapter have been centered primarily on a phenomenological analysis of the meson resonances. Thus it is of interest to make a direct examination of the experimental ππ cross section data. These data are shown in the rather remarkable plot of Fig. 13.8, which is taken from the $\pi^+ p \to \Delta^{++}\pi^+\pi^-$ experiment of Ref. 7 (Fig. 13.8 is Fig. 1c of Ref. 7, with the boxes added by the present author). Figure 13.8 is remarkable in that it illustrates most of the resonances we have been discussing in this chapter. The $\pi^+\pi^-$ spectrum shown in Fig. 13.8 is dominated primarily by the ρ peak, and secondarily by the f peak. The ω meson is clearly in evidence in the form of a dip at the top of the ρ. There is also a small enhancement at the position of the η_N (compare Figs. 13.6 and 13.8). And the data in

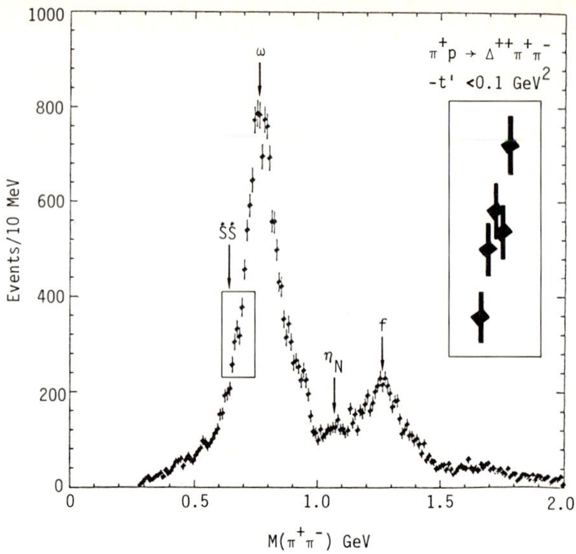

Figure 13.8. The ππ cross section data of Ref. 7. The ρ-ω interference dip (see Fig. 8.1a) and the η_N and f peaks (Fig. 13.6) are clearly in evidence, as well as an indication of a small $\dot{S}\dot{S}$ peak (see Figs. 8.1b and 13.9).

the box show evidence of a small peak near 660 MeV, which logically corresponds to the $\dot{S}\dot{S}$ bandhead of the ρ.

The peak shown near 660 MeV in Fig. 13.8 is actually of marginal statistical significance†. However, evidence of this peak also appears in Fig. 13.9, which is taken from Fig. 15 in Ref. 7, and which shows the ππ cross section data of Fig. 13.8 broken down into components of the momentum transfer $|t'|$. As can be seen in Fig. 13.9, each of the three $|t'|$ regions exhibits some kind of anomaly in the region of 660 MeV. If an $\ddot{S}\ddot{S}$ spinor pair occurs with spin S = 1 in its isotopic spin I = 0

---

†A statistical analysis of this peak indicates that it is centered at 665 MeV and has a significance of about four standard deviations. This analysis was carried out by N. Smiriga and J. Karush, statisticians at Lawrence Livermore Laboratory, and it consisted essentially of performing a variance-stabilizing square-root transformation and then using a curve-fitting procedure, with the points in question removed, to determine the background.

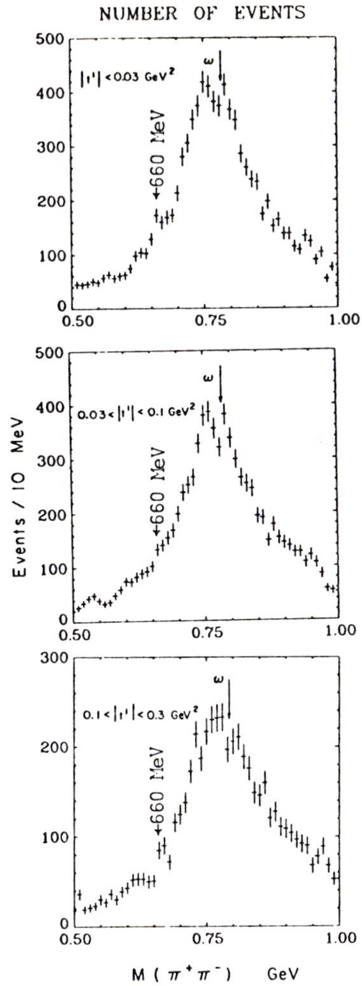

Figure 13.9. The $\pi\pi$ cross section data of Ref. 7 broken down into components of the momentum transfer $|t'|$. As can be seen, an anomaly occurs at 660 MeV which shows a dependence on momentum transfer. From the relationship $|\vec{r} \times \vec{p}| = \sqrt{\ell(\ell+1)}\hbar$, where $\vec{r}$ = impact parameter $\sim$ 1 fermi, $\vec{p}$ = linear momentum transfer, and $\hbar$ = 197 MeV/c fermi, the value $\ell \simeq 1$ corresponds to $|\vec{p}| \sim 0.3$ GeV/c. Hence $|t'| \sim |\vec{p}|^2 \sim 0.1\ (\text{GeV/c})^2$ sets the scale for the magnitude of $|t'|$ values over which changes in the production of the (spin 1) 660 MeV enhancement should be expected to occur.

mode, as is suggested in Chapter 14, then the anomalous behavior observed near 660 MeV in Fig. 13.9 arises from a spin 1 enhancement; and, as is explained in the caption to Fig. 13.9, a spin 1 interfrence effect would have about the observed dependence on $|t'|$ (where $t' \propto \vec{p}^2$, with $\vec{p}$ the linear momentum transfer). Other evidence for a ππ resonance at 660 MeV is supplied by an analysis of Gutay and coworkers[14] (see Fig. 8.1b). This analysis reveals a four-standard-deviation dip in the isotropic component of the ππ amplitude right at 660 MeV, for the I = 0 amplitude only, and it again suggests a spin 1 interference effect. Figures 13.8 and 13.9 illustrate both the positions and the magnitudes of the effects produced by the $\ddot{S}S$, ρ, ω, η_N and f mesons.[15] The two S-wave rotationless resonances ($\ddot{S}S$ and ω) have narrow widths, but the P-wave (ρ) and D-wave (η_N and f) resonances have broad widths, corresponding to their identification as rotational excitations.

13.10 Summary and Implications for Theory

In the present chapter we have attempted to delineate the differences between rotationless and rotational meson resonances, with the low-mass ω and ρ mesons serving as the clearest examples of these two types of excitation. We first showed that the light-quark cluster $\ddot{S}S\pi$ in a spin 1 S-state configuration reproduces the main features of the narrow-width ω meson. Then we used the decay modes shown in Eqs. (13.1) and (13.2) to deduce that the broad-width ρ meson must be an $\ddot{S}S$ spinor pair in a P-state ℓ = 1 rotational mode. From the 655 MeV $\ddot{S}S$ bandhead mass of the ρ, it follows that the ρ must have a rotational energy of 118 MeV in order to account for its peak energy of 773 MeV. We then went on to show that this value for the rotational energy of the ρ is in agreement with: (1) the value that we extrapolate from hadron rotational levels; (2) the value obtained by calculating the moment-of-inertia of an $\ddot{S}S$ spinor pair; and (3) the value obtained directly from the experimental ππ phase shifts. If the ρ is in fact a rotational excitation, as these results all suggest, then its broad distribution of energies corresponds to a distribution of ℓ-values, and we investigated some of the properties of these ℓ-distributions — not only in the ρ, but also in the f and η_N mesons.

The results presented in this chapter are important both because they are of general interest and because they resolve some puzzling features that have emerged from analyses of the experimental data. However, another reason for presenting these results in such detail is that the present interpretation of the ρ and ω resonances is in direct opposition to the interpretation offered by the quark model of Gell-Mann and Zweig. Gell-Mann and Zweig treat the ρ and ω mesons on an equal footing, with the difference in their widths attributed entirely to final-state phase space effects, and not to their assignments as rotational (ρ) and rotationless (ω) excitations. This of course is not the first time that the present results have conflicted with those of Gell-Mann and Zweig: we construct the low-mass resonances π, K, η, and η' entirely of spinless quanta M = 70 MeV, whereas Gell-Mann and Zweig deny that the quantum M even exists. This conflict between the present light-quark model and the Gell-Mann and Zweig quark model extends throughout the spectrum of meson and baryon resonances, but it assumes its most dramatic form in the very-low-mass states — namely, in the states μ, π, K, η, ρ, ω, and η' that have masses less than 1 GeV/c^2. It is very exciting to go charging off to the new frontier of high-mass New Particle states, as is presently occurring (see the quotation by R. Tripp at the end of Chapter 15), but the main underpinning for quark models must come from the manner in which the low-mass states are reproduced. If a theory fails right at its foundations, then any apparent successes with higher-mass and more complicated systems are probably illusory.

References for Chapter 13.

1. B. T. Feld, Models of Elementary Particles, Blaisdell, Waltham (1969), p. 362.

2. R. Y. Cusson and L. P. Staunton, Nuovo Cimento 17A, 303 (1973).

3. Particle Data Group, "Review of Particle Properties", Rev. Mod. Phys. 48, No. 2, Part II, April (1976).

4. S. Hagopian et al., Phys. Rev. Lett. 25, 1050 (1970).

5. See M. H. Mac Gregor, Phys. Rev. D9, 1259 (1974), Table XXII and the accompanying discussion.

6. See Sec. 8.4 in Chapter 8.

7. The data shown in the histogram in Fig. 13.2 are from M. Alston-Garnjost et al., Phys. Lett. 36B, 152 (1971); they were kindly supplied to the author in tabular form by S. Protopopescu.

8. S. Protopopescu et al., Phys. Rev. D7, 1279 (1973). The phase shift values shown in Figs. 13.2 - 13.5 of the present book are from Table VI of this paper, as modified in Table XIII of the same paper. The phase shift notation is δ_ℓ^I, where I and ℓ are the isotopic spin and the angular momentum, respectively.

9. M. David et al., Phys. Rev. D16, 2027 (1977).

10. B. Y. Oh et al., Phys. Rev. D1, 2494 (1970).

11. See Particle Data Group, "Review of Particle Properties", Phys. Lett. 33B, No. 1, Aug. (1970). Of the two experiments used in this reference to obtain the parameters of the f' (see p. 65), one contained only 17 events in the f' peak.

12. C. Whitehead et al., Nucl. Phys. 48B, 365 (1972).

13. See the comments about the ρ on pages 7 and 43 of Ref. 11.

14. L. J. Gutay et al., Nucl. Phys. 27B, 486 (1971); see Sec. 8.4 in Chapter 8.

15. The f'(1522) resonance discussed in Sec. 13.7 is not in evidence in Fig. 13.8, since the f' has a very small (1.4%) branching ratio into $\pi\pi$ [see A. J. Pawlicki et al., Phys. Rev. Lett. 37, 971 (1976)].

CHAPTER 14. TRANSITIONS BETWEEN SPINLESS AND SPINNING MASSES; SPIN QUANTIZATION

14.1 Introduction

The basic quark models of Gell-Mann and Zweig and of Fermi and Yang, and the variations of these models that have been set forth, all feature sets of spinning quarks; these models do not include spinless basis states. However, as we have described in detail in the preceding chapters of the present book, we find ample evidence for a set of spinless states that are formed from the basic spinless mass quantum M ≃ 70 MeV; these spinless basis states, which we denote as "cabers" (linear arrays of quanta M) occur as states that contain 1, 3, 4, or 7 subquanta M. We also find substantial evidence for a spinning basis state, the spinor S ≃ 330 MeV. Furthermore, by applying the equations of relativity† to rotating systems, we were able to demonstrate (Chapter 8) that the spinor S is a relativistically-spinning form of the caber 3, so that $S \equiv (3)_{\text{spinning}}$. This brings up the question of transitions between the caber 3 and the spinor S. These transitions do occur, in both directions, and the present chapter is written to summarize the phenomenological properties of these transitions. We do not have a comprehensive theory for the transitions 3 ↔ S, but the mass values that are obtained from the present light-quark approach to particle structure are accurate enough that, by using the characteristic mass shifts produced in these transitions, we can tell a posteriori when these transitions have taken place. Thus by writing down the basis state configurations for a series of reactions, both in producing hadron resonances and in the decays of these resonances, we can demonstrate the manner in which transitions between spinless and spinning states occur.

Once we consider the possibility of transitions between spinless and spinning basis states, we also open up another possibility — namely, the possiblity of explaining the spin values that occur for spinning states. We know empirically that many elementary particles have spins S = 1/2 ħ, and we believe that quark substates exist which also have spins S = 1/2 ħ. But up to now we have had no clues whatsoever as to why this particular spin quantization value occurs — other than to say (for example) that it appears magically from the Dirac equations (which of course is no

---

†As we described in Chapter 6, both special relativity and general relativity lead to the same equations for relativistically-rotating systems in the inertial frame of reference.

explanation, since the Dirac equations were specifically invented so as to reproduce, among other things, the spin value $S = 1/2\ \hbar$ for the electron). If we can observe spinless to spinning transitions as they take place in hadronic interactions, then maybe we can learn why the spinning states have the spin values that they do. The present microscopic approach to elementary particle structure does not provide all of the answers to this question, but it does lead to some interesting ideas. In Sec. 14.2 we discuss the question of spin quantization *per se*. Then in Sec. 14.3 we discuss a possible relationship between spins and isotopic spins. In Sec. 14.4 we give some examples of transitions between spinless and spinning basis states as they occur in the formation of hadron resonances; and in Sec. 14.5 we give some examples that involve the decays of these resonances. In Sec. 14.6 we again mention the topic of excited states of the spinor S (see Sec. 8.7 in Chapter 8). Section 14.7 concludes Part II of these studies.

14.2 Spin Quantization

The muon is known to have an observable spin $S_z = 1/2\ \hbar$, and the nucleon is also known to have an observable spin $S_z = 1/2\ \hbar$. What can we say about the spins of these particles from the point-of-view of the present representational approach to particle structure? Consider first the muon. In Chapter 7 we described a mathematical model that reproduces the properties of the muon. This model consists of a relativistically-spinning sphere with an equatorial charge distribution. If the radius of the sphere is set equal to its Compton wavelength, $R = \hbar/mc$, and if the sphere is spinning at the full relativistic limit, with its equator moving at the velocity of light, then the calculated spin angular momentum of the spinning sphere is $S = 1/2\ \hbar$. Thus the reasons that the sphere has spin $S = 1/2\ \hbar$ are that (1) it has a radius equal to its Compton wavelength, and (2) it is spinning at the full relativistic limit. This brings up another question: why is the sphere spinning at its full relativistic limit? The model again suggests at least a partial answer to this question. A sphere that is spinning at the full relativistic limit is half again as massive as it was at rest. Thus if three spinless quanta combine together so as to produce two spinning quanta (with the annihilation of one of the original three quanta), then the two spinning quanta, in order to conserve energy, must spin at the full relativistic limit. For example, we can have the decay process

$$3 \equiv 3M = 210 \text{ MeV} \rightarrow \mu^+(105 \text{ MeV}) + \mu^-(105 \text{ MeV}), \tag{14.1}$$

with the quantization of the μ^+ and μ^- spins following from the conservation of energy in the $3 \rightarrow \mu^+\mu^-$ decay process. However, this cannot be the whole story, since we also have decays of the type

$$\pi^{\pm} = 140 \text{ MeV} \rightarrow \mu^{\pm}(105 \text{ MeV}) + \nu, \tag{14.2}$$

where the muon $\mu^{\pm}$ and the neutrino ν are both spin 1/2 particles. In this example, the muon is again spinning at the full relativistic limit, but this limit is not imposed by the requirement of energy conservation, at least in any obvious manner. Thus, to summarize, we can assert that the spin quantization of the muon is due to the fact that it always spins at the full relativistic limit, and we can further suggest that the relativistic mass increase of 3/2 for spinning systems and the requirement of energy conservation combine to explain why it is that pair-produced muons [Eq. (14.1)] always spin at this relativistic limit, but we have no ready explanation for the reason that singly-produced muons [Eq. (14.2)] also spin at this same limit.

Now how do these remarks apply to the nucleon? The nucleon, in contrast to the muon, appears to be a composite structure which is formed from three spinning quarks. Thus the spin of the nucleon follows from the spins of the constituent quarks. Hence we must ask another question: why does the nucleon quark have spin $S = 1/2\,\hbar$? We can't really answer this question in detail until we discuss the geometry of the nucleon quark, which we do in Chapter 18. However, the answer is essentially the same for the nucleon quark as it was for the muon: the nucleon quark, which in the present model is the spinor $S \simeq 330$ MeV, has an axial radius equal to its Compton wavelength, and it is spinning at the full relativistic limit. As we show in Chapter 18, a number of interlocking properties of the nucleon and the spinor S combine to lead us to this conclusion. But the spinor S, in contrast to the muon, does not seem (at least from some of its properties) to have a spherical geometry; it appears essentially as a cylinder, with an axial radius of about 0.6 fermi and an overall length of 2 fermis.

14.3 A Possible Relationship Between Spins and Isotopic Spins

We now discuss an idea that is deduced directly from the phenomenology of the present light-quark model. It seems to apply to the lower-mass meson resonances, but not to all of the higher-mass resonances, so that it may or may not be correct. The idea is the following. As we stated above, the spinor S is a spinning form of the spinless caber 3. We know experimentally that spinors S are never observed singly. Thus direct transitions of the type

$$3 \leftrightarrow S \tag{14.3}$$

do not occur. What we do find are transitions of the type

$$333 \leftrightarrow SS, \tag{14.4}$$

in which spinors S are produced in pairs and decay in pairs. The basic production reaction for nucleons is the multiple transition

$$333 \cdot 333 \cdot 333 \leftrightarrow S\bar{S}S \cdot \bar{S}S\bar{S} \equiv N\bar{N}, \tag{14.5}$$

in which three pairs of spinors $S\bar{S}$ are formed and then split in half to form two spinor triplets, $N \equiv S\bar{S}S$ and $\bar{N} \equiv \bar{S}S\bar{S}$. This appears to be the only process in which a spinor pair $S\bar{S}$ is ever separated, and it leads to the stable configurations $S\bar{S}S$ and $\bar{S}S\bar{S}$, which evidently can't decay because (1) they don't allow the transition $S\bar{S} \rightarrow 333$ to take place, and (2) they don't allow the annihilation $[S\bar{S}]$ to occur.

The process we wish to discuss in the present section is the basic transition of Eq. (14.4), $\dot{3}\dot{3}\dot{3} \leftrightarrow \dot{S}\dot{S}$, which is almost isoergic (the fact that it is not quite isoergic has some important phenomenological ramifications, as we will see below). Let us consider this transition in detail. As we showed in Table 13.1 of the last chapter, the excitation $\dot{3}\dot{3}\dot{3}$ = 630 MeV is a fundamental meson excitation. Thus in a collision between two particles in which the available excitation energy exceeds 630 MeV, we can have the transition

$$\text{Kinetic Energy} \rightarrow \dot{3}\dot{3}\dot{3} \simeq 630 \text{ MeV}. \tag{14.6}$$

If we then have the subsequent transition

$$\dot{3}\dot{3}\dot{3} \rightarrow \dot{S}\dot{S}, \tag{14.4}$$

this must occur so as to conserve angular momentum. Now since $\dot{S} \equiv (\dot{3})_{\text{spinning}}$, Eq. (14.4) implies that we must actually be observing the transition

$$\dot{3}\dot{3}\dot{3} \rightarrow \dot{3}[\dot{3}]\dot{3} \rightarrow \dot{S}\dot{S}, \tag{14.7}$$

in which $[\dot{3}]$ represents the annihilation of one of the original cabers $\dot{3}$. As is illustrated in Fig. 14.1, the effect of this annihilation is to set the two remaining $\dot{3}$'s spinning in opposite directions, with local conservation of angular momentum. Thus the basic transition $\dot{3}\dot{3}\dot{3} \rightarrow \dot{S}\dot{S}$ produces a pair of spinors $\dot{S}\dot{S}$ that have total spin S = 0. Now if the initial state which leads to the transition $\dot{3}\dot{3}\dot{3} \rightarrow \dot{S}\dot{S}$ has angular momentum J = 0, then the final state will also have angular momentum J = 0. However, if the initial state has J = 1, then (and here is the point of the present discussion) we have two main possibilities[†]: <u>(1)</u> the $\dot{S}\dot{S}$ spinor pair can appear as an overall

---

[†]A third possibility is a J = 1, ℓ = 1, S = 1 configuration, which we will not consider here. Also, with a J = 0 initial state, the configuration J = 0, ℓ = 1, S = 1 can be produced; this is an excitation that may occur in the ε-type resonances.

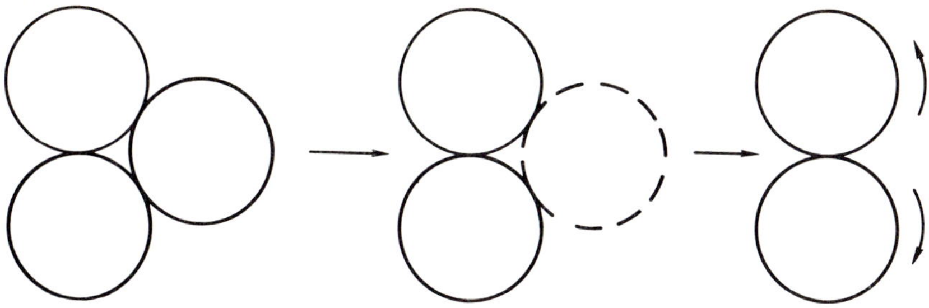

Figure 14.1. A schematic diagram of the basis-state transition $333 \rightarrow 3[3]3 \rightarrow \dot{S}\dot{S}$, in which a spinless basis state $3 \equiv 3M = 210$ MeV annihilates and sets the other two 3's spinning in opposite directions in the form of an $\dot{S}\dot{S}$ spinor pair, where $\dot{S} = (3)_{\text{spinning}}$.

rotational excitation, with $J = 1$, $\ell = 1$, $S = 0$ (as in the ρ meson); (2) the $\dot{S}\dot{S}$ pair can undergo a spin-flip process that reverses the spin of one of the spinors $\dot{S}$, so that the spinor pair $\dot{S}\dot{S}$ then has spin $S = 1$, and it appears as the rotationless S-state excitation $J = 1$, $\ell = 0$, $S = 1$ (as in the ω meson). Now the essential point here is that for the second possibility to occur, there must be a spin 1 "virtual particle" which couples the $J = 1$ angular momentum of the initial state to the $\dot{S}\dot{S}$ pair to produce the spinor-flipping transition $\Delta S = 1$, and the logical condidate for this "virtual particle" is the spin 1 photon. But for the photon to couple to the $\dot{S}\dot{S}$ spinor pair, it seems natural to require that the two spinors be in the charge states $\dot{S}^+$ and $\dot{S}^-$; that is, the $\dot{S}\dot{S}$ spinor pair must be in an isotopic spin $I = 0$ configuration. Hence we arrive at the following phenomenological conclusion about meson resonance formation:

The basic transition $\dot{3}\dot{3}\dot{3} \rightarrow \dot{S}\dot{S}$ occurs with spin $S = 0$ (see Fig. 14.1); (14.8)

If the $\dot{S}\dot{S}$ spinor pair is in an isotopic spin $I = 0$ mode ($\dot{S}^+\dot{S}^-$), then a simultaneous spin-flip can occur which gives spin $S = 1$ for the $\dot{S}\dot{S}$ pair. (14.9)

If we make the further assumption that the reaction (14.9) always occurs when it is allowed to occur, we obtain the following rule:

$\ddot{S}S$ spinor pairs have spin and isotopic spin (S = 1, I = 0) or else (S = 0, I = 1). (14.10)

As an empirical test of Rule (14.10), we show in Table 14.1 a series of narrow-width meson resonances (Fig. 5.3) whose spins and isotopic spins are known, and also the narrow-width $A_{1.5}$ meson† and two broad-width low-mass meson resonances — the I = 1 ρ(773)Γ152 meson and the "strange" I = 1/2 K*(892)Γ50 meson. As can be seen in Table 14.1, all of these resonances except the $A_{1.5}$ follow Rule (14.10),

Table 14.1. A series of meson and kaon resonances whose spin S and isotopic spin I are known. The resonances whose S and I values obey Eq. (14.11) appear as easily-observed states, whereas the resonance which violates this rule does not.

Meson resonance	I	S	ℓ	Follows Eq. (14.10)	Easily observed
ρ	1	0	1	Yes	Yes
ω	0	1	0	Yes	Yes
K*	1/2	0	1	Yes	Yes
φ	0	1	0	Yes	Yes
$A_{1.5}$	1	1	0	No	No
D	0	1	0	Yes	Yes
E	0	1	0	Yes	Yes

which may explain why the $A_{1.5}$ is not well-established or particularly easy to observe. As we will see in Chapter 16, the K* resonance shown in Table 14.1 is an $\dot{S}\dot{S}\pi$ bandhead in an ℓ = 1 rotational mode; thus, from the example of the K*, we can make an extension of Rule (14.10) and obtain the final result of this section:

$\ddot{S}S$ spinor pairs have spin and isotopic spin (S = 1, I = 0) or else (S = 0, I = 1/2 or 1). (14.11)

†The $A_{1.5}$ meson state was not included in Fig. 5.3 both because it is a controversial resonance and because it does not appear to serve as a bandhead for an overlying meson rotational band (see the discussion of this state in Appendix B).

Rule (14.11) is not offered here as a well-determined empirical observation, but rather as an example of the type of deduction that can be obtained by making a detailed microscopic analysis of meson production mechanisms. We continue with this analysis by showing some typical examples of $\dot{3}\dot{3}\dot{3} \to \dot{S}\dot{S}$ transitions as they occur in meson production reactions (Sec. 14.4) and in meson decays (Sec. 14.5). A basic clue to the occurrence of these transitions is provided by the fact that the spinless triplet $\dot{3}\dot{3}\dot{3}$ and the spinor pair $\dot{S}\dot{S}$ have a characteristic energy difference of about 25 MeV.

14.4 Examples of $\dot{3}\dot{3}\dot{3} \to \dot{S}\dot{S}$ Transitions in Meson Production Processes

Transitions $\dot{3}\dot{3}\dot{3} \to \dot{S}\dot{S}$ occur in a variety of meson production reactions. In these transitions, the spinless $\dot{3}\dot{3}\dot{3}$ excitation unit typically has a mass of 9 x 70 = 630 MeV, and the $\dot{S}\dot{S}$ spinor pair typically has a mass of 655 MeV. Thus the transition

$$\dot{3}\dot{3}\dot{3}(630\ \text{MeV}) + 25\ \text{MeV K.E.} \to \dot{S}\dot{S}(655\ \text{MeV}) \tag{14.12}$$

occurs with an energy shift of 25 MeV. Since these mixed basis states bind together with essentially zero binding energy, and since the mass values for the narrow-width meson resonances are given to an accuracy that is in general much better than 25 MeV, we can work backwards from the resonance mass values and deduce the number of $\dot{3}\dot{3}\dot{3} \to \dot{S}\dot{S}$ transitions that have taken place. The best example of this procedure in the low-mass meson resonances is afforded by the D meson and the resonance that produces the small interference dip in the A_2 meson†:

$$\begin{aligned} &\dot{3}\dot{3}\dot{3} \cdot \dot{3}\dot{3}\dot{3} \to \dot{S}\dot{S} \cdot \dot{3}\dot{3}\dot{3} = 655 + 630 = 1285\ \text{MeV},\ J^P = 1^+,\ D(1285\pm10\ \text{MeV}); \\ &\dot{3}\dot{3}\dot{3} \cdot \dot{3}\dot{3}\dot{3} \to \dot{S}\dot{S}\dot{S}\dot{S} = 655 + 655 = 1310\ \text{MeV},\ J^P = 2^+,\ A_2\text{dip}(1310\pm3\ \text{MeV}). \end{aligned} \tag{14.13}$$

These same mass calculations also apply in a very striking manner to the high-mass New Particle resonances. If we start with the basic $\bar{N}N$ excitation (Eq. 14.5),

$$(\dot{3}\dot{3}\dot{3})^3 \to \bar{S}S\bar{S} \cdot S\bar{S}S \equiv \bar{N}N = 1795\ \text{MeV}^{\dagger\dagger} \leftrightarrow \bar{p}n(1794.5\pm1.4\ \text{MeV}), \tag{14.14}$$

---

†See Appendices B and D for references to the data values given in Secs. 14.4 and 14.5.

††See Eq. (8.9) in Chapter 8.

we have

$$(\dot{3}\dot{3}\dot{3})^5 \to \bar{N}N \cdot \dot{S}\dot{S}\dot{S}\dot{S} = 1795 + 1310 = 3105 \text{ MeV} \leftrightarrow J/\psi(3098 \pm 3 \text{ MeV}), \qquad (14.15)$$

where all of the $\dot{3}\dot{3}\dot{3}$ excitation units are converted into spinors, and we have

$$(\dot{3}\dot{3}\dot{3})^6 \to \bar{N}N \cdot (\dot{3}\dot{3}\dot{3})^3 = 1795 + 3 \times 630 = 3685 \text{ MeV} \leftrightarrow \psi'(3684 \pm 4 \text{ MeV}), \quad (14.16)$$

where only half of the $\dot{3}\dot{3}\dot{3}$ excitation units are converted into spinors. The J/ψ and ψ' basis state configurations shown in Eqs. (14.15) and (14.16) were deduced from the experimental mass values; however, as we show in Sec. 14.5 and in Chapter 17, these basis-state assignments are verified by the decay modes of the resonances.

A systematic procedure for cataloging $\dot{3}\dot{3}\dot{3} \to \dot{S}\dot{S}$ transitions is developed in Part III. In the remainder of the present section, we merely give a few more examples which illustrate the general characteristics of these transitions. For the low-mass narrow-width meson resonances we have the production reactions

$$\begin{aligned}
\dot{3}\dot{3}\dot{3}\pi &\to \dot{S}\dot{S}\pi = \omega \to \pi\pi\pi \quad (J=1,\ell=0,S=1,I=0,P=-),\\
\dot{3} \cdot \dot{3}\dot{3}\dot{3}\pi &\to \dot{3}\dot{S}\dot{S}\pi = \phi \to \pi\pi\pi \quad (J=1,\ell=0,S=1,I=0,P=-),\\
\dot{3}\dot{3} \cdot \dot{3}\dot{3}\dot{3}\pi &\to \dot{3}\dot{3}\dot{S}\dot{S}\pi = A_{1.5} \to \pi\pi\pi \quad (J=1,\ell=0,S=1,I=1,P=-),\\
7 \cdot \dot{3}\dot{3}\dot{3} &\to K\dot{S}\dot{S}\pi = K\pi\pi \to K\pi\pi \quad (J=1,\ell=0,S=1,I=1/2,P=+).
\end{aligned} \qquad (14.17)$$

We also have the low-mass broad-width production reactions

$$\begin{aligned}
\dot{3}\dot{3}\dot{3} &\to \dot{S}\dot{S} = \rho \to \pi\pi \quad (J=1,\ell=1,S=0,I=1,P=-),\\
\dot{4}7 &\to \dot{S}\dot{S}\pi = K^* \to K\pi \quad (J=1,\ell=1,S=0,I=1/2,P=-)^{\dagger}.
\end{aligned} \qquad (14.18)$$

And we have the intermediate-mass narrow-width production reactions

$$\begin{aligned}
\dot{3}\dot{3}\dot{3} \cdot \dot{3}\dot{3}\dot{3} &\to \dot{S}\dot{S} \cdot \dot{3}\dot{3}\dot{3} = D \to KK\pi \quad (J=1,\ell=0,S=1,I=0,P=+),\\
\dot{3}\dot{3}\dot{3} \cdot \dot{3}\dot{3}\dot{3}\pi &\to \dot{S}\dot{S} \cdot \dot{3}\dot{3}\dot{3}\pi = E \to K^*K \quad (J=1,\ell=0,S=1,I=0,P=+),\\
\dot{3}\dot{3}\dot{3}\pi \cdot \dot{3}\dot{3}\dot{3}\pi &\to \dot{S}\dot{S}\pi \cdot \dot{3}\dot{3}\dot{3}\pi = F_1 \to K^*K \quad (J=1,\ell=0,S=1,I=1,P=+).
\end{aligned} \qquad (14.19)$$

---

† It isn't clear just how the strangeness quantum number $S = \pm 1$ is carried in the K^*, but the overall basis-state assignment is well-determined (Chapter 16).

From these examples, it can be seen that the transition $\dot{3}\dot{3}\dot{3} \to \dot{S}\dot{S}$ plays a dominant role in all but the lowest-mass meson production reactions. In Chapter 17 we trace the manner in which this transition systematically occurs in the high-mass New Particle resonances. We now go on in Sec. 14.5 to consider this transition from the standpoint of meson decay modes.

14.5 Examples of $\dot{S}\dot{S} \to \dot{3}\dot{3}\dot{3}$ Transitions in Meson Decays

In Sec. 14.4 we gave a number of examples where the light-quark basis-state transition $\dot{3}\dot{3}\dot{3} \to \dot{S}\dot{S}$ occurs in meson production reactions. In the present section we examine the decays of these resonances. These decays usually involve the inverse transition $\dot{S}\dot{S} \to \dot{3}\dot{3}\dot{3}$, but there are some interesting examples where the transition $\dot{3}\dot{3}\dot{3} \to \dot{S}\dot{S}$ occurs.† As the reader may have noticed, we are concentrating in the present chapter on the production and decay reactions for <u>meson</u> resonances only. This is because <u>baryon</u> resonances are produced in most cases by adding spinless excitation quanta to an existing $S\bar{S}S$ spinor core, so that the basis state transition $\dot{3}\dot{3}\dot{3} \to \dot{S}\dot{S}$ is not observed in these baryon decays.

Meson decays are produced in almost all cases by the annihilation of some of the internal subquanta M = 70 MeV. These annihilations can occur in many different ways, as is obvious from the wide range of final states that are observed in meson decays. If spinor pairs $\dot{S}\dot{S}$ are present in a resonance, the annihilation process in the decay of the resonance may or may not be accompanied by transitions $\dot{S}\dot{S} \to \dot{3}\dot{3}\dot{3}$. In the present discussion, we do not attempt any systematic study of meson decays; instead, we give a number of examples which illustrate the variety of meson decay modes that are observed experimentally.

We start with the lowest-mass meson states. The decay mode of the charged pion was shown in Eq. (14.2); this decay involves the annihilation of a single subquantum M. The neutral pion has the decay mode $\pi^0 \to \gamma\gamma$, which involves the annihilation of both subquanta M. The K meson ($K \equiv 7$) and the η meson ($\eta \equiv 4\bar{4}$) each have a variety of different decay modes, which in turn correspond to a variety of annihilations of subquanta M.

The lightest meson states which involve $\dot{S}\dot{S}$ spinor pairs are the ρ and ω mesons. For both of these mesons, the decay process consists of the transition $\dot{S}\dot{S} \to \dot{3}\dot{3}\dot{3}$

---

†In the meson decays, some of the final states may involve unmixed cabers 3 and $\bar{3}$ instead of mixed cabers $\dot{3}$; however, we will for simplicity denote all of these as $\dot{3}$ states. This remark also applies to some of the excitations shown in Sec. 14.4.

followed by the annihilation $[\dot{3}\pi]$:†

$$\rho \equiv \dot{S}\dot{S}_{\ell=1} \to \dot{3}\dot{3}\dot{3} = \dot{3}\pi\pi\pi \to [\dot{3}\pi]\pi\pi,$$
$$\omega \equiv \dot{S}\dot{S}\pi \to \dot{3}\dot{3}\dot{3}\pi = \dot{3}\pi\pi\pi\pi \to [\dot{3}\pi]\pi\pi\pi. \tag{14.20}$$

This same $[\dot{3}\pi]$ annihilation process carries over to the M and η' mesons (see Eqs. (5.8) - (5.11) in Chapter 5) in a very interesting manner. The M meson is produced only in reactions of the type A = B → A + B + M, and it consequently has perfect internal particle-antiparticle symmetry. The M decays by the one-step process

$$M \equiv 7\bar{7} = \dot{3}\dot{3}\dot{3}\dot{3}\pi = \dot{3}\dot{3}\pi\pi\pi\pi \to [\dot{3}\pi][\dot{3}\pi]\pi\pi\gamma, \tag{14.21}$$

where 95% of the final-state $\pi\pi$ pairs <u>are not</u> in the ρ band. The η' meson, on the other hand, is produced in reactions of the type $\bar{K} + p \to \Lambda + \eta'$, which have an asymmetrical character, and the η' has the two-step decay mode

$$\eta' \equiv 7\bar{7} = \dot{3}\dot{3}\dot{3}\dot{3}\pi \to \dot{S}\dot{S}_{\ell=1}[\dot{3}\pi]\gamma, \tag{14.22a}$$

$$\dot{S}\dot{S}_{\ell=1} \to \dot{3}\dot{3}\dot{3} = \dot{3}\pi\pi\pi \to [\dot{3}\pi]\pi\pi, \tag{14.22b}$$

where 95% of the final-state $\pi\pi$ pairs <u>are</u> in the ρ band. Experimentally, the decay modes (14.21) and (14.22) are cleanly separated, and these decay modes indicate in a very convincing manner that the M and η' resonances are distinct entities. In addition to the differences in their production and decay modes, the M(953) and $\eta'(958)$ mesons also have a 5 MeV difference in their mass values. We emphasize these differences between the M and η' resonances, because the M overfills the $J^P = 0^-$ SU(3) meson nonet in the same manner that the η_N overfills the $J^P = 2^+$ meson nonet (see Sec. 13.7 in Chapter 13). The decay of the η' meson is also of interest in that it illustrates the occurrence of a $\dot{3}\dot{3}\dot{3} \to \dot{S}\dot{S}$ transition in a decay mode rather than in the usual production mode.

When we move on to the higher-mass meson resonances, we find an interesting decay mode in the $\phi(1020)$ meson:

$$\phi \equiv \dot{3}\dot{S}\dot{S}\pi \to \dot{3}\dot{3}\dot{3}\dot{3}\pi = \dot{3}\dot{3}\pi\pi\pi\pi \to 3\pi\pi + 3\pi\pi = K + K. \tag{14.23}$$

(See the discussion of this $\phi \to KK$ decay mode in Ref. 7 of Chapter 10.) This ϕ decay mode is triggered by the transition $\dot{S}\dot{S} \to \dot{3}\dot{3}\dot{3}$, and it is an example of a strong meson

---

†We write $\dot{3}\dot{3} = \pi\pi\pi$ in some cases to show the decay modes in detail.

decay where no internal subquanta M are annihilated. The New Particle decay mode $D^* \to D + \pi$ is another example of such a decay (see Fig. 5.4). The ϕ meson has another decay mode that is similar to a decay mode in the E(1416) meson:

$$\phi \equiv \dot{3}\dot{S}\dot{S}\pi \to \dot{3}\dot{3}\pi\pi\pi\pi \to [\dot{3}\dot{3}\pi]\pi\pi\pi;$$

$$E \equiv \dot{3}\dot{3}\dot{3}\dot{S}\dot{S}\pi \to \dot{3}\dot{3}\dot{3}\dot{3}\pi\pi\pi\pi \to [\dot{3}\dot{3}\pi]\eta\pi\pi. \tag{14.24}$$

Two other decay modes of the E meson are

$$E \equiv \dot{3}\dot{3}\dot{3}\dot{S}\dot{S}\pi = \dot{3}\pi\pi\pi\dot{S}\dot{S}\pi \to 3\pi\pi[\pi]SS\pi_{\ell=1} = KK^*,$$

$$E \equiv \dot{3}\dot{3}\dot{3}\dot{S}\dot{S}\pi \to \dot{3}\dot{3}\dot{3}\dot{3}\dot{3}\dot{3}\pi = \dot{3}\dot{3}\dot{3}\dot{3}\pi\pi\pi\pi \to [\pi\pi]KK\pi. \tag{14.25}$$

The D(1285) meson has the decay modes

$$D \equiv \dot{3}\dot{3}\dot{3}\dot{S}\dot{S} \to \dot{3}\dot{3}\dot{3}\dot{3}\dot{3}\dot{3} = \dot{3}\dot{3}\dot{3}\dot{3}\pi\pi\pi \to [\pi]KK\pi,$$

$$\to [\dot{3}\dot{3}]\eta\pi\pi, \tag{14.26}$$

$$D \equiv \dot{3}\dot{3}\dot{3}\dot{S}\dot{S} = \dot{S}\dot{S}\dot{3}\pi\pi\pi \to \dot{S}\dot{S}_{\ell=1}[\dot{3}\pi]\pi\pi = \rho\pi\pi;$$

and the $A_{1.5}$(1197) meson has the decay mode

$$A_{1.5} \equiv \dot{3}\dot{3}\dot{S}\dot{S}\pi \to \dot{S}\dot{S}_{\ell=1}[\dot{3}\dot{3}]\pi = \rho\pi. \tag{14.27}$$

As these examples illustrate, many different combination of subquanta M can be annihilated in meson decays, and these annihilations are often, but not always, accompanied by transitions $\dot{S}\dot{S} \to \dot{3}\dot{3}\dot{3}$ or $\dot{3}\dot{3}\dot{3} \to \dot{S}\dot{S}$.

For the last examples of meson decays, we briefly consider the high-mass New Particle resonances. The $J/\psi \equiv \bar{N}N\dot{S}\dot{S}\dot{S}\dot{S}$ resonance has many decay modes to low-mass final states, including the leptonic decays $J/\psi \to \mu^+\mu^-$ and $J/\psi \to e^+e^-$ that involve the annihilation or transformation of all of the hadronic subquanta M. The $\psi' \equiv \bar{N}N(\dot{3}\dot{3}\dot{3})^3$ resonance also has decays to low-mass final states. However, the most interesting decays from the present point-of-view are the ψ' decays to high-mass final states. These are decays in which only a few of the subquanta M are annihilated, and the levels involved have narrow widths, so that the mass intervals can be accurately determined experimentally. These high-mass ψ' decays thus offer us a unique opportunity to check the workings of the present light-quark model, which was formulated long before the J/ψ and ψ' New Particles were discovered.

The $\psi'(3684)$ meson has radiative decays into the three levels $\chi_1(3414)$, $\chi_2(3503)$, and $\chi_3(3551)$[1]. The energy differences $\psi' - \chi_3 = 133$ MeV and $\psi' - \chi_1 = 270$ MeV indicate clearly that these decays involve the annihilations $[\pi]$ and $[\pi\pi]$, respectively [note the similar annihilation modes in the E meson decays of Eq. (14.25)]. Thus we have

$$\psi' \equiv \bar{N}N(\dot{3}\dot{3}\dot{3})^3 = \bar{N}N(\dot{3}\dot{3}\dot{3})^2\dot{3}\pi\pi\pi \rightarrow \bar{N}N(\dot{3}\dot{3}\dot{3})^2\dot{3}\pi[\pi\pi] \rightarrow \gamma\chi_1, \tag{14.28a}$$

$$\rightarrow \bar{N}N(\dot{3}\dot{3}\dot{3})^2\dot{3}\pi\pi[\pi] \rightarrow \gamma\chi_3, \tag{14.28b}$$

in which no $\dot{3}\dot{3}\dot{3} \rightarrow \dot{S}\dot{S}$ transitions occur. However, the states χ_1 and χ_3 have subsequent decays down to the J/ψ level, so that these decay modes are

$$\chi_1 \equiv \bar{N}N(\dot{3}\dot{3}\dot{3})^2\dot{3}\pi \rightarrow NN\dot{S}\dot{S}\dot{S}\dot{S}[\dot{3}\pi] \rightarrow \gamma J/\psi, \tag{14.29a}$$

$$\chi_3 \equiv \bar{N}N(\dot{3}\dot{3}\dot{3})^2\dot{3}\pi\pi \rightarrow \bar{N}N\dot{S}\dot{S}\dot{S}\dot{S}[\dot{3}\pi\pi] \rightarrow \gamma J/\psi, \tag{14.29b}$$

in which the transition $(\dot{3}\dot{3}\dot{3})^2 \rightarrow \dot{S}\dot{S}\dot{S}\dot{S}$ takes place. In contrast to the decay modes of (14.28), the energy difference of 181 MeV in the $\psi' \rightarrow \chi_2(3503)$ decay indicates that the transition $(\dot{3}\dot{3}\dot{3})^2 \rightarrow \dot{S}\dot{S}\dot{S}\dot{S}$ has accompanied a $[\dot{3}]$ annihilation; this assumption is borne out by the fact that the $\chi_2 - J/\psi$ energy difference is 405 MeV, which is just the sum of the $\psi' - \chi_1$ and $\psi' - \chi_3$ energy differences, and which is a characteristic spinless excitation interval (see Fig. 5.2). Thus we have the sequential decays

$$\psi' \equiv \bar{N}N(\dot{3}\dot{3}\dot{3})^3 = \bar{N}N(\dot{3}\dot{3}\dot{3})^2\dot{3}\dot{3}\dot{3} \rightarrow \bar{N}N\dot{S}\dot{S}\dot{S}\dot{S}\dot{3}\dot{3}[\dot{3}] \rightarrow \gamma\chi_2, \tag{14.30a}$$

$$\chi_2 \equiv \bar{N}N\dot{S}\dot{S}\dot{S}\dot{S}\dot{3}\dot{3} \rightarrow \bar{N}N\dot{S}\dot{S}\dot{S}\dot{S}[\dot{3}\dot{3}] \rightarrow \gamma J/\psi. \tag{14.30b}$$

The ψ' radiative decays of Eqs. (14.28) - (14.30) are illustrated in Fig. 14.2. The ψ' also has two interesting non-radiative decays:

$$\psi' \equiv \bar{N}N(\dot{3}\dot{3}\dot{3})^3 = \bar{N}N(\dot{3}\dot{3}\dot{3})^2\dot{3}\dot{3}MMM \rightarrow \bar{N}N\dot{S}\dot{S}\dot{S}\dot{S}\dot{3}\dot{3}\pi[M] = J/\psi + \eta, \tag{14.31}$$

in which a single quantum $M = 70$ MeV is annihilated; and

$$\psi' \equiv \bar{N}N(\dot{3}\dot{3}\dot{3})^3 = \bar{N}N(\dot{3}\dot{3}\dot{3})^2\dot{3}\pi\pi\pi \rightarrow \bar{N}N\dot{S}\dot{S}\dot{S}\dot{S}[\dot{3}\pi]\pi\pi = J/\psi + \pi\pi, \tag{14.32}$$

in which the annihilation $[\dot{3}\pi]$ occurs. The decay mode of Eq. (14.31) is unusual in that it has a large branching ratio in spite of its very small energy release; and the decay mode of Eq. (14.32) is interesting in that the two pions are blown off in opposite directions, with the J/ψ remaining essentially at rest, which has the effect

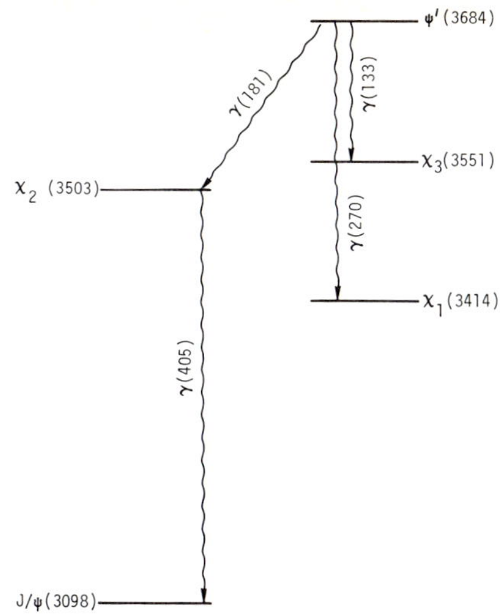

Figure 14.2. The $\chi_1(3414)$, $\chi_2(3503)$, and $\chi_3(3551)$ radiative decay levels of the $\psi'(3684)$ New Particle state. The $\gamma(133)$ and $\gamma(270)$ decays correspond to the annihilation of two and four subquanta M, respectively [Eqs. (14.28b) and (14.28a)]. The $\gamma(181)$ decay corresponds to the annihilation of three subquanta M accompanied by a spinor transition [Eq. (14.30a)], and the $\gamma(405)$ decay corresponds to the annihilation of six subquanta M [Eq. (14.30b)]. Note the set of related equations $(\psi' - \chi_3) + (\psi' - \chi_1) = (\chi_2 - J/\psi)$; [2M] + [4M] = [6M]; 133 MeV + 270 MeV $\simeq$ 405 MeV.

that the final-state $\pi\pi$ energy distribution is skewed in the direction of high mass values (these results are discussed in Chapter 17).

If we study the decay modes of Eqs. (14.31), (14.28b), (14.30a), (14.28a), (14.29a), (14.30b), and (14.29b), we observe an interesting result. These decays involve the annihilation of 1, 2, 3, 4, 5, 6, and 7 subquanta M, respectively, and in the [1], [3], [5], and [7] annihilations the transition $(\dot{3}\dot{3}\dot{3})^2 \to \dot{S}\dot{S}\dot{S}\dot{S}$ occurs, whereas in the [2], [4], and [6] annihilations it does not. Thus we see a pattern emerging from the systematics of these high-mass decays. This line of thought is developed in more detail in Chapter 17. The main point we wish to bring out here is that the decays of the high-mass New Particle resonances occur in the same general manner as do the decays of the low-mass meson resonances.

From the viewpoint of the present light-quark model, meson resonances are produced by first transforming kinetic energy into spinless hadronic excitation quanta, and then converting some of these spinless quanta into spinor pairs via transitions of the type $\dot{3}\dot{3}\dot{3} \rightarrow \dot{S}\dot{S}$; the decays of these meson resonances follow the same path, but in a more-or-less reversed order, and with sequential decays of the hadronic basis states. The annihilations that we have shown in these decay modes represent the sudden conversion of hadronic matter back into kinetic energy, and the important point about these annihilations is the quantized manner in which they occur. As we pointed out in Chapter 10, it is these annihilations which trigger the decays of the elementary particle states; and the onset of these annihilations is determined, at least for the long-lived "particles" that have lifetimes $\tau > 10^{-21}$ sec, by electromagnetic considerations (as is evidenced by the scaling of the "particle" lifetimes in powers of $\alpha = e^2/\hbar c$). In the present chapter we have tried to illustrate the many types of hadronic production and decay reactions that are observed experimentally. In spite of the complexity of these reactions, and the richness of the observed spectrum of elementary particle resonances, they are all based on some very simple transitions between a small and well-defined set of light-quark basis states.

14.6 Excited States of the Spinor S

In Sec. 8.7 of Chapter 8 we briefly discussed the possibility that the spinor S may occur in higher excited states, as seems to be required by the small moments of inertia for the higher-mass hadron rotational bands. The only comment we wish to add here is the observation that if we can form the transition $\dot{3}\dot{3}\dot{3} \rightarrow \dot{S}\dot{S}$, with the absorption (annihilation) of one spinless caber $\dot{3}$, then maybe we can form the transition $\dot{3}\dot{3}\dot{3}\dot{3} \rightarrow (\dot{S}\dot{S})'$, with the absorption of two (or more) spinless cabers. This suggests the additional possibility that we can form the transition $\dot{3} + S\bar{S}S \rightarrow (S\bar{S}S)'$, where the (invariant) spinor triplet $S\bar{S}S$ is excited to a higher-energy mode. This type of excitation is phenomenologically analogous to the vibrational excitations of nuclear physics, although the actual mechanism may be completely different. This topic is mentioned again in Chapters 15 and 16, where we map the experimental baryon and meson rotational bands, and where we obtain numerical information about the moments of inertia (and hence the geometries) of these hadron rotational bands. It is also discussed in Chapter 18, where we assemble the information that we have obtained about the geometry of the elementary particle.

14.7 Conclusion

In Part I of the present book we started with the assumptions that quarks exist and that they have small binding energies, and we then used the experimental data to deduce the mass values for a set of light-quark basis states. In Part II we went on to investigate the angular momentum properties of these basis states; we found many similarities between hadron resonances and atomic nuclei, but also some essential differences. The main task still remaining before us is to determine the patterns of the observed resonance excitations. Clearly, many more combinations of spinless quanta M = 70 MeV and spinors S = 330 MeV can be formed than are actually observed. Thus we must start with the observed specturm of elementary particle resonances, and then work backwards to determine which basis-state combinations are required in order to reproduce these resonances. This is the task which occupies Part III of the book. This procedure is in agreement with our general policy in this book of using the experimental elementary particle data as a guide in the development of a theory for these particles. In the preceding chapters we have of course already discussed many of the elementary particle resonances. But in Part III we consider the totality of the observed resonances (insofar as we can at this time define this totality), and we confront the light-quark model with the task of reproducing all of these resonances — and only these resonances.

References for Chapter 14.

1. The mass values for the χ_1, χ_2, and χ_3 radiative decay levels of the ψ' are taken from W. Tanenbaum et al., Phys. Rev. D (March 1, 1978).

PART III. EXCITATION MECHANISMS: MAPPING THE ELEMENTARY PARTICLE SPECTRUM

INTRODUCTION

In part I of the present book we studied the mass quantization of the spectrum of elementary particles. From these studies, we deduced the existence of a fundamental set of light-quark basis states — 1,3,4,7,S — which are held together with small binding energies, and which are formed as spinless (1,3,4,7) and spinning (S) combinations of the fundamental mass quantum $1 \equiv M = 70$ MeV. The properties of these basis states are summarized in Appendix C. The lowest-mass spinless meson resonances are composed entirely of spinless quanta M, and the higher-mass meson resonances are formed as combinations of spinless quanta with spinor pairs $\ddot{S}S$ or with nucleon-antinucleon pairs $N\bar{N}$. The lowest-mass baryon states are the proton and neutron, which are spinor triplets $S\bar{S}S$; and baryon excitations are formed by adding spinless excitation quanta to these (invariant) spinor triplets.

In Part II of the book we studied the angular momentum properties of the hadron resonances, including the decomposition of the total angular momentum $\vec{J}$ of a resonant state into its spin $\vec{S}$ and orbital $\vec{\ell}$ components. Due to the small masses of these hadron resonances, orbital rotational energies are very large. By studying a few specific examples, we showed that hadron orbital rotations are quite similar to orbital rotations in atomic nuclei, and we showed in particular that the energies of the hadron rotational levels follow an $\ell(\ell+1)$ interval rule. However, there is one major difference between hadron rotations and the rotations which occur in nuclear physics: the angular momentum barriers that are so important for the low-energy nuclear decays are unimportant for the very energetic hadronic decays. As a consequence of this result, hadron rotational levels have broad widths and short lifetimes, whereas nuclear rotational levels have narrow widths and long lifetimes. Thus the rotationless S-state levels in the hadron resonances characteristically have narrower widths than do the higher-spin rotational levels, which is just the opposite of the situation that occurs in nuclear physics.

In Part III of the book we come to the heart of the present studies — namely, the problem of accounting for the complete spectrum of observed resonances. Up to this point we have examined the properties of some individual resonances, and we have analyzed the systematics of selected sets of resonances. But now we must attempt to combine <u>all</u> of the observed states together into one overall scheme. Once this is accomplished, we can then judge the completeness and accuracy of the results — both from the standpoint of experiment and from that of theory.

When we actually proceed to map the observed spectrum of hadron resonances, we find that they divide naturally into three groups: (1) the baryon and hyperon resonances; (2) the low-mass meson and kaon resonances (which are mostly below the $N\bar{N}$ threshold); (3) the high-mass New Particle meson and kaon resonances (which are all above the $N\bar{N}$ threshold). Historically, the present light-quark model was developed entirely from the systematics of the first two of these three groups. Thus the fact that the third group — the New Particle resonances — can be accurately reproduced by this same model should be regarded as a major success of the model. In the next three chapters we successively examine these three groups of resonances. Since the systematics of these resonances tell their own story in a rather clear-cut manner, we can keep the discussion of these results to a minimum, and concentrate instead on graphic displays of the data. In judging the results that are displayed in these three chapters, the reader should keep in mind that we strive always for simplicity, overall unity, and a connection with known principles in other areas of physics.

As a final comment in this Introduction to Part III, we emphasize again that the conclusions we reach here are not the generally-accepted explanations for the classification of these elementary particle resonances. We find that the resonances which occur in elementary particle physics are in many respects similar to the resonances which occur in nuclear physics, but with modifications that arise mainly from the quantization of the excitation processes, the smallness of the elementary particle masses, and the large energy release in the decays of these particles. However, the conventional explanation for the spectra of these elementary particle families is in terms of SU(3) symmetries, which have nothing to do with nuclear physics, nor with atomic or molecular physics. SU(3) symmetries have led to some important elementary particle successes, and the manner in which these symmetries correlate isotopic spins and strangeness quantum numbers is very striking. However, the SU(3) approach has also led, at least in the opinion of the present author, to some equally striking failures. For example, we mentioned at the end of Chapter 9 some of the difficulties that are encountered by SU(3) in the calculation of hadron mass values. Also, as we point out in Sec. 15.8 of Chapter 15, SU(3) symmetries do a rather poor job of reproducing the observed spectrum of baryon and hyperon resonances. And, as is mentioned in Chapter 16, there are some problems for SU(3) in the mapping of the meson resonances. Furthermore, as we discuss in Chapter 19 (see Table 19.6), calculations of baryon magnetic moments within the framework of SU(6) do not match the experimental data. Thus, in spite of its wide acceptance at the present time, the SU(3) approach to elementary particle structure is, in the opinion of the author, very far from what would seem to constitute an acceptable theory for these particles. The New Particle resonances that are discussed in Chapter 17 are widely interpreted as heralding the existence of "charm", where the charmed quark has beed added as the fourth fundamental quark in the quark model of Gell-Mann and Zweig. However, the validity of the concept of charm depends on the validity of the Gell-Mann and Zweig quark model, which in turn is inexorably bound up with the validity of SU(3). Hence any problems that exist for SU(3) are also problems that exist for the charm model, and the totality of these theories must be considered altogether. A building, or a physical theory, is only as strong as its foundations.

We now proceed to the mapping of the baryon and hyperon resonances; the low-mass meson and kaon resonances; and the high-mass New Particle meson and kaon resonances.

CHAPTER 15. MAPPING THE BARYON RESONANCES

15.1 The Experimental Data

> "Apart from the properties of the stable baryons, we have, in my judgment, very little accurate-enough experimental information which is direct in the sense that it is model independent. The extraction of resonance parameters from phase shifts, and the phase shifts themselves, involve theoretical assumptions. Thus most of the so-called experimental information is an unholy or non-kosher mixture of implicit theoretical considerations and experiments, making it difficult to judge the validity of a particular theoretical model."
>
> K. C. Wali, invited talk on "Current Status of Baryon Spectroscopy" given at the ZGS Summer Symposium on New Directions in Hadron Spectroscopy, Argonne National Laboratory, Argonne, Illinois, July (1975)

The experimental baryon and hyperon resonances that have been identified to date are summarized in Table 15.1. In this table we have grouped the well-established resonances at the top, the less-well-established resonances in the middle, and some reported narrow-width resonances at the bottom (the significance of the bottom group of resonances will become apparent in Sec. 15.3). Most of the information about the baryon and hyperon resonances of Table 15.1 comes from phase shift analyses of πN and $\bar{K}N$ scattering experiments.[1] One of the problems encountered in these analyses is that a resonance which appears in one reaction channel may not appear in another channel, or it may appear in the second channel at a different energy or with a different width. And sometimes different groups of workers, using basically the same set of data but slightly different models to represent the phase shifts, arrive at different values for some of the resonances. The reason for this ambiguity in the determination of the baryon resonances can be traced to the fact that the πN and $\bar{K}N$ scattering data have two serious defects: (1) the data are known to be manifestly incomplete; (2) the resonance parameters that are extracted from the data are a mixture of theory and experiment.[2] The incompleteness of the experimental data arises from the fact that four different types of spin-space scattering measurements are required at each energy in order to completely specify the πN or $\bar{K}N$ scattering matrix; namely,

Table 15.1. Experimentally-identified baryon and hyperon resonances. The data for these resonances are listed in Tables B2 - B6 of Appendix B. Group I in Table 15.1 includes resonances which have been identified as "well-established" by the members of the Particle Data Group (Ref. 1). Group II includes resonances which are not yet regarded as well-established, plus some recently-reported levels. Group III is a listing of narrow-width resonances, some of which may be redundant with resonances that appear at the same mass values in Groups I or II but with broader widths; these Group III resonances are listed here because of their phenomenological significance (see Figs. 15.1 and 15.9). In Groups I and II, the rotationless S-state excitations are placed at the top of each column. Resonances which have masses larger than 2500 MeV have generally been excluded from Table 15.1 because they have no spin identification. For the sake of internal consistency, the N and Δ resonance values were based in large part on the phase shift analyses of AYED 76 and BARBOUR 74 in Ref. 1, and the Λ and Σ resonance values were based in large part on RLIC 76.

N	Δ	Λ	Σ	Ξ
Group I. Well-established resonances (**** and ***)[†]				
N(939)Γ0	$\Delta(1623)S_{31}\Gamma161$	Λ(1116)Γ0	Σ(1193)Γ0	Ξ(1321)Γ0
$N(1511)S_{11}\Gamma104$	--	$\Lambda(1402)S_{01}\Gamma38$	$\Sigma(1770)S_{11}\Gamma60$	Ξ(1535)Γ10
$N(1674)S_{11}\Gamma187$	$\Delta(1232)P_{33}\Gamma115$	$\Lambda(1519)D_{03}\Gamma15$	--	Ξ(1819)3/2Γ19
--	$\Delta(1676)D_{33}\Gamma207$	$\Lambda(1672)S_{01}\Gamma33$	$\Sigma(1385)P_{13}\Gamma38$	Ξ(1956)Γ43
$N(1413)P_{11}\Gamma187$	$\Delta(1869)F_{35}\Gamma255$	--	$\Sigma(1670)D_{13}\Gamma50$	
$N(1517)D_{13}\Gamma130$	$\Delta(1900)P_{31}\Gamma200$	$\Lambda(1690)D_{03}\Gamma60$	$\Sigma(1774)D_{15}\Gamma130$	
$N(1674)D_{15}\Gamma159$	$\Delta(1928)F_{37}\Gamma237$	$\Lambda(1822)F_{05}\Gamma81$	$\Sigma(1920)D_{13}\Gamma300$	
$N(1680)F_{15}\Gamma123$	$\Delta(2406)H_{3,11}\Gamma289$	$\Lambda(1825)D_{05}\Gamma94$	$\Sigma(1920)F_{15}\Gamma130$	
$N(1720)P_{11}\Gamma137$		$\Lambda(1881)P_{03}\Gamma101$	$\Sigma(2040)F_{17}\Gamma190$	
$N(1722)P_{13}\Gamma132$		$\Lambda(2110)G_{07}\Gamma250$	$\Sigma(2210)G_{19}\Gamma60$	
$N(2132)G_{17}\Gamma247$		$\Lambda(2360)H_{09}\Gamma120$	Σ(2455)Γ220	
$N(2183)H_{19}\Gamma364$				
$N(2650)I_{1,11}\Gamma350$				

(continued on the next page)

[†]See Table B1 in Appendix B for a definition of the baryon resonance confidence level ratings that are used in Table 15.1.

Table 15.1. (Continued)

N	Δ	Λ	Σ	Ξ
Group II. Less-well-established resonances (**,*,†, and new)				
$N(2100)S_{11}\Gamma 200$	$\Delta(2070)S_{31}\Gamma 100$	$\Lambda(1827)S_{01}\Gamma 150$	$\Sigma(1476)\Gamma 34$	$\Xi(1617)\Gamma 28$
--	--	--	$\Sigma(1582)D_{13}\Gamma 11$	$\Xi(2024)\Gamma 16$
$N(\sim 1230)P_{11}\Gamma\sim 200$	$\Delta(1550)P_{31}\Gamma 110$	$\Lambda(1573)P_{01}\Gamma 147$	$\Sigma(1623)S_{11}\Gamma 45$	$\Xi(2121)\Gamma 24$
$N(1540)P_{13}\Gamma 200$	$\Delta(1680)P_{33}\Gamma 220$	$\Lambda(1747)P_{01}\Gamma 120$	$\Sigma(1955)S_{11}\Gamma 170$	$\Xi(2295)\Gamma<30$
$N(1710)D_{13}\Gamma 100$	$\Delta(1889)D_{35}\Gamma 121$	$\Lambda(2010)D_{03}\Gamma 130$	--	
$N(1986)F_{17}\Gamma 214$	$\Delta(2174)G_{39}\Gamma 205$	$\Lambda(2020)F_{07}\Gamma 160$	$\Sigma(1676)P_{11}\Gamma 120$	
$N(1989)F_{15}\Gamma 179$	$\Delta(2200)G_{37}\Gamma 293$	$\Lambda(2120)F_{05}\Gamma 170$	$\Sigma(1840)P_{13}\Gamma 120$	
$N(2029)D_{13}\Gamma 116$			$\Sigma(1954)P_{11}\Gamma 213$	
$N(2100)D_{15}\Gamma 220$			$\Sigma(2070)F_{15}\Gamma 140$	
$N(2142)G_{19}\Gamma 297$			$\Sigma(2090)G_{17}\Gamma 103$	
			$\Sigma(2100)P_{13}\Gamma 200$	
Group III. Other reported narrow-width resonances				
$N(1532)P_{11}\Gamma 89$		$\Lambda(1327)\Gamma<25$	$\Sigma(1680)P_{11}\Gamma 25$	$\Xi(1871)\Gamma 17$
		$\Lambda(1687)D_{03}\Gamma 30$	$\Sigma(1697)S_{11}\Gamma 66$	$\Xi(2203)\Gamma<20$
		$\Lambda(1690)P_{01}\Gamma 22$	$\Sigma(1760)P_{11}\Gamma 77$	
		$\Lambda(1710)P_{03}\Gamma 20$	$\Sigma(1925)P_{13}\Gamma 65$	
		$\Lambda(1808)G_{09}\Gamma 27$	$\Sigma(1945)D_{13}\Gamma 65$	
		$\Lambda(1817)P_{01}\Gamma 12$		

a single-scattering differential cross section measurement σ, a double-scattering polarization measurement P, and two triple-scattering measurements R and A. However, only a few of the very difficult R and A measurements have been carried out, so that the πN and $\bar{K}N$ phase shift analyses have been based almost exclusively on σ and P data. Furthermore, when inelastic scattering channels are open, which is generally the case, the analysis becomes even more involved. As a way of overcoming some of these difficulties, energy-dependent parameters are introduced which link the scattering amplitudes from different energies together, and various models are used to represent inelastic effects. This use of theoretical information to bolster the incomplete experimental information is the second defect listed above, and it has led to the situation which is aptly summarized by K. C. Wali in the quotation given at the beginning of this section. One of the dangers inherent in this approach is that the "experimental data", which in reality are "experimental data plus theoretical data", tend to favor the theoretical data which have been inserted (see the discussion in Section B1 of Appendix B).

Before moving on to an examination of the experimental baryon and hyperon resonances, we first make a comparison between these resonances and the resonances that are observed in the area of nuclear physics.

15.2 The Analogy to Nuclear Physics

The baryon resonances are the excited states of the proton and neutron. As a possible guide to the form that these excited states might take, we can study the excited states of atomic nuclei, since these nuclei are formed from protons and neutrons, and since the masses of the very light nuclei are roughly comparable to the masses of the baryon excitations. We have already seen in Sec. 12.6 of Chapter 12 that hadronic rotational excitations extrapolate smoothly into the rotational excitations of light nuclei. Thus it seems reasonable to investigate this idea in more depth, and to see if the observed baryon excitations are in fact analogous to the observed nuclear excitations. This is in line with the Golden Rule of Scientific Research that we enunciated at the end of Chapter 1: try the simple solutions first, and go on to more complicated solutions only if the simple solutions can't be made to work.

The observed excitations in atomic nuclei follow a distinctive pattern: (1) atomic nuclei exist as a series of stable or almost-stable ground states; (2) bombardment of these ground states with various projectiles (photons, electrons, nucleons, heavy ions, etc.) results in a sequence of excited states that are characterized as vibrational excitations; (3) the ground states and the excited vibrational states can in addition be excited into rotational modes, so that rotational bands are observed superimposed on these states. Now, the spectrum of baryon and hyperon resonances that is given in Table 15.1 bears certain similarities to the spectrum of nuclear states that we have just described. In particular, there are six distinct families of baryon and hyperon excitations — N, Δ, Λ, Σ, Ξ, and Ω — , and each of these families has a stable or metastable ground state (the N and Δ families share the same ground state). Also, the N, Λ, and Σ families, whose properties are the most completely known, all have several identified S-state levels. Since these S-states represent rotationless ($\ell=0$) excitations, they are analogous to the vibrational excitations of atomic nuclei, and they can serve as the bandheads for rotational excitations. (In nuclear physics, the lowest state in a rotational band is commonly denoted as the "bandhead".) Hence, if we wish to complete this analogy between nuclear excitations and baryon excitations, we must identify the broad-width P, D, F, G, and H baryon resonances as rotational levels and group them with the appropriate S-states. As we show in the present chapter, this can be done, but there are four difficulties here that are not present in the nuclear case: (1) baryon rotational energies are expected to be large (a few hundred MeV), and in fact comparable to the separation of the S-state bandheads (see Table 15.1), so that the rotational levels for the various rotational bands will be overlapping rather than cleanly separated as in nuclear physics; (2) some of the baryon rotational levels are missing (and also some of the S-state bandheads); (3) the parities of the states are not as clearly defined for these baryon resonances as they are for the nuclear resonances, so that, for example, we have (as we will demonstrate) J = 1/2 S-states and narrow-width P-states appearing at the same mass values as manifestations of the same rotationless excitations; (4) the number of rotational levels in a hadron rotational band is limited by relativistic constraints, with some rotational bands having only a single rotational level (and sometimes an unobserved S-state bandhead). In view of these difficulties, the only way that we can untangle the maze of observed baryon resonances is to study the entire spectrum of observed resonances, obtain a systematic method for locating and identifying the S-state bandheads, and then deduce a method for estimating moments of inertia and hence rotational energies. If we can pinpoint the S-states, and if we can approximate the rotational energies, then we can assign the baryon resonances to rotational bands.

In carrying out the task of mapping the baryon rotational bands, we fortunately have four pieces of information which tend to offset the four difficulties cited above. These four "known" pieces of information are the following: (1) Rotational energies for quantum mechanical systems are known to generally follow an $\ell(\ell+1)$ interval rule, where ℓ is the orbital angular momentum quantum number, as we discussed in Sec. 12.4 of Chapter 12.† (2) Rigid-body estimates for the moments-of-inertia I of atomic nuclei give values for very light nuclei that are quite accurate (see Table 12.1), and they give values for heavy nuclei that are generally too large by just about a factor of two. This same result seems to occur in baryon rotations (although for quite a different reason), so that we can obtain approximate values for baryon

---

†Under certain conditions, which were specified in Chapter 12, the orbital angular momentum ℓ in an atomic nucleus may couple to the intrinsic spin of the nucleus so as to produce a $J(J+1)$ interval rule rather than an $\ell(\ell+1)$ interval rule, where J is the total angular momentum of the nucleus. But since these specialized conditions are strongly violated in baryon and meson rotations, we expect to find $\ell(\ell+1)$ energy intervals (which we do) rather than $J(J+1)$ energy intervals (which we do not).

rotational energies from rigid-body estimates. (3) From the observed quantization of meson and baryon excitation energies in units of M = 70 MeV (Chapter 5), and from the further fact that the metastable hyperon "ground state" resonances are formed by the addition of $\bar{3} \equiv 3\bar{M}$ = 210 MeV and $\bar{4} \equiv 4\bar{M}$ = 280 MeV excitation quanta to the basic nucleon spinor triplet $S\bar{S}S$ (Chapter 9), it seems plausible that baryon and hyperon S-state bandheads are spaced by characteristic energy intervals of 210 MeV and 280 MeV; empirically, this assumption turns out to be correct, as we have already demonstrated in Sec. 5.4 of Chapter 5. (4) Rotationless baryon excitations characteristically have narrower widths than do rotational levels, as we discussed at some length in Sec. 12.2 of Chapter 12.

If we combine the information from all of the above sources, and if we correlate the results from the various baryon families, then we can carry out the program of mapping the resonances of Table 15.1 into baryon rotational bands. In doing this, we first locate the S-states (Sec. 15.3); then estimate the rotational energies (Sec. 15.4); then identify the individual rotational bands (Sec. 15.5); and, finally, attempt to account for observed and missing levels in terms of "doorway excitations" and relativistic constraints (Sec. 15.6), and also in terms of rotation invariance and overlapping levels (Sec. 15.7). These results are summarized in Sec. 15.8, where they are compared briefly with the SU(3) mapping of these same resonances.

The mapping of the basic N, Λ, Σ, Ξ, and Ω "ground states" was discussed in Chapter 9. In the present chapter we are studying the "excited states" of these baryon and hyperon families.

15.3 Mapping the Baryon S-States

As we discussed in Sec. 15.1, the experimental πN and $\bar{K}N$ scattering data are known to be incomplete. One indication of the ambiguities that are inherent in these data is provided by the systematics of the narrow-width rotationless baryon and hyperon excitations, which we now discuss.

Let us assume that the various baryon and hyperon states each contain three spin 1/2 quarks, as is generally agreed to be correct. Then these baryon states can have intrinsic spin values S = 1/2 or S = 3/2, depending on the alignment of the quark spins. If, as we have stated above, rotationless (ℓ=0) excitations have narrower widths than do rotational (ℓ>0) excitations, then we expect to find some J = 1/2 and J = 3/2 baryon resonances that have narrow widths (these are resonances where the angular momentum values J are due to the intrinsic spins S, so that ℓ = 0 for these states), but we do not expect to find any J ≥ 5/2 baryon resonances that have narrow widths (since ℓ > 0 orbital components must necessarily be present in these resonances). This is a result that we have already demonstrated experimentally in Sec. 12.3 of Chapter 12 (see Fig. 12.1). However — and this is a new aspect that we introduce here — , if the parity P of a resonance is not pinned down experimentally, then we may see both $J^P = 1/2^-$ (S-state) and $J^P = 1/2^+$ (P-state) resonances† appearing at the same mass value as manifestations of the same resonance enhancement. And, if the energy difference between S = 1/2 and S = 3/2 spin configurations is small (as seems empirically to be the case), we may see narrow-width S-, P-, and D-state J = 1/2 and J = 3/2 resonances all appearing at the same mass value.

The force of these remarks is contained in an important empirical observation that is based on the experimental compilation of baryon resonances given in Table 15.1. In Table 15.1, which contains all of the reported baryon and hyperon resonances, we have divided the identified resonances into two groups: Group I includes the well-established **** and *** resonances as rated by the Particle Data Group[1], and Group II includes the not-so-well established resonances that have ** and * ratings plus some newly-reported resonances. (In each of these groups, the rotationless ($\ell = 0$) excitations are placed at the tops of the columns.) However, in addition to these two groups of identified baryon and hyperon resonances, we have also included a third group, Group III, at the bottom of Table 15.1. This third group is a collection of narrow-width peaks which have been reported in the literature, and which are contained in the data compilation of the Particle Data Group[1] (see Appendix B for references to these data), but which in most cases have presented a puzzle, either because they don't fit in with the established set of resonances (some of these narrow peaks have mass values that coincide with those of Group I or II resonances), or else because they are seen only in certain reaction channels and not in other expected reaction channels. In fact, one of these resonances, $\Lambda(1327)\Gamma<25$, has been officially declared "dead" by the Particle Data Group[1], although, as we stated at the end of Sec. 5.4 in Chapter 5, it is possible that this burial may have been premature. The empirical observation that we make here — an observation that is crucial to the results of the present chapter — has to do with this third group of resonances in Table 15.1, the group of "renegade" levels, which exhibit the following interesting behavior: when these narrow-width anomalies are plotted in the form of an energy level diagram above the basic N, Λ, and Σ "ground states", as we have done in previous chapters, it turns out that they do not appear at random locations, but rather at the same locations as those at which the S-state resonances appear. Hence they seem to be manifestations of the same rotationless baryon excitations that give rise to the S-states. This result is illustrated in Fig. 15.1, which is the energy level diagram for these resonances.

---

†The notation used to classify these states is summarized in Appendix A.

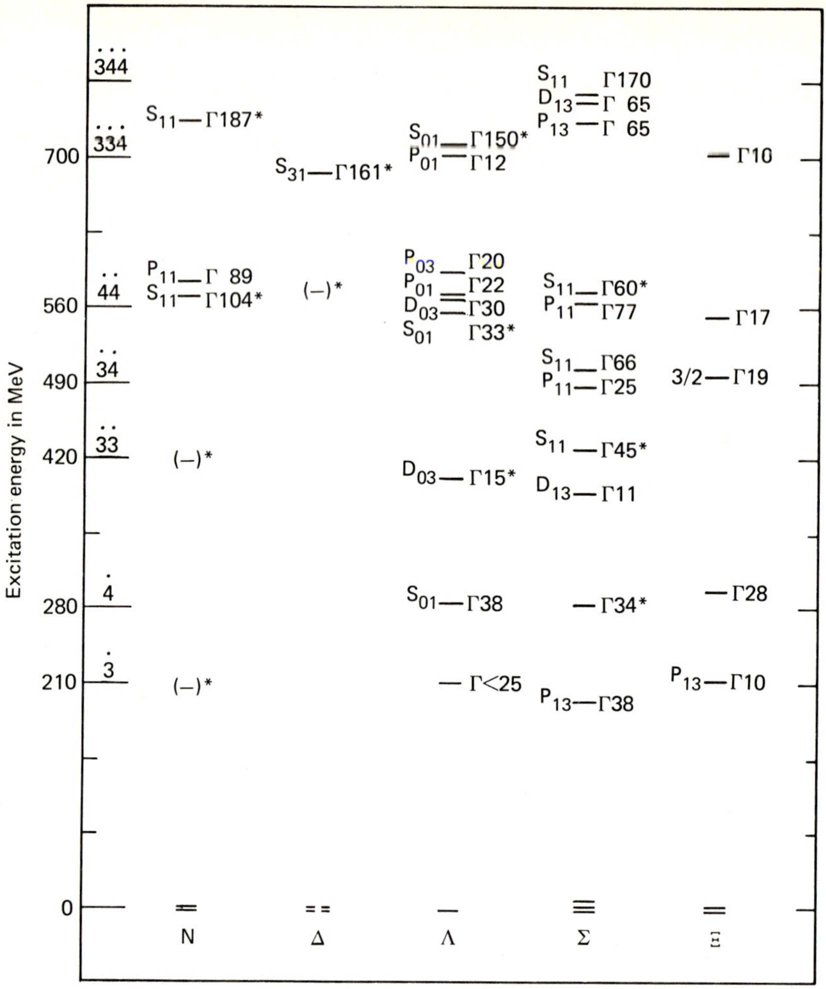

Figure 15.1. The narrow-width and S-state baryon and hyperon resonances. The states labeled with asterisks serve as the bandheads for overlying rotational levels. The levels enclosed in parentheses are unobserved bandheads that can be identified on the basis of their observed rotational levels. As can be seen in the figure, the resonances occur at the quantized excitation energies which are indicated along the ordinate. Furthermore, there is a tendency for S-states and narrow-width J = 1/2 P-states and J = 3/2 P-states and D-states to appear at the same mass values as manifestations of the same (rotationless) baryon excitation.

In studying the systematics of Fig. 15.1, we first note that the narrow-width and S-state resonances occur in discrete excitation bands, and that the excitation-band energies are in multiples of $\dot{3}$ = 210 MeV and $\dot{4}$ = 280 MeV, in agreement with our speculation in Sec. 15.2 above. In fact, it is useful, and physically significant, to use these $\dot{3}$ and $\dot{4}$ excitation multiples as labels for the excitation bands. (In Appendix E we demonstrate that the $\dot{3}$ and $\dot{3}\dot{3}$ excitations have quite different properties from the $\dot{4}$ and $\dot{4}\dot{4}$ excitations.)

We next examine the Λ and Σ resonances of Fig. 15.1 in detail, starting with the lowest-mass levels. The $\Lambda\dot{3}$ level and the $\Sigma\dot{4}$ level, which are $N\overline{3}\dot{3}$ and $N\overline{4}\dot{4}$ excitations, respectively,† have no spin identification other than their narrow widths, but the $\Sigma\dot{4}$ level serves as the bandhead for overlying rotational levels (as we have indicated by the asterisk). The other low-lying level, the $\Lambda(1402)S_{01}\Gamma 38$ resonance, while nominally a member of the Λ family, is actually a K^-p bound state.† It has no associated rotational levels.

The $\Lambda\dot{3}\dot{3}$ excitation — the $\Lambda(1519)D_{03}\Gamma 15$ level — must be a rotationless excitation, both from its narrow width and from its identification with overlying rotational levels (these levels are shown in Fig. 12.5). Although no S-state is in evidence near this D-state, we note that the corresponding $\Sigma\dot{3}\dot{3}$ excitation band has both an identified S-state and a very-narrow-width D-state that resembles the Λ-band D-state. Thus the analogy between these Λ and Σ excitations serves to confirm the interpretation that we have given here to the Λ-band D-state. Further confirmation is supplied by the fact (which is described in Appendix E) that the $\dot{3}$ and $\dot{3}\dot{3}$ excitations are typically difficult to observe as S-states.

In the $\Lambda\dot{4}\dot{4}$ excitation band, we find the interesting result that very-narrow-width S_{01}, P_{01}, P_{03}, and D_{03} resonances all appear at about the same mass value, and with comparable widths. The corresponding $\Sigma\dot{4}\dot{4}$ excitation band also features a pair of matching levels — S_{11} and P_{11}. Thus these resonances illustrate the important fact, which we anticipated in the discussion above, that a single rotationless excitation can appear with the spin values S = 1/2 or S = 3/2, and with positive or negative parity. This is a result that has not been noted before anywhere in the literature, and it brings into question the uniqueness of the parity assignments for these resonances. The S-states in both of these $\dot{4}\dot{4}$ excitation bands have identified overlying rotational levels.

Continuing on to the $\Lambda\dot{3}\dot{3}\dot{4}$ excitation band, we encounter another pair of matching levels — S_{01} and P_{01}, but whereas the S-state has the broad width (Γ = 150 MeV)

---

†It was shown in Chapter 9 that $\Lambda \equiv N\overline{3}$, $\Sigma \equiv N\overline{4}$, $\Xi \equiv N\overline{33}$, and $\Lambda(1402) \equiv N\overline{7} = pK^-$.

that seems to be typical for S-states at the higher mass values, the P-state has the very narrow width Γ = 12 MeV. The experimenter who reported this P-state remarked[3] that it is "a surprisingly narrow resonance", and he went on to comment that "this narrow effect needs further study as to its origin in the data".[†] From the viewpoint of the present light-quark model, the narrow width of this state arises from its identification as a rotationless excitation. The neighboring Σ resonances, at a slight higher mass value, feature three matching states — S_{11}, P_{13}, and D_{13} — , with the P- and D-states having narrower widths (Γ = 65 MeV) than the S-state (Γ = 170 MeV). Thus here again we have an example where results in the Σ resonances corroborate results that we have obtained in the Λ resonances. Another pair of matching S- and P-states can be seen in the $\Sigma\ddot{34}$ excitation band, and once again the P-state has a much narrower width (Γ = 25 MeV) than the S state (Γ = 66 MeV). These $\Sigma\ddot{34}$ excitations have no counter part in the Λ excitations, and they do not correspond to identified overlying rotational levels, but they match one of the well-established Ξ excitations — the $\Xi(1819)\Gamma 19$ resonance — , as can be seen in Fig. 15.1.

If we found only a single pair of J = 1/2 S- and P-states appearing at the same mass value in Fig. 15.1, we might regard it as a coincidence. But the number of times that this result occurs in the Λ and Σ resonances of Fig. 15.1 removes it from this category. Since Table 15.1 is a reasonably-complete compilation of <u>all</u> of the observed baryon resonances, the regularities shown in Fig. 15.1 represent a general result for <u>narrow-width</u> and S-state resonances, and not just a result that follows from a carefully-selected subset of data. This result is an important one from the standpoint of the present light-quark model in that it reinforces our ideas about the difference between rotationless and rotational excitations: rotationless excitation levels are in general narrower than rotational levels, which is just the opposite of the situation that occurs in nuclear physics. Thus it seems clear, as we argued in Chapter 11, that the concept of angular momentum barriers does <u>not</u> apply to the energetic decays of these baryon resonances. This conclusion has important theoretical ramifications in that it calls into question a series of SU(3) decay-rate calculations which feature the use of explicit nuclear-physics-type angular momentum barrier factors.[4]

Moving on in our study of Fig. 15.1 from the Λ and Σ resonances to the N, Δ, and Ξ resonances, we notice two interesting facts: <u>(1)</u> The N and Δ resonances are

---

[†]There is a narrow-width $\Lambda(1808)G_{09}\Gamma 27$ enhancement listed in Table 15.1 that also appears at about this same energy. The very high spin value reported for this narrow-width state (which comes from a single unconfirmed experiment) makes it somewhat suspect, and it may simply be a reflection of the sharp discontinuity manifested by the nearby $\Lambda(1817)P_{01}\Gamma 12$ resonance.

systematically broader than the Λ and Σ resonances, which in turn are systematically broader than the Ξ resonances (as can be verified in Table 15.1). (2) The N and Δ resonances have fewer S-states than do the Λ and Σ resonances; and the Ξ resonances, from their narrow widths, seem to be mostly S-states (as is shown by a study of the Ξ resonances in Table 15.1). Now, there is a logical connection between these two facts, as we demonstrate in the following explanation. The strangeness S = 0 N and Δ resonances are typically produced by pion-nucleon collisions which tend to occur in a peripheral manner that enhances the formation of rotational levels as opposed to rotationless S-state levels, and which leads to broad widths for these levels. The strangeness S = -1 Λ and Σ resonances, on the other hand, arise from kaon-nucleon interactions, and they require a rather direct impact in order to produce the basic S = -1 "bound state" excitations $\Lambda \equiv N\bar{3}$ and $\Sigma \equiv N\bar{4}$; these "direct-impact" interactions lead to an increased number of rotationless S-state excitations, and they also result in a sharper definition (narrower width) for the rotational levels. The strangeness S = -2 Ξ resonances, in which the basic excitation $\Xi \equiv N\bar{3}\bar{3}$ must be produced, require an even more direct knock-on type of interaction than do the Λ and Σ resonances, so that the majority of the observed Ξ resonances (see Table 15.1) have the very narrow widths that we associate with rotationless excitations. The strangeness S = -3 Ω hyperon represents the extreme limit in this sequence of interactions: it requires such a direct impact in order to produce the $\Omega \equiv N\bar{3}\bar{4}\bar{4}$ excitation that to date only the "ground state" form of this resonance has been observed.† From this impact parameter argument, we conclude that the N and Δ resonances typically have broad widths and a scarcity of $\ell = 0$ levels; the Λ and Σ resonances have narrower widths and more $\ell = 0$ levels; and the Ξ resonances have very narrow widths and (probably) very many $\ell = 0$ levels.

Since N and Δ S-states seem characteristically difficult to produce, we need some means other than direct observation for locating all of the bandheads of the N and Δ rotational bands. It turns out that by intercomparing all of the rotational bands, and by extrapolating from the results of the Λ and Σ resonances, we can determine where unseen N and Δ bandheads must be, as we will demonstrate later on in the chapter. These locations are indicated in Fig. 15.1 by the lines enclosed in parentheses — namely, the $N\dot{3}$ and $N\dot{3}\dot{3}$ excitations, which correspond to unseen N bandheads, and the $N\dot{4}\dot{4}$ excitation, which corresponds to an unseen Δ bandhead. The information on the Ξ resonances is not sufficient to permit the identification of rotational levels, but the narrow-width Ξ resonances in Fig. 15.1 match the rotationless excitation bands that are populated by the N, Δ, Λ, and Σ resonances.

---

†The tendency for baryon production reactions to become more strongly peaked along the direction of the incident beam as the strangeness of the state increases is a result that can be directly observed in the experimental data.

When we combine all of the above information, we can pretty well pinpoint the locations of the bandheads for the various baryon rotational families. These are the levels in Fig. 15.1 that are labeled with asterisks. As can be seen in Fig. 15.1, the lower rotational bands are separated by characteristic intervals of 210 MeV, and the higher-mass rotational bands are separated by intervals of 140 MeV. In Chapter 16 we will see this result repeated for the rotational bands of the meson resonances. One of the most important results to be drawn from Fig. 15.1 is that the S-state excitation systematics are similar for all of these baryon and hyperon families. Later on in the chapter, we will demonstrate some striking similarities between the rotational bands of the various families. These similarities indicate that the rotational motion of these states is essentially a mechanical phenomenon, and it does not depend, for example, on the strangeness values of the constituent basis states.

As the first step in the assignment of rotational levels to these bandheads, we must determine the magnitude of the baryon rotational energies. This is the task that is described in the next section.

15.4 How Large are Baryon Rotational Energies?

If we wish to sort the baryon resonances into rotational bands, the magnitude of the baryon rotational energies must be deduced. One way to deduce the rotational energy of a baryon is to try to calculate it. In the quark model of Gell-Mann and Zweig, orbital rotations are ascribed to the motion of individual quark pairs[5], and the masses of the quarks are generally regarded as unknown, so that no definitive statements about baryon rotational energies can be made on the basis of this model. However, if we adopt the viewpoint of the present light-quark model, and attribute orbital energies to a nuclear-physics-type rotation of the entire nucleus, as was discussed in Chapter 12, then the energies of the rotational levels are given by the equations

$$E(\ell) = E_o + E_{rot} \times \ell(\ell+1) \tag{15.1}$$

and

$$E_{rot} \equiv \hbar^2/2I, \tag{15.2}$$

where E_o is the S-state energy of the rotational band (as given in Fig. 15.1), and where I is the moment-of-inertia of the S-state bandhead. In Sec. 13.4 of Chapter 13, we made a classical estimate for the moment-of-inertia of the rho meson (see Fig. 13.1) which turned out to be empirically correct. If we picture the nucleon, $N \equiv S\bar{S}S$, in the same manner (see Fig. 18.17), then we can make the same estimate for the moment of inertia of this three-spinor configuration. This estimate gives

$$I_{nucleon} \simeq 3 \times (2/5\ MR^2 + 4/3\ MR^2) = 646\ \text{MeV fm}^2, \tag{15.3}$$

where we set M = 313 MeV (one-third of the nucleon mass) and R = 0.63 fm (the Compton wavelength that corresponds to the mass M = 313 MeV). This value for I then gives

$$E_{rot} = \hbar^2/2I_{nucleon} \simeq 30\ \text{MeV}. \tag{15.4}$$

Thus we expect baryon rotational levels, if they are analogous to meson rotational levels, to have energies that are roughly characterized by this value for E_{rot}.

Another way to estimate baryon rotational energies is to study the rotational levels that have been observed experimentally. If we insert the value for E_{rot} from Eq. (15.4) into Eq. (15.1), then, for example, $\ell = 2$ D-states should be separated from $\ell = 3$ F-states by about 6 x 30 = 180 MeV, and these D-states should be separated from $\ell = 5$ H-states by about 24 x 30 = 720 MeV. To check this result, we show in Fig. 15.2 the mass values for all of the well-established baryon resonances (the Group I resonances of Table 15.1), plotted on an absolute mass (or energy) scale. As can be seen in Fig. 15.2, the G, H, and I resonances have very large mass values; the D and F resonances have intermediate mass values; and the S and P resonances, while extending up to the higher mass values, also include all of the lowest-mass states. The separation energies between these three angular momentum groupings amount to several hundred MeV, which indicates rather directly that baryon rotational energies are large. Furthermore, the lowest-mass broad-width D, F, and H resonances that appear in Fig. 15.2 are the N excitations $N(1517)D_{13}\Gamma 130$, $N(1680)F_{15}\Gamma 123$, and $N(2183)H_{19}\Gamma 364$ of Table 15.1. The F - D separation energy for these levels is 163 MeV, and the H - D separation energy is 666 MeV. These values agree closely with the F - D and H - D estimates of 180 MeV and 720 MeV given above, and they indicate that these particular levels belong to an N rotational band. If we thus assign these D-, F-, and H-levels to this rotational band, then the F - D mass difference† gives a value for E_{rot} of 163 ÷ 6 = 27.2 MeV, and it indicates that the bandhead mass for this rotational band should be 163 MeV (the calculated D - S mass difference for E_{rot} = 27.2 MeV) below the mass of the D-level, or at 1354 MeV. Since the $N\dot{3}\dot{3}$ excitation energy in Fig. 15.1 is 939 + 420 = 1359 MeV, it seems clear that the $N\dot{3}\dot{3}$ excitation serves as the (unobserved) bandhead for this N rotational band. The P-state level for this rotational band should then appear with a mass that is 2 x 27.2 ≃ 54 MeV above the S-state bandhead mass, or at a mass value of 1359 + 54 = 1413 MeV.

---

†The very broad width of the H-level suggests that it may have contributions from more than one rotational band.

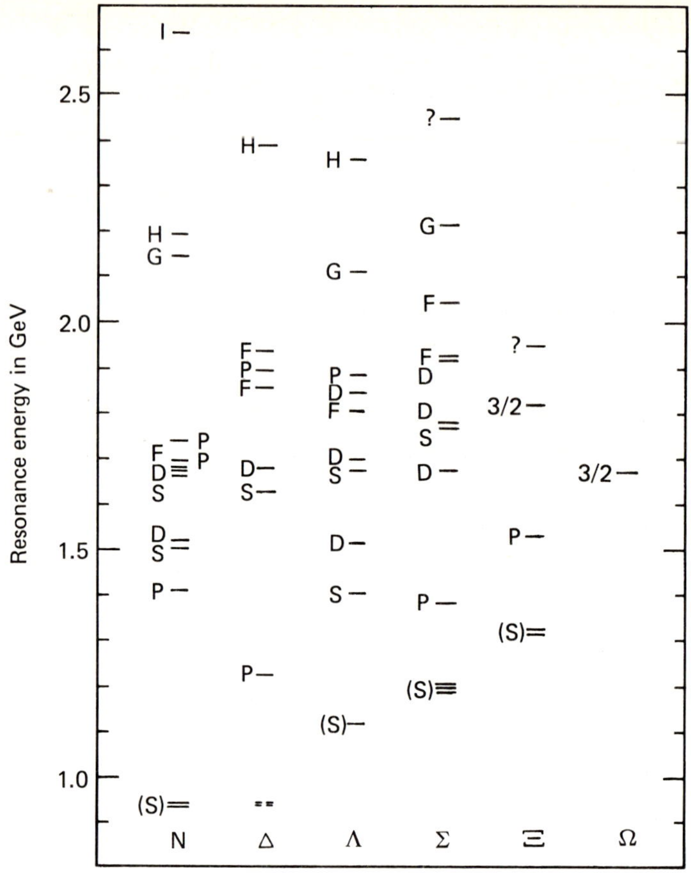

Figure 15.2. A plot of all of the well-established baryon and hyperon resonances, using an absolute mass or energy scale. As can be seen, the resonances below 1.6 GeV are all S-, P-, and D-states; the F- and G-states fall in the range 1.7 - 2.2 GeV; and the H- and I-states fall in the range 2.2 - 2.7 GeV. This indicates in a direct manner that the rotational energies for these levels are very large (several hundred MeV for ℓ = 3 or 4, and $\sim$1 GeV for ℓ = 5 or 6).

This mass assignment matches the N(1413)$P_{11}\Gamma 187$ resonance that is shown in Fig. 15.2 and Table 15.1. Thus we have identified the P, D, F, and H members of this N rotational band; the G-level is missing, and so is the S-level.

There is one final aspect of this study of baryon rotational energies that we must still consider — namely, the manner in which the moment-of-inertia I of a baryon excitation depends on the mass of the bandhead. And here we encounter a rather surprising result. If the dimensions of the nucleon bandhead $S\bar{S}S$ stay constant as excitation quanta are added (possibly in the form of vibrational excitations), then the moment-of-inertia I will vary inversely with the mass M; and if the dimensions of the bandhead increase with increasing mass, then I will fall off faster than 1/M. However, what we find empirically is that the moment-of-inertia I is almost constant over the entire range of bandhead mass values, so that I is falling off more slowly than 1/M. Thus it appears that the effective radius R of the rotating system actually <u>decreases</u> as the mass <u>increases</u>. At first sight this seems to be a paradoxical situation. However, it must be recalled that when we constructed specific models for the muon μ and the spinor S (see Fig. 7.3), we assigned radii that are proportional to their Compton wavelengths, $R \propto \hbar/mc$, so that the radii R_μ and R_S are inversely proportional to the masses in question. This is a situation that we also encounter in the meson resonances, and it is discussed in more detail in Chapter 18. We can state here as a general result that for any system in which the spin angular momentum is attributed to the rotation of an extended mass distribution, an increase in the mass of the spinning system must necessarily be accompanied by either a decrease in radius or a decrease in angular velocity, or both, if spin angular momentum is to be conserved (as it must be). Thus this observed behavior of the moments of inertia for the baryon rotational bands is not unreasonable, and it fits into a consistent pattern with the moments of inertia of the meson resonances, as we will see in Chapter 16.

As one final comment in this section, we estimated in Eqs. (15.3) and (15.4) that the rotational energy parameter for the nucleon if it serves as a baryon rotational bandhead (which it does not) is $E_{rot} \simeq 30$ MeV. Now, most of the baryon bandheads shown in Fig. 15.1 have masses that are roughly twice as large as the nucleon mass. Thus our first guess for the rotational energy parameters of these bandheads would be $E_{rot} \simeq 15$ MeV. However, as we will see, the majority of the observed values for E_{rot} fall in the range $E_{rot} \simeq 26 - 30$ MeV. Thus we superficially have the same situation here as occurs in nuclear physics, wherein the classical (rigid body) estimates for the moments of inertia of the heavy nuclei are too high by a factor of two. But the discrepancy in the case of atomic nuclei is attributed to the fluid nature of large nuclei, whereas the discrepancy in the hadronic case seems to arise from the necessity of conserving spin angular momentum.

We now go on to display the various baryon and hyperon rotational bands.

15.5 The Baryon Rotational Bands

Guided by the discussions in the preceding sections, we can summarize the data on the baryon and hyperon rotational bands in a rather succinct manner. The task here is to sort the broad-width P-, D-, F-, G-, and H-resonances of Groups I and II in Table 15.1 into rotational bands. In presenting the results of this task, we display the rotational bands in groupings which bring out certain aspects of the rotational systematics.

Figure 15.3 shows the lowest-mass baryon rotational levels. These are the very prominent Δ(1232) level and the very weak N(∿1230) level, which both correspond to the unseen N$\dot{3}$(1149) bandhead. An estimate for the rotational energy of the Δ(1232) level was given in Sec. 12.5 of Chapter 12. The difficulty in observing the N$\dot{3}$ excitation directly as an S-state resonance is a common characteristic of $\dot{3}$ excitations, as we demonstrate in Appendix E.

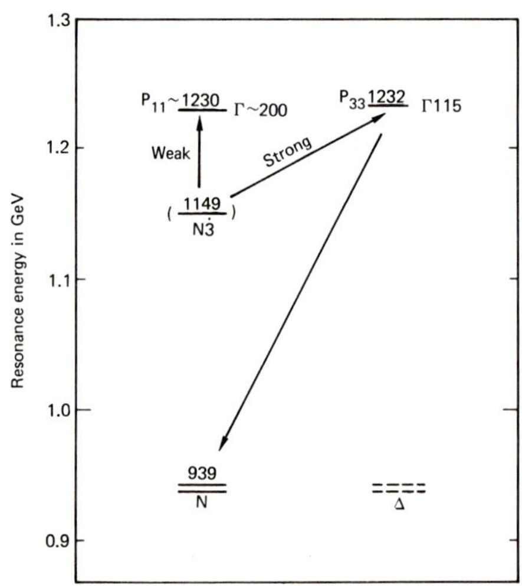

Figure 15.3. Rotational levels for the lowest-mass baryon rotational band. The unseen baryon excitation N$\dot{3}$ serves as the bandhead for both the prominent Δ(1232) rotational level and the weak N(∿1230) rotational level.

Figure 15.4 shows the rotational bands for the $\dot{3}\dot{3}$ and $\dot{3}\dot{3}\dot{4}$ excitations of Fig. 15.1. The N$\dot{3}\dot{3}$ rotational band in Fig. 15.4 is the rotational band that was identified in the discussions of the preceding section; as can be seen, both the S-level

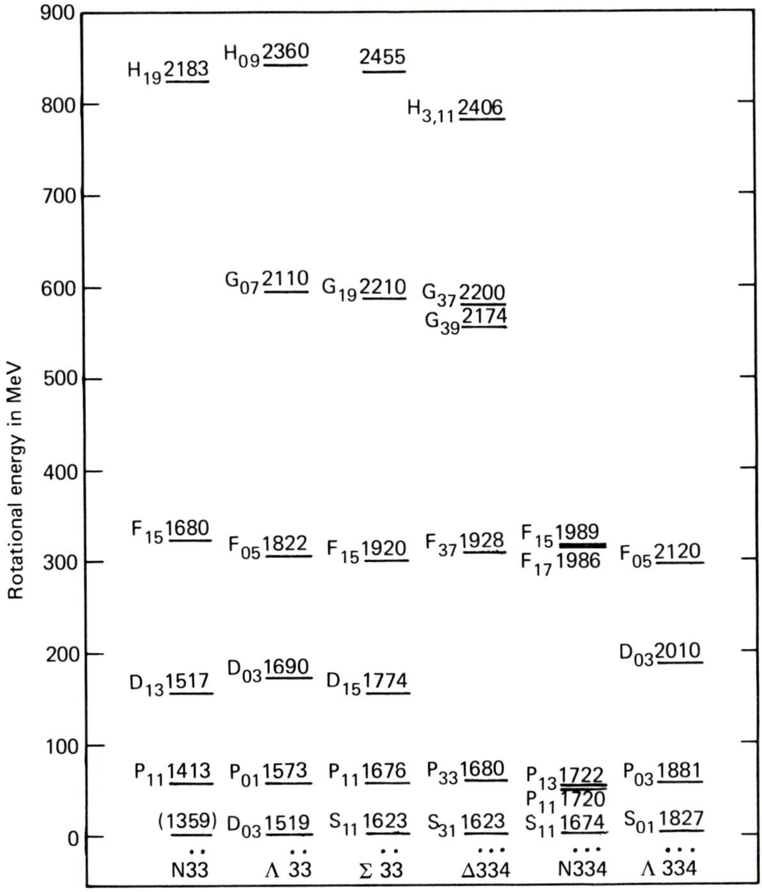

Figure 15.4. The baryon rotational bands that correspond to the $\dot{3}\dot{3}$ and $\dot{3}\dot{3}\dot{4}$ excitations of Fig. 15.1. As can be seen, the rotational energies are essentially independent of the strangeness and isotopic spin quantum numbers. The level in parentheses is an unobserved S-state bandhead. In general, the bandheads for these $\dot{3}\dot{3}$ and $\dot{3}\dot{3}\dot{4}$ rotational bands do <u>not</u> correspond to easily-identified S-states (contrast this to the situation shown in Fig. 15.5). Tables H1 to H4 in Appendix H give numerical values for the rotational energy parameters of the baryon and hyperon rotational bands that are shown in Fig. 15.4.

and the G-level are missing from this band, and the very broad H-level probably includes contributions from a higher-mass rotational band. The $\Lambda\dot{3}\dot{3}$ rotational band in Fig. 15.4 was previously shown in Fig. 12.5 of Chapter 12, where it was demonstrated that this rotational band follows a fairly accurate $\ell(\ell+1)$ interval rule. Both the $\Lambda\dot{3}\dot{3}$ band and the neighboring $\Sigma\dot{3}\dot{3}$ and $\Delta\dot{3}\dot{3}\dot{4}$ bands contain the G-level that is missing in the $N\dot{3}\dot{3}$ band. We note the following systematic results with respect to the S-state bandheads of the $\dot{3}\dot{3}$-type rotational bands in Fig. 15.4: the $N\dot{3}\dot{3}$ band is missing the S-state level; the $\Lambda\dot{3}\dot{3}$ band has this level, but in the form of a narrow-width D-state rather than an S-state; and the $\Sigma\dot{3}\dot{3}$ band has both an S-state level plus an associated narrow-width D-state level (as shown in Fig. 15.1), with the S-state level (which, it should be noted, was predicted[6] on the basis of the present systematics) not yet receiving acceptance by the Particle Data Group as a well-established resonance (it appears in.Group II of Table 15.1 rather than in Group I). These examples illustrate the difficulty in observing S-states for the $\dot{3}\dot{3}$-type excitations; in Fig. 15.3 we found a similar situation for a $\dot{3}$-type excitation. The $\Delta\dot{3}\dot{3}\dot{4}$ and $N\dot{3}\dot{3}\dot{4}$ rotational bands in Fig. 15.4 are missing the D-levels, and the $\Lambda\dot{3}\dot{3}\dot{4}$ band has the D-level at an anomalously high value. The highest-mass rotational bands are rather fragmentary, and the mass values for many of the levels in these bands are not well-established experimentally. In general, the rotational bands which extend up to the highest angular momentum values are the best-determined bands, and most of the levels in these bands correspond to well-established resonances.

Figure 15.5 contains the rotational bands for the $\dot{4}$ and $\dot{4}\dot{4}$ excitations.† In contrast to the rotationless $\dot{3}\dot{3}$ excitations of Fig. 15.4, the rotationless $\dot{4}\dot{4}$ excitations in Fig. 15.5 all correspond to well-established S-state resonances. The P-level is missing from the $\Sigma\dot{4}$ rotational band, and it appears at somewhat anomalous positions in the other rotational bands. In general, the positions of the P-state resonances have been difficult to pin down experimentally. Part of this difficulty may stem from the fact, which was demonstrated in Fig. 15.1, that many of the rotationless $\ell = 0$ S-state enhancements also include corresponding P-state components.

In Figs. 15.4 and 15.5, we see several examples where the $J = \ell - 1/2$ and $J = \ell + 1/2$ states that correspond to a particular ℓ-value both resonate at about the same mass value. These examples illustrate the fact, which was discussed in Sec. 15.2 above, that baryon rotational energies depend essentially just on ℓ and not on J [see Eqs. (12.6) - (12.11) in Chapter 12].

---

†The lowest-mass $\dot{4}$-type excitation, $\Lambda\dot{4} = \Lambda(1402)S_{01}\Gamma 38$, is actually a K^-p bound state, and it does not appear to have overlying rotational bands.

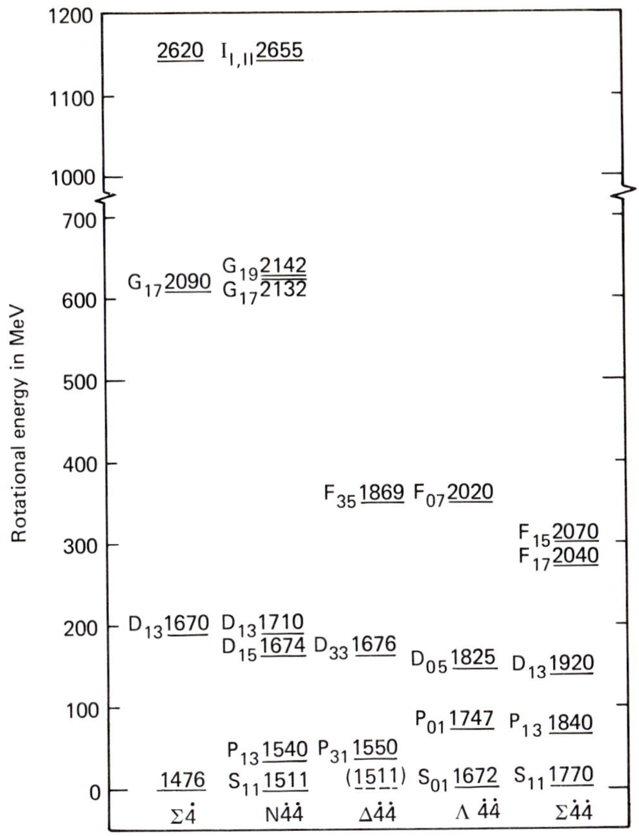

Figure 15.5. The baryon rotational bands that correspond to the $\dot{4}$ and $\dot{4}\dot{4}$ excitations of Fig. 15.1. In contrast to the bandheads shown in Fig. 15.4, the $N\dot{4}\dot{4}$, $\Lambda\dot{4}\dot{4}$, and $\Sigma\dot{4}\dot{4}$ bandheads of Fig. 15.5 correspond to prominent S-state resonances. Tables H1 to H4 in Appendix H give numerical values for the rotational energy parameters of the baryon and hyperon rotational bands that are shown in Fig. 15.5.

Figures 15.3 - 15.5 contain all of the identified baryon and hyperon rotational bands, although from the appearance of scattered S-, P-, and D-levels at higher energies it seems clear that, at least in principle, these rotational bands extend on up to higher mass values. In Fig. 15.6 we combine all of the rotational bands from Figs. 15.3 - 15.5 together in one plot, with the rotational bands arranged according to the mass of the bandhead. As can be seen, these rotational bands form a reasonably unified pattern (some possible explanations for missing levels in this pattern are discussed in the next section).

One of the interesting aspects of the rotational levels shown in Fig. 15.6 is the manner in which the ℓ-values and J-values appear. As we discussed above, the energies of these levels depend essentially just on the ℓ-values. However, these levels of course also have associated J-values, where $\vec{J} = \vec{\ell} + \vec{S}$, and S is the intrinsic spin of the bandhead. A study of the rotational levels in Fig. 15.6 reveals that the J-values for a particular ℓ-value occur in a systematic manner: the J-values alternate with increasing mass, and the lowest J-value appears first (with one important exception). To illustrate this result, let us introduce the notation $\ell_- \equiv (J = \ell - 1/2)$ and $\ell_+ \equiv (J = \ell + 1/2)$, so that, for example, P_- denotes $\ell = 1$, $J = 1/2$ and P_+ denotes $\ell = 1$, $J = 3/2$. Using this notation, we have plotted in Table 15.2 the rotational bands of Fig. 15.6, grouped according to baryon and hyperon families. In addition to the - and + subscripts on the ℓ-values, we have also included superscripts in Table 15.2 that denote the confidence level which has been assigned to each resonance by the Particle Data Group[1] (e.g., the superscript 4 denotes a top-rated **** resonance). As can be seen in Table 15.2, the D_-, D_+ levels, F_-, F_+ levels, and G_-, G_+ levels for the various baryon families appear in a regular alternating pattern. The P_-, P_+ levels also generally follow this pattern, although the lowest-mass P-state, which belongs to the $\Delta\dot{3}$ family (see Fig. 15.3), is a P_+ level rather than a P_- level (this is the exception cited above). The $N\dot{3}\dot{3}$ and $\Lambda\dot{3}\dot{3}$ rotational bands, which feature top-rated resonances, include only ℓ_- levels; but the $\Sigma\dot{3}\dot{3}$ rotational band, which also features top-rated resonances, has some ℓ_+ levels due to the presence of the lower-mass $\Sigma\dot{4}$ rotational band (which has preempted some of the ℓ_- levels). In the Δ family, the $\Delta\dot{3}$ and $\Delta\dot{3}\dot{3}\dot{4}$ rotational bands are ℓ_+ levels, and the $\Delta\dot{4}\dot{4}$ band includes only ℓ_- levels. One of the important conclusions to be drawn from a table such as Table 15.2 is that the regularities exhibited by the data are an indication that the groupings displayed in the table possess real physical significance: the groupings that appear in Table 15.2 were obtained on the basis of energies and ℓ-values, and <u>not</u> on the basis of J-values, but the groupings thus obtained show interesting regularities with respect to the J-values.†

---

†Although the bandhead of the $\Lambda\dot{3}\dot{3}$ rotational band is manifested experimentally as a $J = 3/2$ D state, it seems probable from the systematics of Table 15.2 that this bandhead appears as a $J = 1/2$ state in its rotational modes.

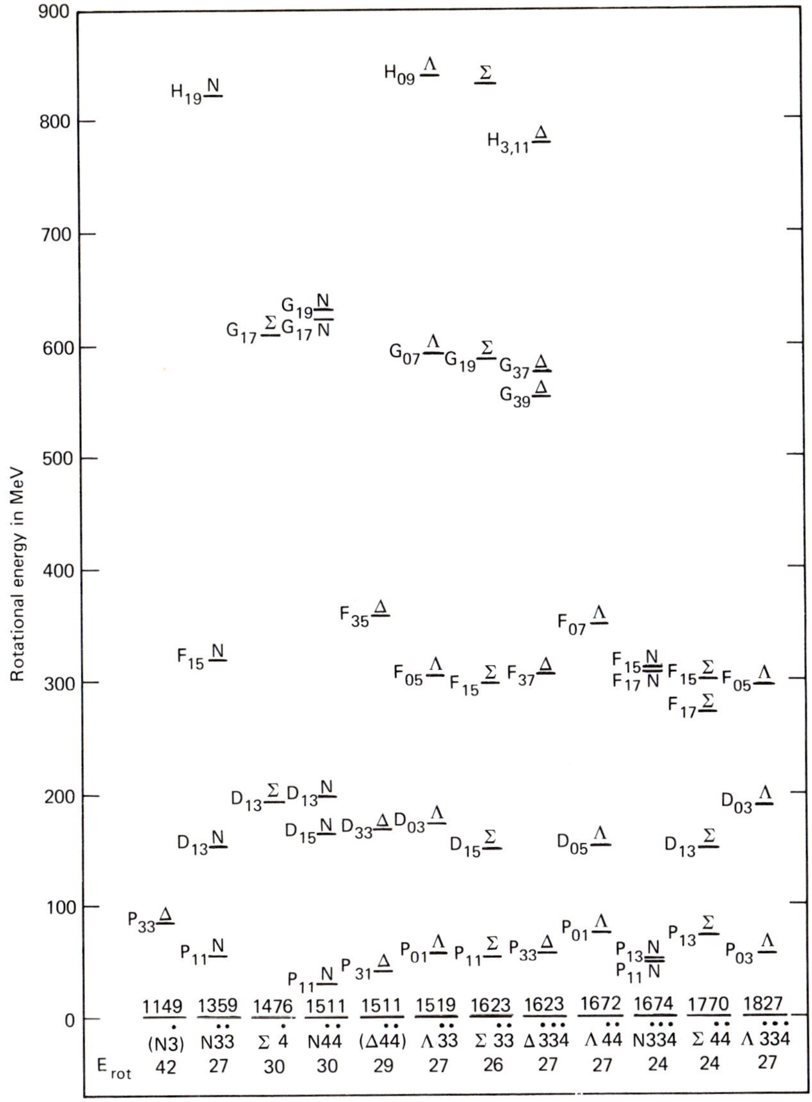

Figure 15.6. The baryon and hyperon rotational levels of Figs. 15.3 - 15.5 combined together into one overall level diagram (some high-spin levels with rotational energies > 900 MeV are not included). As can be seen in the figure, these baryon rotational bands form a consistent pattern, with the rotational energies being about the same for each bandhead.

Table 15.2. The ℓ_- and ℓ_+ variation of the baryon resonances, where $\ell_- \equiv (J = \ell - 1/2)$ and $\ell_+ \equiv (J = \ell + 1/2)$, and where $\ell = 0,1,2,3,4,5,6$ corresponds to S,P, D,F,G,H,I waves. The baryon rotational levels shown in Table 15.2 are the ones that are displayed graphically in Fig. 15.6. In general, the ℓ_- amplitude resonates at the lowest mass value, and the ℓ_+ and ℓ_- amplitudes then resonate alternately with increasing mass. The left superscript on each $\ell_\pm$ state indicates the confidence-level rating assigned to this state in Table 15.1 (4 = ****, etc.).

$N\dot{3}$	--	$^{0}P_-$					
$N\dot{3}\dot{3}$	--	$^{4}P_-$	$^{4}D_-$	$^{4}F_-$		$^{3}H_-$	
$N\dot{4}\dot{4}$	^{4}S	$^{0}P_+$	$^{4}D_+$		$^{3}G_-$		$^{3}I_-$
			$^{2}D_-$		$^{3}G_+$		
$N\dot{3}\dot{3}\dot{4}$	^{4}S	$^{3}P_-$		$^{3}F_+$			
		$^{3}P_+$		$^{1}F_-$			
$\Delta\dot{3}$	--	$^{4}P_+$					
$\Delta\dot{4}\dot{4}$	--	$^{0}P_-$	$^{3}D_-$	$^{3}F_-$			
$\Delta\dot{3}\dot{3}\dot{4}$	^{4}S	$^{1}P_+$		$^{4}F_+$	$^{2}G_+$	$^{3}H_+$	
					$^{0}G_-$		
$\Lambda\dot{3}\dot{3}$	$^{4}D_-$	$^{1}P_-$	$^{4}D_-$	$^{4}F_-$	$^{4}G_-$	$^{4}H_-$	
$\Lambda\dot{4}\dot{4}$	^{4}S	$^{2}P_-$	$^{4}D_+$	$^{1}F_+$			
$\Lambda\dot{3}\dot{3}\dot{4}$	^{2}S	$^{3}P_+$	$^{2}D_-$	$^{2}F_-$			
$\Sigma\dot{4}$	"S"		$^{4}D_-$		$^{2}G_-$		
$\Sigma\dot{3}\dot{3}$	^{2}S	$^{2}P_-$	$^{4}D_+$	$^{4}F_-$	$^{4}G_+$		
$\Sigma\dot{4}\dot{4}$	^{3}S	$^{1}P_+$	$^{3}D_-$	$^{4}F_+$			
				$^{1}F_-$			

At the bottom of each rotational band in Fig. 15.6 is shown the average value of E_{rot} in MeV for the levels in that band, where E_{rot} is defined in Eq. (15.2). As can be seen, the experimental values for E_{rot}, while showing a slight tendency to decrease with increasing mass values, are remarkably constant over the whole range of bandhead masses. In Chapter 16 we compare these results with the results obtained for the meson rotational bands; and in Chapter 18, where we discuss the geometry of the elementary particle, we try to see what conclusions can be drawn from these data. The baryon rotational bands shown in Fig. 15.6 are summarized in tabular form in Appendix H.

As the final result in this section, we demonstrate some interesting similarities between the N and Λ rotational bands, and also between the Δ and Σ rotational bands. The similarities between the N and Λ rotational bands are illustrated in Fig. 15.7, and the similarities between the Δ and Σ rotational bands are illustrated in Fig. 15.8; the rotational bands for these figures are taken from Fig. 15.6, but they are replotted on absolute mass scales in order to more clearly illustrate the nature of the similarities.

Figure 15.7 shows a pair of matching N and Λ rotational bands which occur a little above 1500 MeV, and another pair of matching bands which occur at about 1670 MeV – roughly 140 MeV higher. Figure 15.8 similarly shows a pair of matching Σ and Δ rotational bands which occur at about 1500 MeV, and another pair of matching bands which occur at 1623 MeV – again roughly 140 MeV higher. Some of the rotational bands shown in these figures have missing levels, but the matching (N,Λ) and (Σ,Δ) rotational bands have, in addition to their similar bandhead mass values, at least one well-determined level in common, and the occupation of the levels in the matching pairs tends to be complementary.

In the next section we study the completeness of the rotational bands in more detail.

15.6 Doorway Excitations and Relativistic Constraints

It is apparent from a study of Fig. 15.6 that the baryon rotational bands have some missing levels, and also that there seem to be some overall constraints which limit the number of allowed levels. The occurrence of missing levels may be at least partially explained by the concept of "doorway excitations", which is another idea that we take over from nuclear physics. The origin of an overall constraint on rotational levels, on the other hand, may follow from a limitation imposed by special relativity which does not come into play in the case of nuclear physics.

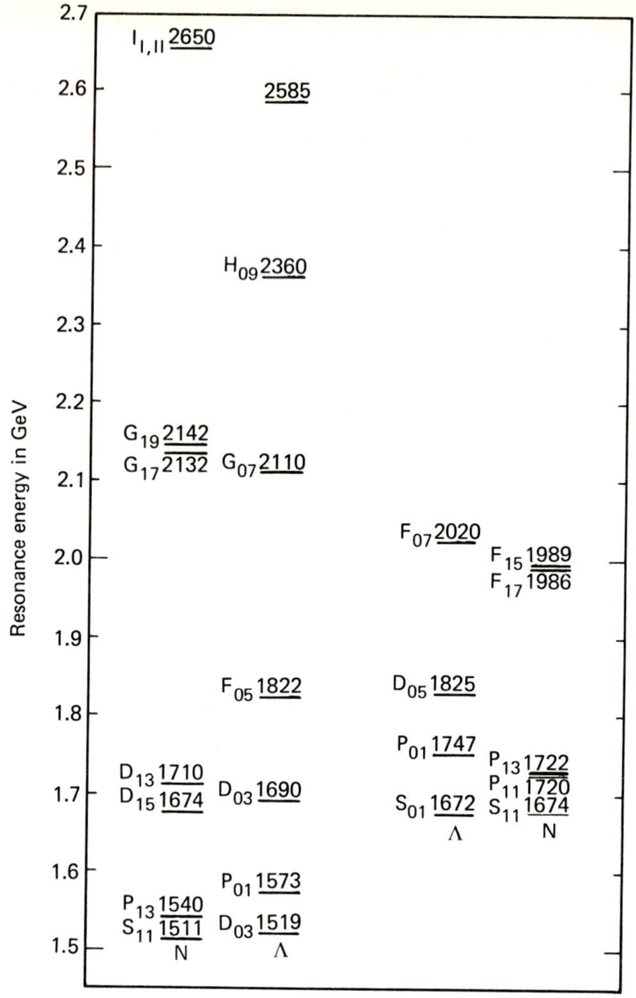

Figure 15.7. A comparison between N and Λ rotational bands. As can be seen, there are two closely-matched pairs of rotational bands, with almost identical rotational energies in each N and Λ pair. The filling of the levels in the matching pairs tends to be complementary.

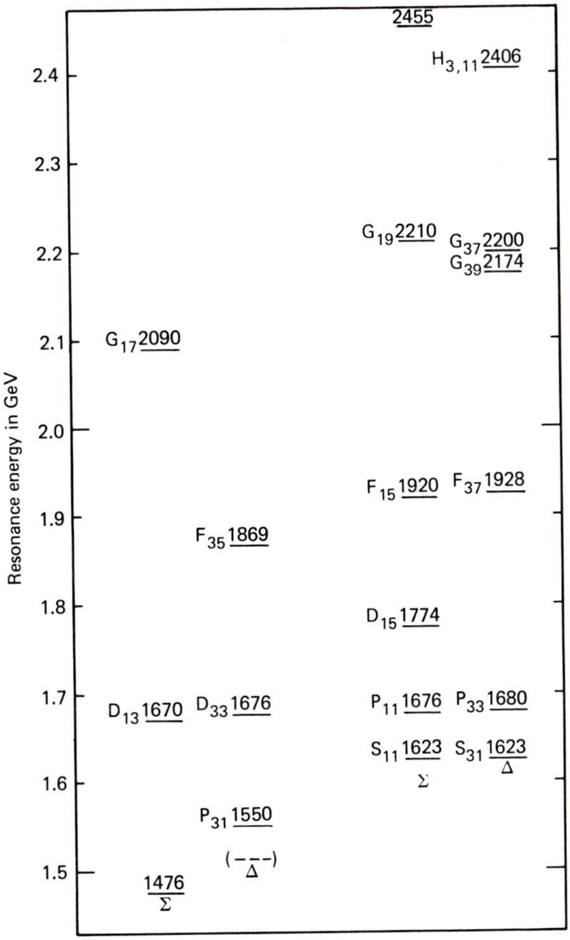

Figure 15.8. A comparison between Σ and Δ rotational bands. As can be seen, there are two closely-matched pairs of rotational bands, with almost identical rotational energies in each Σ and Δ pair. The filling of the levels in the matching pairs tends to be complementary.

In nuclear physics, a "doorway excitation" is a real or virtual excitation which presumably mediates the transition from an incident single particle state to an intermediate compound nucleus state; the doorway excitation may or may not be an observable phenomenon. It is possible that a similar process mediates baryon excitations. In order to investigate the idea of "baryon doorway excitations", we should plot the baryon resonances on an absolute mass scale rather than in the form of an energy level diagram, since all of the baryon and hyperon resonances represent excited states of the same initial ground state – the nucleon. Such a plot is shown in Fig. 15.9, where we have included only the N and Λ families, and where we have taken only the rotationless excitations of Fig. 15.1 plus the rotational D-, F-, and G-levels (which are well-determined experimentally) from these families. As can be seen in Fig. 15.9, the N and Λ families both have all of the narrow-width S-states and broad-width D-, F-, and G-states appearing at the same rather-sharply-quantized mass values. Since Fig. 15.9 includes all of the resonances in these categories which have been reported, it seems as if there are "excitation doorways" at mass values of about 1520 MeV, 1690 MeV, 1820 MeV, 2020 MeV, and 2100 MeV through which all of these N and Λ resonances can be excited. If these doorways exist, they could account for the absence of certain levels in the N and Λ rotational bands.

The "doorway excitation" band at about 1520 MeV in Fig. 15.9 is logically associated with an N + η type of excitation, and the band at about 1690 MeV is logically associated with an N + ρ type of excitation, where η and ρ are the eta and rho mesons, respectively. The 1670 - 1700 MeV "doorway excitation" is especially dominant, and it includes many prominent Δ and Σ resonances, and also the strangeness S = -3 Ω hyperon. The Ξ resonances that are listed in Table 15.1 also show a correlation with these same excitation bands.

As a final comment in connection with Fig. 15.9, we call attention to the pairs of resonances $N(1674)D_{15}\Gamma 159$, $N(1680)F_{15}\Gamma 123$ and $\Lambda(1825)D_{05}\Gamma 94$, $\Lambda(1822)F_{05}\Gamma 81$. Both of these pairs represent $J^P = 5/2^-$ and $5/2^+$ enhancements which are observed at the same mass value and with about the same width. This again raises the question, which was raised initially in connection with the matching S- and P-states in Fig. 15.1, as to how well the parities of these states are actually determined by the data.

The last topic we discuss in this section is that of relativistic constraints on rotational motion. These constraints arise from the fact that the periphery of the rotating system cannot exceed the velocity of light c. It is easy to demonstrate by first-order estimates that these relativistic constraints must be important, but we cannot calculate the actual limits with any accuracy since we don't know how to pinpoint the location of the periphery. However, one interesting result which emerges here is that when we study these relativistic constraints, we are forced to the same conclusion that we reached earlier on the basis of the observed moments of

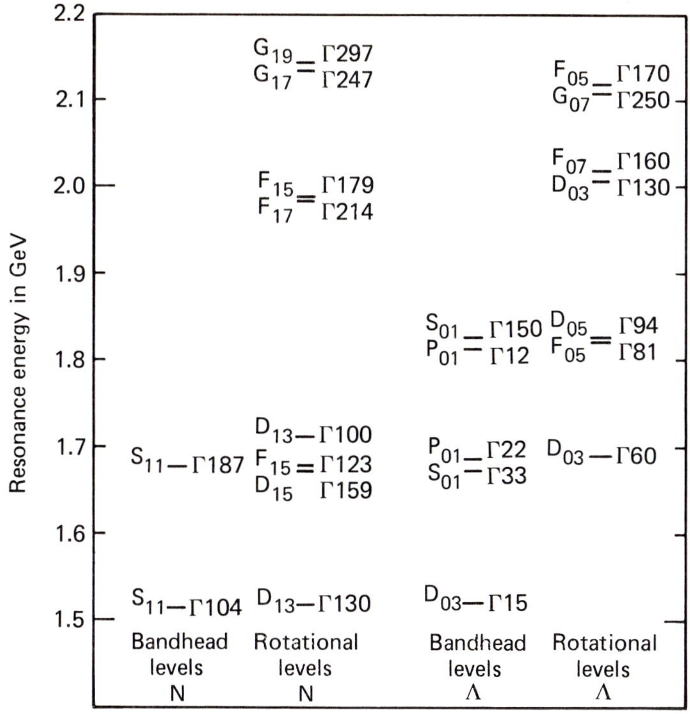

Figure 15.9. A plot of the N and Λ rotationless excitations of Fig. 15.1 together with the broad-width D, F, and G resonances for the N and Λ families (these are the resonances whose masses and widths are most reliably determined). As can be seen, both the bandhead levels and the broad rotational levels occur at the same quantized mass values. This suggests the existence of "excitation doorways" which are common to the N and Λ resonances, and which lead to the excitation of both rotationless bandheads and rotational levels within the mass band of the doorway.

inertia for the baryon resonances — namely, that as the mass of the rotational bandhead increases, its radius necessarily decreases.

Let us make an estimate for the relativistic limit on the rotation of a nucleon. Using the same nucleon model as we did in connection with the calculations of Eqs. (15.3) and (15.4), we picture the nucleon as a three-quark cluster that is analogous to the two-quark rho meson cluster shown in Fig. 13.1. In this picture of the nucleon, if R is the radius of a quark, then $R + \sqrt{4/3}\, R = R_p$ is the radius of the periphery of the nucleon. Choosing R = 0.63 fm, as in Eq. (15.3), we obtain R_p = 1.36 fm. Now the limit on angular velocities is $\omega_{max} = c/R_p$, so that, setting c = 1, we have $\omega_{max} = 0.74\ \text{fm}^{-1}$. Then, since $I_{nucleon} = 646\ \text{MeV fm}^2$ [Eq. (15.3)], the maximum possible angular momentum for the nucleon is given by the equation $\sqrt{\ell_{max}(\ell_{max}+1)}\hbar = I_{nucleon}\omega_{max} = 476$ MeV-fm. Finally, inserting the value $\hbar$ = 197 MeV-fm (in these units), we obtain $\ell_{max}(\ell_{max} + 1) = 2.41$, which gives $\ell_{max} \simeq 2.0$. Thus if the nucleon itself had rotational excitations (which it apparently does not), it would be limited to P-states and D-states. If we repeat this calculation for the rho meson configuration of Fig. 13.1, we obtain $\ell_{max} \simeq 1.0$, which explains why the rotating $\ddot{S}\dot{S}$ bandhead of the rho appears strongly as a P-state resonance, but does not appear as a D-state resonance (see Chapter 13).

If we now extend these calculations to the baryon bandheads of Fig. 15.6, we encounter a difficulty. From Eq. (15.4), the calculated value of E_{rot} for the nucleon is $E_{rot} \simeq 30$ MeV, which, as can be seen in Fig. 15.6, is roughly the same as the E_{rot} values for the higher-mass baryon rotational bands. Thus, from Eq. (15.2), the bandheads for the various baryon rotational bands all have moments-of-inertia I which are comparable to that of the nucleon. Since the rotational limit ℓ_{max} depends only on I and R_p, and since the values of I are roughly the same for all of the baryon rotational bands as for the nucleon, then if R_p is the same for these higher-mass bandheads as it is for the nucleon, we obtain $\ell_{max} \simeq 2$ as the calculated limit for all of the baryon rotational bands, which is obviously incorrect. However, in order to account for the constancy of the E_{rot} values in Fig. 15.6, we have already been forced to the conclusion that R_p, the radius of the periphery, must decrease as the bandhead mass increases; the present analysis of relativistic limitations on baryon rotations substantiates that conclusion.

In order to account for the observed constancy of the E_{rot} values in Fig. 15.6, the effective radii of the higher-mass bandheads relative to that of the nucleon must be reduced by a factor which is typically 0.8. This decrease in radius leads to the calculated value $\ell_{max} \simeq 3$ as the relativistic limit on rotations for these bandheads, and in fact many of the rotational bands in Fig. 15.6 extend only up to F states. However, some of the bands in Fig. 15.6 extend up to ℓ = 5 H levels and ℓ = 6 I levels. In analyzing this situation, we encounter another conceptual problem. As we have concluded above, increasing the mass of a bandhead leads to a decrease in the overall radius of the bandhead. Now, the highest rotational levels in Fig. 15.6 have rotational energies which amount to as much as 1 GeV — over one-third of the total energy of the resonance. On the basis of the above line of reasoning, it appears as if the rotational energy of a resonance should also contribute to the shrinking of the overall size of the resonance. Thus, from this point of view, the higher-ℓ resonances in a rotational band have smaller effective sizes than the lower-ℓ resonances, which is a result that seems to be required from our calculations of the relativistic constraints imposed on rotations. If this is true, then we can logically inquire if the concept of a baryon rotational band even makes any sense. However, the constancy of the E_{rot} values in Fig. 15.6 now comes to our rescue. From the empirical E_{rot} values in Fig. 15.6, we see that as the mass of a bandhead increases, the radius shrinks in such a manner as to keep E_{rot} approximately constant. Hence if large rotational energies cause an additional shrinking, the value for E_{rot} should still remain approximately constant. Since the definition of a rotational band is a set of levels which have a common bandhead mass and a common value for E_{rot}, these systematically-shrinking resonances still fit this definition. In the area of nuclear physics, rotational energies are very small, so that these large relativistic rotations do not come into play. However, even for nuclear rotational bands, the moment-of-inertia I of the band is often observed to vary as the ℓ-value is increased, so that rotational bands with variable moments of inertia are an observed phenomenon.

In the present studies, we do not attempt to calculate relativistic constraints on hadron rotations in any detail. However, on the basis of the estimates that have been presented here, we can understand why it is that the lowest-mass rotational bands extend only up to $\ell = 1$ P waves, whereas higher-mass rotational bands extend systematically up to higher ℓ-values.

It seems to be a universal result in the baryon resonances, and also (as we will see in Chapter 16) in the meson resonances, that the "size" of a resonance structure varies inversely with the mass of the structure. This is the same result that we deduced for the muon model of Chapter 7: in order to be in agreement with experiment, the radius R_μ of the muon must be proportional to its Compton wavelength — $R_\mu \propto \hbar/m_\mu c$ — for all values of the mass m_μ, so that as m_μ increases, R_μ decreases. This same result may also be in evidence in hadron production processes, since it has been found[7] that "the intrinsic production cross section (σ) of a particle of mass m varies as $1/m^2$" (which, as a consequence of the fact that $\sigma \propto R^2$, leads to the result that $R \propto 1/m$); in particular, this $1/m^2$ scaling law "correctly accounts for the production yields of K^-, $\bar{p}$, Λ, and Σ particles"[7], and "the total ψp cross section ... is also consistent with a $1/m^2$ dependence"[7]. Finally, going to a completely different scale of phenomena, we note that this same result is also found in astronomical systems, where normal stars, white dwarfs, and neutron stars exhibit successively increasing densities[8].

15.7 Rotation Invariance and Overlapping Levels

The mapping of the baryon and hyperon rotational bands that is shown in Fig. 15.6 reveals that these bands form a reasonably-consistent overall pattern. However, there are a number of missing levels in these bands; also, there appears to be an "envelope" which limits the maximum number of levels in a band. In Sec. 15.6 we discussed the idea of "doorway excitations", which may account for some of the missing levels; and we also discussed the limitations imposed on orbital rotations by special relativity (the periphery of the rotating structure cannot exceed the velocity of light), which may account for the shape of the "envelope". In the present section we briefly mention two other factors that may account for missing levels – namely, rotation invariance and overlapping levels.

Rotation invariance is another concept that we borrow from nuclear physics.[9] If an axially-symmetric rotor has fore-and-aft symmetry (e.g., is not pear-shaped), so that a rotation of 180° around a perpendicular to the axis of symmetry leaves the system unchanged, then the rotor is said to be "rotationally invariant", which leads to the result that the rotational band contains only even or odd ℓ-values.[9] Strictly speaking, rotation invariance applies only to spin zero rotors. However, it may also apply to the highly nonadiabatic hadron rotations, since the very rapid rotations tend to decouple the spin from the orbital angular momentum, particularly at the higher ℓ-values[†]. As can be seen in Figs. 15.4 - 15.8, there is a tendency in some of the rotational bands for the levels to be populated only with alternate ℓ-values (e.g., see the $\Sigma\dot{4}$ and $N\ddot{4}\ddot{4}$ rotational bands in Fig. 15.5); this may indicate that these bandheads or rotors are rotationally invariant, which would provide an explanation for the missing levels.

One final factor that can account for missing levels is the tendency for hadron rotational bands to overlap. We saw in Fig. 15.1 that the bandheads in a baryon or hyperon family are separated in mass by about 140 or 210 MeV. And we saw in Fig. 15.2 that baryon rotational energies are very large – as much as 1000 MeV or more. Thus the rotational levels from the various rotational bands in a baryon

---

[†]Evidence for this result is indicated by the occurrence of matching pairs of $J = \ell + 1/2$ and $J = \ell - 1/2$ resonances which appear with about the same mass values (see Fig. 15.6).

family are interleaved. Furthermore, since the values for E_{rot} decrease slightly with increasing bandhead mass, as is shown in Fig. 15.6, there is a tendency for the higher-ℓ levels in two adjacent rotational bands to coincide. This result is brought out empirically in Tables H1 and H4 of Appendix H, where it is demonstrated that we can place certain G, H, and I levels into either of two different rotational bands and obtain consistent fits to the S, P, D, and F levels in these bands. Thus the assignment of these high-ℓ levels is phenomenologically ambiguous, and they may in some cases reflect contributions from more than one rotational band. In particular, the broad widths of the N-family G-levels in Table H1, and the fact that the G-level values for E_{rot} are somewhat high in the $N\dot{4}\dot{4}$ rotational band and somewhat low in the $N\dot{3}\dot{3}\dot{4}$ rotational band in Table H1, suggest that these very broad G-levels may actually be common to both of these rotational bands.

The mapping considerations that we have discussed in Sec. 15.6 and 15.7 will remain at the level of conjectures until specific mathematical models are developed for the rapidly-rotating baryon and hyperon bandheads. However, doorway excitations and rotation invariance are known to be of importance in the area of nuclear physics, and we have repeatedly stressed the similarities between these hadron rotations and nuclear rotations. Also, it seems clear that hadron rotations are rapid enough to bring relativistic constraints into operation. Thus, although we do not yet have a mathematical formalism which gives a detailed accounting of the rotational bands that are shown in Figs. 15.3 - 15.8, these rotational bands seem to be in reasonable agreement with the results that we expect by extrapolating our knowledge from other areas of physics. In this context, the intrinsic incompleteness of the πN and $\bar{K}N$ data sets which underlie the resonances of Table 15.1 must be continually kept in mind. In spite of the relative stability of the reported πN and $\bar{K}N$ phase shift results in recent years, new resonances such as the $\Lambda(1817)P_{01}\Gamma 12$ level[3] appear from time to time to remind us that the πN and $\bar{K}N$ scattering matrices are not yet unambiguously determined. The smooth variation of the ℓ_+ and ℓ_- scattering amplitudes shown in Table 15.2 may be an indication of the powerful constraint imposed by the requirement of energy continuity when an incomplete data set is used. And if a model-dependent form is superimposed on the energy dependence, then it becomes a very difficult task to extricate experiment from theory, as K. C. Wali reminds us in the quotation given at the beginning of Sec. 15.1. In the opinion of the present author[2], the question of what constitutes a complete spectrum of baryon and hyperon resonances is far from being a resolved matter at the present time — either experimentally or theoretically.

Our final task in the present chapter is to summarize the efficiency of our mapping procedure for the baryon and hyperon resonances, and to briefly compare it to the efficiency which has been achieved in the more conventional SU(3) approach to the problem. This task is carried out in the next section.

15.8 Summary and a Comparison with SU(3)

In the preceding sections we discussed the problems and ramifications that are associated with the task of mapping the baryon resonances into nuclear-physics-type rotational bands, and we made specific assignments of the observed resonances into these rotational bands. Having done this, it is important that we now make an *a posteriori* assessment of the efficacy of this procedure. In Chapter 19 we will discuss the accuracy of the mass values for these baryon resonances, so our evaluation here has to do just with the number of resonances that were actually mapped versus the number of resonances that have been observed experimentally.

We cannot really evaluate the mapping efficiency until we know that we have a "complete set" of experimental measurements. In this regard, the data on baryon resonances (as summarized in Table 15.1) have pretty well stabilized in recent years, so that, generally speaking, all phase shift analyses agree on the same overall set of resonances, although they may differ in the locations and widths of some of these resonances. However, as we have mentioned in the discussions above, the πN and $\bar{K}N$ experimental data are known to be incomplete: the specification of the scattering matrix for the scattering of a spin 0 particle off a spin 1/2 particle requires *four* independent spin measurements at each energy; but, with just a few exceptions, only *two* measurements are available. The linking together of experiments at different energies by means of an assigned energy dependence for the phase shifts removes some of the ambiguities that are inherent in the data, but it may also tend to bias the data. The unexpected appearance recently of some narrow-width P waves[3] is an indication that the baryon resonance data may not actually be as complete as they appear to be, and the problems with parity ambiguities that we discussed above may be another indication of this same situation. With this caveat, we now proceed to a discussion of the mapping efficiency that has been achieved for the baryon resonances.

The baryon and hyperon resonances that have been identified experimentally are listed in Table 15.1, and detailed references to these data are given in Appendix B. In evaluating the efficiency of our mapping procedure for these resonances, we exclude the N, Λ, and Σ "ground state" resonances because they fall into a somewhat different category, and we exclude the Ξ resonances because they have no spin identification (these resonances, as we have seen, can be accounted-for within the present light-quark formalism). In addition, we exclude the peaks above 2500 MeV because they have no spin identification (the N(2650)$I_{1,11}$ resonance is a possible exception), and we exclude a few identified S-, P-, and D-resonances which occur above 1900 MeV; these states can in principle be mapped, but the information on them is fragmentary, so that they do not constitute a good test for theoretical models. Finally, we exclude the controversial Group III resonances of Table 15.1, although, as we have seen in Fig. 15.1, they fit in naturally with the present formalism for the rotationless

S-states. After all of these exclusions, we are left with 34 N, Δ, Λ, and Σ resonances from Group I in Table 15.1, and 24 resonances from Group II; these 58 resonances constitute a reasonable data base against which to test theoretical models.

Of the 58 baryon resonances in this selected experimental data base, 57 are mapped by the S-state levels of Fig. 15.1 and the rotational levels of Fig. 15.6 (the $\Delta(1889)D_{35}$ resonance is the only one that does not fit naturally into these rotational bands). Thus the mapping efficiency is 57 ÷ 58 = 98%. Of course, some of the rotational levels – especially P-states – appear with anomalous mass values, but the mass values for many of these states are not well known experimentally, and the mass values given by these rotational bands compare favorably with the mass values that are given (for example) by SU(3). As a way of taking these mass anomalies into account, let us claim an overall mapping efficiency of (say) 90% rather than 98%.

In addition to the mapping efficiency of the observed levels, the other question to determine is the number of levels which are predicted but not seen experimentally – that is, the predictive efficiency. (As we discussed above, "doorway excitations" may be responsible for some missing levels, and relativistic constraints serve to set limits for ℓ_{max} in each rotational band.) As a way of estimating the predictive efficiency of the rotational bands in Fig. 15.6, we simply count the number of missing levels (including S-states). This procedure gives 8 levels, including 2 S-states, which would be required in order to fill out all of the observed rotational bands shown in Fig. 15.6. Thus if 57 + 8 = 65 resonances constitute a "complete set" of baryon rotational levels, the predictive efficiency is 57 ÷ 65 = 88%. Hence the mapping and predictive efficiencies for the baryon rotational bands of Fig. 15.6 are 90% and 88%, respectively.† Of course, these are only crude estimates, but they serve as a basis for similar comparisons to other models.

As a comparison to the present light-quark model and its nuclear-physics-type rotational bands, let us examine the original SU(3) symmetry model, whose fits to the experimental baryon data are summarized in a recent Review of Particle Properties[10]. In this summary, all of the SU(3) baryon multiplets are listed which have two or more identified members (a complete multiplet consists of four or five members). These "established" SU(3) multiplets include 18 of the 34 Group I resonances listed in Table 15.1, but they include none of the 24 Group II resonances in Table 15.1 (see Section B1 in Appendix B for a comment on this result; this situation is especially serious in view of the quotation by K. C. Wali: the theory which is to be "proved"

†The present light-quark model also accounts for 8 or 9 of the 12 reported Ξ resonances as S-states, and it accounts for 11 of the 12 Group III resonances.

is used for extracting parameters from the data). The SU(3) mapping efficiency is 18 ÷ 58 = 31%, as compared to 90% or so for the systematics of Fig. 15.6.

The predictive efficiency for the SU(3) model is more difficult to pin down than the mapping efficiency. There are 40 Group I and Group II resonances in Table 15.1 which are not mapped into the "established"[10] SU(3) baryon multiplets. With respect to these 40 unmapped resonances, we have three possibilities: (1) we can say that these 40 resonances do not fit into SU(3) multiplets, which would constitute a disaster for SU(3); (2) we can say that each of these 40 resonances is the only observed member of an SU(3) multiplet, which would constitute even more of a disaster, since in this latter case there are at least 3 x 40 = 120 other resonances which are predicted but not observed; (3) we can choose some combination of (1) and (2), which is actually the case experimentally. As an additional complication for SU(3), it should be pointed out that the SU(3) assignments for the experimental baryon resonances are made chiefly on the basis of their J^P spin and parity values and not on the basis of their mass values, since the calculation of baryon mass values in (broken) SU(3) is in general not very accurate.

The SU(3) symmetrists are aware of these difficulties in mapping the baryon resonances, and they agree[10] that it is necessary to go to a higher symmetry (e.g., SU(6)) in order to accommodate the observed spectrum of baryon resonances. However, whereas the shift to higher symmetries helps with the mapping efficiency (the observed resonances can all be mapped), it ruins the predictive efficiency, which was already bad enough in SU(3), since the increased dimensionality of the higher symmetries greatly increases the number of predicted states. As one worker in the field remarked[11]: "It is clear that one expects a bewilderingly large number of states". Furthermore, there is no unique prescription as to which higher symmetry to choose, and many different ad hoc choices have been investigated. In particular, now that "charm" has come into vogue, the tendency is to replace SU(3) with SU(4), and to use SU(8) symmetries as a guide for mapping the baryon resonances.

The reason we have emphasized SU(3)-type symmetries in our discussion here is that, in spite of the difficulties we have just cited (which tend to be glossed over), the majority of the elementary particle physicists still consider SU(3) or one of its derivatives to be "the theory" for these baryon resonances. We close the present chapter with the following precautionary quotation about SU(3):

"Too much of what we think we know about the relation between baryons and symmetry theory is still based on wishful thinking. Perhaps this is a legitimate attitude in view of the apparent successes which brought us to the present situation. Still, some of us conservatives find it uneasy to see everyone now heading towards a new spectroscopy, lured by the appearance of charm, while reasonable doubts remain concerning the old spectroscopy. Too many easily accessible experiments are still possible."

Robert D. Tripp

in "The Outlook for Baryon Spectroscopy", a talk presented at the Oxford Conference on Baryon Resonances, July, 1976.

References for Chapter 15.

1. T. G. Trippe et al., "Review of Particle Properties", Rev. Mod. Phys., Vol. 48, No. 2, Part II, April (1976); see pages S147 - S165 and S197 - S210.
2. The present author has had experience with $\bar{K}N$ phase shift analyses as a thesis adviser (see D. Bailey, UCRL 50617, 1969, unpublished; values from this report are quoted on pages S214 and S228 of Ref. 1). He has also had extensive experience with nucleon-nucleon phase shift analyses, in which five types of spin-space data are required — σ, P, D, R, and A (see M. H. Mac Gregor et al., Phys. Rev. 182, 1714 (1969), and the references contained therein).
3. R. D. Tripp, Lawrence Berkeley Laboratory report LBL-5502, July (1976).
4. See Ref. 1, page S240.
5. See B. T. Feld, Models of Elementary Particles, Blaisdell, Waltham (1969), Fig. 16.6 on page 382.
6. M. H. Mac Gregor, Coral Gables Conference on Fundamental Interactions at High Energy, Gordon and Breach (1971), Volume 3, Invited Papers, p. 75; see Table A2 on p. 135 and Table B2 on p. 150.
7. T. K. Gaisser et al., Phys. Rev. D12, 1968 (1975).
8. See M. H. Mac Gregor, Lett. Nuovo Cimento 1, 759 (1971).
9. See M. H. MacGregor, Lett. Nuovo Cimento 2, 846 (1971).

10. See Table I on page S241 of Ref. 1.

11. K. C. Wali, COO-3533-61, "Current status of baryon spectroscopy", August (1975), talk given at the ZGS Summer Symposium on New Directions in Hadron Spectroscopy, Argonne National Laboratory, Argonne, Illinois, July, 1975.

CHAPTER 16. MAPPING THE LOW-MASS MESON RESONANCES

16.1 The Experimental Data

"As you well know, the data from formation experiments are far more extensive and complete than from production experiments. This results in the healthier state of baryon spectroscopy as compared to meson spectroscopy."

K. Lanius, in his rapporteur summary "Meson and Baryon Spectroscopy" given at the XVIIIth International Conference on High Energy Physics, Tbilisi, USSR, July (1976)

In the last chapter we dwelled at some length on the intrinsic shortcomings of the experimental baryon and hyperon data. Thus it comes as somewhat of a shock to realize that the meson and kaon data are in even poorer shape. Much of the baryon data consists of formation experiments, in which the phase shifts for amplitude transitions from a well-defined set of initial states to a reasonably-well-defined set of final states are determined. In these formation experiments, the angular momentum components of the scattering matrix at a single energy are all considered simultaneously, and the energy dependence of the phase shifts then reveals the angular momentum components that exhibit resonant behavior. Most of the meson data, however, consists of production experiments, in which a resonance is deduced on the basis of an observed peak in a total cross section or in an invariant center-of-mass plot. In these production experiments, the quantum numbers of the observed peak must be ascertained by measuring the decay products of the peak in a manner that is essentially divorced from the mechanism which produced the peak in the first place. Unless the peak is well separated from other nearby peaks, and unless the experimental statistics are very good, the quantum numbers of the resonance which corresponds to the peak can be difficult to obtain.

There is another difficulty that is encountered in meson production experiments — namely, the task of determining if a peak even exists. The question of the existence of a peak is ultimately a question of reproducibility. But what do we do if a peak appears in one production channel and not in another expected channel? Even more perplexing are cases where an experiment reveals a peak, and then a follow-up experiment that is designed to closely repeat the original experiment, but with better statistics, fails to confirm the peak. We can illustrate these difficulties with three specific examples:

(1) The kappa meson, $\kappa(725)\Gamma 15$. The kappa is a low-mass narrow-width kaon resonance which was seen in a number of early experiments[1], but which has not been observed in recent experiments. One very recent experiment shows some indications of a kappa peak[2], but this peak may be spurious[3]. The kappa meson is of interest theoretically because it is a strange kaon state which appears just 70 MeV above the opening up of a meson channel at 655 MeV (see Sec. 16.2), and it logically echoes the behavior of the New Particle resonances (Chapter 17), in which the "charmed" kaon state $D^o(1863)$ appears just 70 MeV above the opening up of the $\bar{N}N(1795)$ meson channel. The kappa was originally observed in high-momentum-transfer experiments, and there is some question as to whether these original experiments have been properly repeated.

(2) The $\varepsilon(\sim 700)$, $S^*(993)$, and $\varepsilon'(\sim 1200)$ mesons. These are a series of $J^P = 0^+$ resonances whose existence is required in order to comply with the mapping of states in the conventional quark model of Gell-Mann and Zweig. The $\varepsilon(\sim 700)$ is a low-mass $\pi\pi$ enhancement that was listed in the 1974 edition of the Review of Particle Properties[4], but which disappeared in the 1976 edition[5], being replaced by the higher-mass S^* and ε' states. The S^* resonance has been identified on the basis of a number of experiments and phase shift analyses[5], but the question still remains as to whether it is a narrow-width or a broad-width effect[6]. The $\varepsilon'(\sim 1200)$ resonance is also based on an extensive mass of experimental data[5], but there have been some recent experimental disagreements as to the quantum numbers of states observed in this mass region[7]. The reason we discuss these resonances here is that a recent CERN analysis of these experiments[8] suggests that none of the resonances ε, S^*, and ε' may actually exist. To further complicate this situation, the $\delta^{\pm}(976)$ meson state is often assumed to be the $J^P = 0^+$ isotopic spin I = 1 member of the 0^+ SU(3) nonet that contains the S^* and ε'; but the spin of the $\delta^{\pm}$ has not been determined experimentally, and present systematics suggest that it may be a J = 1 state. There have even been suggestions[9] that the δ is not a resonance at all, but just a $K^{\pm}K^o$ threshold effect. Hence the existence of the entire set of $J^P = 0^+$ meson resonances is undecided at the present time.

(3) The CERN missing-mass effects. An early CERN missing-mass experiment[10] provided evidence for — among other things — the following meson effects: (a) a narrow peak, $X^-(1153)\Gamma 16$; (b) a narrow-width spin 2 interference dip, $A_2\text{dip}(1310)\Gamma\sim 10$; (c) a narrow-width high-mass peak, $S(1929)\Gamma\lesssim 35$. The tangled maze of follow-up experiments which appeared subsequent to this CERN experiment not only vividly illustrate the difficulties that are inherent in these meson experiments, but they also involve results that are crucial to the present light-quark approach to elementary particle structure. The $X^-(1153)\Gamma 16$ peak, which is visible in the experimental CERN

data[10], was not in fact claimed as a resonance by the CERN experimenters[10], and the mass and width that we quote here are from a fit to these data by the present author[11]. The existence of this X^- peak was reinforced by the subsequent observation at Argonne[12] of the neutral peak $M^o(1148)\Gamma 15$, which closely matches the peak $X^-(1153)\Gamma 16$. However, a NE/SUNY missing-mass experiment[13], which was designed to reproduce the CERN experiment[10], did not show the X^- peak, and an Argonne follow-up experiment[14] did not show the M^o peak. From a phenomenological point of view, we will demonstrate in Sec. 16.4 that the (M^o,X^-) peak corresponds to the (seen or unseen) bandhead of the most prominent meson rotational band.

The experimental situation becomes even more obscure when we move on to the other two CERN missing-mass effects mentioned above. The deep interference dip in the A_2 meson that was observed in the CERN experiment[10] was also observed in a number of subsequent experiments[5], but it was not observed in several other experiments[5]. In particular, the NE/SUNY experiment[13], which was deliberately patterned after the original CERN experiment, but with improved statistics, did not show this dip. However, three bubble chamber experiments all show the dip[15], and always at the same mass value — 1310 MeV —, and a reanalysis of the original CERN data by Kienzle[16] continues to indicate the presence of the dip — again, right at 1310 MeV (see the discussion of these results in Sec. 8.4 of Chapter 8). The NE/SUNY experiment[13] also failed to confirm the narrow peak $S(1929)\Gamma\leq 35$ that was observed in the CERN experiment[10], but the existence of this peak has now been verified in three recent experiments[17]. As we have already shown in Sec. 5.3 of Chapter 5, this high-mass S peak plays an important role in the systematics of the New Particle resonances (see Fig. 5.4).

The conclusion we draw from these conflicting experimental results is that the production of <u>narrow-width</u> spin 1 or spin 2 meson resonances, which involves the creation of $\dot{S}\dot{S}$ spinor pairs or $\dot{S}\dot{S}\dot{S}\dot{S}$ spinor quartets in rotationless ($\ell = 0$) configurations, is a process that depends sensitively on the momentum transfer conditions in the experiment, so that slight changes in the experimental arrangement can radically alter the probability for producing these resonances. Thus, when a narrow-width meson resonance is seen in one experiment but not in another, it is not necessarily a matter of one experiment being "right" and the other "wrong"; the experiments in general are probably both "right", but the momentum transfer conditions in the two experiments must be somewhat different. From our present viewpoint, the interesting thing about the narrow-width spin 1 meson resonances is that in the experiments where they are observed, they tend to appear at mass locations which are spaced by intervals of 70 MeV and 140 MeV.

In view of the experimental difficulties that we have just cited, it is not possible at this time to define a "complete set" of low-mass meson resonances. In spite of the enormous amount of experimental effort that has gone into the measurements of these low-mass (below 1800 MeV) meson and kaon states, the experimental situation with respect to many of these resonances is still very much up in the air. It is to be hoped that in the rush to the glamorous higher-mass New Particle resonances, physicists will not neglect the problems that still remain in these equally-important low-mass meson states.[†]

In Table 15.1 of Chapter 15 we listed the observed spectrum of baryon and hyperon resonances, and we grouped these resonances according to their confidence-level ratings. However, in the case of the meson resonances, it is not really possible at this time to similarly assign confidence-level ratings to all of the reported meson and kaon resonances: in many cases the actual existence of a meson peak is in dispute, and in other cases the quantum numbers which correspond to a peak are unknown. As we have discussed in some detail in Secs. 5.2 and 5.3 of Chapter 5 and in Chapter 13, one of the most important pieces of information we have about a meson resonance is its width. Accordingly, in Table 16.1, which lists the "observed" spectrum of low-mass meson and kaon resonances, we have grouped the resonances according to width (Groups I, II, and III). For clarity, we have placed a few controversial narrow-width effects into a separate group (Group IV), and we have placed a few very recent (and thus unverified) resonances into another separate group (Group V). We now proceed to a discussion of these resonances. In this discussion, we can draw heavily on our experience with the baryon and hyperon resonances (Chapter 15), since baryons and mesons appear to be quite similar in their nuclear-physics-type rotational aspects.

16.2 Low-Mass Meson Production Thresholds

Before proceding with the mapping of the meson and kaon states of Table 16.1, we first briefly introduce the concept of meson production thresholds. This concept will be of major importance when we come to the New Particle resonances of Chapter 17. Its importance in the present chapter lies mainly in the fact that the identification of low-mass production thresholds paves the way for the subsequent identification of the corresponding high-mass New Particle production thresholds.

The importance of a meson production threshold resides not in its implications for the mesons themselves, but rather in its implications for the underlying substructures which make up these mesons. The occurrence of a meson production threshold

[†]In this connection, we remind the reader of the quotation by R. Tripp at the end of Chapter 15 with respect to the baryon states.

Table 16.1. The low-mass meson and kaon resonances. These are the resonances that are mostly below the $\bar{N}N(1795)$ New Particle threshold. References for these resonances are given in Tables B7 and B8 of Appendix B. Group I in Table 16.1 contains the very-narrow-width spinless resonances, Group II contains the narrow-width spin 1 rotationless resonances, and Group III contains the broad-width rotational levels. Group IV contains a few controversial narrow-width meson states, and Group V contains some recently-reported resonances.

MESONS Resonance	J^P	KAONS Resonance	J^P
Group I. Very-narrow-width resonances			
$\pi^0(135.0)\Gamma\simeq0$	0^-	$K^{\pm}(493.7)\Gamma\simeq0$	0^-
$\pi^{\pm}(139.6)\Gamma\simeq0$	0^-	$K^0(497.7)\Gamma\simeq0$	0^-
$\eta(548.8)\Gamma\simeq0$	0^-		
$M(953\pm2)\Gamma\leq10$			
$\eta'(957.6)\Gamma<1$	0^-		
Group II. Narrow-width resonances			
$\omega(783)\Gamma10$	1^-	$K\pi\pi(1278)\Gamma25$	1^+
$S^*(993)\Gamma40$	0^+	$K\pi\pi(1692)\Gamma26$	
$\phi(1020)\Gamma4$	1^-		
$M^0(1148)\Gamma15$			
$A_{1.5}(1197)\Gamma<29$	1^-		
$D(1285)\Gamma21$	1^+		
$X_1(1415)\Gamma30$			
$E(1416)\Gamma60$	1^+		
$F_1(1542)\Gamma24$			
$X(1681)\Gamma64$			

(continued on next page)

Table 16.1 (Continued)

MESONS Resonance	J^P	KAONS Resonance	J^P
Group III. Broad-width resonances			
$\rho(773)\Gamma152$	1^-	$K^*(892)\Gamma49$	1^-
$\delta^{\pm}(976)\Gamma47$		$Q(1234)\Gamma188$	1^+
$\eta_N(1083)\Gamma98$	2^+	$Q(1368)\Gamma241$	1^+
$A_1(1100)\Gamma\sim300$	1^+	$K_N(1421)\Gamma108$	2^+
$B(1230)\Gamma125$	1^+	$K_N(1673)\Gamma77$	
$\rho'(1250)\Gamma130$	1^-	$L(1765)\Gamma138$	2^-
$f(1271)\Gamma180$	2^+	$K_N(1785)\Gamma100$	3^-
$A_2(1310)\Gamma102$	2^+	$K_N(1840)\Gamma100$	
$\chi_0(1411)\Gamma95$			
$f'(1522)\Gamma62$	2^+		
$\rho'(1600)\Gamma\sim300$	1^-		
$A_3(1640)\Gamma240$	2^-		
$\omega(1667)\Gamma150$	3^-		
$g(1690)\Gamma180$	3^-		
$A_4(1825)\Gamma90$			
$h(2040)\Gamma193$	4^+		
Group IV. Other narrow-width effects			
$\pi\pi(658)\Gamma\sim15$		$\kappa(725)\Gamma15$	
$H(\sim1000)$			
$X^-(1153)\Gamma16$			
$A_2\text{dip}(1310)\Gamma\sim10$	2^+		

(continued on next page)

Table 16.1 (Continued)

MESONS Resonance	J^P	KAONS Resonance	J^P
Group V. Recently-reported resonances			
$\pi\pi(2150)\Gamma 200$	3^-	$K_N(2115)\Gamma 300$	4^+
$\pi\pi(2310)\Gamma 210$	4^+		
$\pi\pi(2480)\Gamma 280$	5^-		

signals the onset of a new type of quark substructure. In particular, if a new class of meson resonances suddenly appears above a certain mass value, this indicates that a new combination of quark substates has become energetically possible at that mass value. By using the threshold mass value plus the associated spin value, we can deduce the properties of this new type of quark substructure.

The meson production thresholds that are observed in the low-mass meson resonances are illustrated in Fig. 16.1. The lowest-mass meson state, the π meson, appears at a mass of 140 MeV, and it signals the opening up of the spin 0 meson production channel. The other very-narrow-width spinless meson states, η and (M,η'), appear at characteristic intervals of 410 MeV above the π meson, and the strange spinless kaon K appears at half the mass of the η' as a manifestation of the bound state $\eta' \equiv \bar{K}K$. These states are all composed of spinless quanta $M \simeq 70$ MeV (see Sec. 5.2 in Chapter 5).

At a mass of roughly 780 MeV in Fig. 16.1, we suddenly have the appearance of two spin 1 meson resonances – the dominant meson states $\rho(773)\Gamma 152$ and $\omega(783)\Gamma 10$. The opening up of this spin 1 meson threshold indicates that a new type of quark substate has become energetically possible – a quark state that is not composed solely of spinless quanta M = 70 MeV. These spin 1 meson and kaon states occur in two quite different forms: (1) the narrow-width resonances ω, ϕ, $A_{1.5}$, $K\pi\pi$, D, E, and F, which have widths $\Gamma \simeq 10 - 25$ MeV; (2) the broad-width resonances ρ, K^*, A_1, B, Q, Q, ..., which have widths $\Gamma \simeq 50 - 150$ MeV. As we have argued in Chapter 13, the broad-width resonances correspond to rotational ($\ell = 1$) states, so that their angular momenta arise from the overall rotational motion; whereas the narrow-width resonances correspond to rotationless ($\ell = 0$) states, so that their spins arise from the intrinsic spins of the quark substates. Thus the narrow-width resonances are the most revealing with respect to their quark substructure. The appearance of the narrow-width spin 1 $\omega(783)\Gamma 10$ meson resonance signals the appearance of a pair of spin 1/2 quark

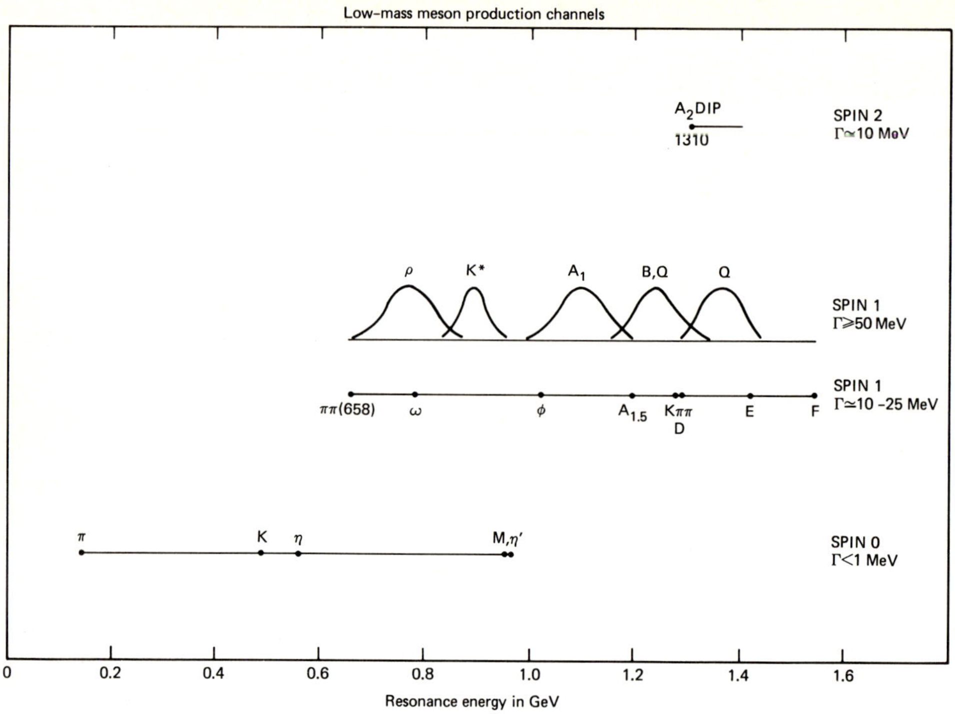

Figure 16.1. Meson production channels for the low-mass meson resonances. The phenomenological significance of these channels is that the occurrence of a meson production threshold signals the onset of a new type of quark substructure. In Fig. 16.1, the opening up of the spin 0 channel signals the appearance of the mass quantum M = 70 MeV in the form $\pi = M\bar{M}$; the opening up of the spin 1 channel signals the appearance of the spinor $\dot{S}$ = 330 MeV in the forms $\dot{S}\dot{S}$(rotationless) = ππ(658) and $\dot{S}\dot{S}$(rotational) = ρ; and the opening up of the spin 2 channel signals the appearance of the spinor quartet $\dot{S}\dot{S}\dot{S}\dot{S}$. These results will take on added importance when we extend this diagram to include the high-mass New Particle meson and kaon resonances, which we do in Fig. 17.1.

substates, which we have denoted in the present book as the spinor pair $\dot{S}\dot{S}$ (see Appendix C). In order to pinpoint the mass of the spinor pair $\dot{S}\dot{S}$, we must pinpoint the mass at which this spin 1 meson channel actually opens up†. This can be done in several different ways: (1) since the three-spinor nucleon has a mass of 939 MeV, a two-spinor meson state should have a mass, apart from small (4%) binding energy effects, of about 2/3 x 939 MeV = 626 MeV; (2) the ρ meson, which is the dominant meson excitation in this energy range, must also be associated with the opening up of this meson channel, and the ρ has a lower-mass threshold (see Figs. 13.2 and 13.8) of about 625 - 650 MeV; (3) there are two narrow-width $\pi\pi$ effects that can be observed at 660 MeV (see Figs. 8.1b, 13.8, and 13.9); (4) if we construct the ω as the light-quark combination $\omega = \dot{S}\dot{S}\pi^0$ (Eq. 13.3), then the $\dot{S}\dot{S}$ spinor pair has a mass (apart from small binding energy effects) of $\omega(783) - \pi^0(135) = 648$ MeV. When all of these effects are analyzed together (see Appendix D), we arrive at an effective $\dot{S}\dot{S}$ spinor pair mass of about 655 MeV, and the appearance of this $\dot{S}\dot{S}$ spinor pair as a component of the meson and kaon resonances is signaled by the opening up of the spin 1 meson production channel at 655 MeV.

As additional confirmation for this result, we note, as is shown in Fig. 16.1, that a narrow-width spin 2 effect, the A_2 interference dip, occurs at 1310 MeV. Now 1310 MeV is just twice the 655 MeV mass of the spin 1 threshold (compare Figs. 8.1b and 8.1c), and it signals the appearance of the spin 2 spinor quartet $\dot{S}\dot{S}\dot{S}\dot{S}$, with the consequent opening up of the spin 2 meson production channel for narrow-width resonances (broad-width spin 2 resonances appear at lower energies as rotational excitations). Unfortunately, we do not have other identified narrow-width spin 2 resonances that can be associated with this production channel. However, the $\dot{S}\dot{S}\dot{S}\dot{S}$ spinor quartet figures prominently as an excitation unit in the New Particle resonances (e.g., see Fig. 8.3).

The real importance of Fig. 16.1 occurs when we extend it to include the opening up of the $\bar{N}N$ channel, which we do in Chapter 17. In fact, the $\bar{N}N$ threshold serves as the dividing line between the low-mass meson resonances of Chapter 16 and the high-mass New Particle resonances of Chapter 17. The sudden onset of the narrow-width New Particle resonances above 1800 MeV signals the appearance of the $\bar{N}N$ configuration as a light-quark substate, just as the onset of the narrow-width spin 1 resonances above 655 MeV signals the appearance of the $\dot{S}\dot{S}$ spinor pair as a light-quark

†In Fig. 5.3 we showed this spin 1 meson channel (for simplicity) as opening up at the mass of the ω; that is, at 783 MeV. However, in Sec. 13.2 of Chapter 13 we presented arguments which indicate that this spin 1 channel actually opens up at a mass of about 655 MeV.

substate. Furthermore, as we will see in Chapter 17, the so-called R-ratio (the ratio of hadron production to muon pair production) has two plateau regions: $R \sim 2.5$ for energies somewhat above 1800 MeV, and $R \sim 5$ for energies somewhat above 3600 MeV. Thus the $\bar{N}N(\sim 1800)$ meson production channel leads to the $R \sim 2.5$ plateau region, and the $\bar{N}N\bar{N}N(\sim 3600)$ meson production channel leads to the $R \sim 5$ plateau region.

We now return to the problem of mapping the low-mass meson resonances.

16.3 Mapping the Meson S-States

The rotationless ($\ell = 0$) low-mass meson and kaon S-states include the very-narrow-width resonances of Group I in Table 16.1 plus the narrow-width resonances of Groups II and IV. The Group I resonances correspond to the spin 0 meson production channel shown in Fig. 16.1, and the Group II and Group IV resonances are associated with the spin 1 and spin 2 meson production channels in Fig. 16.1 (the A_2 dip is the only member of the spin 2 channel).

In terms of the light-quark spinless basis states $1 \equiv M \simeq 70$ MeV, $3 \equiv 3M$, $4 \equiv 4M$, and $7 \equiv 7M$ (Appendix C), the spinless meson and kaon resonances of Group I are reproduced as the light-quark configurations

$$\pi = 1\bar{1}, \quad \eta = 4\bar{4}, \quad (M,\eta') = 7\bar{7}, \quad K = 7, \quad K = \bar{7}. \qquad (16.1)$$

The properties of these resonances were discussed in detail in Sec. 5.2 of Chapter 5, and the mass values that follow from these configurations were summarized in Eq. (9.4) of Chapter 9. When coulomb effects are incorporated (Chapter 19), the light-quark configurations of Eq. (16.1) give absolute mass values for these states that are accurate to within 0.1% (see Table 19.1).

The resonances that correspond to the spin 1 meson production channel of Fig. 16.1 are shown in Fig. 16.2 in the form of an energy level diagram above the basic ground state $\dot{S}\dot{S} = 655$ MeV. Each of these resonances consists of an $\dot{S}\dot{S}$ spinor pair that is combined with 70 MeV-quantized spinless excitation quanta. The ω and (S^*,ϕ) mesons are the light-quark configurations.[†]

$$\omega = \dot{S}\dot{S}\pi \quad \text{and} \quad S^*,\phi = \dot{3}\dot{S}\dot{S}\pi. \qquad (16.2)$$

[†]The H($\sim$1000) meson may correspond to the $\pi\pi\pi$ decay mode of the ϕ meson (see Appendix B); and we identify the S* as the J = 0 form of this configuration.

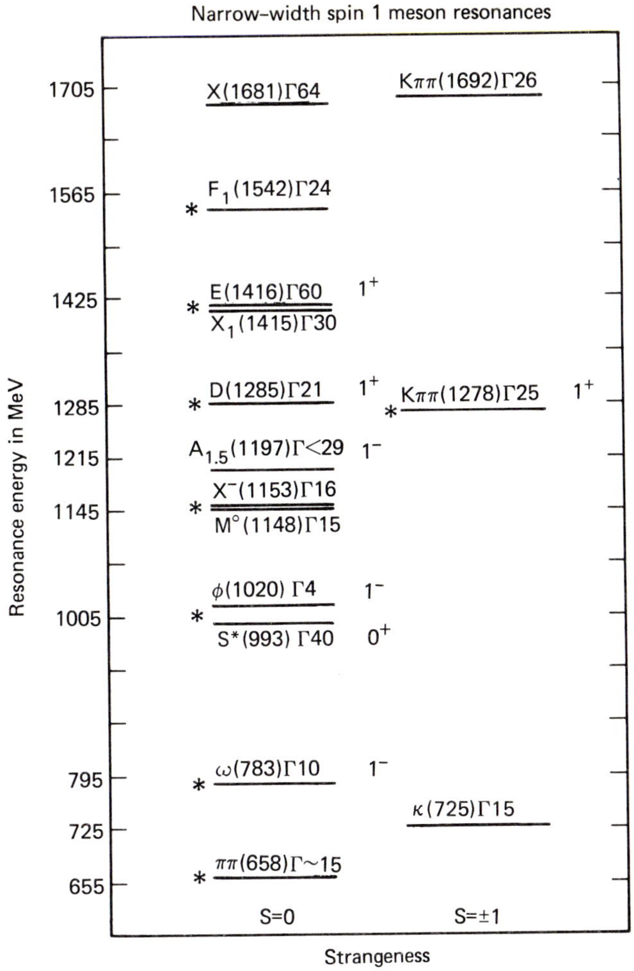

Figure 16.2. The meson resonances that are associated with the narrow-width "spin 1" production channel of Fig. 16.1. These resonances are arranged here in the form of an energy level diagram above the $\ddot{S}\ddot{S}$(658) "ground state" (which is the threshold resonance for this channel in Fig. 16.1). The levels labeled with asterisks serve as bandheads for overlying meson and kaon rotational bands. For energies above 1000 MeV, the meson bandheads appear at regular 140 MeV intervals. As noted in Fig. 16.2, most of the narrow-width resonances shown here are known to be either $J^P = 1^+$ or $J^P = 1^-$ states.

The M(1150), D(1285), E(1416), F_1(1542), and X(1681) levels appear at regular 140 MeV intervals above the (S*,ϕ)(1005) level, thus indicating the successive addition of π = 140 MeV spinless excitation quanta (see Sec. 5.3 in Chapter 5). The levels in Fig. 16.2 that are labeled with asterisks are the levels which serve as bandheads for overlying rotational bands; the weak $A_{1.5}$(1197) level (which may be an $\dot{S}\dot{S}\eta$ configuration) and the high-mass X(1681) level are the only states in Fig. 16.2 that are not identified with rotational bands. [The broad width of the $\delta^{\pm}(976)\Gamma 47$ meson resonance in Table 16.1 suggests that it may be a rotational excitation. Also, the shoulder that can be observed on the high-energy side of the rho meson in Fig. 13.8 suggests the presence of a fairly broad $\pi\pi$ enhancement in this energy range. From the present systematics, these enhancements could be ℓ = 1 rotational levels that correspond to an unseen $\dot{3}\dot{S}\dot{S}$(865) meson bandhead.]

The narrow-width kaon states shown in Fig. 16.2 correspond to the light-quark configurations

$$\kappa(725) = \dot{S}\dot{S}1, \quad K\pi\pi(1278) = K\dot{S}\dot{S}\pi, \quad K\pi\pi(1692) = K\dot{S}\dot{S}\eta. \qquad (16.3)$$

In each of these kaon resonances, the strangeness quantum number S = ±1 is carried by the strange excitation quantum 1 ≡ M or K ≡ 7M ≡ 7. There seems to be a real similarity between the states $\pi\pi$(658), κ(725), ω(783) in Fig. 16.2 and the states $\bar{N}N$(1795), D(1863), $\bar{p}p$(1935), D*(2006) in Fig. 5.4. In each case, a strange narrow-width kaon resonance appears just 70 MeV above the opening up of a meson threshold. Thus the experimental problem of clarifying the status of the κ(725) meson may be an important one from the standpoint of ascertaining the significance of "charm".

The narrow-width meson and kaon states that are shown in Figs. 16.1 and 16.2 include <u>all</u> of the low-mass states that have been reported to date (as summarized in Table 16.1). The important fact to note with respect to the mapping of these narrow-width states is that when they are sorted out into the appropriate meson production channels (as in Fig. 16.2), they exhibit a clear-cut mass quantization that is in intervals or multiples of 70 MeV. This is a result that is expected on the basis of the present light-quark model, but it is not a result that emerges in any obvious manner from the conventional quark model of Gell-Mann and Zweig.

We now move on to an examination of the meson and kaon rotational bands that are associated with these narrow-width levels.

16.4 The Meson Rotational Bands

The rotational ($\ell > 0$) meson and kaon resonances are the broad-width resonances of Groups III and V in Table 16.1. In sorting these resonances into rotational bands, we use the narrow-width resonances from Sec. 16.3 as bandheads, and we assign rotational energies in accordance with the nuclear-physics-type estimates of Chapters 12 and 13 and the empirical results of Chapter 15. In this connection, it should be noted that the very-narrow-width spinless resonances of Group I in Table 16.1 do not seem to have associated rotational bands; the rotational baryon bandheads seem to be restricted to the resonances that contain spinor pairs or spinor triplets.

The identified meson and kaon rotational bands are shown graphically in Fig. 16.3, and they are given in tabular form in Table H5 of Appendix H. As can be seen in Fig. 16.3, the meson rotational bands are quite similar to the baryon rotational bands of Chapter 15, both in their overall shape and in their values for the rotational energy parameter E_{rot} (once the difference in bandhead masses is taken into account). The same light-quark bandhead structures that give rise to the nonstrange meson rotational bands also give rise to the strange kaon rotational bands. In fact, it is only by grouping the meson and kaon rotational bands together, as we have done in Fig. 16.3, that we can obtain a reasonably complete mapping of the rotational levels (compare this to the baryon and hyperon groupings shown in Figs. 15.7 and 15.8).

In order to illustrate the accurate $\ell(\ell+1)$ interval rule that is obeyed by these meson and baryon rotational bands, we have plotted the most complete meson rotational band separately in Fig. 16.4, where it is compared to a set of rotational levels that are calculated from the equation

$$E_\ell = E_o + E_{rot} \times \ell(\ell+1), \qquad (12.13)$$

with E_o = 1150 MeV and $E_{rot} \equiv \hbar^2/2I$ = 44.5 MeV. As the topmost rotational level in this figure, we have included the newly-reported $J^P = 5^-$ level that is shown in Group V of Table 16.1. As can be seen in Fig. 16.4, the agreement between the experimental levels and the calculated levels is very accurate over the entire span of energies included (Fig. 12.5 shows a similar comparison for one of the baryon rotational bands).

Figures 16.1 - 16.3 include essentially all of the low-mass meson and kaon resonances that have been reported to date (Table 16.1). In view of the difficulties with the data that were cited at the beginning of this chapter, we cannot make reliable statements as to the completeness or accuracy of the mapping. In particular,

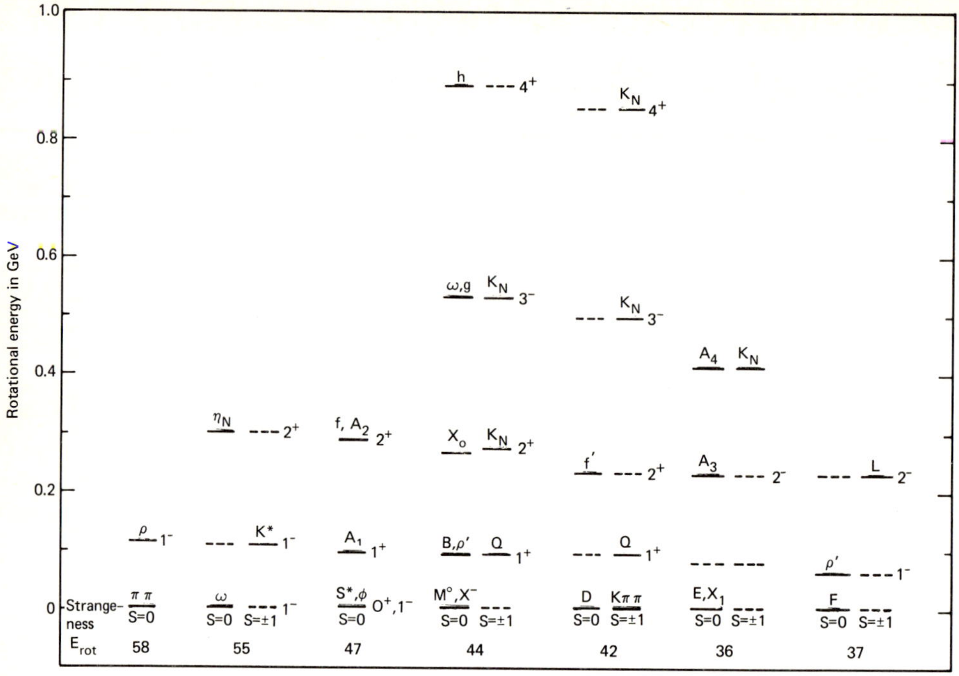

Figure 16.3. The meson and kaon rotational bands that are associated with the bandhead levels of Fig. 16.2 (compare these to the baryon and hyperon rotational bands of Fig. 15.6). Although there are large uncertainties in the data for some of these resonances, the strangeness S = 0 meson resonances and strangeness S = ±1 kaon resonances seem to be associated with the same basic bandheads, and the nonstrange and strange rotational levels tend to occur in a complementary manner (compare this to the paired baryon and hyperon rotational levels in Figs. 15.7 and 15.8). Table H5 in Appendix H gives numerical values for the rotational energy parameters of the meson and kaon rotational bands that are shown in Fig. 16.3. The dominant (M^{o},X^{-}) rotational band in Fig. 16.3 is shown in more detail in Fig. 16.4, where it is extended to a higher mass value.

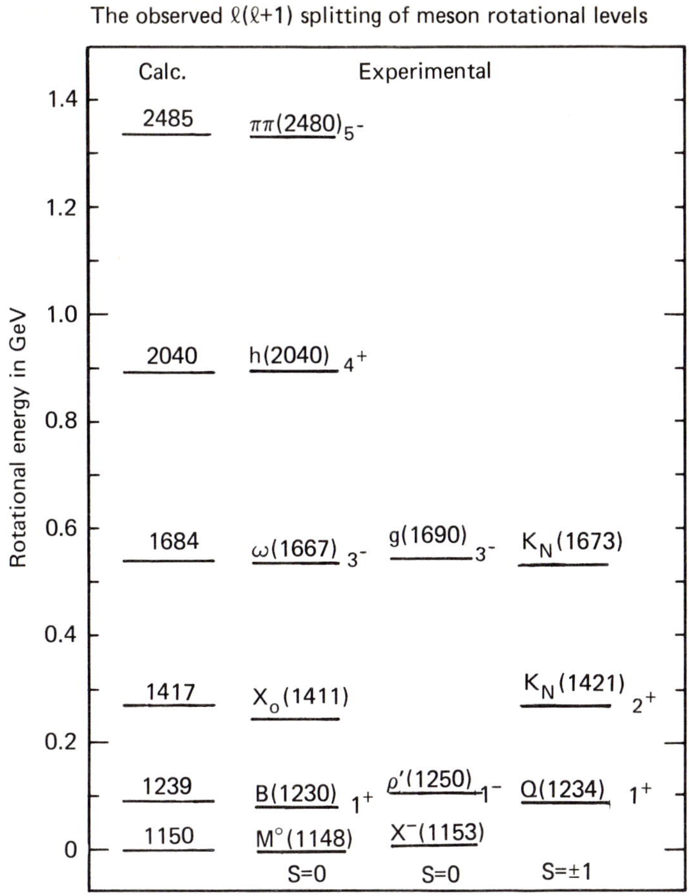

Figure 16.4. A comparison of experimental meson and kaon rotational levels with theoretical levels that are calculated from Eq. 12.13, using the rotational energy parameter E_{rot} = 44.5 MeV. As can be seen in the figure, the calculated $\ell(\ell+1)$ rotational energy rule is very accurately followed by the experimental resonances. A similar result for a prominent baryon rotational band is shown in Fig. 12.5 of Chapter 12.

we have included some levels in Fig. 16.3 whose spins have not yet been determined. However, the levels whose spins are known do fit naturally into the rotational bands of Fig. 16.3.

We can use the present meson systematics to make a few comments about some of the disputed meson and kaon resonances. The $M^o(1148)\Gamma 15$ meson resonance serves as the bandhead for the prominent rotational band shown in Fig. 16.4; thus it seems probable that this resonance actually does exist. The spin 2 $\eta_N(1083)\Gamma 98$ meson resonance appears as the $\ell = 2$ level in the (ω, K^*, η_N) rotational band of Fig. 16.3, so that this resonance probably also exists (see the discussion of this resonance in Sec. 13.7 of Chapter 13 and in Sec. B1 of Appendix B); this is an important result theoretically, because the η_N overfills the 2^+ nonet of the conventional quark model. We continue to predict, on the basis of Fig. 16.1, that the spin 2 interference effect manifested as the A_2 dip is a real effect; and we similarly feel that the $\pi\pi(658)$ dip is also a real effect (see Fig. 8.1). The $\kappa(725)\Gamma 15$ kaon state is a more speculative resonance, but we will demonstrate in the next chapter that it seems to be consistent with the systematics exhibited by the New Particle resonances.

16.5 Comparison to Baryon Rotational Bands

We saw in Chapter 15 that baryon and hyperon rotational bands are very similar to one another, and we saw in the present chapter that meson and kaon rotational bands are also similar to one another. Thus the rotational characteristics of these resonances do not depend on their strangeness values S or isotopic spins I. By comparing the baryon rotational bands with the meson rotational bands, we can now determine if the rotational characteristics depend in an essential manner on the baryon quantum number B.

A visual comparison of Fig. 15.6 with Fig. 16.3 shows that the baryon and meson rotational families are in fact quite similar. In each case an $\ell(\ell+1)$ splitting of the levels is observed, and the numbers of levels in the rotational bands are roughly comparable in the two cases. In order to make a quantitative comparison between the baryon and meson rotational bands, we should study the experimental values that are obtained for the rotational energy parameter E_{rot} in each of these rotational bands. These values for E_{rot} are shown in Fig. 16.5, where they are plotted as a function of the mass of the bandhead. As can be seen in Fig. 16.5, the meson bandhead masses are systematically lighter than the baryon bandhead masses, but the experimental meson and baryon values for E_{rot} form a smoothly continuous set of values. Thus the meson rotational energies are intrinsically comparable to the baryon rotational energies, so that the rotational motion does not depend in an essential manner on the baryon number B, but only on the overall mass of the bandhead.

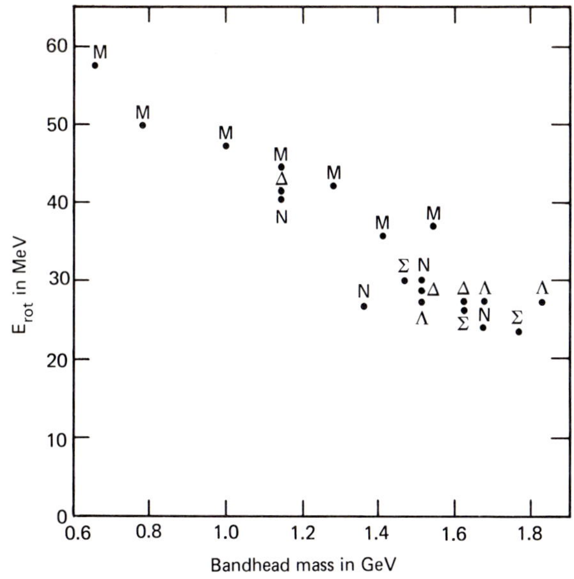

Figure 16.5. Experimental values of the rotational energy parameter E_{rot} in Eq. (12.13) as determined from meson (M) and baryon (N,Δ,Λ,Σ) rotational bands. When plotted as a function of bandhead mass, the E_{rot} values exhibit a rather uniform dependence for all types of resonance, which shows that the moment of inertia of a hadron resonance does not depend on the baryon, strangeness, or isotopic spin quantum numbers — but only on the mechanical properties of the resonance. The values for E_{rot} shown here are used in Chapter 19, Fig. 19.4, to calculate average mass accuracies for all of the hadron rotational levels.

The result that we have just described has an important theoretical ramification. In the quark model of Gell-Mann and Zweig, the quarks in a baryon are given an orbital degree of freedom ℓ; but the orbital motion in the baryon is attributed to a complicated internal rotation of individual quark pairings†, and not to an overall rotation of the three-quark structure. Thus we expect the orbital rotational energy of a Gell-Mann and Zweig baryon to bear little relation to the rotational energy of a

---

†See Fig. 16.6 on page 382 of B. T. Feld's book Models of Elementary Particles, Blaisdell, Waltham (1969).

Gell-Mann and Zweig two-quark meson. However, we find empirically that the rotational energies for baryons and mesons are essentially identical to one another once an allowance has been made for overall differences in mass values, so that both baryons and mesons have nuclear-physics-type rotational excitations which depend only on the moment-of-inertia of the rotating resonance structure. This result does not seem to follow in any obvious manner from the Gell-Mann and Zweig formulation.

We can reinforce this conclusion by extending these results to include rotational motion in very light atomic nuclei, which we do in the next section.

16.6 Comparison to Rotations in Very Light Nuclei

If baryons and mesons have nuclear-physics-type nonadiabatic rotational excitations, as we have argued above, then it is of interest to see if these rotations extrapolate smoothly to match the nonadiabatic rotations that seem to occur in very light atomic nuclei. This is a result that we have already considered in Sec. 12.6 of Chapter 12. Table 12.1 of Chapter 12 lists experimental values for E_{rot} that are deduced from the orbital rotations of the light nuclei. If we plot the E_{rot} values for the lightest of these nuclei together with the E_{rot} values from Fig. 16.5, using a log log scale, we obtain the curve that is shown in Fig. 16.6 (which is similar to Fig. 12.6 in Chapter 12). As can be seen in Fig. 16.6, the rotational energies of mesons, baryons, and light nuclei exhibit a smooth monotonic dependence on the masses of the rotating bandheads, thus indicating that all of these systems have the same fundamental type of rotation. Hence, with respect to their rotational characteristics, baryons and mesons differ from light nuclei mainly in their very small moments of inertia, and their consequently large rotational energies.

16.7 Meson Moments of Inertia

In Chapter 15 we analyzed the rotational systematics of the baryon resonances and came to the rather surprising conclusion that as the mass of a baryon rotational bandhead increases, the overall size of the bandhead decreases. This conclusion was deduced both from the moments of inertia of the bandheads (as obtained from the E_{rot} values shown in Fig. 15.6), and also from the maximum ℓ-values that were observed for the rotational bands (which characterize the peripheral radii, and hence the relativistic limits, that apply to these bands). We find the same situation occurring in the meson rotational systematics of the present chapter. From the values for the rotational energy parameter E_{rot} that are shown in Fig. 16.3, we can calculate that the moments of inertia of the meson bandheads fall off more slowly than $1/m$, where m is the mass of the bandhead. Since a $1/m$ dependence would mean that all bandheads have

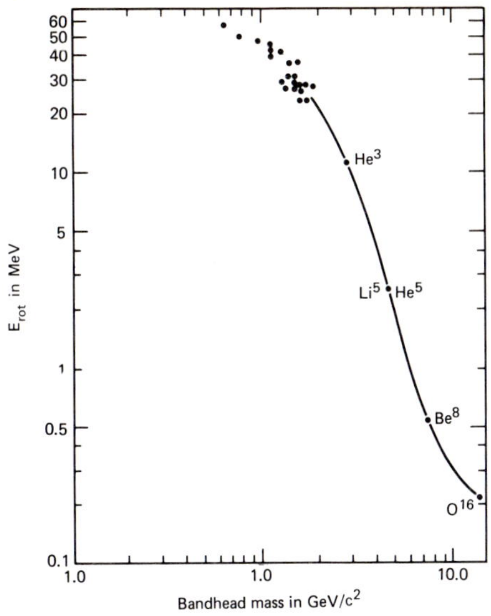

Figure 16.6. Experimental values of the rotational energy parameter E_{rot} for mesons, baryons, and light atomic nuclei, shown on a log log plot. This figure is similar to the moment-of-inertia plot shown in Fig. 12.6, and it indicates that mesons, baryons, and light nuclei all have similar rotational excitations.

the same "size", a dependence that is slower than 1/m means that the size is actually decreasing as the mass increases. Hence we not only have a smooth and continuous behavior for the moments of inertia of the mesons and baryons with increasing mass, as is shown in Fig. 16.5, but we also have a uniform behavior for the "shrinking" of the meson and baryon quark structures with increasing mass. This topic is considered again in Chapter 18, where we discuss the geometry of the elementary particle.

16.8 Summary

With the results in the present chapter, we conclude our study of hadron rotational systematics. We have tried to show, from as many viewpoints as possible, that it makes sense to apply the concepts of nuclear physics also to hadron physics. The systematic results that we obtained by this procedure are internally consistent; they extrapolate to match the results from light atomic nuclei; and they permit us to achieve a reasonably accurate and complete mapping of the observed baryon and meson resonances. This approach is of course quite different from the conventional quark model mapping of these resonances. The conventional quark model, in spite of its early successes, has not led to an accurate mapping of the spectrum of meson resonances; in particular, if all of the resonances that we have shown in Figs. 16.1 - 16.3 actually exist, they overfill the "established" SU(3) meson nonets[5] while still leaving the other meson nonets underfilled.

In the next chapter, the final chapter in Part III, we move from the low-mass meson and kaon resonances to the high-mass New Particle meson and kaon resonances. These New Particle states have very narrow widths, which makes them useful from the standpoint of studying mass quantizations. The narrow widths also mean that the New Particles are rotationless excitations, so that we will be studying a system which is composed more-or-less completely of S-states.

References for Chapter 16.

1. See the review by A. R. Erwin et al., Nucl. Phys. B9, 364 (1969).

2. See Figs. 2(a), 2(c), and 2(d) in the Berkeley report LBL-5534, October, 1976, by G. Goldhaber.

3. The present author was informed by G. Goldhaber (private communication) that this peak has been studied, and that it probably originates from misidentified events (e.g., Ke pairs).

4. V. Chaloupka et al., Phys. Lett. 50B, No. 1, April, 1974.

5. T. G. Trippe et al., Rev. Mod. Phys. 48, No. 2, Part II, April, 1976.

6. S. M. Flatté, Phys. Lett. 63B, 228 (1976).

7. N. M. Cason et al., Phys. Rev. Lett. 36, 1485 (1976), identified a rather narrow-width $K\bar{K}$ S-wave state near 1300 MeV, which they concluded probably has isotopic I = 1. However, A. J. Pawlicki et al., Phys. Rev. Lett. 37, 1666 (1976), while confirming the existence of this state, concluded that it has a much broader width, and that it is an I = 0 enhancement.

8. M. Cerrada et al., Phys. Lett. 62B, 353 (1976).

9. S. M. Flatté, in Proceedings of the Eleventh Rencontre de Moriond, 1976, edited by J. Tran Thanh Van (to be published), Vol. 1. These results have been subsequently discussed by A. C. Irving, Phys. Lett. 70B, 217 (1977).

10. G. Chikovani et al., Phys. Lett. 25B, 44 (1967); H. Benz et al., Phys. Lett. 28B, 233 (1968).

11. The $X^-(1153)\Gamma 16$ values were obtained by fitting a Gaussian peak to the data of Ref. 10. It should be noted here that this X^- peak, which appears in the final form of these data (see page S88 in A. Rittenberg et al., Rev. Mod. Phys. 43, No. 2, Part II (supplement), April, 1971), also appears in earlier subsets of the data; whereas another nearby peak in the final data, $X^-(1080)$, does not show up in the earlier subsets, and hence is probably spurious.

12. R. W. Jacobel et al., Phys. Rev. Lett. 29, 671 (1972).

13. D. Bowen et al., Phys. Rev. Lett. 26, 1663 (1971).

14. M. Buttram et al., Phys. Rev. Lett. 35, 970 (1975).

15. BDNPT Collaboration, Nucl. Phys. B16, 221 (1969); K. W. J. Barnham et al., Phys. Rev. Lett. 26, 1494 (1971); I. J. Bloodworth et al., Nucl. Phys. B37, 203 (1972).

16. W. Kienzle, CERN report "MMS, 1967", 1972 (unpublished).

17. A. S. Carroll et al., Phys. Rev. Lett. 32, 247 (1974); V. Chaloupka et al., Phys. Lett. 61B, 487 (1976); W. Brückner et al., Phys. Lett. 67B, 222 (1977).

CHAPTER 17. MAPPING THE HIGH-MASS "NEW PARTICLE" MESON RESONANCES

"The events of November 1974 have proven to be a watershed in the history of elementary particle physics."

K. Gottfried, invited paper on "The Spectroscopy of the New Particles", presented at the 1977 International Symposium on Lepton and Photon Interactions at High Energies, Hamburg, Germany, August (1977)

17.1 The Situation in the Summer of 1974

The so-called "New Particles" are a series of narrow-width high-mass meson and kaon resonances that were unexpectedly discovered after the spectrum of lower-mass meson and kaon resonances had already been rather completely explored. The first of the New Particles, the spectacular $J/\psi(3098)$ peak, was simultaneously discovered late in 1974 by a Brookhaven group[1] (who named it the J particle) and by a SLAC-LBL group[2] (who named it the ψ particle). The Brookhaven group was studying the reaction $p + Be \rightarrow e^+ + e^- + X$; and the SLAC-LBL group was studying the reaction $e^+ + e^- \rightarrow$ hadrons, e^+e^-, and $\mu^+\mu^-$. The second New Particle, the $\psi'(3684)$, was discovered shortly afterwards by the SLAC-LBL team[3] (who had the sole privilege of naming it, since the ψ' is not produced in the Brookhaven reaction). The exciting aspect of these New Particles is their unique combination of a high mass with a long lifetime. As we have seen from the hadron systematics described in Chapters 10, 15, and 16, the higher-mass states characteristically have shorter lifetimes (and higher spins) than do the lower-mass states, which corresponds naturally to the fact that a higher-mass excitation is inherently more unstable than a lower-mass excitation. Thus the fact that the New Particles, although they have even higher masses than do the resonances discussed in Chapters 15 and 16, also have relatively long lifetimes (and low spin values) clearly indicates that these New Particles are built from a new type of quark substructure — a substructure which is evidently too massive to be excited at lower energies. The question then is: what is this new type of quark substructure?

Since we are discussing elementary particle structure from the viewpoint of the present light-quark model, it is important that we review the situation with respect to the development of this light-quark model as it appeared in the Summer of 1974, just prior to the discovery of the New Particles. (Appendix J summarizes the papers and reports that have been written by the present author on the subject of the light-quark model.) The basic idea behind the light-quark model — that quark binding energies are very small (a few percent at most) — was deduced in the year 1969 from a study of the mass systematics of the then-known baryon and meson resonances. Two years later, with the aid of the subsequently-discovered $\bar{p}n(1795)$ bound state resonance, the systematics of the light-quark model had been worked out in essentially its present form. That is, the fundamental light-quark basis states 1,3,4,7,S were identified; the dichotomy of 4% binding energies for unmixed states and ∿0% binding energies for mixed states was recognized; the general rules for handling strangeness quantum numbers and electromagnetic effects were worked out; and the identification of meson and baryon resonances as sets of 70 MeV-quantized S-states with superimposed rotational bands was written down in detail. This work is summarized in a long paper that was submitted to Physical Review D late in 1971, and was finally published in mid-1974, just prior to the appearance of the New Particles. Due to the great length of this paper, the editors of Physical Review D split it into two parts, which we denote here as Paper I and Paper II (refs. 4 and 5 in this chapter).

The meson resonances that were treated in Papers I and II were limited mainly to energies below 2 GeV, because information on meson states above 2 GeV was both sparse and inconclusive. However, even from these low-energy meson states, it was possible to deduce that the meson excitation series $\dot{3}\dot{3}\dot{3}^n$ (where $\dot{3}\dot{3}\dot{3}$ = 630 MeV is a fundamental spinless caber grouping) serves as the generator for high-mass meson excitations. The properties of this $\dot{3}\dot{3}\dot{3}^n$ meson excitation series are described in Tables XXII and XXIII of Paper I. At that time, only the terms n = 1, 2, and 3 of the $\dot{3}\dot{3}\dot{3}^n$ series were filled with identified resonances, but the expectation was noted in Paper I that this series should extend to higher terms (although the dramatic properties of the n = 5 and n = 6 J/ψ and ψ' terms were certainly not anticipated). The n = 3 term in this series is the $\bar{p}n(1795)$ resonance, which was singled out as an especially important meson state [see Nuovo Cimento Letters 3, 197 (1972) and 5, 135 (1972), and see Particles and Nuclei 3, 173 (1972) and 5, 69 (1973), as summarized in Appendix J]. Also, a strong argument was made in Paper I for the reality of the narrow-width S(1935) meson state (see Refs. 42 and 43 in that paper); and this argument was advanced at a time when the experimental evidence was very much in doubt (and indeed largely negative), and when the conventional quark-model theorists had no reason (which they still do not have) for anticipating the existence of this state. We cite these particular results here because they are results that are germain to the subsequent appearance of the New Particles.

In physics, it is the experimentalists and not the theorists who ultimately determine which theoretical approach is correct — or at least most nearly correct. In Papers I and II it was established that a light-quark weak-binding-energy approach was *sufficient* to account for the known spectrum of elementary particles. But was this approach necessary? The answer to this question could only come from new experimental results. And these were not long in coming. In November 1974 the Brookhaven and SLAC-LBL groups simultaneously announced[1,2] the discovery of the J/ψ(3098) meson, with an enormous cross section and an extremely narrow width (which is why it hadn't been found earlier), and the world of elementary particle physics hasn't been quite the same ever since.

17.2 The Main Implication of the New Particles

In the remainder of the present chapter we will discuss the New Particles as they are reproduced by the light-quark basis states 1,3,4,7,S. This is a very specialized viewpoint. However, from any viewpoint whatsoever, there seems to be one main conclusion that emerges from a study of the spectrum of New Particle excitations: the New Particles are formed from quarks which are held together with small binding energies. This is a conclusion that is gradually (if somewhat unconsciously) being accepted by the theorists who are involved with this problem. For example, the J/ψ meson, with a mass of about 3098 MeV, is currently regarded by many elementary particle physicists as being composed of a pair of "charmed" quarks, so that $J/\psi \equiv c\bar{c}$. Detailed calculations have been made of "charmonium" energy levels, and the mass that is commonly quoted for the quark c in these calculations is in the range of 1550[6] to 1600[7] MeV. Thus, from this "charm" viewpoint, the $c\bar{c}$ quark-antiquark binding energy in the J/ψ is no more than a few percent. De Rújula, Georgi, and Glashow,[8] who pinpointed the mass of the charmed quark c as 1660 ± 30 MeV, commented that "We expect the dominant contribution to its (the J/ψ) mass to be the rest mass of the constituent quarks"; and they commented elsewhere in their paper that "In our framework, the masses of the quarks have a realistic meaning".

This assumption about weak quark binding energies is also trickling down to the lower-mass states. Lichtenberg et al.[6] and others[9] (see Sec. 8.6 in Chapter 8) now quote the masses of the Gell-Mann and Zweig u and d (nucleon) quarks as u,d ≃ 390 MeV; and they quote the mass of the s (strange) quark as s ≃ 510 MeV. De Rújula, Georgi, and Glashow[8] quote the mass values u,d ≃ 336 MeV and s ≃ 540 MeV. With these mass values, the Gell-Mann and Zweig quark binding energies for the low-mass states are mostly in the range of 10% or less (with the glaring exception of the π meson). From these recent examples, we can see that the concept of very massive Gell-Mann and Zweig u, d, and s quarks, with masses in the range of 5 GeV or more, is gradually being done away with.

We have emphasized in as much detail as possible throughout the present book that if quark binding energies are small, so that the mass of a resonance arises mainly from the masses of the constituent quarks,[8] then we are forced to consider the mass quantum M = 70 MeV as a basis state. This quantum appears in meson resonances in the form $\pi = M\bar{M} \simeq 140$ MeV, as we discussed in detail in Sec. 5.3 of Chapter 5. In support of this viewpoint (which was advanced before the advent of the New Particles), we can cite the New Particle decay $D^* \rightarrow D + \pi$ (see Fig. 5.5 in Chapter 5), which has a Q value of only a few MeV, and which graphically illustrates both the small magnitude of quark binding energies and the reality of the π as a meson basis state. Furthermore, the alternate New Particle decay mode $D^* \rightarrow D + \gamma$ (see Fig. 5.5) demonstrates the fact that the π substate in the D^* can spontaneously annihilate instead of being emitted as a final-state particle. This internal annihilation of one or more π substates can also be observed in the radiative decays of the ψ' New Particle (see Fig. 5.5). No matter what assumptions are made about the constituent quarks, the systematics of the spectrum of New Particle resonances indicates in a very clear-cut manner that quark binding energies are weak.

An important point to keep in mind with respect to the New Particles is their very narrow widths. Narrow widths not only mean that the mass values are accurately known, so that we can observe the mass quantization of the level spacings directly; but these narrow widths also imply that the New Particles are rotationless excitations, so that we do not have the large rotational energies which serve to obscure the mass quantization of the lower-mass Old Particle resonances.

We now proceed to an examination of the New Particles, starting first with the J/ψ and ψ' states, and with the transitions between these states.

17.3 The J/ψ and ψ' Mesons, and the $\psi' \rightarrow J/\psi$ Decays

The light-quark basis states M = 70 MeV and S = 330 MeV are relatively small building blocks. Thus, when the J/ψ New Particle state appeared with a mass of about 3100 MeV, the immediate question posed for the present light-quark model was to come up with a logical grouping of basis states to account for this resonance. A natural starting point for this task is the $\dot{3}\dot{3}\dot{3}^n$ meson excitation series that was mentioned in Sec. 17.1 as being the generator of high-mass meson resonances (see Sec. 17.6). The n = 5 term in this series, $\dot{3}\dot{3}\dot{3}^5$, contains fifteen spinless cabers $\dot{3}$; and its calculated mass of 15 x 210 = 3150 MeV is close to the mass of the J/ψ. Moreover, as we show in Sec. 17.6, and as was illustrated in Table 13.1 of Chapter 13, the odd-n terms in this series correspond to excitations which are composed entirely of spinors S. These spinors S are produced by basis-state transformations of the type $\dot{3}\dot{3}\dot{3} \rightarrow SS$ (see Chapter 14). Thus the fifteen spinless cabers $\dot{3}$ in the $\dot{3}\dot{3}\dot{3}^5$ term of the series are transformed into ten spinors S. We can associate six of these ten spinors with the $N\bar{N}$ bound state, $\bar{p}n(1795) \equiv \bar{S}S\bar{S}\cdot S\bar{S}S$, that serves as the fundamental "ground state" for the New Particles; and we can associate the other four spinors with the $\dot{S}\dot{S}\dot{S}\dot{S} = 1310$

MeV meson excitation that is discussed in Sec. 8.4 of Chapter 8 and in Table D2 of Appendix D. These two basis-state groupings combine together with essentially zero binding energy, so that we have

$$J/\psi \equiv N\bar{N}\cdot\dot{S}\dot{S}\dot{S}\dot{S} = 1795 + 1310 = 3105 \text{ MeV}. \tag{17.1}$$

This basis state assignment for the J/ψ is not quite as arbitrary as it might seem from the above discussion. The J/ψ production process

$$e^+ + e^- \rightarrow \text{hadron-antihadron} \rightarrow J/\psi \tag{17.2}$$

suggests that an energetic e^+e^- pair annihilates and is transformed into a pair of excited spin 1/2 hadron states which bind together to form the J/ψ. These excited spin 1/2 hadron states are baryon excitations, which were seen in Chapter 15 to be formed by adding spinless meson excitation quanta $\bar{3},\bar{4},\dot{3},\dot{4}$ to an $S\bar{S}S$ nucleon spinor core. But since the J/ψ is a new type of meson, its existence suggests that a new type of baryon excitation has been formed, and an obvious choice is to combine an $S\bar{S}S$ spinor core with the dominant $\dot{S}\dot{S}$ meson excitation that, in a rotational mode, corresponds to the ρ meson. Thus we have (at least schematically) the light-quark excitation

$$(S\bar{S}S\cdot\dot{S}\dot{S})\cdot(\dot{S}\dot{S}\cdot\bar{S}S\bar{S}) = (N\cdot\rho)\cdot(\rho\cdot\bar{N}) = J/\psi = N\bar{N}\cdot\dot{S}\dot{S}\dot{S}\dot{S}\ . \tag{17.3}$$

The key point to note about this excitation is that the basis-state transformation $\dot{3}\dot{3}\dot{3} \rightarrow \dot{S}\dot{S}$ does not produce the $\dot{S}\dot{S}$ spinor pair in a rotationless S-state, but only in a rotational mode as the ρ meson (see Table 13.1; and see Appendix E). Thus the transformation $\dot{3}\dot{3}\dot{3} \rightarrow \dot{S}\dot{S}$ does not operate to produce single baryon excitations of the type $N\cdot\dot{S}\dot{S}$. But if a matching pair of baryon excitations are simultaneously produced, as in (17.3), then the parity (?) constraint against single S-states is satisfied, and the $\dot{S}\dot{S}\cdot\dot{S}\dot{S}$ pair of excitations can appear together in an S-state resonance. The J/ψ is the lowest-mass $N\bar{N}$ resonance in which this basis-state combination can appear, and the large cross section and long lifetime of the J/ψ attest to the importance of this combination.

The n = 6 term in the $\dot{3}\dot{3}\dot{3}^n$ meson excitation series, which we identify here with the $\psi'(3684)$ New Particle state, has a calculated spinless caber mass of 18 x 210 = 3780 MeV. Now, it was shown in Table 13.1 of Chapter 13 that whereas the odd-n terms in the $\dot{3}\dot{3}\dot{3}^n$ series are composed entirely of (transformed) spinors S, the even-n terms are composed of (transformed) spinors S combined together with (untransformed) spinless caber clusters $\dot{3}\dot{3}\dot{3}$. If we associate the transformed spinors S with the $N\bar{N}$ bound state, as we did above for the J/ψ New Particle state, then the n = 6 term in the $\dot{3}\dot{3}\dot{3}^n$ series appears as the light-quark combination

$$\psi' \equiv N\bar{N}\cdot\dot{3}\dot{3}\dot{3}^3 = 1795 + 1890 = 3685 \text{ MeV}, \qquad (17.4)$$

which correctly reproduces the mass of the ψ' in a manner similar to that of the J/ψ configuration of Eq. (17.1).

When the J/ψ and ψ' New Particles first appeared, it was possible on the basis of the above line of reasoning to immediately identify them as being the n = 5 and n = 6 members of the fundamental $\dot{3}\dot{3}\dot{3}^n$ meson excitation series, and a paper to this effect was published.[10]

If the basis-state configurations $J/\psi \equiv N\bar{N}\cdot\dot{S}\dot{S}\dot{S}\dot{S}$ and $\psi' \equiv N\bar{N}\cdot\dot{3}\dot{3}\dot{3}^3$ correspond to physical reality, then we should be able to make use of these configurations in the interpretation of the six $\psi' \to J/\psi$ transitions that have been observed experimentally. These transitions include two <u>single-step</u> decays,

$$\psi' \to J/\psi + \eta \quad \text{and} \quad \psi' \to J/\psi + \pi + \pi\,, \qquad (17.5)$$

and four <u>two-step</u> radiative decays, which are of the form

$$\psi' \to \chi + \gamma\,, \quad \chi \to J/\psi + \gamma\,. \qquad (17.6)$$

Four different intermediate states, $\chi(3551)$, $\chi(3503)$, $\chi(3455)$, and $\chi(3414)$, have been experimentally identified in the radiative decay channels of (17.6); these correspond to the four two-step radiative decays. In order to transform the ψ' into the J/ψ, the basis-state transformation

$$\dot{3}\dot{3}\dot{3}^2 \to \dot{S}\dot{S}\dot{S}\dot{S} \qquad (17.7)$$

must necessarily take place in each of these $\psi' \to J/\psi$ decays. The properties of this transformation were discussed in detail in Chapter 14. Since we have the basis-state mass values $\dot{3}\dot{3}\dot{3}^2 \equiv \dot{3}\dot{3}\dot{3}\cdot\dot{3}\dot{3}\dot{3} \simeq 6 \times 210 = 1260$ MeV and $\dot{S}\dot{S}\dot{S}\dot{S} \simeq 4 \times 330 \simeq 1310$ MeV (see Table D2 in Appendix D), it can be seen that the transformation $\dot{3}\dot{3}\dot{3}^2 \to \dot{S}\dot{S}\dot{S}\dot{S}$ is endoergic by approximately 50 MeV. In fact, we can calculate this mass difference empirically by using the accurately-measured $\psi' - J/\psi$ mass difference of 588.6 ± 0.8 MeV[11], as follows:

$$\psi' - J/\psi = 588.6 \text{ MeV} = N\bar{N}\cdot\dot{3}\dot{3}\dot{3}^3 - N\bar{N}\cdot\dot{S}\dot{S}\dot{S}\dot{S} = \dot{3}\dot{3}\dot{3} + (\dot{3}\dot{3}\dot{3}^2 - \dot{S}\dot{S}\dot{S}\dot{S}), \qquad (17.8a)$$

which gives

$$\dot{S}\dot{S}\dot{S}\dot{S} - \dot{3}\dot{3}\dot{3}^2 = 630 \text{ MeV} - 588.6 \text{ MeV} = 41.4 \text{ MeV}. \qquad (17.8b)$$

The phenomenological significance of this 41 MeV mass shift lies in the fact that the two-step radiative decays of (17.6) can occur with the $\dot{3}\dot{3}\dot{3} \to \dot{S}\dot{S}\dot{S}\dot{S}$ transformation taking place either in the first step or else in the second step of the decay, and the 41 MeV mass shift enables us to distinguish between these two possibilities. Using this method, we identify two of the radiative decays in (17.6) as having the $\dot{3}\dot{3}\dot{3}^2 \to \dot{S}\dot{S}\dot{S}\dot{S}$ transition occurring in the <u>second</u> step:

$$\psi' \to \chi(3551) + \gamma \,, \quad \Delta m = 133 \text{ MeV} \qquad \chi(3551) \to J/\psi + \gamma \,, \tag{17.9}$$

$$\psi' \to \chi(3414) + \gamma \,, \quad \Delta m = 270 \text{ MeV} \qquad \chi(3414) \to J/\psi + \gamma \,; \tag{17.10}$$

and the other two radiative decays as having the $\dot{3}\dot{3}\dot{3}^2 \to \dot{S}\dot{S}\dot{S}\dot{S}$ transition occurring in the <u>first</u> step:

$$\psi' \to \chi(3503) + \gamma \,, \qquad \chi(3503) \to J/\psi + \gamma \,, \quad \Delta m = 405 \text{ MeV} \tag{17.11}$$

$$\psi' \to \chi(3455) + \gamma \,, \qquad \chi(3455) \to J/\psi + \gamma \,. \quad \Delta m = 357 \text{ MeV} \tag{17.12}$$

The essential point here is that the Δm values shown in (17.9) - (17.12) accurately correspond to the annihilation of 2, 4, 6, and 5 subquanta M = 70 MeV, respectively, so that these particular decays are <u>not</u> accompanied by the $\dot{3}\dot{3}\dot{3}^2 \to \dot{S}\dot{S}\dot{S}\dot{S}$ transformation with its characteristic 41 MeV mass shift. By noting the mass difference

$$\chi(3455) - \chi(3414) = 41 \text{ MeV}, \tag{17.13}$$

which is precisely the mass difference that is shown in (17.8), we can see that the reactions (17.10) and (17.12) both correspond to the same decay channel, but with the basis-state transformation $\dot{3}\dot{3}\dot{3}^2 \to \dot{S}\dot{S}\dot{S}\dot{S}$ occurring in step two in (17.10) and in step one in (17.12).

The really interesting aspect of these $\psi' \to J/\psi$ decays is revealed when we classify them on the basis of the number of subquanta M that are annihilated in each step of the decay. This result is illustrated in Table 17.1, where we have included all of the decay modes that are associated with the decay channels of (17.5) and (17.6). From the systematics shown in Table 17.1, we can deduce the following rule about these high-mass New Particle decays:

The decays in which an odd number of subquanta M are annihilated

Table 17.1. Light-quark basis-state configurations for the decay modes that occur in the $\psi' \to J/\psi$ transitions of Eqs. (17.5) and (17.6). $M \simeq 70$ MeV, $\pi \simeq 140$ MeV, and $\dot{3} \simeq 210$ MeV are spinless substates, and $\dot{S} \simeq 330$ MeV and $N \simeq 939$ MeV are spin 1/2 substates. $N\bar{N} = 1795$ MeV is the nucleon-antinucleon bound state, and $\dot{3}\dot{3}\dot{3} \simeq 630$ MeV and $\dot{S}\dot{S}\dot{S}\dot{S} \simeq 1310$ MeV are fundamental meson excitation quanta. The brackets [ ] denote the annihilation of some of the constituent subquanta. As is shown in Eq. (17.8), the basis-state transition $\dot{3}\dot{3}\dot{3}^2 \to \dot{S}\dot{S}\dot{S}\dot{S}$ is accompanied by a 41 MeV energy shift, which we can use to identify decay modes in which this transition has occurred. The results shown in Table 17.1 demonstrate the fact that decays in which an odd number (1,3,5,7) of subquanta M are annihilated are associated with the $\dot{3}\dot{3}\dot{3}^2 \to \dot{S}\dot{S}\dot{S}\dot{S}$ transition, whereas decays in which an even number (2,4,6) are annihilated are not. The only exceptions to this rule in Table 17.1 are the $\psi' \to \chi(3455)$ and $\chi(3455) \to J/\psi$ decays that are enclosed in parentheses, and experimentally these decay modes are very strongly discriminated against. Table 13.1 in Chapter 13 and Table 17.3 in the present chapter show that this same general rule applies to production modes as well as to decay modes.

Observed decay	Basis-state transformation	No. of quanta M annihilated	$\dot{3}\dot{3}\dot{3}^2 \to \dot{S}\dot{S}\dot{S}\dot{S}$ transition
$\psi' \to J/\psi + \eta$	$N\bar{N}\cdot\dot{3}\dot{3}\dot{3}^3 \to N\bar{N}\cdot\dot{S}\dot{S}\dot{S}\dot{S} + 3\bar{3}\pi[M]$	1	yes
$\psi' \to \chi(3551) + \gamma$	$N\bar{N}\cdot\dot{3}\dot{3}\dot{3}^3 \to N\bar{N}\cdot\dot{3}\dot{3}\dot{3}^2\cdot\dot{3}\pi\pi[\pi] + \gamma$	2	no
$\psi' \to \chi(3503) + \gamma$	$N\bar{N}\cdot\dot{3}\dot{3}\dot{3}^3 \to N\bar{N}\cdot\dot{S}\dot{S}\dot{S}\dot{S}\cdot 3\bar{3}[\dot{3}] + \gamma$	3	yes
$(\psi' \to \chi(3455) + \gamma)$	$N\bar{N}\cdot\dot{3}\dot{3}\dot{3}^3 \to N\bar{N}\cdot\dot{S}\dot{S}\dot{S}\dot{S}\cdot\dot{3}\pi[\pi\pi] + \gamma$	(4)	(yes)
$\psi' \to \chi(3414) + \gamma$	$N\bar{N}\cdot\dot{3}\dot{3}\dot{3}^3 \to N\bar{N}\cdot\dot{3}\dot{3}\dot{3}^2\cdot\dot{3}\pi[\pi\pi] + \gamma$	4	no
$\chi(3414) \to J/\psi + \gamma$	$N\bar{N}\cdot\dot{3}\dot{3}\dot{3}^2\cdot\dot{3}\pi \to N\bar{N}\cdot\dot{S}\dot{S}\dot{S}\dot{S}[\dot{3}\pi] + \gamma$	5	yes
$(\chi(3455) \to J/\psi + \gamma)$	$N\bar{N}\cdot\dot{S}\dot{S}\dot{S}\dot{S}\cdot\dot{3}\pi \to N\bar{N}\cdot\dot{S}\dot{S}\dot{S}\dot{S}[\dot{3}\pi] + \gamma$	(5)	(no)
$\psi' \to J/\psi + \pi + \pi$	$N\bar{N}\cdot\dot{3}\dot{3}\dot{3}^3 \to N\bar{N}\cdot\dot{S}\dot{S}\dot{S}\dot{S}[\dot{3}\pi] + \pi + \pi$	5	yes
$\chi(3503) \to J/\psi + \gamma$	$N\bar{N}\cdot\dot{S}\dot{S}\dot{S}\dot{S}\cdot 3\bar{3} \to N\bar{N}\cdot\dot{S}\dot{S}\dot{S}\dot{S}[3\bar{3}] + \gamma$	6	no
$\chi(3551) \to J/\psi + \gamma$	$N\bar{N}\cdot\dot{3}\dot{3}\dot{3}^2\cdot\dot{3}\pi\pi \to N\bar{N}\cdot\dot{S}\dot{S}\dot{S}\dot{S}[\dot{3}\pi\pi] + \gamma$	7	yes

are accompanied by the transition $\dot{3}\dot{3}\dot{3}^2 \to \dot{S}\dot{S}\dot{S}\dot{S}$, whereas the decays in which an even number of subquanta M are annihilated are not accompanied by this transition. (17.14)

The only exception to this rule in Table 17.1 is the decay sequence $\psi' \to \chi(3455) + \gamma$, $\chi(3455) \to J/\psi + \gamma$ that is set off with parentheses; experimentally, this decay mode is very weak, with only a handful of events being recorded. Thus the correlation that is expressed in (17.14) serves as a powerful constraint on decay modes: decays that violate this rule have very small cross sections, whereas decays that are in agreement with this rule have much larger cross sections. A case in point is the $\psi' \rightarrow J/\psi + \eta$ decay shown at the top of Table 17.1, which is in agreement with the rule; as one physicist has remarked[12], this reaction "has everything working against it", including a very small Q-value, and yet it has a large cross section that accounts for 4% of all of the ψ' decays.

Rule (17.14) applies to meson decay modes. However, as is demonstrated in Table 13.1 of Chapter 13 and in Table 17.3 of the present chapter, this same general correlation also applies to production modes: meson excitations that correspond to the creation of an odd number of subquanta M appear as fully-transformed spinors, whereas meson excitations that correspond to the creation of an even number of subquanta M are only partially transformed into spinors. We have already made use of this rule in the present chapter, in connection with the derivation of Eqs. (17.1) and (17.4).

From a theoretical point-of-view, the $\psi' \rightarrow J/\psi$ decays that are currently of most interest are the $\psi' \rightarrow \chi \rightarrow J/\psi$ channels that involve the intermediate states $\chi(3551)$, $\chi(3503)$, and $\chi(3414)$. In the "charm" model, these three states are regarded as being the $^3P_2$, $^3P_1$, and $^3P_0$ levels, respectively, of a positronium-like $c\bar{c}$ pair of charmed quarks; and, indeed, the spin values of these states, although not yet completely determined experimentally, seem to be in remarkable agreement with this hypothesis. However, the mass splittings $^3P_2 - {}^3P_1$ and $^3P_2 - {}^3P_0$ of these levels, which were calculated by De Rújula, Georgi, and Glashow to be 6 MeV and 14 MeV,[8] have turned out instead to be 48 MeV and 137 MeV, and no completely convincing "charm model" explanation has been forthcoming to explain this order-of-magnitude discrepancy in the level spacings.

In the charm model, the states $\chi(3551)$, $\chi(3503)$, and $\chi(3414)$ are all on an equal footing. However, this is not true in the light-quark model. From Eqs. (17.9) to (17.11), and from the systematics of Table 17.1, we see that

$$\chi(3551) \equiv N\bar{N}\cdot\dot{3}\dot{3}\dot{3}^2\cdot\dot{3}\pi\pi \ , \qquad (17.15a)$$

$$\chi(3503) \equiv N\bar{N}\cdot\dot{S}\dot{S}\dot{S}\dot{S}\cdot 3\bar{3} \ , \qquad (17.15b)$$

$$\chi(3414) \equiv N\bar{N}\cdot\dot{3}\dot{3}\dot{3}^2\cdot\dot{3}\pi \ . \qquad (17.15c)$$

Thus the $\chi(3551)$ and $\chi(3414)$ states are essentially $\psi' \equiv N\bar{N}\cdot\dot{3}\dot{3}\dot{3}^3$ configurations in which one and two subquanta π, respectively, have annihilated; whereas the $\chi(3503)$ state is a $J/\psi\cdot 3\bar{3}$ configuration in which the transition $\dot{3}\dot{3}\dot{3}^2 \rightarrow \dot{S}\dot{S}\dot{S}\dot{S}$ has already taken place. Experimentally, the $\chi(3503)$ state has always appeared to be somewhat different from the $\chi(3551)$ and $\chi(3414)$ states, and it was originally denoted as the P_c level; in particular, this level has a larger branching ratio for decaying into the J/ψ (23%) than do the $\chi(3551)$ (16%) and $\chi(3414)$ (3%) levels, which is a result that seems in line with the basis-state configurations shown in (17.15). Since the $\chi(3503)$ is produced in the reaction

$$\psi' \equiv N\bar{N}\cdot\dot{3}\dot{3}\dot{3}^3 \;\rightarrow\; N\bar{N}\cdot\dot{S}\dot{S}\dot{S}\dot{S}\cdot 3\bar{3}[\dot{3}] + \gamma \;\equiv\; \chi(3503) + \gamma\ , \tag{17.16}$$

we would expect it to have a mass value of 3684 - 210 + 41 = 3515 MeV. The somewhat lower mass value of 3503 MeV that is observed (3508 MeV in RPP78) indicates that the $3\bar{3}$ excitation pair in the $\chi(3503) \equiv J/\psi\cdot 3\bar{3}$ configuration has an appreciable binding energy; this conclusion is reinforced by the fact that the $\chi(3503)$ appears 405 MeV above the mass of the J/ψ, and not 420 MeV as we would expect for a $\dot{3}\dot{3}$ excitation.

There is one recent experimental result that also singles out the $\chi(3503)$ or P_c state for special attention. In a $p + p \rightarrow e^+ + e^- + X$ experiment at CERN, J. H. Cobb <u>et</u> al.[13] studied the production of J/ψ particles accompanied by photons, and they concluded that 43 ± 21% of the J/ψ's are produced via the photonic decay of one of the χ levels. Although their experiment did not have sufficient resolution to pinpoint the mass of the χ level, the experimenters guessed that it is probably "the $\chi/P_c(3.51)$ state which has the largest branching ratio into J/ψ's plus photon". Thus the J/ψ seems to be as readily produced in the form of the excited $J/\psi\cdot 3\bar{3}$ configuration as it is directly in the J/ψ ground state. As a somewhat analogous result, we note that the η meson appears in the light-quark model as the excitation $\pi\cdot 3\bar{3}$, and the η' appears as the excitation $\eta\cdot 3\bar{3}$ (see Fig. 5.2 in Chapter 5).

If the $\chi(3551)$, $\chi(3503)$, and $\chi(3414)$ spin values that are predicted by the "charm" model turn out to be correct, then this must be regarded as an impressive accomplishment. However, the light-quark type of mass quantization that appears in these levels is an equally-impressive result that must also be taken into account. Furthermore, the $\chi(3455)$ level appears naturally in the present light-quark model as a $\chi(3414)$ state in which the transition $\dot{3}\dot{3}\dot{3}^2 \rightarrow \dot{S}\dot{S}\dot{S}\dot{S}$ has taken place (thus increasing its mass by 41 MeV); but the "charm" model currently has no explanation for the existence of this $\chi(3455)$ state[14]. Thus, while the successes of the "charm" model appear at this point to be very impressive, they are by no means complete.

As a final comment with respect to the $\psi' \rightarrow J/\psi$ transitions shown in Table 17.1, we note the interesting decay $\psi' \rightarrow J/\psi + \pi + \pi$, which historically was the first decay in Table 17.1 to be observed. In this decay, which is the light-quark sequence

$$\psi' \;\equiv\; N\bar{N}\cdot\dot{3}\dot{3}\dot{3}^3 \;\rightarrow\; N\bar{N}\cdot\dot{S}\dot{S}\dot{S}\dot{S}[\dot{3}\pi]\pi\pi \;\rightarrow\; J/\psi + \pi + \pi\ , \tag{17.17}$$

the two pions are blown off in opposite directions, with the J/ψ remaining essentially at rest. The kinematic effect of this type of annihilation is to distort the final-state $\pi\pi$ phase space, so that the two pions carry away more energy on the average than would normally be expected.[15] This decay reaction, Eq. (17.17), demonstrates the explosive nature of the $[\dot{3}\pi]$ annihilation inside the ψ', in which 350 MeV is suddenly liberated as kinetic energy. This type of decay is typical for hadrons, but it is of a completely different nature from the decays that occur in atomic nuclei. Thus, as we suggested in Chapter 11, it seems that nuclear-physics-type angular momentum barriers will play only a modest role in these violent hadronic annihilations.

17.4 New Particle Data and New Particle Production Thresholds

The so-called New Particles are in fact mesons and kaons of the same general type as the well-known Old Particle mesons and kaons, and the 1978 edition of the Review of Particle Properties groups them together. For our purposes here, we define the New Particles to be the narrow-width resonances that appear with masses greater than or equal to 1795 MeV, where $\bar{p}n(1795)$ is the nucleon-antinucleon bound state that serves as the threshold for New Particle production. These New Particle states are listed in Table B9 of Appendix B, where references to the data are given. These same states are shown here in Table 17.2, where we have grouped them according to the manner in which they are observed. At the present time, it is difficult to judge the completeness of the New Particle data set listed in Table 17.2. New experimental results continue to appear, but very intensive peak-hunting searches have now been carried out over most of this energy range. Thus, while additions will certainly be made to the data shown in Table 17.2, it seems unlikely that radical improvements will occur in the near future.

The theoretical significance of the New Particles is that they signify the onset of a new type of quark substructure. In particular, the opening up of a New Particle threshold signifies that a new quark combination has become energetically possible. This is a result that we have already discussed in Sec. 16.2 of Ch. 16, in connection with the low-mass meson and kaon resonances; and we now extend that discussion to include the higher-mass New Particle meson and kaon resonances.

Figure 16.1 in Chapter 16 illustrated the meson production channels for: the very-low-mass spinless mesons; the low-mass "spin 1" mesons; and the narrow-width spin 2 mesons. The thresholds for the spin 1 and spin 2 channels are manifested only by the difficult-to-observed $\pi\pi(658)$ and A_2dip(1310) interference effects (see Sec. 8.4 in Chapter 8), which correspond to the production of $\dot{S}\dot{S}$ spinor pairs and $\dot{S}\dot{S}\dot{S}\dot{S}$

Table 17.2. High-mass New Particle resonances, grouped according to the manner in which they are observed. Groups I - IV are meson and kaon states, and Group V lists two hyperon states. These resonances are documented in Tables B9 and B10 of Appendix B.

Group I. R-ratio peaks			
$J/\psi(3098)\Gamma 0.067$	$J^P = 1^-$	$\psi'''(4414)\Gamma 33$	$J^P = 1^-$
$\psi'(3684)\Gamma 0.228$	$J^P = 1^-$	$\Upsilon(9400)$	$J^P = 1^-$
$\psi''(3772)\Gamma 28$	$J^P = 1^-$	$\Upsilon(10010)$	
$\psi(4028)$		$\Upsilon(10400)$	
$\psi(4100)$			
Group II. $\bar{N}N$ peaks			
$\bar{p}n(1795)\Gamma\leq 8$		$\bar{p}p(2660)$	
$M(1817)\Gamma 30$		$\bar{N}N(2850)\Gamma\leq 30$	
$\bar{p}p(1935)\Gamma\leq 4$		$\bar{p}p\pi(2950)\Gamma<15$	
$\bar{p}p(2020)\Gamma 24$		$\bar{N}N(3050)\Gamma<20$	
$\bar{p}p(2204)\Gamma 16$			
Group III. Final-state enhancements and levels			
$D^o(1863.3)\Gamma<2.4$	$J^P = 0^-$	$I(2600)\Gamma\leq 19$	
$D^+(1868.3)\Gamma<2.4$	$J^P = 0^-$	$X(2830)$	
$D^{*o}(2006)\Gamma<2$	$J^P = 1^-$	$\chi(3340)$	
$D^{*+}(2008.6)\Gamma<2$	$J^P = 1^-$	$\chi(3414)$	$J^P = 0^+$
$F(2030)$		$\chi(3455)$	
$F^*(2140)$		$\chi(3503)$	$J^P = 1^+$?
$(K\pi)^o(2290)$		$\chi(3551)$	$J^P = 2^+$?
$(K\pi)^o(2420)$			
Group IV. $(K\pi)^o$ enhancements recoiling against the $D(\sim 1865)$			
$(K\pi)^o_R(\sim 1860)$	$(K\pi)^o_R(\sim 2005)$	$(K\pi)^o_R(\sim 2430)$	
Group V. Narrow-width hyperon enhancements			
$\bar{\Lambda}(2260)\Gamma\leq 40 \rightarrow \bar{\Lambda}\pi\pi\pi$		$\Lambda(2426) \rightarrow \Lambda\pi\pi\pi\pi$	

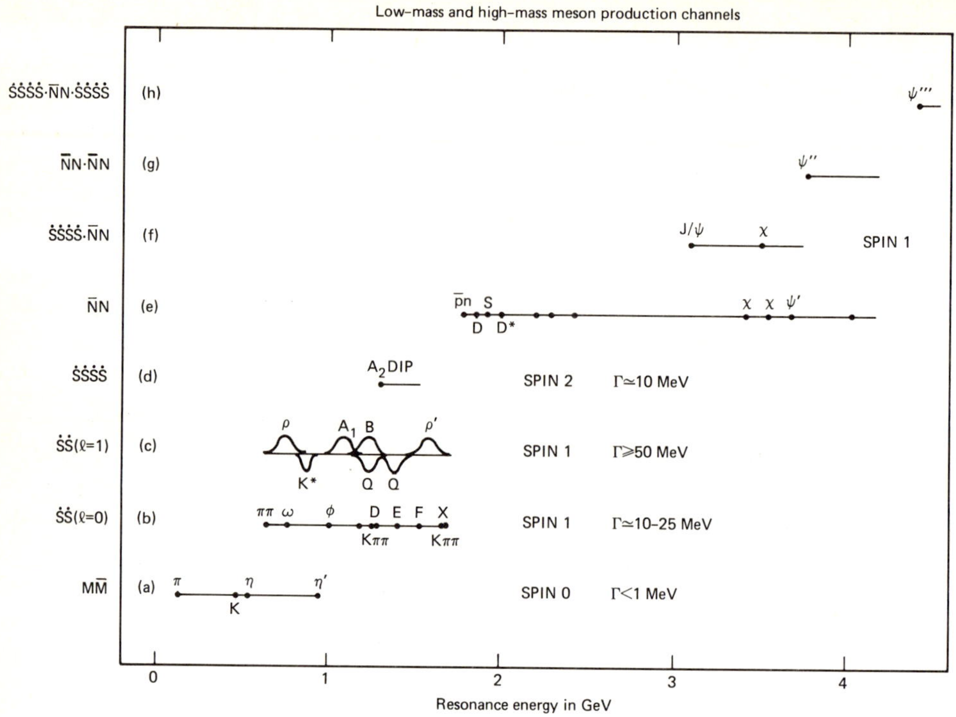

Figure 17.1. Meson production channels. This is an extension of Fig. 16.2 to include the New Particle resonances, which first appear at the N̄N threshold. Thus the N̄N combination of light-quark basis states is the new type of quark substructure which gives rise to the unique properties of the New Particles (their combination of large masses and narrow widths). Channels (a) - (d) in Fig. 17.1 are the low-mass Old Particle production channels that were shown in Fig. 16.2, and channels (e) - (h) are the high-mass New Particle production channels. As is indicated at the left in Fig. 17.1, combinations of the basis-state units N̄N and ṠṠṠṠ generate the production channels (e) - (h), with a prominent New Particle state appearing as the threshold resonance in each of these channels.

spinor quartets, respectively. Figure 17.1 extends the systematics of Fig. 16.1 up into the higher-mass realm of the New Particles. Channels (a) - (d) in Fig. 17.1 are the low-mass meson production channels that were shown in Fig. 16.1. Channel (e) is the N̄N channel, and it includes those New Particle states which are formed by adding spinless excitation quanta to the p̄n(1795) "ground state"; these states include the p̄p(1935) or S(1935) meson, the D(1865) and D*(2007) "charmed" kaons, the χ(3414) and χ(3551) radiative decay levels, the ψ'(3684) meson, and several other possible states that are indicated in the figure by dots. Channel (f) is the $\bar{N}N \cdot \dot{S}\dot{S}\dot{S}\dot{S}$ channel, and it includes the J/ψ(3098) threshold resonance, the very weak χ(3455) radiative decay level (which is not shown in the figure), and the χ(3503) or P_c radiative decay level. Channel (g) is the $\bar{N}N \cdot \bar{N}N$ channel, and its threshold resonance is the ψ''(3772) meson. Channel (h) is the $\bar{N}N \cdot \dot{S}\dot{S}\dot{S}\dot{S}^2$ channel, and its threshold resonance is the ψ'''(4414) meson.

The rationale behind the systematics of Fig. 17.1 is that meson resonances are formed by adding spinless excitation quanta to a set of fundamental "ground states". For example, the π meson, $\pi = M\bar{M}$, forms the ground state for the spinless mesons, whose excitations are illustrated in Fig. 5.2 of Chapter 5. Also, the $\dot{S}\dot{S}$ spinor pair forms the ground state for the "spin 1" meson resonances, whose excitations are illustrated in Fig. 16.2 of Chapter 16. With the exception of the spinless mesons, the "ground states" for the various production channels seem to be formed entirely of spinors; these are the spinor configurations which are shown at the left in Fig. 17.1. Thus the opening up of a new meson production channel means that a new type of spinor configuration has become energetically possible.

Since the spectrum of New Particle resonances becomes very complex above 4 GeV, it is not possible at this time to sort out or even identify all of the resonance peaks in this energy region. Thus we cannot delineate the meson production channels above 4 GeV with any confidence. The ψ' → J/ψ transitions that we studied in the preceding section involve meson states which decay from one production channel to another; that is, they decay from one "ground state" spinor configuration to another.

There is another way that we can study the meson production channels, using direct experimental information. This involves the use of the so-called R-ratio in e^+e^- annihilations, which is defined as

$$R \equiv (e^+e^- \rightarrow \text{hadrons}) \,/\, (e^+e^- \rightarrow \mu^+\mu^-) \, . \qquad (17.18)$$

The reaction $e^+e^- \rightarrow$ hadrons in (17.18) serves to indicate the rate at which meson resonances are being produced, and the reaction $e^+e^- \rightarrow \mu^+\mu^-$ serves as a normalization factor for the incident e^+ and e^- colliding beams. Thus an increase in the R-ratio signals the appearance of a hadron resonance which decays into hadronic final states. Experimental values of the R-ratio[16] are shown in Fig. 17.2, with the contributions of the dominant J/ψ and ψ' peaks removed. The R-ratio has a value of about 2.5 in the energy region from 1.8 to 3.6 GeV, where it reflects the production of hadron resonances in the $\bar{N}N$ channel. There is considerable structure in the R-ratio near 4 GeV, which reflects resonance formation associated with the opening up of the $\bar{N}N \cdot \bar{N}N$ channel. Above 4.5 GeV, the R-ratio settles down to a value of about 5, which is just twice the value it had in the lower-energy region. The R-ratio thus gives a

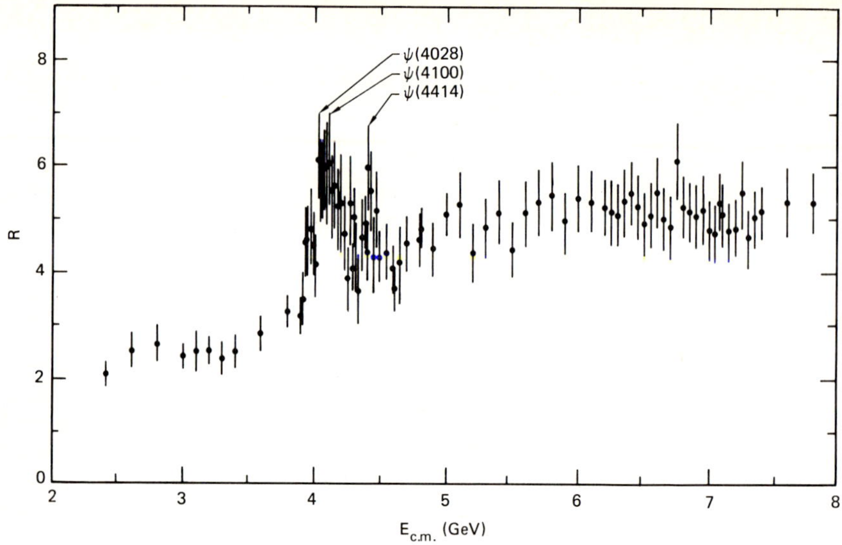

Figure 17.2. SPEAR measurements (Ref. 16) of the R-ratio, which is the ratio of hadron production to $\mu^+\mu^-$ pair production [Eq. (17.18)] in e^+e^- annihilations. In this figure, contributions from the J/ψ and ψ' resonances have been removed. As can be seen, R has the value R ∿ 2.5 in the 2.5 - 4 GeV region that lies above the $\bar{N}N$(∿1.8) threshold, and it has the value R ∿ 5 in the 5 - 8 GeV region that lies above the $\bar{N}N \cdot \bar{N}N$(∿3.6) threshold. The New Particle states ψ(4028), ψ(4100), and ψ'''(4414) appear as peaks in the R-ratio.

graphic illustration of the contributions from the various meson channels to hadron production; in particular, it singles out the importance of the $\bar{N}N$ and $\bar{N}N \cdot \bar{N}N$ channels for forming New Particle resonances. The peaks ψ(4028), ψ(4100), and ψ'''(4414) shown in Fig. 17.2 are R-ratio peaks that seem to be associated with identifiable meson resonances.

As we have outlined above, the $\bar{N}N$ bound state serves as the threshold for the production of the New Particles. Some of the most interesting New Particle resonances — the "charmed" D and D* kaons — appear right above the $\bar{N}N$ threshold. We discuss these resonances in the next section.

17.5 The "Charmed" D and D* Kaons

In Sec. 17.3 we discussed the J/ψ(3098) and ψ'(3684) mesons and the χ(3414), χ(3503), and χ(3551) radiative decay levels. From the viewpoint of the Gell-Mann and Zweig quark model, as extended by Glashow and others, these states are regarded as being formed from a pair of charmed quarks, $c\bar{c}$. From our present light-quark viewpoint, the interesting thing about these states is that, regardless of interpretation, they are quantized in precisely the manner that we would expect from the systematics of the light-quark model.

Speaking again from the viewpoint of the "charm" model, if the J/ψ is formed from a pair of charmed quarks, then it should be possible to form a "charmed" meson that contains one high-mass charmed quark and one low-mass ordinary quark. States that correspond to this prediction have been discovered by Goldhaber *et al.*[17]; they are the D^o(1863.3), D^+(1868.3), D^{*o}(2006), and D^{*+}(2008.6) kaon resonances listed in Table 17.2. From the systematics of the charm model[18], the D and D* kaons are expected to be spin 0 and spin 1 states, respectively; and the experimental data, while not yet conclusive, favor these assignments[17]. Also, the charm model predicts specific decay modes for these states, and the predictions seem to be verified experimentally. However, the interesting aspect of these states from our present viewpoint is that, again regardless of interpretation, their mass values are quantized in precisely the manner that we expect from the systematics of the light-quark model.

We have already discussed the D and D* kaons in Sec. 5.3 of Chapter 5, because they provide one of the clearest examples of a mass quantization that is in units or multiples of 70 MeV. As was illustrated in Fig. 5.4, the D* levels appear just 140 MeV above the D levels; and a narrow-width meson level, the $\bar{p}p$(1935) or S(1935) state, appears halfway between the D and D* levels and just 140 MeV above the $\bar{p}n$(1795) "ground state" level. As a result, the strange D and D* pair of levels are offset in mass from the nonstrange $\bar{p}n$ and $\bar{p}p$ pair of levels by precisely 70 MeV. In terms of light-quark basis states, we have the configurations

$$\bar{p}n(1795) = \bar{S}S\bar{S}\cdot S\bar{S}S \ , \tag{17.19}$$

$$D(\sim 1865) = \bar{S}S\bar{S}\cdot S\bar{S}S\cdot M \ , \tag{17.20}$$

$$\bar{p}p(1935) = \pi\cdot\bar{S}S\bar{S}\cdot S\bar{S}S \ , \tag{17.21}$$

$$D^*(\sim 2007) = \pi\cdot\bar{S}S\bar{S}\cdot S\bar{S}S\cdot M \ . \tag{17.22}$$

Since a single quantum M = 70 MeV is strange (see Appendix C), the D and D* configurations are strange, whereas the $\bar{p}n$ and $\bar{p}p$ configurations are nonstrange. Furthermore, the π substate in the D* is directly emitted as a final-state particle in the decay $D^* \to D + \pi$, thus illustrating the reality of the configuration shown in (17.22).

It turns out experimentally that the D and D* kaons are not produced singly, but only in the form of DD, DD*, and D*D* pairs. Since $D \equiv \bar{N}N\cdot M$ and $D^* \equiv \pi\cdot\bar{N}N\cdot M$, a resonance that decays into a pair of D kaons must be above the threshold for the $\bar{N}N\cdot\bar{N}N$ channel. The peaks $\psi''(3772)$, $\psi(4028)$, and $\psi'''(4414)$ are all above this threshold, and they have all been studied experimentally as resonances that decay into pairs of "charmed" kaons. By analyzing the decays of these three resonances into D and D* pairs, we can obtain interesting information about their light-quark substructures, as we now demonstrate.

Let us assume that the $\psi''(3772)$ peak has the light-quark configuration

$$\psi''(3772) \quad = \quad \bar{N}N\cdot\bar{N}N\cdot X \ , \tag{17.23}$$

where X denotes an unknown spinless excitation quantum. From the decay mode $\psi'' \rightarrow D + \bar{D} \simeq 3730$ MeV, we see that about 42 MeV of kinetic energy is liberated in this decay process. If the decay is accompanied by the internal annihilation of a $\pi \simeq 140$ MeV substate, this means that the $(\bar{N}N)\cdot(\bar{N}N)$ binding energy must be $140 - 42 = 98$ MeV. In this case, we have $X = \pi\pi$ in (17.23), so that $\psi''(3772) \equiv S\cdot S$, where $S(1935) = \bar{N}N\cdot\pi$ is the $\bar{p}p$ meson state shown in (17.21). (The calculated mass value for the ψ'' is $1935 + 1935 - 98 = 3772$ MeV.) The decay mode of the ψ'' is then

$$\psi'' \quad \equiv \quad \bar{N}N\cdot\pi\pi\cdot\bar{N}N \quad \rightarrow \quad \bar{N}N\cdot M[\pi]\bar{M}\cdot\bar{N}N \quad \rightarrow \quad D + \bar{D} \ , \tag{17.24}$$

where the annihilation $[\pi]$ splits the resonance into a strangeness $S = +1$, $S = -1$ kaon-antikaon pair.

The decay of the $\psi(4028)$ peak into $D + \bar{D}$, $D + \bar{D}^*$, and $D^* + \bar{D}^*$ pairs illustrates the mass quantization of the internal annihilations in a remarkable manner. From its mass value, the $\psi(4028)$ is the light-quark configuration

$$\psi(4028) \quad \equiv \quad \psi'\cdot\dot{3}\pi \quad = \quad N\bar{N}\cdot\dot{3}\dot{3}\dot{3}^3\cdot\dot{3}\pi \ . \tag{17.25}$$

In the decay of the $\psi(4028)$ into D's and D*'s, the basis-state transformation $\dot{3}\dot{3}\dot{3}^3 \rightarrow N\bar{N}\cdot M$ must take place. This transformation is exoergic, with a calculated energy release of about $3 \times 630 - 1865 = 25$ MeV. We can obtain a more accurate value for this energy release by noting that the decay $\psi(4028) \rightarrow D^*(2007) + \bar{D}^*(2007)$ has a kinetic energy of about 14 MeV, which we can attribute to the $\dot{3}\dot{3}\dot{3}^3 \rightarrow N\bar{N}\cdot M$ transformation. Since this transformation leads to a strange excitation, it is clear that the rest of the resonance structure must be transformed into the opposite strangeness, so that the D and D* kaons are produced in matching kaon-antikaon pairs. In the decay mode $\psi(4028) \rightarrow D + \bar{D}^*$, a total energy of $4028 - 1865 - 2007 = 156$ MeV is liberated. If we attribute 14 MeV to the internal basis-state transformation, then $156 - 14 = 142$ MeV corresponds to the internal annihilation of a π substate in the $\psi(4028)$. This annihilation energy of 142 MeV can be compared to the annihilation energy of 142.7 MeV that is observed in the decay $D^{*0}(2006) \rightarrow D^0(1863.3) + \gamma$ (see Fig. 5.5 in Chapter 5). In the decay mode $\psi(4028) \rightarrow D + \bar{D}$, a total energy of 4028 - 1865 - 1865 = 298 MeV is liberated. Again subtracting 14 MeV, we obtain 284 MeV as the energy liberated by the internal annihilation of two π substates in the $\psi(4028)$. We can compare these $[\pi]$ and $[\pi\pi]$ annihilations in the $\psi(4028)$ to the $[\pi]$ and $[\pi\pi]$ annihilations in the $\psi'(3684)$ that were described in Sec. 17.3, as follows:

$$\psi'(3684) \rightarrow \chi(3551) + \gamma\ , \quad \Delta m = 133 \text{ MeV} \qquad \psi'(3684) \rightarrow \chi(3414) + \gamma\ , \quad \Delta m = 270 \text{ MeV} \tag{17.26}$$

$$\psi(4028) \rightarrow D(1865) + \bar{D}^*(2007)\ , \quad \Delta m^* = 142 \text{ MeV} \qquad \psi(4028) \rightarrow D(1865) + \bar{D}(1865)\ , \quad \Delta m^* = 284 \text{ MeV} \tag{17.27}$$

where Δm^* denotes the fact that we have attributed 14 MeV of the observed total energy to the exoergic basis-state transformation $\dot{3}\dot{3}\dot{3}^3 \rightarrow N\bar{N}{\cdot}M$. Thus the $\psi'(3684)$ and $\psi(4028)$ mesons, by their matching $[\pi]$ and $[\pi\pi]$ decays, reveal that they have similar light-quark substructures.

The $\psi'''(4414)$ peak, which has also been used to study the production of D and D* kaons, is the light-quark configuration

$$\psi''' \equiv N\bar{N}{\cdot}\dot{S}\dot{S}\dot{S}\dot{S}^2\ . \tag{17.28}$$

In the decay mode $\psi''' \rightarrow D^* + \bar{D}^*$, the $\dot{S}\dot{S}\dot{S}\dot{S}^2 \simeq 2620$ MeV spinor configuration is transformed into the configuration $N\bar{N}{\cdot}\dot{3}\dot{3}\dot{3} \simeq 1795 + 630 = 2425$ MeV, which releases an energy of 195 MeV. Also, the annihilation $[\dot{3}]$ takes place, which liberates another 210 MeV. Thus the total calculated energy release is 405 MeV. Experimentally, this value is 4414 - 2 x 2007 = 400 MeV. In the decay mode $\psi''' \rightarrow D + \bar{D}^*$, an additional subquantum π is annihilated; and in the decay mode $\psi''' \rightarrow D + \bar{D}$, two additional subquanta π are annihilated. All of these decays are in agreement with rule (17.14).

Probably the clearest examples of 70 MeV mass quantizations in all of hadron physics are provided by the $\psi(4028)$ peak and its associated decays. The $\psi(4028)$ is 344 MeV (5 x 70 MeV) above the $\psi'(3684)$, which in turn is 1889 MeV (3^3 x 70 MeV) above the $\bar{p}n(1795)$ New Particle "ground state". The $\psi(4028)$ decays into $D + \bar{D}^*$ with a (corrected) annihilation Q-value of 142 MeV (2 x 70 MeV), and it decays into $D + \bar{D}$ with a (corrected) annihilation Q-value of 284 MeV (4 x 70 MeV). The D*(2007) is 212 MeV (3 x 70 MeV) above the mass of the $\bar{p}n(1795)$ ground state, and the D(1865) is 70 MeV above the mass of the $\bar{p}n(1795)$ ground state. Thus all of these levels appear at accurate multiples of 70 MeV above the $\bar{p}n(1795)$ New Particle threshold.

The main conclusion we would like to draw from this study of the D and D* "charmed" kaons and their associated production peaks is that, irrespective of any quark interpretation which is applied to explain their existence, they occur with mass intervals that are accurately quantized in units of 70 MeV. Also, the D and D* production peaks undergo basis-state transformations which can be understood in terms of the spin 1/2 spinor S = 330 MeV and the spinless excitations M = 70 MeV, π = 140 MeV, and $\dot{3}$ = 210 MeV. Thus, whatever macrostructure is ascribed to these "charmed" kaon states, their microstructure seems to be clearly formed from the light-quark basis states M = 70 MeV and S = 330 MeV.

17.6 The Fundamental Meson Excitation Series $\dot{3}\dot{3}\dot{3}^n$

In Sec. 13.3 of Chapter 13, we discussed the properties of the fundamental meson excitation series $\dot{3}\dot{3}\dot{3}^n$ that is shown in Table 13.1. The reason this topic was introduced in Chapter 13 was because we wanted to illustrate the manner in which the dominant ρ meson excitation appears as the first term in the series.

The historical importance of the series $\dot{3}\dot{3}\dot{3}^n$, which serves as the generator of the high-mass meson excitations, is that it was recognized and written down prior to the discovery of the New Particles — at a time when only the first three terms in the series could be identified with observed meson resonances (see Tables XXII and XXIII in Paper I). As soon as the J/ψ and ψ' resonances were discovered, it became apparent that they correspond to the n = 5 and n = 6 terms in this series, and a paper describing this fact was published.[10] In a later paper,[19] which was devoted mainly to the radiative decays of the ψ', the ψ'''(4414) peak was identified as the n = 7 member of this series.

In the discussion of the $\dot{3}\dot{3}\dot{3}^n$ series in Chapter 13, we were concerned mainly with the rationale behind the importance of the $\dot{3}\dot{3}\dot{3}$ excitation unit, and with the manner in which this series accurately reproduces the mass values of the observed resonances. In the present section, we wish to briefly emphasize two other aspects of this series: (1) the difference between the basis-state transformations $\dot{3}\dot{3}\dot{3} \rightarrow$ SS as they occur in the odd-n terms of the series, and as they occur in the even-n terms of the series; (2) the extension of the series to include higher-mass resonances. Table 17.3 illustrates these aspects of the $\dot{3}\dot{3}\dot{3}^n$ excitation series.

In Table 17.3, the first seven terms in the $\dot{3}\dot{3}\dot{3}^n$ series have the following phenomenological properties:

n = 1. The basis-state transformation $\dot{3}\dot{3}\dot{3} \rightarrow \dot{S}\dot{S}$ takes place, with the $\dot{S}\dot{S}$ spinor pair being observed in an ℓ = 0 mode as the very weak spin 1 $\pi\pi$(658) interference effect[20], and in an ℓ = 1 mode as the dominant ρ(773) meson (see Chapter 13).

Table 17.3. The fundamental meson excitation series $\dot{3}\dot{3}\dot{3}^n$. The first seven terms in this series are also shown in Table 13.1, where the mass values for these terms are calculated. As can be seen in Table 17.3, the odd-n terms correspond to complete $\dot{3}\dot{3}\dot{3} \to SS$ spinor transformations, whereas the even-n terms correspond to partial $\dot{3}\dot{3}\dot{3} \to SS$ transformations. This result, which is for the creation of subquanta M = 70 MeV, is analogous to the result shown in Table 17.1 [see rule (17.14)], which is for the annihilation of subquanta M. The extension of the $\dot{3}\dot{3}\dot{3}^n$ series to include higher-mass terms is discussed in the text.

Excitation term	Observed configuration	Observed resonance	Spin state	$\dot{3}\dot{3}\dot{3} \to SS$ transitions
$\dot{3}\dot{3}\dot{3}^1$	$\dot{S}\dot{S}(\ell = 0)$	$\pi\pi(658)$ interference dip	1	complete
	$\dot{S}\dot{S}(\ell = 1)$	ρ meson	1	complete?
$\dot{3}\dot{3}\dot{3}^2$	$\dot{S}\dot{S}\cdot\dot{3}\dot{3}\dot{3}$	D(1285) meson	1	partial
	$\dot{S}\dot{S}\dot{S}\dot{S}$	A_2dip interference effect	2	complete
$\dot{3}\dot{3}\dot{3}^3$	$S\bar{S}S\cdot\bar{S}S\bar{S} \equiv N\bar{N}$	$\bar{p}n(1795)$ bound state		complete
$\dot{3}\dot{3}\dot{3}^4$	$N\bar{N}\cdot\dot{3}\dot{3}\dot{3}$	$(K\pi)^o(\sim 2425)$		partial
$\dot{3}\dot{3}\dot{3}^5$	$N\bar{N}\cdot\dot{S}\dot{S}\dot{S}\dot{S}$	$J/\psi(3098)$	1	complete
$\dot{3}\dot{3}\dot{3}^6$	$N\bar{N}\cdot\dot{3}\dot{3}\dot{3}^3$	$\psi'(3684)$	1	partial
$\dot{3}\dot{3}\dot{3}^7$	$N\bar{N}\cdot\dot{S}\dot{S}\dot{S}\dot{S}^2$	$\psi'''(4414)$	1	complete
⋮				
$\dot{3}\dot{3}\dot{3}^{10}$	$(J/\psi)^2$	T(5970)???		
⋮				
$\dot{3}\dot{3}\dot{3}^{15}$	$(J/\psi)^3$	T(9400)		

n = 2. The partial transformation $\dot{3}\dot{3}\dot{3}^2 \to \dot{S}\dot{S}\cdot\dot{3}\dot{3}\dot{3}$ corresponds to the dominant spin 1 D(1285) meson excitation; and the complete transformation $\dot{3}\dot{3}\dot{3}^2 \to \dot{S}\dot{S}\dot{S}\dot{S}$ corresponds to the very weak spin 2 A_2dip(1310) interference effect[20].

n = 3. The complete transformation $\dot{3}\dot{3}\dot{3}^3 \to S\bar{S}S\cdot\bar{S}S\bar{S} \equiv N\bar{N}$ corresponds to the fundamental $\bar{p}n(1795)$ bound state resonance, which serves as the "ground state" for the New Particle excitations.

n = 4. The partial transformation $\dot{3}\dot{3}\dot{3}^4 \to N\bar{N}\cdot\dot{3}\dot{3}\dot{3} \to \pi\cdot N\bar{N}\cdot\dot{3}\dot{3}\cdot M$ corresponds to the strange kaon excitations $(K\pi)^o(2420)$ and $(K\pi)^o_R(\sim 2430)$ (see Table 17.1). Note the similarity between this basis-state configuration and the D*(2007) $\equiv \pi\cdot N\bar{N}\cdot M$ configuration that is shown in Eq. (17.22). The D*(2007) appears 140 MeV above the D(1865), and the $(K\pi)^o(\sim 2425)$ appears 3 x 140 = 420 MeV above the D*(2007).

n = 5. The complete transformation $\dot{3}\dot{3}\dot{3}^5 \rightarrow N\bar{N}\cdot\dot{S}\dot{S}\dot{S}\dot{S}$ corresponds to the J/ψ(3098) meson. The reasons for the unusual strength of this excitation are discussed in connection with Eq. (17.3).

n = 6. The partial transformation $\dot{3}\dot{3}\dot{3}^6 \rightarrow N\bar{N}\cdot\dot{3}\dot{3}\dot{3}^3$ corresponds to the ψ'(3684) meson.

n = 7. The complete transformation $\dot{3}\dot{3}\dot{3}^7 \rightarrow N\bar{N}\cdot\dot{S}\dot{S}\dot{S}\dot{S}^2$ corresponds to the ψ'''(4414) meson.

As can be seen from these observed excitations, the n = 3, 5, and 7 terms correspond to complete transformations, in which the basic $\dot{3}\dot{3}\dot{3}$ excitation quanta are entirely transformed into spinors S; whereas the n = 4 and 6 terms correspond to partial transformations, in which the $\dot{3}\dot{3}\dot{3}$ excitation quanta are only partially transformed into spinors. The n = 2 term has both a partial and a complete transformation mode, but the partial mode is more prominent experimentally than the complete mode. The n = 1 term corresponds to a complete transformation in its weak ππ(658) interference mode, and it may or may not correspond to a complete transformation in its dominant ρ meson rotational mode (see Chapter 13). From these systematics, it can be seen that the odd-n terms in the series, which correspond to the creation of an odd number of subquanta M = 70 MeV, have a tendency to completely transform the spinless excitation quanta into spinors; whereas the even-n terms, which correspond to the creation of an even number of subquanta M, tend to transform the $\dot{3}\dot{3}\dot{3}$ excitation quanta only partly into spinors. This result, which is for the creation of subquanta M, seems closely analogous to the result displayed in Table 17.1, wherein the annihilation of an odd number of subquanta M is accompanied by a $\dot{3}\dot{3}\dot{3}^2 \rightarrow \dot{S}\dot{S}\dot{S}\dot{S}$ spinor transformation, whereas the annihilation of an even number of subquanta M is not accompanied by this transformation. Both of these results suggest the conservation of a fundamental quantum number during the creation or annihilation of subquanta M, and the parity quantum number P seems a logical candidate. This dependence of the $\dot{3}\dot{3}\dot{3} \rightarrow SS$ transition on the evenness or oddness of the change in the number of subquanta M also seems to be related to the results described in Appendix E, wherein $\dot{3}$-type excitations are shown to have different symmetry properties from $\dot{4}$-type excitations.

The last topic to discuss in connection with Table 17.3 is one about which we can actually say very little at the present time — namely, the extension of the series $\dot{3}\dot{3}\dot{3}^n$ to higher mass values. From the linear manner in which the light-quark basis states are created, it seems that higher-energy excitations should simply be linear combinations of the lower-energy excitations. Thus, from the dominance of the J/ψ resonance, we might expect to find higher-mass resonances appearing at roughly 6200 and 9300 MeV as multiples of the 3100 MeV mass of the J/ψ. These would correspond to the terms n = 10 and n = 15 in Table 17.3. As is shown in Table 17.3, an upsilon resonance was in fact reported at about 6000 MeV[21], but it has unfortunately subsequently disappeared. However, a new upsilon resonance has been discovered at 9400 MeV, which is just about where we would expect to find it on the basis of these systematics.

The exact (hadronic) nature of the T(9400) has not yet been verified, so that we have little to go on other than its mass value. The T(9400) exhibits considerable structure, and in Table 17.1 we have listed the three levels that have (possibly) been identified to date. The spacing of these levels seems similar to spacings observed at lower energies, but the data are not yet precise enough to permit a meaningful analysis.

We have one final comment to make with respect to the systematics of Table 17.3. If we see resonances which appear as multiples of some basic mass value, this suggests that the resonances may have a compound quark structure. The series $\dot{3}\dot{3}\dot{3}^n$ is of course one illustration of this result. In this connection, the observation has been made[22] that the $\phi(1020)$ meson, $J/\psi(3098)$ meson, and T(9400) meson have masses that increase by factors of three, which suggests that the constituent quarks may be similarly compounded. In this regard, we recall the suggestion that was discussed in Sec. 8.7 of Chapter 8, in Sec. 14.6 of Chapter 14, and in Chapters 15 and 16, to the effect that it may be possible to form excited states of the spinor S which have higher masses than 330 MeV (this topic is discussed in some detail in Sec. 18.7 of Chapter 18). Now, in Chapter 16 we reproduced the ϕ meson as the light-quark cluster $\phi \equiv \dot{S}\dot{S}\dot{3}\pi$. However, the experimental properties of the ϕ are so distinctive that the ϕ may actually represent a new type of light-quark cluster, $\phi \equiv S'S'$, where $S' \sim 510$ MeV. If this is true, then a logical extension of this idea is to form the clusters $J/\psi \equiv (S'S')^3$ and $T \equiv (S'S')^9$. But it is necessary to trace the consequences of this assumption through the entire collection of $\sim$200 meson and baryon resonances before its usefulness can be ascertained. This example illustrates the difficulties that are inherent in working with a physical system whose basic building blocks are hidden from view: the evidence that we can produce is always circumstantial, and it takes a great deal of circumstantial evidence to provide a convincing case.

17.7 Mapping the Spectrum of New Particles

In the previous sections of the present chapter, we have given a number of examples that illustrate the manner in which the light-quark basis states M = 70 MeV, π = 140 MeV, $\dot{3}$ = 210 MeV, and S = 330 MeV can be used to accurately reproduce certain groupings of the New Particle resonances. We have also demonstrated the manner in which the fundamental hadronic excitation series $\dot{3}\dot{3}\dot{3}^n$ serves as the generator of the high-mass meson resonances. Our remaining task in this chapter is to attempt to account for the New Particle states that are listed in Table 17.2. Although the light-quark basis states that we have developed in the preceding chapters seem adequate for handling these New Particle states, it would be very difficult to use these basis states in a completely *a priori* manner to predict the unique properties of the New Particles. Thus we limit ourselves in the present section to giving *a posteriori* explanations for the New Particle resonances that have been observed to date. As we will see, we can accurately reproduce the mass values of most of the New Particle levels. However, the determination of spin values seems beyond the present scope of the model. We will handle the mapping of the New Particle resonances by discussing in turn the various groupings of related New Particle states.

17.7A The J/ψ, ψ', ψ'', and ψ''' Mesons

The $J/\psi(3098)$, $\psi'(3684)$, and $\psi'''(4414)$ mesons were the first New Particles to be identified experimentally. As is shown in Tables 13.1 and 17.3, these states accurately correspond to the n = 5, 6, and 7 terms in the $\dot{3}\dot{3}\dot{3}^n$ excitation series. The light-quark configurations $J/\psi \equiv N\bar{N}\cdot\dot{S}\dot{S}\dot{S}\dot{S}$, $\psi' \equiv N\bar{N}\cdot\dot{3}\dot{3}\dot{3}^3$, and $\psi''' \equiv N\bar{N}\cdot\dot{S}\dot{S}\dot{S}\dot{S}^2$ follow from the mass values shown in Table 13.1 and the excitation systematics shown in Table 17.3. As we discussed in Sec. 17.3, the J/ψ and ψ' light-quark configurations are also fundamental to an understanding of the four χ radiative decay levels of the ψ' meson.

The $\psi''(3772)$ meson, which sits just above the threshold for producing a pair of "charmed" D(1865) kaons, appears to be an $N\bar{N}\cdot N\bar{N}$ type of excitation (see Fig. 17.1). In Eqs. (17.23) and (17.24), we assumed the light-quark configuration $\psi'' \equiv \pi\cdot N\bar{N}\cdot N\bar{N}\cdot\pi$, where $N\bar{N}\cdot\pi \equiv S(1935)$ is an observed meson excitation, and we thereby obtained a calculated $(N\bar{N})\cdot(N\bar{N})$ binding energy of 98 MeV. However, the configuration $\psi'' \equiv N\bar{N}\cdot N\bar{N}\cdot\dot{3}$ could also be chosen, which gives a binding energy of 28 MeV. With only this one example, we cannot pin down the light-quark configuration of the ψ'', although we can reasonably account for its existence.

17.7B The "Charmonium" Resonances

The resonances most closely associated with "charmonium" (which supposedly parallels the levels found in positronium) are the J/ψ and ψ' mesons discussed above, the $\chi(3414)$, $\chi(3455)$, $\chi(3503)$, and $\chi(3551)$ radiative decay levels of the ψ', and the X(2830) radiative decay level of the J/ψ. In Sec. 17.3 we discussed all of these levels except the X(2830), which we discuss here.

To briefly recapitulate the results of Sec. 17.3, the $\chi(3551)$ and $\chi(3414)$ levels are 133 MeV and 274 MeV below the $\psi'(3684)$, so that they correspond to the internal annihilations $[\pi]$ and $[\pi\pi]$, respectively, in the ψ'. The $\chi(3503)$ or P_C level is 405 MeV <u>above</u> the $J/\psi(3098)$, so that it corresponds to the internal annihilation $[\dot{3}]$ in the ψ' plus an associated $\dot{3}\dot{3}\dot{3}^2 \to \dot{S}\dot{S}\dot{S}\dot{S}$ basis-state transformation (see Table 17.1). The $\chi(3455)$ level is 41 MeV above the $\chi(3414)$ level, so that it corresponds to the same decay channel as the $\chi(3414)$, but with an associated $\dot{3}\dot{3}\dot{3}^2 \to \dot{S}\dot{S}\dot{S}\dot{S}$ transformation [see Eq. (17.8)]; since this decay mode violates rule (17.14), the cross section for producing the $\chi(3455)$ is very small.

The X(2830) level is an important one for us to consider, because the "charmonium" model has no natural way of accounting for the existence of this state[14]. The X(2830) is a level that was observed at DESY in the decay channel $J/\psi \rightarrow X + \gamma$, $X \rightarrow 2\gamma$. The DESY experiment placed the mass of the X at 2820 ± 20 MeV.[23] A recent Serpukov experiment[24], using the reaction channel $\pi^- + p \rightarrow X + n$, $X \rightarrow 2\gamma$, has singled out an $X \rightarrow 2\gamma$ resonance peak at 2880 ± 60 MeV, which they identify with the X seen in the CERN experiment. Thus the X appears experimentally with a mass of about 2850 MeV. In attempting to reproduce this state within the context of the present light-quark model, we first note that the strongest radiative decay channel of the ψ' is the $\psi' \rightarrow P_c + \gamma$, $P_c \rightarrow J/\psi + \gamma$ channel, which corresponds to the light-quark transformations

$$\psi' \;\equiv\; N\bar{N}\cdot\dot{3}\dot{3}\dot{3}^3 \;\rightarrow\; N\bar{N}\cdot\dot{3}\dot{3}\dot{3}^2\cdot\dot{3}\dot{3}[\dot{3}] \;\rightarrow\; N\bar{N}\cdot\dot{S}\dot{S}\dot{S}\dot{S}\cdot\dot{3}\dot{3} + \gamma \equiv P_c + \gamma\;, \qquad (17.29a)$$

$$P_c \;\rightarrow\; N\bar{N}\cdot\dot{S}\dot{S}\dot{S}\dot{S}[\dot{3}\dot{3}] \;\rightarrow\; N\bar{N}\cdot\dot{S}\dot{S}\dot{S}\dot{S} + \gamma \;\equiv\; J/\psi + \gamma\;. \qquad (17.29b)$$

These $\dot{3}$ annihilations accompanied by the emission of γ rays are in line with the systematics described in Appendix E, wherein it is demonstrated that $\dot{3}$-type excitations are characteristically associated with radiative transitions. Since $J/\psi \rightarrow X(2830) + \gamma$ is the only radiative level of the J/ψ that has yet been observed, it seems natural that this decay must be associated with a $[\dot{3}]$ annihilation in the J/ψ, which in turn requires that the $J/\psi \rightarrow X + \gamma$ decay must include the basis-state transformation $\dot{S}\dot{S}\dot{S}\dot{S} \rightarrow \dot{3}\dot{3}\dot{3}^2$. Since this transformation is exoergic by 41 MeV [Eq. (17.8)], and since the annihilation $[\dot{3}]$ liberates an energy of 210 MeV, we expect to find the X(2830) appearing 41 + 210 = 251 MeV below the J/ψ; hence the X(2830) has a calculated mass of 3098 - 251 = 2847 MeV, which is in good agreement with experiment. Thus the decay $J/\psi \rightarrow X + \gamma$ corresponds to the light-quark transformations

$$J/\psi \;\equiv\; N\bar{N}\cdot\dot{S}\dot{S}\dot{S}\dot{S} \;\rightarrow\; N\bar{N}\cdot\dot{3}\dot{3}\dot{3}^2 \;\rightarrow\; N\bar{N}\cdot\dot{3}\dot{3}\dot{3}\cdot\dot{3}\dot{3}[\dot{3}] \;\rightarrow\; X + \gamma\;. \qquad (17.30)$$

The transition $\dot{S}\dot{S}\dot{S}\dot{S} \rightarrow \dot{3}\dot{3}\dot{3}^2$ shown in (17.30) is the inverse of the transition $\dot{3}\dot{3}\dot{3}^2 \rightarrow \dot{S}\dot{S}\dot{S}\dot{S}$ shown for the ψ' in Table 17.1, and it is in accord with the general systematics of rule (17.14). Furthermore, as we discuss in Sections 17.7F and 17.7G, the light-quark configuration $N\bar{N}\cdot\dot{3}\dot{3}\dot{3}\cdot\dot{3}\dot{3}$ is also associated with the narrow-width $N\bar{N}(2850)$ resonance. Thus both the existence and the mass value of the X(2830) are in accurate agreement with the workings of the light-quark model.

Table 17.2 also lists the radiative decay level $\chi(3340)$. This level appears as an alternate to the $\chi(3455)$ level in the decay channel $\psi' \rightarrow \chi + \gamma$, $\chi \rightarrow J/\psi + \gamma$ (depending on whether the higher-energy γ is emitted in the first or second step of the decay). Since the $\chi(3340)$ is 344 MeV below the mass of the $\psi'(3684)$, we interpret this level as a ψ' internal annihilation $[\dot{3}\pi]$ that is unaccompanied by a $\dot{3}\dot{3}\dot{3}^2 \rightarrow \dot{S}\dot{S}\dot{S}\dot{S}$ transition, which is in violation of rule (17.14) [as is the transformation which produces the $\chi(3455)$]. Thus the $\chi(3340)$ can be accommodated naturally within the framework of the light-quark model, although there is no experimental indication that the levels $\chi(3340)$ and $\chi(3455)$ both exist. We note in passing that $\psi(4028) - \psi'(3684) = \psi'(3684) - \chi(3340) = 344$ MeV.

17.7C The $\psi(4028)$ and $\psi(4100)$ R-ratio Peaks

The $\psi(4028)$ and $\psi(4100)$ R-ratio peaks are illustrated in Fig. 17.2. The $\psi(4028)$ peak was discussed at some length in Sec. 17.5 in connection with the "charmed" D and D* kaons. The interesting feature about the $\psi(4100)$ peak from the present light-quark point-of-view is of course the fact that it appears just 70 MeV above the $\psi(4028)$ peak as another example of a New Particle mass quantization that is in units of 70 MeV. The $\psi(4028)$ and $\psi(4100)$ peaks correspond to the light-quark configurations

$$\psi(4028) \equiv \psi' \cdot \dot{3}\pi \quad \text{and} \quad \psi(4100) \equiv \psi' \cdot \dot{3}\dot{3} \qquad (17.31)$$

[see Eq. (17.25)], which we can compare to

$$\chi(3455) \equiv J/\psi \cdot \dot{3}\pi \quad \text{and} \quad \chi(3503) \equiv J/\psi \cdot \dot{3}\dot{3} \qquad (17.32)$$

[see Table 17.1 and Eq. (17.29)]. Thus, while we have no explanation as to why the particular excitations shown in (17.31) should result in prominent resonance peaks, we can see these same excitations occurring elsewhere in the New Particle spectrum.

17.7D The "Charmed" Kaons and Matching "Uncharmed" Mesons

The "charmed" kaons are the D($\sim$1865) and D*($\sim$2007) resonances that were discussed at some length in Sec. 17.5. The matching "uncharmed" mesons are the narrow-width $\bar{p}n(1795)$ and $\bar{p}p(1935)$ resonances that were also discussed in Sec. 17.5. As we have illustrated in Fig. 5.4 of Chapter 5, these four resonances are separated by very accurate 70 MeV intervals; and as we demonstrated in Eqs. (17.19) - (17.22) of the present chapter, the strangeness assignments for these states follow from their light-quark configurations. There are two points we would like to make with respect to this $\bar{p}n(1795)$, D(1865), $\bar{p}p(1935)$, D*(2007) sequence of resonances: (1) In view of the many examples of mass quantizations in units or multiples of 70 MeV that we have been able to produce in the present chapter and in Chapters 5, 15, and 16, the 70 MeV mass intervals exhibited by these four resonances cannot be regarded as accidental. (2) The "charm" model, with its u,d,s,c set of quarks, has led to the prediction of the D and D* kaons, but it has no logical way of accounting for the $\bar{p}n$ and $\bar{p}p$ mesons, or of anticipating their existence. Thus the "charm" model approach to this set of resonances is, in spite of its apparent successes with the D and D* kaons, manifestly incomplete.

In addition to the D and D* kaons, the "charm" model also unambiguously predicts the existence of two additional kaon resonances — the F and F* kaons, which are expected to appear with masses of 1975 and 2061 MeV, respectively[8]. In Table 17.2 the levels F(2030) and F*(2140) are shown. The evidence for these two levels is very meagre, and they may or may not have the properties that are ascribed to them by the charm model. The interesting thing about these levels from the present point-of-view is that they fit reasonably well into the 70 MeV and 140 MeV mass intervals that are exhibited by the other kaon resonances in this mass region, as we demonstrate in the next section.

17.7E The π = 140 MeV Kaon Excitation Tower

In Fig. 5.3 of Chapter 5, we demonstrated the manner in which narrow-width meson resonances occur at accurate 140 MeV intervals, and in Fig. 5.6 we demonstrated this same result for narrow-width kaon resonances. If we think of the D(∿1865) kaon as constituting a fundamental New Particle kaon "ground state", then we can successively add π = 140 MeV excitation quanta to this ground state so as to generate a "kaon excitation tower". Matching the levels in this excitation tower to the observed narrow-width kaon resonances gives the following results:

Excitation level in MeV	Observed resonances	
1865	D(∿1865)	
2005	D*(∿2007), F(2030)	
2145	F*(2140)	(17.33)
2285	$(K\pi)^{o}(2290)$	
2425	$(K\pi)^{o}(2420)$, $(K\pi)^{o}_{R}(\sim 2430)$	
2565	I(2600)	

The D, D*, F, and F* resonances in this kaon excitation tower were discussed in Sec. 17.7D. The $(K\pi)^{o}(2290)$ and $(K\pi)^{o}(2420)$ peaks are from a Fermilab experiment.[25] Of these two peaks, the $(K\pi)^{o}(2290)$ is prominent enough (a "4σ" effect) that it was specifically singled out in the paper[25] as an enhancement. The $(K\pi)^{o}(2420)$ peak, on the other hand, is small enough (a "3σ" effect) that it was only mentioned in a footnote (see Ref. 10 in Ref. 25). However, this $(K\pi)^{o}(2420)$ peak accurately matches the SLAC $(K\pi)^{o}_{R}($ 2430) recoil peak[26] that is listed next to it in the kaon excitation tower; hence it is probably a real effect. In the SLAC experiment[26], the recoil peaks $(K\pi)^{o}_{R}(\sim 1860)$ and $(K\pi)^{o}_{R}(\sim 2005)$ that are listed in Table 17.2 were also observed; these clearly correspond to the D^{o} and D^{*o} kaons. The I(2600) kaon state that is shown in (17.33) may or may not exist experimentally (see Ref. 83 in Appendix B).

The manner in which these narrow-width kaon resonances appear at regular 140 MeV intervals in the form of a kaon excitation tower suggests that they have similar light-quark structures. The interesting theoretical aspect of this kaon excitation tower is the fact that it naturally accommodates the "charmed" kaon states D, D*, F, and F*. Although we have commented throughout the text that hadronic quark substates cannot be observed directly, it may be that in the decay mode $D^* \to D + \pi$ we are actually witnessing the ejection of a quark substate from the D*.

17.7F The $\dot{3}$ = 210 MeV $\bar{N}N$ Excitation Tower

The pion state π = 140 MeV is a fundamental meson and kaon building block, as we demonstrated in the kaon excitation tower of (17.33), and also in the systematics displayed in Sec. 5.3 of Chapter 5. Another fundamental meson building block is the spinless excitation unit $\dot{3}$ = 210 MeV, as is apparent from the systematics of the $\dot{3}\dot{3}\dot{3}^{n}$ excitation series shown in Table 17.3, and also from the decay modes $\psi' \to P_c \to J/\psi$ shown in (17.29) and $J/\psi \to X(2830) + \gamma$ shown in (17.30). If we use $\dot{3}$ excitations to form a "meson excitation tower" based on the fundamental $\bar{p}n(1795)$ "ground state", we find an interesting correlation with a sequence of observed $\bar{N}N$ resonances, as is demonstrated in the following table:

Excitation level in MeV	Observed resonance	
1795	$\bar{p}n(1795)$	
2005	$\bar{p}p(2020)$	
2215	$\bar{p}p(2204)$	
2425		(17.34)
2635	$\bar{p}p(2660)$	
2845	$\bar{N}N(2850)$	
3055	$\bar{N}N(3050)$	

The $\bar{p}n(1795)$ resonance that we have used as the "ground state" for this $\dot{3}$ = 210 MeV meson excitation tower is of course the $\bar{p}n$ bound state that serves as the threshold for all of the New Particle resonances. The $\bar{p}p(2020)$ and $\bar{p}p(2204)$ levels were both observed in the same experiment[27], which also confirmed the existence of the $\bar{p}p(1935)$ peak that we discussed in Sec. 17.7D. The $\bar{p}p(2660)$ peak is a very weak experimental effect that is mentioned only incidentally in Ref. 10 of Ref. 25; however, this peak seems to fit in naturally with the systematics shown in (17.34). The $\bar{N}N(2850)$ and $\bar{N}N(3050)$ peaks are a pair of enhancements that were observed in the same experiment; we will have more to say about these states in the next section.

The main point we are trying to bring out in a display such as (17.34) is that the light-quark basis states M = 70 MeV, π = 140 MeV, $\dot{3}$ = 210 MeV, and S = 330 MeV can be combined together in many different ways, and the spectrum of observed resonances indicates that systematic groupings of these excitation clusters really are being produced in the various New Particle experiments.

17.7G The $\bar{N}N(2850)$ and $\bar{N}N(3050)$ Resonances

The $\bar{N}N(2850)$ and $\bar{N}N(3050)$ mesons shown in (17.34) are narrow peaks that are produced in $\bar{p}d$ interactions.[28] The $\bar{N}N(3050)$ peak, in particular, is a very sharp resonance, with a single-bin width of less than 20 MeV, and with a quoted mass value of 3050 ± 10 MeV[28]. The significance of the $\bar{N}N(3050)$ meson lies in its close proximity to the $J/\psi(3098)$ meson. In Sec. 17.3 we reproduced the J/ψ as the light-quark configuration $J/\psi = \bar{N}N\cdot\dot{S}\dot{S}\dot{S}\dot{S}$ [Eq. (17.1)]; and in this same section we used the measured $\psi' - J/\psi$ mass difference of 588.6 MeV to deduce the light-quark basis-state mass difference $\dot{S}\dot{S}\dot{S}\dot{S} - \dot{3}\dot{3}\dot{3}^2$ = 41 MeV [Eq. (17.8)]. Thus if we could produce the J/ψ as the configuration $J/\psi = \bar{N}N\cdot\dot{3}\dot{3}\dot{3}^2$ (compare this to the configuration $\psi' = \bar{N}N\cdot\dot{3}\dot{3}\dot{3}^3$) rather than $J/\psi = \bar{N}N\cdot\dot{S}\dot{S}\dot{S}\dot{S}$, we would obtain a resonance with a mass that is 41 MeV lower than that of the J/ψ; that is, at 3098 - 41 = 3057 MeV. This mass value coincides with the measured mass of the $\bar{N}N(3050)$ enhancement. Furthermore, we can see from the excitation systematics of the meson excitation tower shown in (17.34) that the $\bar{N}N(3050)$ has the light-quark structure $\bar{N}N(3050) = \bar{N}N\cdot\dot{3}\dot{3}\dot{3}^2$.

The $\bar{N}N(2850)$ meson is equally significant, because it coincides with the X(2830) level that is observed in the radiative decay $J/\psi \to X + \gamma$, $X \to 2\gamma$. As we showed in Eq. (17.30), the radiative decay of the J/ψ takes place by means of the transition $\dot{S}\dot{S}\dot{S}\dot{S} \to \dot{3}\dot{3}\dot{3}^2$ and the internal annihilation $[\dot{3}]$; these two decay processes lower the mass of the J/ψ by 41 + 210 = 251 MeV, which places the mass of the radiative decay level X at a value of 3098 - 251 = 2847 MeV. This matches the quoted mass value of 2850 ± 5 MeV[28] for the $\bar{N}N(2850)$ resonance. Also, the light-quark structure $\bar{N}N(2850) = \bar{N}N\cdot\dot{3}\dot{3}\dot{3}\cdot\dot{3}\dot{3}$ that is dictated by the systematics of (17.34) is the same configuration that was deduced in Eq. (17.30).

The important point we are trying to make in these examples is that the light-quark method of classifying New Particle resonances has a significance that goes well beyond just reproducing a set of observed mass values. The transitions that take place between these resonances serve to link them together, and this cross-linking can be followed in detail by means of the internal basis-state transformations that occur. The New Particles are especially advantageous for this kind of analysis because their mass uncertainties are much smaller than the 70 MeV mass quantization of the basis states, and also because many of their decays involve the internal annihilation of only a small part of the resonance structure.

17.7H The Upsilon Peaks

The Υ(9400), Υ(10010), and Υ(10400) peaks were mentioned briefly in Sec. 17.6 in connection with the fundamental meson excitation series $\dot{3}\dot{3}\dot{3}^n$. Since the experimental nature of these upsilon peaks has not yet been clarified, about the only relevant observations we can offer here are that these peaks appear at three times the mass of the J/ψ, and that they have suggested level spacings which seem of the same magnitude as the level spacings observed for the lower-mass New Particle states.

17.7I The M(1817) Peaks

The meson state that we have denoted as M(1817)Γ30 in Table 17.2 is actually an average value of the peaks M(1821)Γ31, M(1819)Γ24, and M(1812)Γ34 that were reported in three closely-related Frascati experiments (see Ref. 74 in Appendix B). These peaks are hadron production enhancements that were observed in e^+e^- colliding-beam experiments. We have stressed throughout the present book that the $\bar{p}n(1795)$ bound state represents the fundamental light-quark configuration which underlies all of the New Particle resonances. If this is true, then, since the New Particles are copiously produced in e^+e^- annihilations, we logically expect to find some kind of an e^+e^- effect occurring near 1795 MeV. The narrow-width M(1817) enhancements are suggested here as possible manifestations of this effect.

17.7J The "Charmed" Baryons

The two "charmed" baryons that have been reported to date are the states $\bar{\Lambda}(2260)$ and Λ(2426) shown in Table 17.2. With respect to these states, we note the following mass values:

$$\begin{aligned} \bar{p}\cdot\dot{S}\dot{S}\dot{S}\dot{S} &= 939 + 1310 = 2249 \text{ MeV}, && \bar{\Lambda}(2260); \\ \Lambda\cdot\dot{S}\dot{S}\dot{S}\dot{S} &= 1116 + 1310 = 2426 \text{ MeV}, && \Lambda(2426); \\ \bar{N}N\cdot\dot{S}\dot{S}\dot{S}\dot{S} &= 1795 + 1310 = 3105 \text{ MeV}, && J/\psi(3098). \end{aligned} \tag{17.35}$$

The mass relationships shown in (17.35) suggest that the excitation unit $\dot{S}\dot{S}\dot{S}\dot{S}$, which is of such importance in the New Particle resonances, is also fundamental to these "charmed" baryon enhancements. However, the manner in which the strangeness quantum number is accounted for in the case of the $\bar{\Lambda}(2260)$ baryon state is not immediately apparent from these configurations.

17.7K Summary

The discussions in Sections 17.7A - 17.7J include all of the New Particle resonances listed in Table 17.2 except the $\bar{p}p\pi(2950)$ resonance, for which no ready explanation has emerged.[+] We have been able to reproduce the mass values of these resonances to mass accuracies of roughly ±10 MeV, using light-quark basis states which are in multiples of M = 70 MeV and S = 330 MeV. Also, we have been able to identify basic light-quark clusters which are common to groups of New Particle excitations. Thus, while we have not attempted to provide comprehensive rules for the ways in which the light-quark basis states combine together to produce New Particle resonances (except in the case of the $\dot{3}\dot{3}\dot{3}^n$ excitation series shown in Table 17.3), it seems apparent that these rules do exist. By discussing the New Particle states of Table 17.2 in naturally-related groupings, as we have done here, we have tried to bring out the numerous relationships and cross-relationships which exist among these data sets. It might be argued that we are using such small quark basis states for these New Particles that we can fit anything, but this argument will not stand up under a critical analysis if the results that are displayed in the present chapter are studied carefully.

As in the cases of the baryon resonances in Chapter 15 and the low-mass meson resonances in Chapter 16, there seems to be a reasonable correspondence between the number of light-quark combinations that can be formed and the spectrum of observed resonances. Also, the systematics of the light-quark excitations is more-or-less the same for all of these hadron states. Although we have not attempted to extract spin information from these New Particle light-quark combinations, this is a task that must eventually be carried out. The determination of the spins of these resonances is a difficult problem experimentally, but one that is of obvious importance, particularly from the viewpoint of the "charm" model.

Perhaps the main point to be noted here is that in reproducing the spectrum of New Particle resonances, we have not been forced to modify the light-quark basis set, which was completely defined before any of the New Particle resonances were reported. The New Particles have some unique properties, but this uniqueness stems from the new light-quark basis-state combinations they contain (in particular, the $\bar{N}N$ cluster), and not from the appearance of new types of basis states. Thus we find no need to invoke a new quantum number (charm) in order to account for the existence of the New Particles.

[+]Recent experimental results, with twenty times the data, do not confirm this peak.

17.8 Concluding Comments

With this study of the New Particles, we have essentially concluded our basic examination of the spectrum of strongly-interacting (hadronic) elementary particles. In Part I of this book we examined the mass spectra of selected sets of hadrons in order to deduce their quark substates. Then, having concluded from this examination that the binding energies between quarks are weak (so that the quarks have small masses), and having thereby arrived at a suitable set of light-quark basis states, we went on in Part II to examine the angular momentum properties of the hadrons. We found that it seems to make sense to sort the hadrons into the same types of rotational bands that occur in nuclear physics. Finally, in Part III we examined the entire spectrum of identified hadron resonances (Tables 15.1, 16.1, and 17.2). We discovered that, although baryons and mesons have certain characteristic differences, they also have many similarities; in particular, the same light-quark basis states that appear in baryons also appear in mesons, but in somewhat different combinations. And, in the chapter just concluded, we found that the New Particles utilize these same basis states, but again with some characteristic differences in their combinations. In assessing the significance of the present studies, this last result is especially important, because the light-quark basis states were formulated without any knowledge of the New Particles. The fact that it has not been necessary to modify this established formalism in order to accommodate the spectrum of New Particle resonances is one of the most encouraging results which has yet emerged with respect to this light-quark viewpoint of elementary particle structure.

In the remaining three chapters, which form Part IV of the book, we discuss a number of ramifications of this work – both in the area of elementary particle physics and in the related area of nuclear physics. Our approach in all of these studies is unabashedly phenomenological; however, given the vast amount of experimental data which is now available to us, the phenomenological avenue of approach may well be the most reliable road to success.

References for Chapter 17.

1. J. J. Aubert et al., Phys. Rev. Lett. 33, 1404 (1974).
2. J.-E. Augustin et al., Phys. Rev. Lett. 33, 1406 (1974).
3. G. S. Abrams et al., Phys. Rev. Lett. 33, 1453 (1974).
4. M. H. Mac Gregor, Phys. Rev. D9, 1259 (1974), denoted here as Paper I.
5. M. H. Mac Gregor, Phys. Rev. D10, 850 (1974), denoted here as Paper II.

6. D. B. Lichtenberg et al., Phys. Rev. Lett. 39, 1592 (1977).

7. E. Eichten et al., Phys. Rev. Lett. 34, 369 (1975).

8. A. De Rújula et al., Phys. Rev. D12, 147 (1975).

9. W. K. H. Panofsky, SLAC-PUB-2025, Oct. (1977).

10. M. H. Mac Gregor, Phys. Rev. D12, 1492 (1975). This paper was submitted to Physical Review D in Feb., 1975, three months after the discovery of the J/ψ and ψ' New Particle states.

11. The mass difference ψ' - J/ψ = 588.6 ± 0.8 MeV is listed in RPP78.

12. G. J. Feldman, SLAC-PUB-1851, Dec. (1976), p. 57.

13. J. H. Cobb et al., Phys. Lett. 72B, 497 (1978).

14. E. Eichten, talk given at Am. Phys. Soc. meeting, San Francisco, Calif., Jan. (1978); K. Gottfried, Cornell Univ. report CLNS-376, Nov. (1977), pp. 8 - 12.

15. See, for example, R. Kögerler et al., Phys. Lett. 65B, 441 (1976), Fig. 3.

16. Fig. 17.2 is taken from Ref. 12.

17. For an early review of these states, see G. Goldhaber, LBL-5534, Oct. (1976). A more recent review is given by G. J. Feldman in SLAC-PUB-2068, Dec. (1977).

18. See the second reference in each of Refs. 14 and 17.

19. M. H. Mac Gregor, Phys. Rev. D14, 1463 (1976).

20. See Fig. 8.1 in Sec. 8.4 of Chapter 8, and see Appendix D.

21. D. C. Hom et al., Phys. Rev. Lett. 36, 1236 (1976) and 37, 1374 (1976); D. Eartly et al., Phys. Rev. Lett. 36, 1355 (1976).

22. R. A. Carrigan, Jr., "Compound Quarks as an Explanation for the Apparent Quark Mass Spectrum", FERMILAB-78/13, January (1978).

23. W. Braunschweig et al., Phys. Lett. 67B, 243 (1977).

24. W. D. Apel et al., Phys. Lett. 72B, 500 (1978).

25. M. A. Abolins et al., Phys. Rev. Lett. 37, 417 (1976).

26. G. Goldhaber, Lawrence Berkeley Laboratory report LBL-5534, Oct. (1976).

27. P. Benkheiri et al., Phys. Lett. 68B, 483 (1977).

28. H. Braun et al., Phys. Lett. 60B, 481 (1976).

PART IV. REFINEMENTS AND RAMIFICATIONS

INTRODUCTION

In Part IV, the concluding section of the book, we take as an ansatz the light-quark basis states that were deduced from the studies described in Parts I through III. We then use these basis states, plus the implications that can be drawn from their existence, as a guide in extending and broadening our investigation of the elementary particle.

In Chapters 18 and 19 we delve more deeply into the description of the strongly interacting hadronic particles, and we investigate their geometries and their electromagnetic properties; these studies culminate in the phenomenological calculation of precision mass values for the basic long-lived baryons and mesons. In Chapter 20, the final chapter of the book, we move from the domain of elementary particles to the domain of nuclear physics. We have borrowed heavily from the concepts of nuclear physics in the studies of the preceding sections – particularly in Parts II and III. We can now repay this debt by demonstrating the manner in which the structure of the elementary particle affects the structure of the atomic nucleus. The answer we obtain here is rather surprising, and it leads to a model of the nucleus which simply and accurately accounts for the observed symmetrical and asymmetrical fission modes of the heavy nuclides. Chapter 20 concludes with a brief comment on what is probably the most arcane of all the elementary particles – the photon.

A series of appendices amplify some of the discussions in the text, and they furnish additional background information about this light-quark approach to the structure of the elementary particle.

CHAPTER 18. THE GEOMETRY OF THE ELEMENTARY PARTICLE

"One picture is worth ten thousand words."

18.1 What Should we Expect to See?

Man is by nature a visual creature, and he likes to understand the ultimate nature of something by "seeing how it is" or "seeing how it works". One of the frustrations in elementary particle physics is that we are forced to deal with systems which are so small that we can't literally see them. This brings up the question as to the observational nature of the elementary particle: if we were able to make ourselves progressively smaller and smaller, would we discover that molecules, atoms, atomic nuclei, and nucleons are in turn just smaller and smaller versions of the same systems that we see in the macroscopic world; or would we find out that the ultimate microscopic world is of a completely different nature – perhaps populated by a collection of fuzzy circular objects which continually exchange parts of themselves with one another (this is the field-theory concept of the elementary particle).

On the molecular level, which is a level that we can now view with the aid of the electron microscope, we have discovered that molecules – both organic and inorganic – have clearly-ordered geometries. At the atomic level, however, which is still beyond the range of our vision, the initial concept of the Bohr atom as a small solar system has been replaced by the quantum-mechanical concept of the atom as a small but massive nucleus surrounded by a set of diffuse "electron clouds". And the application of field theory to the nucleon has led to the viewpoint that the nucleon consists of a more-or-less spherical core region which is surrounded by a more-or-less spherical pion cloud region. In this field-theory viewpoint, two adjacent nucleons exchange pions, ρ mesons, ω mesons, etc. back and forth, with this exchange providing the nuclear force that holds the nucleons together. Since the masses of 140 MeV π mesons, 773 MeV ρ mesons, and 783 MeV ω mesons are comparable to the mass of the 939 MeV nucleon, the total mass in the pion exchange cloud can in principle be comparable to the masses of the residual nucleon cores, so that a collection of nucleons becomes, in this field-theory context, a pion-nucleon soup.

Now, the Heisenberg Uncertainly Principle suggests that we have no hope of seeing an elementary particle: in order to have dimensions that are on the order of one fermi (the size of the nucleon), a photon has to have an energy

$$E_{\gamma} = h\nu = \hbar c/\lambda = 197 \text{ MeV-fm}/1 \text{ fm} = 197 \text{ MeV},$$

which is so large that it would perturb the system it is trying to investigate. However, even if we cannot hope to "see" an elementary particle, we can do experiments on it to measure its static and dynamic properties, and we can then devise mathematical models which reproduce the results of the experiments. These mathematical models may in turn tell us something about the geometry of the elementary particle. In Chapter 7 we have already taken a step in this direction by devising a mathematical muon model which is based on a specific geometrical configuration for the muon — namely, Fig. 7.3. In the present chapter we attempt to extend this result, insofar as we are able to do so, to apply also to the nucleon and the other hadrons, and hence to their constituent quarks.

The usefulness of assigning a specific geometry to an elementary particle rests solely on the mathematical conclusions that we can thereby draw with respect to that particle. In this context, it should be pointed out that the field-theory picture of the pion-nucleon soup which we described above has not led, in spite of intensive theoretical efforts, to the calculation of any of the static properties of the nucleon; nor, except in a very qualitative way, to calculations of the nuclear force which are not dominated by the use of arbitrary parameters. In contrast to the essentially structureless field theory picture of the nucleon, we will see in the present chapter that some very useful results can be obtained by assigning specific geometries to the nucleon and other hadrons. Furthermore, as an extension of these results, we will demonstrate in Chapter 20 that by using the geometry of the nucleon to deduce a corresponding geometry for the atomic nucleus, which is composed of nucleons, we can obtain some very striking results which pertain to nuclear mass systematics and nuclear fission systematics.

The question of the geometry of the hadron ultimately becomes a question of the geometry of the quark. In this connection, the weak binding energies that we have found for the hadronic quarks; the clear-cut mass quantization of these quarks, in which they are all constructed as spinless and spinning combinations of one fundamental mass quantum — the quantum M = 70 MeV; and the complex but universal manner in which the same few quark states occur in all of the baryon and meson resonances, combine together to suggest that these quarks have well-defined geometries, so that the hadrons which are formed as clusters of quarks also have well-defined geometries.

We begin this study of the geometry of the elementary particle with the only particle whose "size" has been directly measured — the nucleon.

18.2 The Geometry of the Nucleon

The conventional wisdom in nuclear and particle physics used to be that all subatomic particles are more-or-less spherical, and all spin zero subatomic particles are strictly spherical. Thus it came as quite a surprise when it was discovered that the spin zero U^{238} nucleus is markedly prolate in shape, with a "length" that is about 30% larger than its "width". Many atomic nuclei, both heavy and light, are now known to have shapes that are markedly aspherical, and there is no reason why this same situation doesn't also apply to the elementary particles. In particular, the nucleon appears to be constructed of three lightly-bound equal-mass quarks, and this type of substructure does not suggest a spherical configuration.

We know something about the overall size of the nucleon from the many electron-proton and electron-neutron scattering experiments that have been carried out, starting with the pioneering work of Hofstadter and coworkers at Stanford. These experiments probe both the electric charge distribution and the magnetic field distribution in the proton and in the neutron. In the case of the proton, the interesting result was obtained that the spatial distribution of the electric charge and of the magnetic field are equal to one another; they each correspond to a root-mean-squared radius for the proton of about 0.8 fermi[†]. This equality of the electric and magnetic spatial distributions in the proton is very important, because it suggests that the magnetic field arises from a rotation of the electric charge (and not from the presence of magnetic monopoles). In the neutron, the equally interesting result was obtained that the electric charge distribution is very small, but not identically zero; and the magnetic field distribution has a spatial extent which is the same as that of the proton – that is, an rms radius of 0.8 fermi. From this neutron result, we conclude that the neutron contains equal numbers of internal + and - electric charges, and that the neutron and proton are particles which have very nearly, and perhaps exactly, the same "size". Thus it seems clear from the electron scattering experiments that the proton and neutron are two different forms of the same basic state – the nucleon. This, of course, is the same conclusion that we have already reached on the basis of our studies in Part I. This conclusion is reinforced by the observation of the beta decay transitions $n \to p + e^- + \bar{\nu}$ and $p \to n + e^+ + \nu$, in which neutrons and protons are transformed into one another.

If the nucleon has a falloff radius of 0.8 fermi, then as a first approximation we can calculate the "volume" of a nucleon as that of a sphere with an

---

[†] The electric charge and magnetic field distributions in the proton are characterized by rather uniform central regions plus exponential tails, and the rms radius defines a suitable midpoint in the fall-off region of the distribution.

overall radius of (say) 1 fermi. This gives a volume $V = 4\pi/3\ fm^3 = 4.2\ fm^3$. However, we can obtain a better answer than this from nuclear physics. In large atomic nuclei, where two hundred or so nucleons are packed together, optical model calculations have shown that the radius of a spherical nucleus is given by the equation $R_{nucleus} = 1.25\ A^{1/3}$, where R is in fermis and A is the total number of nucleons in the nucleus. Thus the volume of a single nucleon is $V_{nucleon} = V_{nucleus}/A = (4\pi/3) \times (1.25)^3 = 8.2\ fm^3$. Hence the effective radius of a nucleon, taken to be a spherical object, is 1.25 fermi. Now, it turns out phenomenologically that we can define the geometry of the nucleon more accurately than this, and in particular we can delineate its nonspherical nature. But in order to do this, we must study the geometry of the nucleon quark, which is the spinor S = 330 MeV. This study is carried out in the next section.

18.3 The Geometry of the Spinor S

As we have seen from the studies in Chapter 8 of Part I, the nucleon is composed of three Fermi-Yang-type spinors S, so that $N = S\bar{S}S$ and $\bar{N} = \bar{S}S\bar{S}$. Hence, in this light-quark approach to nucleon structure, the spinor S = 330 MeV is the counterpart of the nucleon quark of Gell-Mann and Zweig. Now, what can we determine about the geometry of the spinor S?

We can deduce the overall dimensions of the spinor S by determining its radius R_S and its volume V_S. The radius R_S can be obtained in several different ways: (1) We showed in Chapter 6 that if a sphere is set into rotation at the full relativistic limit (with its equator moving at the velocity of light), and if the sphere is assigned a radius equal to its Compton wavelength, $R = \hbar/mc$, then the sphere has a calculated spin angular momentum $S = 1/2\ \hbar$. Since the nucleon quark is known to have a spin $S = 1/2\ \hbar$, this suggests that we should set the radius of the spinor S equal to its Compton wavelength, so that $R_S = \hbar/m_S c = 0.60$ fm, where $m_S = 330$ MeV. (2) In the Gell-Mann and Zweig formulation of the quark model, the intrinsic magnetic moment of a quark (the magnetic moment per integral electric charge Q) is assumed to be the same as that of the proton — $\mu = 2.79$ nuclear magnetons[1]. If we assume that this magnetic moment arises from a rotating current loop, as given by the equation $\mu = eR_S/2$ [Eq. (7.2) in Chapter 7], then the value $\mu = 2.79$ n. m. corresponds to the spinor radius $R_S = 0.59$ fm. (3) If the nucleon, with an overall rms electromagnetic radius of 0.8 fermi, is composed of three equal-size substates, then a simple calculation shows that each of these substates must have an electromagnetic radius of just about 0.6 fermi. We thus have three independent reasons for pinpointing the radius of the spinor S at the value $R_S \simeq 0.60$ fermi.

We have previously concluded from the relativistic calculations of Chapters 6 and 7 and from the basis-state transitions of Chapter 14 that the spinor S is composed of three subquanta M = 70 MeV in a relativistically-spinning configuration: $S \equiv (3M)_{\text{spinning}}$. We have further seen from the mass and binding energy calculations of Chapter 9 and of Sec. 5.2 in Chapter 5 that the subquanta M inside a basis state tend to combine in the form of linear arrays (cabers). Hence the spinor S does not necessarily have a spherical shape. In fact, as we now demonstrate, S appears to have an overall shape which is essentially that of an axially-symmetric cylinder. Suppose that the atomic nucleus consists of a set of closely-packed nucleons, and that each nucleon consists of three spinors S which have radii R_S and overall "lengths" L_S. The volume occupied by a spinor is $V_S = (2R_S)^2 \times L_S$, so that the total volume occupied by a nucleon is $V_{\text{nucleon}} \simeq 3\,V_S = 12\,R_S^2 L_S$. If we now set $V_{\text{nucleon}} = 8.2\ \text{fm}^3$ and $R_S = 0.6$ fm, as discussed above, we obtain $L_S = 1.9$ fm. Thus the "length" of the spinor S, and hence the "height" of the nucleon N, is just about 2 fermis. As we will see in Chapter 20, this height for the (aspherical) nucleon is just the value that is required in order to account quantitatively for the asymmetric fission of the uranium nucleus.

Now let us synthesize the above observations in a systematic manner, using the information that we have developed in the preceding chapters. To construct a singly-charged spinor $S^{\pm}$, we first align three subquanta M in a linear array (see Fig. 18.1a) as the spinless caber $3^{\pm}$:

Figure 18.1. The spinor S as it appears with (a) spherical subquanta M, and (b) flattened (oblate spheroidal) subquanta M. The spherical subquanta would result in a nucleon that is too large to be in agreement with experiment. By flattening the subquanta M as shown in (b), we simultaneously improve the calculated values for the nucleon volume, the spinor mass, and the spinor magnetic moment, as is illustrated in Fig. 18.2.

$$3^{+} \equiv M^{o}M^{+}M^{o} \qquad \text{and} \qquad 3^{-} \equiv \bar{M}^{o}M^{-}\bar{M}^{o} . \tag{18.1}$$

The calculated mass of the caber $3^{\pm}$ is 70.0 + 74.6 + 70.0 = 214.6 MeV. We then set the caber $3^{\pm}$ spinning at the full relativistic limit about its axis of symmetry to produce the spinor

$$S^{\pm} \equiv (3^{\pm})_{\text{spinning}}. \tag{18.2}$$

If we assume that each subquantum M in the spinor S has a spherical shape, we can apply the relativistic mass increase of 3/2 for rotating systems [Eq. (6.8) in Chapter 6] to obtain

$$m_s^{\pm} = 3/2 \times 214.6 \text{ MeV} = 321.9 \text{ MeV} \tag{18.3}$$

as the calculated mass of the spinor $S^{\pm}$. The radius of <u>each spherical subquantum M</u> is then assigned the value

$$R_s = \hbar/m_s c = 0.613 \text{ fm}, \tag{18.4}$$

where m_s = 321.9 MeV is the <u>total mass</u> of the spinor $S^{\pm}$. With this value for R_s, each subquantum M has a calculated spin angular momentum S = 1/6 ħ, and the spinor $S^{\pm}$ has a total spin angular momentum S = 1/2 ħ. The overall length of the spinor $S^{\pm}$ (see Fig. 18.1) is

$$L_s = 6 \times R_s = 3.68 \text{ fm}. \tag{18.5}$$

Finally, we allow the (point) electric charge on the charged subquantum $M^{\pm}$ to move freely; this charge is forced to the equator of the spinning sphere by the action of the induced magnetic field, where it produces a calculated magnetic moment

$$\mu_s = eR_s/s = 2.92 \text{ n. m.} \tag{18.6}$$

If we now compare these calculations with the above discussion, we discover that the calculations are deficient in three respects: <u>(1)</u> The calculated mass of the spinor $S^{\pm}$ in Eq. (18.3) is too low; it should be 330 MeV rather than 321.9 MeV. <u>(2)</u> The length L_s of the spinor $S^{\pm}$ in Eq. (18.5) is too large; it should be about 2 fermis rather than 3.68 fermis. <u>(3)</u> The magnetic moment of the spinor $S^{\pm}$ in Eq. (18.6) is too large; it should be 2.79 n. m. rather than 2.92 n. m. The only quantity we have calculated accurately is the spin angular momentum S = 1/2 ħ. However, the remarkable fact which now emerges is that by making one basic change in the geometry of the spinor $S^{\pm}$, we can simultaneously correct all three of these deficiencies. The change we make is to flatten the spherical subquanta M of the Spinor $S^{\pm}$ into oblate spheroids, as shown in Fig. 18.1b. The relativistic equations for an oblate spheroid are given in Eqs. (6.14) - (6.17) of Chapter 6, and they are best evaluated numerically on a computer. If the radius R_S is held fixed, the effect of flattening the subquanta M is to slightly increase both the relativistic mass m [Eq. (6.16)] and the relativistic moment-of-inertia I [Eq. (6.17)], which in turn slightly increases the spin angular momentum S. If we then decrease R_S [Eq. (18.4)] slightly so as to maintain the calculated spin angular momentum at the value $S = 1/2\,\hbar$, this will decrease the calculated value [Eq. (18.6)] of the magnetic moment μ. And, of course, flattening the subquanta M into oblate spheroids decreases the overall length L_S of the spinor $S^{\pm}$. Thus if we flatten the subquanta M of the spinor $S^{\pm}$ into spheroids and adjust R_S to preserve the spin angular momentum, we change all three of the erroneously-calculated quantities — mass, volume, and magnetic moment — in the proper direction. To make a numerical evaluation of the accuracy of this procedure, we express the relativistic mass, the relativistic moment of inertia, the magnetic moment, and the volume of the spinor $S^{\pm}$ (or of the nucleon) as functions of the deformation parameter ε [Eq. (6.15)], and we plot these quantities together with a common ε abscissa. This plot is shown in Fig. 18.2. From the numerical values shown in Fig. 18.2, it can be ascertained that a deformation parameter $\varepsilon \simeq 0.65$ gives the proper value for the magnetic moment μ, a deformation parameter $\varepsilon \simeq 0.7$ gives the proper volume for the nucleon, and a deformation parameter $\varepsilon \sim 1$ gives the proper mass for the spinor $S^{\pm}$. The important point here is not the exact value for ε that is obtained in each of these calculations, but rather the fact that a single value of ε, $\varepsilon \simeq 0.7$, gives accurate numerical results for all of these quantities ($\varepsilon = 0.7$ is the value that is illustrated in Fig. 18.1b). Thus we have a sound phenomenological basis for choosing the flattened configuration for the spinor $S^{\pm}$ that is shown in Fig. 18.1b, and the quantities which relate to this geometry can all be calculated in an internally-consistent manner. Hence the use of a specific geometry leads to mathematically-significant results, and these results justify its usage.

There are two main conclusions to be drawn from the detailed calculations that we have just discussed: (1) by assigning a specific geometry to the spinor S, it is possible to make accurate numerical calculations of the mass, spin angular momentum, magnetic moment, and overall volume of the spinor S; (2) the spinor S has an overall shape which is essentially that of a cylinder of radius $R_S = 0.6$ fm and

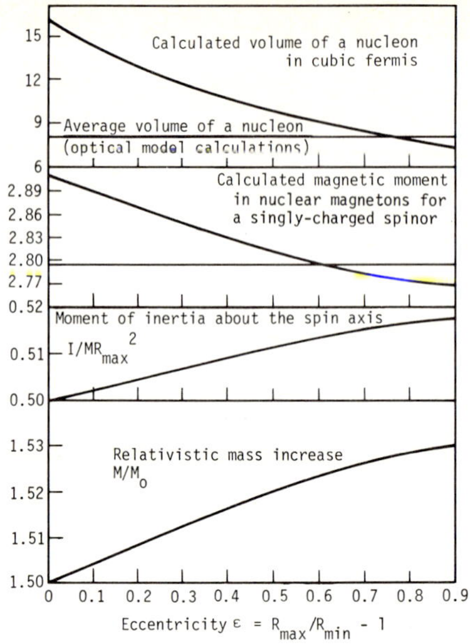

Figure 18.2. Calculated quantities for a relativistically-spinning spinor S (see Fig. 18.1) as a function of the deformation parameter $\varepsilon = R_{max}/R_{min} - 1$, where R_{max} and R_{min} are the equatorial and axial radii of the subquanta M [see Eq. (6.15)]. If the spherical subquanta M in Fig. 18.1a are flattened, and the radius R_{max} is simultaneously decreased so as to maintain the spin angular momentum value $S = 1/2\ \hbar$, then the nucleon volume, mass, and magnetic moment approach their proper values when $\varepsilon \simeq 0.7$, which is the deformation parameter that is illustrated in Fig. 18.1b.

length $L_S = 2.0$ fm†. The implications of the first conclusion are rather self-evident; they suggest that we can construct a microscopic model for the nucleon which is based on more-or-less classical macroscopic concepts. The implications of the second conclusion are not so self-evident, but they are very far-reaching — not only for the structure of the nucleon, but also for the structure of the nucleus. In the nucleon, these cylindrically-shaped spinors bind together in a side-by-side manner. Thus, as we showed in Sec. 8.2 of Chapter 8, we can choose the mass values

---

†When the deformation parameter ε is increased in value from $\varepsilon = 0$ to $\varepsilon = 0.7$ (see Figs. 18.1a and 18.1b), and the radius R_S is accordingly decreased by 3% to preserve the spin angular momentum value $S = 1/2\ \hbar$, the overall calculated length L_S of the spinor $S^{\pm}$ becomes $L_S = 2.0$ fm rather than the previously-estimated value in the text of $L_S = 1.9$ fm.

$S = \bar{S} = 330$ MeV and the (4%) binding energies $SS = \bar{S}\bar{S} = 0$ and $S\bar{S} \simeq -26$ MeV, and thereby calculate the mass of the nucleon as

$$N = \bar{S}^{S}_{S} = 3 \times 330 - 2 \times 26 = 938 \text{ MeV}, \qquad (18.7)$$

which is the correct mass experimentally. Furthermore, by binding a nucleon N and an antinucleon $\bar{N}$ together in a side-by-side manner, we can calculate the $\bar{N}N$ mass as

$$\bar{N}N = S^{\bar{S}}_{\bar{S}}S_{\bar{S}}{}_{S}{}^{S} = 6 \times 330 - 7 \times 26 = 1798 \text{ MeV}, \qquad (18.8)$$

which compares to the experimental $\bar{p}n$ bound state mass of 1795 MeV. (In Chapter 19 we will repeat these calculations with coulomb effects added in.) The calculations of Eqs. (18.7) and (18.8) are so simple and so accurate that they must be physically significant. And the crucial point here is that it is the side-by-side binding of the nucleon N with the antinucleon $\bar{N}$ which leads to the fact that there are precisely seven $S\bar{S}$ bonds in the $\bar{N}N$ cluster. As a continuation of these results, we have already demonstrated in Eqs. (9.9) - (9.18) of Chapter 9 that the quarks in a hyperon resonance are clustered together in a side-by-side manner; and we will demonstrate in Chapter 20 not only that the four nucleons in an alpha particle are bound together in a side-by-side manner, but also that the α particles themselves bind together side by side to form a series of two-fermi-thick nucleon layers in the nucleus. It is these (Ising) nucleon layers in the atomic nucleus which account for the symmetry properties of the thermal fission process: an even number of layers leads to symmetric fission, and an odd number of layers leads to asymmetric fission. We now consider the geometries of the spinless quark states.

18.4 The Geometry of the Kaon

Because of the wealth of available experimental data on the properties of the nucleon, it has been possible to determine the geometry of the nucleon quark — the spinor S — in considerable detail, as we discussed in the preceding section. However, the information that we can gather together with respect to the geometries of the spinless cabers $1 \equiv M$, $3 \equiv 3M$, $4 \equiv 4M$, and $7 \equiv 7M$ is considerably less. Moreover, these spinless cabers do not have the geometrical information which comes from the dynamical aspects of spin angular momenta and magnetic moments. Hence we are limited mainly to mass values and binding energies when we attempt to delineate the geometries of these spinless cabers. Nevertheless, we can draw some conclusions about the geometries of these spinless basis states, as we discuss in this section and the next.

The geometry of the kaon, or K meson, is of interest because of the clear-cut manner in which it can be deduced from the mass values of the charged and neutral kaons. The kaon is composed of seven spinless quanta M, as we have seen in Eq. (9.4) of Chapter 9. The charged kaons K^+ and K^- each contain six neutral quanta $M^o = 70.0$ MeV and one charged quantum $M^{\pm} = 74.6$ MeV. If we assume that the binding energies among these seven subquanta are negligible, we obtain

$$K^{\pm} = 6\ M^o + M^{\pm} = 6 \times 70.0 + 74.6 = 494.6 \text{ MeV} \tag{18.9}$$

as the calculated $K^{\pm}$ mass; the experimental $K^{\pm}$ mass is 493.7 MeV†. The neutral kaons K^o and $\bar{K}^o$ each contain five neutral quanta M^o and two charged quanta $M^{\pm}$, so that they are more massive than the charged kaons (in contradiction to the prediction of the Gell-Mann and Zweig quark model). The calculated K^o mass is

$$K^o = 5\ M^o + 2\ M^{\pm} = 4 \times 70.0 + 2 \times 74.6 = 499.2 \text{ MeV}, \tag{18.10}$$

as compared to the experimental K^o mass of 497.7 MeV. Now, the crucial point here is that the calculated $K^o - K^{\pm}$ mass difference shown in Eqs. (18.9) and (18.10) is 4.6 MeV, whereas the experimental $K^o - K^{\pm}$ mass difference is 3.99 ± 0.13 MeV[2]. This discrepancy of 0.6 MeV between the calculated and experimental $K^o - K^{\pm}$ mass differences can be attributed to the fact that in the calculation of Eq. (18.10) we have not taken into account the coulomb attraction between the internal + and - charges in the K^o. This internal coulomb attraction evidently lowers the K^o mass by 0.6 MeV. But if this is the case, then we can calculate the separation distance r between these two charges: r is given by the equation

$$e^2/r = 0.6 \text{ MeV} \rightarrow r = 2.40 \times 10^{-13} \text{ cm}. \tag{18.11}$$

Thus the + and - charges in the K^o have a separation distance of 2.4 fermis!

We don't know the geometries of the spinless subquanta M in the kaon, but we do know that the spinning subquanta M_s in the spinor S have the shape of oblate spheroids (see Fig. 18.1b), with equatorial radii $R_s = 0.6$ fm, and with axial lengths $\ell_s = 2/3$ fm (so that the spinor S has a length $L_s = 3\ \ell_s = 2$ fm). If we assume these same dimensions for the spinless subquanta, then a caber $K = 7 \equiv 7M$ that is constructed as a linear array of subquanta M (see Fig. 18.3) will have an overall length of $7 \times 2/3 = 4.7$ fm. If we then construct the K^o as the charge

---

†Since the experimental $K^{\pm}$ mass is 0.9 MeV smaller than the calculated mass, the seven subquanta M in Eq. (18.9) must have a total internal binding energy of 0.9 MeV. This internal binding energy is small enough that we neglect it here; its effect is to reduce all of the calculated kaon masses by the same small constant factor.

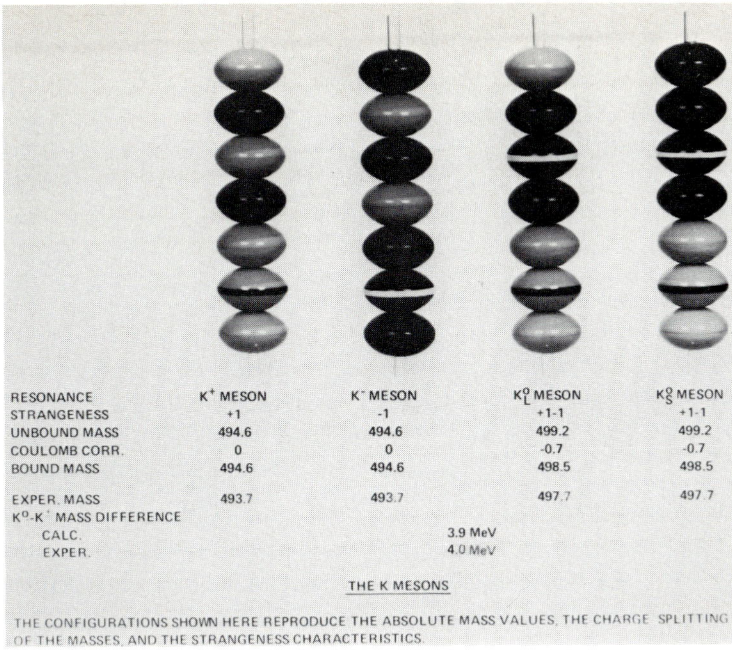

Figure 18.3. Linear light-quark configurations that reproduce the main properties of the K mesons.

configuration

$$K^0 = (M^0M^+M^0)(M^0M^-M^0)M^0 \ , \tag{18.12}$$

for example, the + and - charges will have a separation distance of about 2 fm, in rough agreement with the "experimental" separation distance of 2.4 fm given in Eq. (18.11). This suggests that the linear configuration shown in Fig. 18.3 is a reasonable representation for the neutral kaon. Hence, neutral kaons, and presumably also charged kaons, occur in the form of linear arrays of subquanta M, which we denote as cabers. Some lifetime relationships that pertain to these linear kaon configurations are illustrated in Fig. 10.10 of Chapter 10.

We will obtain confirmatory evidence for this linear kaon structure from the binding energy systematics discussed in the next section. For example, the fact that the η' meson is constructed as a K^+K^- bound state [see Eq. (9.4) in Chapter 9] suggests from the binding energy systematics that the K^+ and K^- mesons have linear structures. But the example of the kaon is so simple and so clear-cut that it can stand on its own merits.

Having now shown that both the spinor S and the spinless kaon K contain linear arrays of subquanta M, our next task is to study clusters which contain combinations of quark substates. To do this, we first relate the geometrical configurations of the quarks to their binding energies (Sec. 18.5), and we then illustrate the manner in which these quark states combine together in the basic meson and baryon resonances (Sec. 18.6).

18.5 Quark Geometries and Quark Binding Energies

In Sec. 18.3 we studied the geometry of the spinor S as it appears in the nucleon, and we concluded that S is composed of three oblate spheroids (spinning subquanta M_s) in a linear array (Fig. 18.1b). Then in Sec. 18.4 we showed that if we assume the same geometrical shape for the nonspinning quantum M = 70 MeV as for its spinning counterpart M_s, and if we place seven of these subquanta M in a linear array (Fig. 18.3), we can reproduce both the absolute masses and the mass difference of the K^o and $K^{\pm}$ mesons. In the present section we extend these results to include all of the light-quark basis states, and we demonstrate that if all of the subquanta in these basis states are assumed to have the same geometry (that of an oblate spheroid), then the various combinations of basis states accurately correlate the quark-quark binding energies that we have deduced from the resonance mass values.

The oblate spheroidal geometry of the quantum M is shown in Fig. 18.4. M has an axial radius of 0.6 fm and a total "height" of 2/3 fermi. It occurs in the particle states M^+ and M^o and in the antiparticle states $\bar{M}^o$ and M^-, corresponding to its isotopic spin I = 1/2. The mass difference of 4.6 MeV between $M^{\pm}$ = 74.6 MeV and M^o, $\bar{M}^o$ = 70.0 MeV is chosen to match the 4.6 MeV mass difference between the $\pi^{\pm}(139.6)$ and $\pi^o(135.0)$ mesons. (A <u>spinning</u> quantum $M_s^{\pm}$ has a mass of about 110 MeV.)

The light-quark basis states 1,3,4,7,S that are necessary in order to reproduce the observed spectrum of meson and baryon resonances are illustrated in Fig. 18.5. In this figure, the fundamental spinless cabers 1 ≡ M, 3 ≡ 3M, 4 ≡ 4M, and 7 ≡ 7M and the basic spinor S ≡ $(3M)_{\text{spinning}}$ are shown at the left, and a possible quark substructure for these cabers is illustrated at the right. The masses of the constituent subquanta M in these cabers are assumed for simplicity to add linearly (apart from small electromagnetic effects). As we have seen in the case of the K meson (Fig. 18.3), this assumption of zero intercaber binding energy is accurate to within about 0.2% in absolute mass.

One of the principal advantages of the <u>linear</u> quark structures shown in Fig. 18.5 is the natural way in which they account for the observed quark-quark

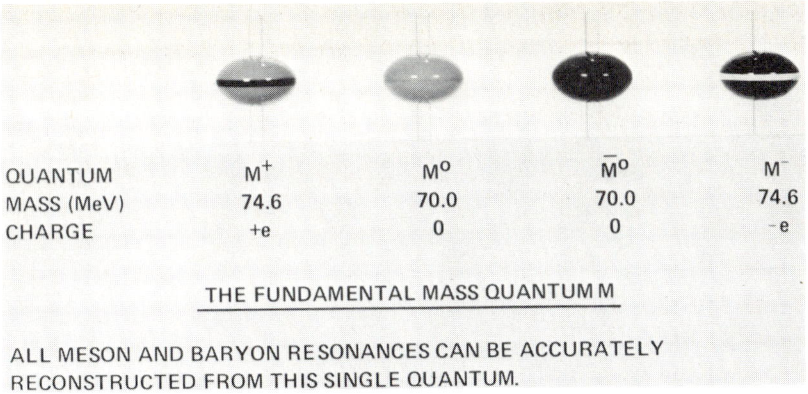

Figure 18.4. The fundamental hadronic mass quantum M.

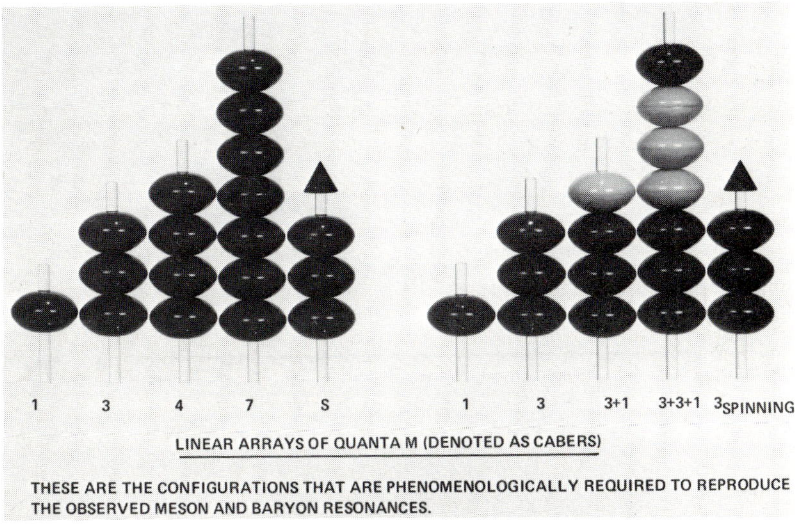

Figure 18.5. The fundamental light-quark basis states 1,3,4,7,S. These linear cabers form a necessary and sufficient basis set for reproducing the spectrum of meson and baryon resonances. A possible substructure for these cabers is suggested by the configurations shown at the right in the figure.

binding energies. These binding energies are delineated in Fig. 18.6. As can be

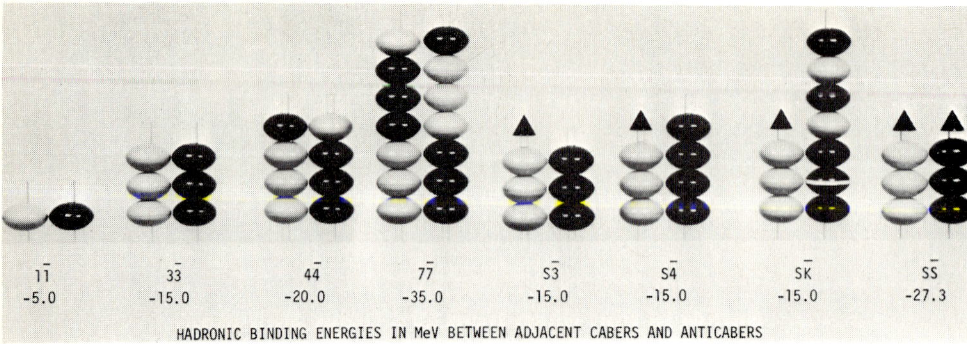

Figure 18.6. The linear systematics of the ∿4% hadronic binding energies. These very-short-ranged binding energies operate only between adjacent cabers and anticabers in a cluster of cabers.

seen in Fig. 18.6, the $1\bar{1}$, $3\bar{3}$, $4\bar{4}$, and $7\bar{7}$, spinless caber-anticaber combinations have binding energies which are strictly proportional to their masses, and which amount in each case to about 4% of the unbound masses[†]. Also, the corresponding $3\bar{3}$ (spinless) and $S\bar{S}$ (spinning) quark pairings have comparable (∿4%) binding energies. Finally, and most interestingly, the $S\bar{3}$, $S\bar{4}$, and $S\bar{7}$ combinations of spinning and spinless quarks <u>all have precisely the same binding energy</u>; this is a result that is clearly required by the experimental baryon mass values, as we have already shown in Eqs. (9.9) - (9.19) of Chapter 9, and will demonstrate with more accuracy in Chapter 19, and it is the strongest reason we have for assigning the same geometrical shape to both the spinning and the nonspinning subquanta M.

In addition to the "unmixed" basis states of Fig. 18.6, which all have ∿4% binding energies, there are also "mixed" basis states that contain both M and $\bar{M}$ subquanta scrambled together, and which have ∿0% binding energies (see Appendix C). The mixed and unmixed spinless cabers $3,\dot{3},\dot{\bar{3}},\bar{3}$ are illustrated (perhaps schematically) in Fig. 18.7, and the corresponding mixed and unmixed spinors $S,\dot{S},\dot{\bar{S}},\bar{S}$ are illustrated in Fig. 18.8. The unmixed spinless cabers 3 and $\bar{3}$ are "strange" (they carry the strangeness quantum numbers S = +1 and S = -1, respectively), whereas the mixed spinless cabers $\dot{3}$ and $\dot{\bar{3}}$ are "nonstrange" (they carry the strangeness quantum number S = 0). As we discussed in Sec. 9.4 of Chapter 9, the 4% binding energies of the

---

[†]The magnitudes of the quark binding energies were discussed in connection with Eqs. (4.6) - (4.9) of Chapter 4 and Eqs. (8.3) - (8.9) of Chapter 8, and they are formally defined in Appendix C.

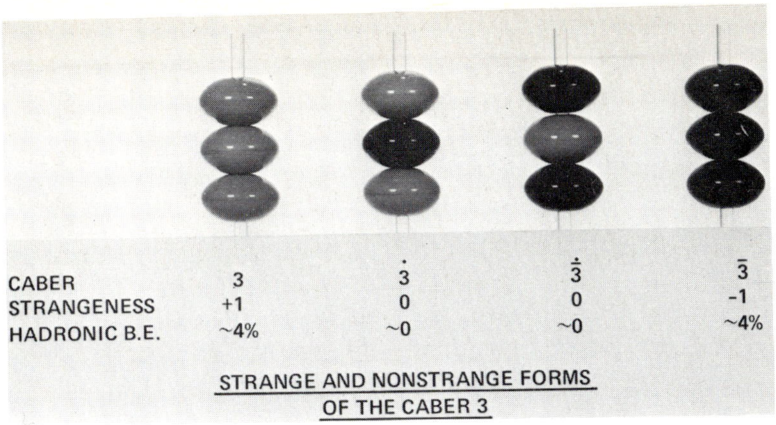

Figure 18.7. Strange and nonstrange forms of the spinless caber 3 (compare Figs. 18.7 and 18.8). Note the correlation between strangeness quantum numbers and hadronic binding energies.

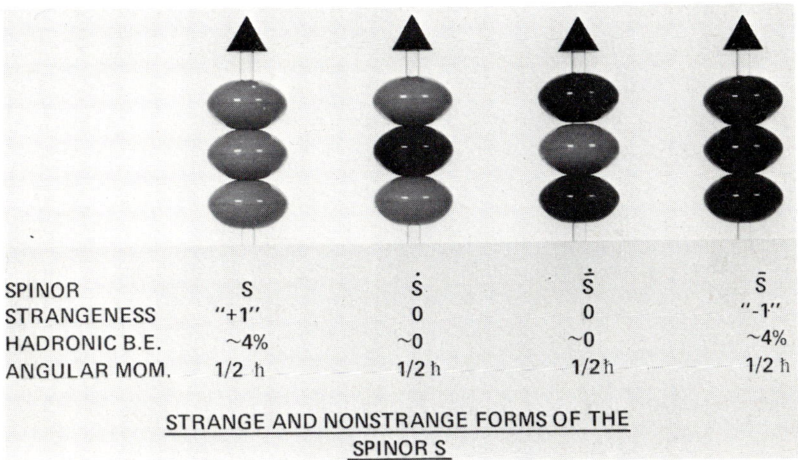

Figure 18.8. Strange and nonstrange forms of the spinor S. The "strangeness" of the spinor S, which is established here on the basis of the hadronic binding energies (see Fig. 18.7), is an "invariant strangeness" that does not enter into the bookkeeping of the process of Associated Production, as is discussed in the text.

unmixed spinors S and $\bar{S}$ suggest that these spinors are "strange"; however, due to the invariant nature of the spinor triplets $S\bar{S}S$ that occur in the baryon resonances, the strangeness of the spinor S does not enter into the bookkeeping of the process of Associated Production (see Appendices A and I). Although it is not illustrated here, the spinless caber 4 also occurs in both unmixed $(4,\bar{4})$ and mixed $(\dot{4})$ forms, with ~4% and ~0% binding energies, and with the strangeness quantum numbers (S=+,-) and (S=0), respectively [see Eqs. (4.7) and (4.8) in Chapter 4].

Having defined the geometries of these fundamental light-quark basis states, we now demonstrate the manner in which they are combined together in the meson and baryon resonances. This is sketched in the next section, and it is developed in more detail (with electromagnetic effects included) in Chapter 19.

18.6 Quark Geometries of some Basic Meson and Baryon Resonances

The lowest-mass and stablest – and therefore the most basic – meson resonances are the very-narrow-width states π, K, η, and (M,η'); similarly, the lowest-mass and stablest baryon resonances are the long-lived states N, Λ, Σ, Ξ, and Ω. Any quark model which purports to represent the structures of the hadrons must be able to account for these fundamental meson and baryon resonances in a natural and accurate manner. In Sec. 9.2 of Chapter 9 we have already written down the equations which represent the light-quark configurations for these resonances, and we have demonstrated the accurate mass values and strangeness quantum number assignments that are obtained from these configurations. In the present section we display the overall geometries of these light-quark clusters. We will treat these geometries in more detail in Chapter 19, where we use them as the basis for introducing electromagnetic effects.

Figure 18.9 shows the lowest-mass hadron resonances – the pi mesons. The mass values of the quanta M^o = 70.0 MeV and $M^{\pm}$ = 74.6 MeV, together with the binding energies of Fig. 18.5, were chosen so as to exactly reproduce the masses of these pion states. If we postulate that each hadron resonance must contain at least one neutral quantum M^o, then we can account for the isotopic spin I = 1 of the pion†. Although the π^o meson is constructed here as a neutral $M^o\bar{M}^o$ bound state, the facts that (1) the π^o decay is into two photons, and (2) the $\pi^{\pm}$ and π^o lifetimes differ by a factor of α^4 (see Chapter 10), both indicate that the neutral substates M^o and $\bar{M}^o$ are closely related to electromagnetic phenomena (as is the neutral neutron, for example).

---

†Note that the muon $\mu^{\pm}$, which is a 74.6 MeV mass quantum $M^{\pm}$ in a relativistically-spinning (and expanded) configuration (Fig. 7.3), does not contain a neutral quantum M^o, and it does not interact hadronically.

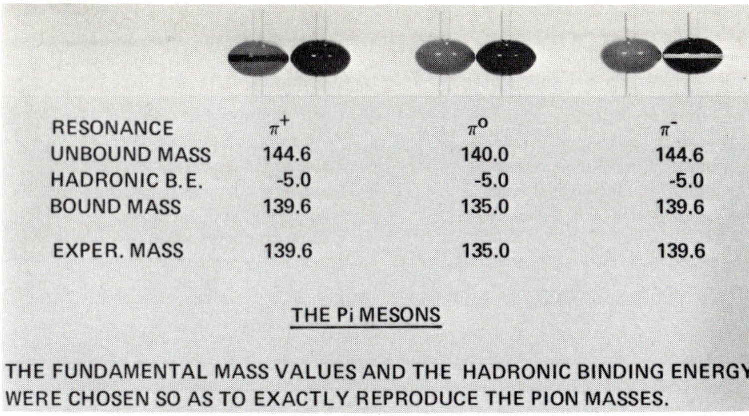

Figure 18.9. The pi mesons — the lowest-mass hadronic states.

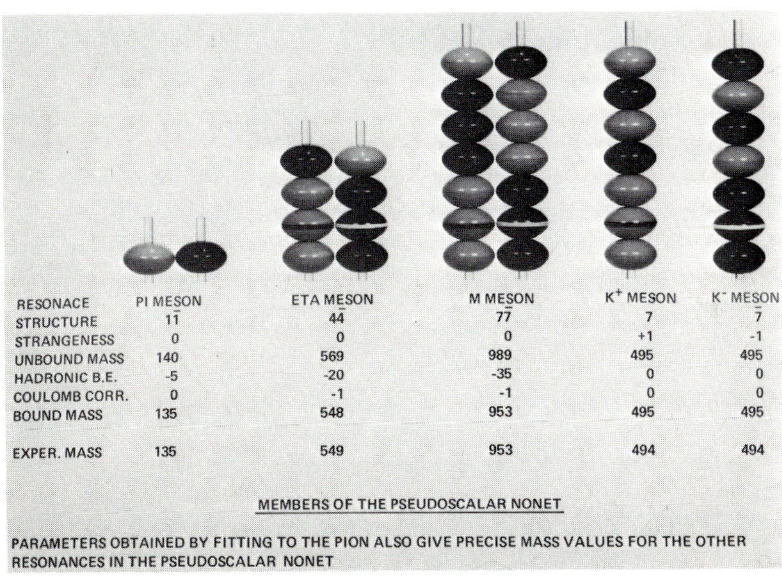

Figure 18.10. The π, η, M, K^+, and K^- mesons. The η' meson is shown in Fig. 18.11, and the K_L^o and K_S^o mesons are shown in Fig. 18.3.

Figure 18.10 extends the results of Fig. 18.9 to include the low-mass pseudoscalar meson states η, M, K^+, and K^-. The basis-state mass values and binding energies that were tailored to give an exact fit to the pion masses of Fig. 18.9 also give accurate fits to the masses of these related higher-mass resonances, which is a very stringent test of this light-quark phenomenology. The K^o mesons that correspond to the K^+ and K^- mesons of Fig. 18.10 were shown in Fig. 18.3. The accurate mass values that are obtained for the kaons follow of course from the fact that the kaons are single (unbound) cabers; this may seem to be a minor point of the phenomenology, but it is precisely the result that is needed to make all of these calculations fit together. The π, K, η, and (M,η') meson states span a range of masses from 135 MeV to 958 MeV, and to devise a simple model which gives all of these masses to an absolute accuracy of 1 MeV (as we will demonstrate in Chapter 19) is a nontrivial accomplishment that must have real physical significance.

The distinction between the M(953) meson and the nearby η'(958) meson is brought out schematically in Fig. 18.11. The η', from its asymmetric production and decay modes [see Eqs. (5.8) - (5.11) in Chapter 5], has a mismatched internal MM pair, which lowers its binding energy and raises its observed mass by 5 MeV. The mass difference between M and the η' thus becomes a measure of the $M\bar{M}$ binding energy.

The most fundamental baryon resonances — the nucleon N, the antinucleon $\bar{N}$, and the $N\bar{N}$ bound state — are illustrated in Fig. 18.12. The nucleon N is the fundamental "ground state" for the baryon and hyperon excitations of Chapter 15, and the $N\bar{N}$ bound state is the fundamental "ground state" for the New Particle resonances of Chapter 17. The calculation of the mass values for these configurations was discussed in Chapter 8 [Eqs. (8.1) - (8.9)], reviewed in Chapter 9 [Eqs. (9.5) - (9.8), and reviewed again in the present chapter [Eqs. (18.7) and (18.8)]. In Chapter 19 we will add electromagnetic effects to these calculations, and thereby demonstrate that we can reproduce not only the absolute mass values of the p, n, and $\bar{p}n$ resonances, but also the small n - p mass difference, which has never emerged from any basic theoretical calculation.

The one point we want to bring out visually in Fig. 18.12 is the side-by-side manner in which the spinors S in the nucleon and antinucleon cluster together. It is this side-by-side clustering that makes the N and $N\bar{N}$ mass calculations "work", just as it was the zero binding energy that made the kaon mass calculations of Fig. 18.10 "work". The side-by-side clustering of the nucleon and antinucleon follows both from the overall cylindrical geometry of the spinor S, and also from the fact that these spinors bind together with very-short-range forces that are essentially contact (nearest-neighbor) interactions. The $N\bar{N}$ bound state also provides the clearest experimental example of the 4% hadronic binding energy that occurs between these particle and antiparticle states and substates.

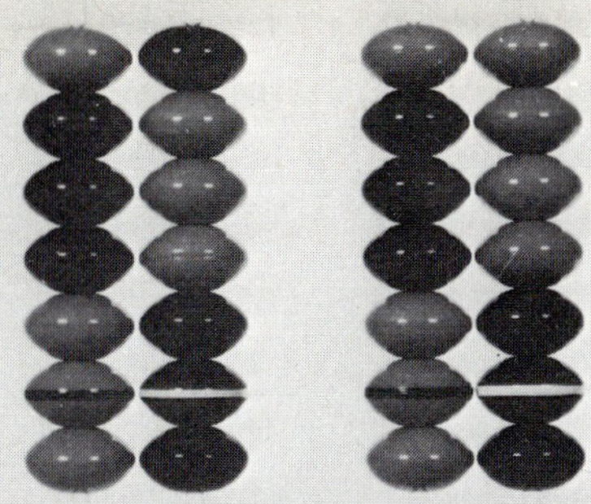

RESONANCE	M MESON	η' MESON
PRODUCTION PROCESS	A+B→A+B+RES.	A+B→C+RES.
UNBOUND MASS	989	989
HADRONIC B.E.	-35	-30
COULOMB CORR.	-1	-1
BOUND MASS	953	958
EXPER. MASS	953	958

Figure 18.11. The M(953) and η'(958) mesons. The relationship between these two closely-related meson states is described in detail in Sec. 5.2 of Chapter 5 [see Eqs. (5.8)-(5.11)]. The M meson is produced in reactions of the type A + B → A + B + M, so that it necessarily has perfect internal particle-antiparticle symmetry; whereas the η' is produced in reactions of the type $K^- + p \to \eta' + \Lambda$, so that it does not have this internal symmetry constraint. Also, the M decays via the one-step decay mode M → ππγ (ππ not in the ρ band); whereas the η' decays via the two-step decay mode η' → ργ, ρ → ππ. If we attribute the 5 MeV difference between the M and η' masses to a mismatch in the M = 70 MeV substates in the η', as is suggested in Fig. 18.11, then the η' - M mass difference becomes an experimental measurement of the $M\bar{M}$ binding energy. Although the lifetimes of these two states are not known precisely (see Chapter 10 and Appendix B), the lifetime of the symmetric M meson appears to be just about one power of $\alpha = e^2/\hbar c$ shorter than the lifetime of the asymmetric η' meson.

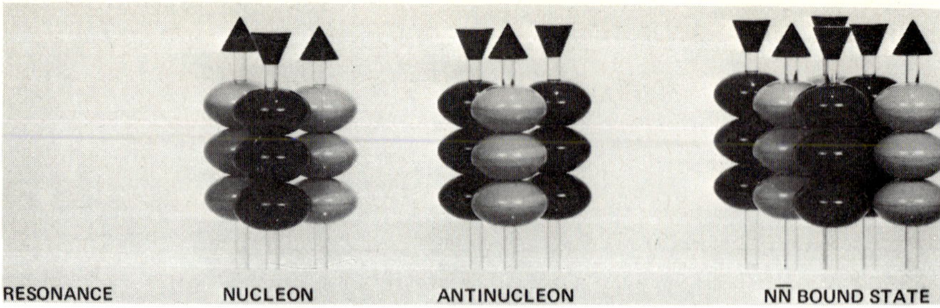

Figure 18.12. The nucleon N, the antinucleon N̄, and the nucleon-antinucleon bound state NN̄. The spinor configurations shown here accurately reproduce the masses of these states [see Eqs. (18.7) and (18.8)]. The NN̄ bound state serves as the fundamental "ground state" for the New Particle resonances, as is discussed in Chapter 17.

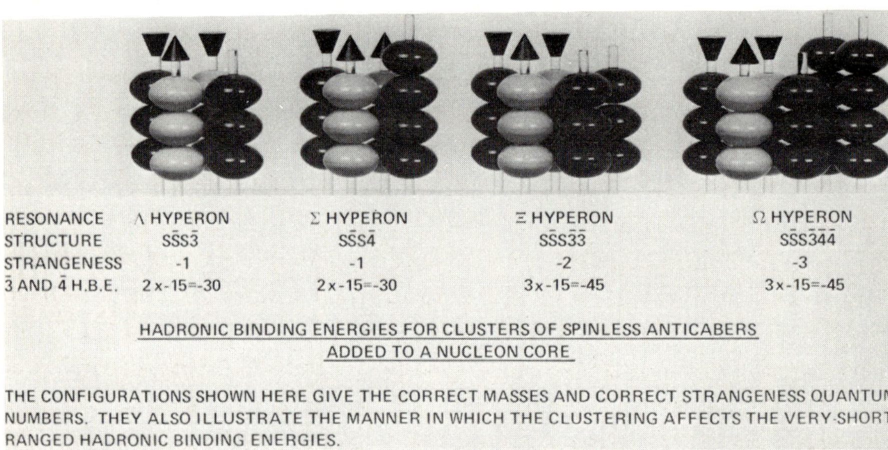

Figure 18.13. Light-quark configurations for the Λ, Σ, Ξ, and Ω hyperons. The strangeness quantum numbers of these hyperon states are carried by the spinless cabers 3̄ and 4̄ that are bound to the invariant SS̄S spinor core, and the baryon quantum number is carried by the invariant spinor core itself. The values shown here for the binding energies of the spinless cabers 3̄ and 4̄ to the SS̄S spinor core result from the binding energy systematics described in Fig. 18.6. The light-quark basis-state configurations displayed here accurately reproduce the mass values of all of these states [see Eqs. (9.9)-(9.16) in Ch. 9].

Figure 18.13 illustrates the light-quark configurations for the fundamental Λ, Σ, Ξ, and Ω hyperon excitations. In each case, spinless anticabers $\bar{3}$ or $\bar{4}$ are added to the invariant $\bar{S}SS$ spinor core, and it is these spinless excitations that carry the strangeness quantum numbers. The mass values for these configurations were given in Eqs. (9.9) - (9.16) of Chapter 9. The mass value of the K^-p bound state, $S\bar{S}S\bar{7}$, is also obtained in this same manner [Eqs. (9.18) and (9.19) of Chapter 9]. The key point to observe in Fig. 18.13 is again the side-by-side clustering of these light-quark basis states. In these hyperon excitations, the first spinless anticaber can bind hadronically to two spinors S (in the Λ and Σ hyperons), but the second spinless anticaber can bind to only one spinor S (in the Ξ hyperon), and the third spinless anticaber does not seem to bind at all (in the Ω hyperon; see the discussion of this point in connection with Eq. (9.17) of Chapter 9). We will return to these resonances in Chapter 19, where we include electromagnetic effects and thereby attempt to reproduce their magnetic moments and the charge splitting of their masses.

We close the present section by mentioning three other interesting results that can be portrayed by these geometrical light-quark representations. Figure 18.14 illustrates the fundamental (isoergic) basis state transition 333 → 3[3]3 → $\dot{S}\dot{S}$, in which one spinless caber 3 annihilates and sends the other two 3's spinning in opposite directions to form an $\dot{S}\dot{S}$ spinor pair — the ρ meson (see Chapter 13, and see Sec. 14.4 in Chapter 14). Figure 18.15 illustrates the reverse transition $\dot{S}\dot{S}$ → 3{3}3 → 333, in which a spinor pair $\dot{S}\dot{S}$ is transformed back into a spinless triplet 333 in the $\phi \rightarrow K\bar{K}$ decay; the spinless quanta in the final state combine together to form two kaons, which by their noncollinear exit trajectories carry away the angular momentum J = 1 of the φ (see Sec. 14.5 in Chapter 14). Finally, Fig. 18.16 illustrates an Associated Production reaction in which a π meson is incident on a proton, the kinetic collision energy is transformed into an η meson excitation, and the resultant excited state splits apart asymmetrically to form a positive strangeness kaon and a negative strangeness Λ hyperon (see Appendix I).

18.7 The Geometries of Rotating Hadrons

In the studies of the present chapter, we have been investigating the geometry of the elementary particle: the geometry of the spinor S = 330 MeV was deduced from its angular momentum and magnetic moment properties, and from the overall geometry of the nucleon; the geometry of the nucleon, $N = S\bar{S}S$, was deduced from the results of electron scattering experiments, from the volume of a nucleon as manifested in heavy atomic nuclei, and from the geometry of the spinor S; and the geometry of the spinless quantum M = 70 MeV was inferred from the mass systematics of the π, K, η, (M,η') mesons, from the charge splitting of the masses in the kaon, and from the

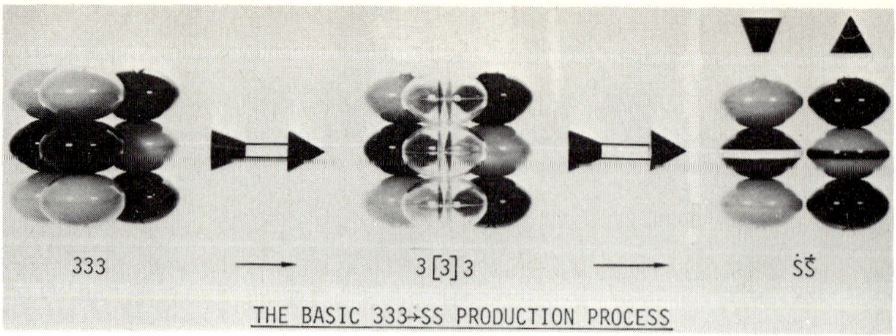

Figure 18.14. The basic spinor production process 333 → 3[3]3 → $\dot{S}\dot{S}$, in which a spinless caber 3 annihilates and sets the other two 3's spinning in opposite directions. A "top view" of this production process is shown in Fig. 14.1. The inverse process $\dot{S}\dot{S}$ → 333 is illustrated in Fig. 18.15.

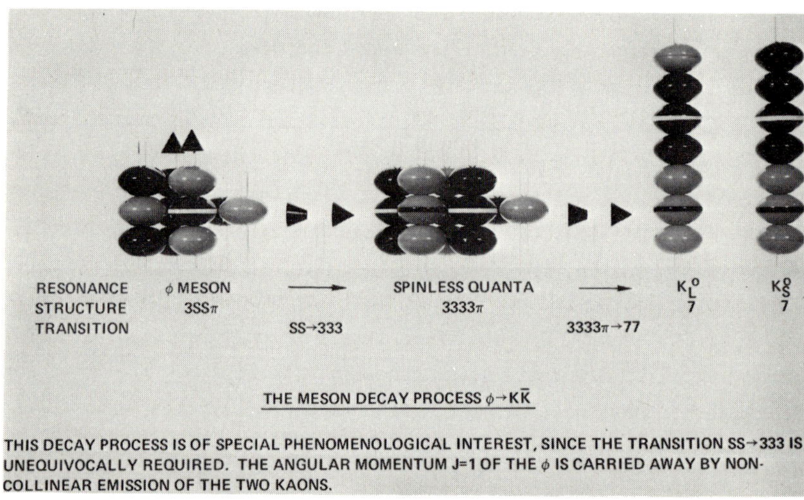

Figure 18.15. The meson decay process ϕ → KK. This decay is the light-quark sequence $\phi \equiv 3\dot{S}\dot{S}\pi \rightarrow 3333\pi \rightarrow K + K$, in which the basis-state transition $\dot{S}\dot{S}$ → 3{3}3 = 333 is unequivocally required. The spin 1 of the ϕ is carried away by noncollinear emission of the two kaons.

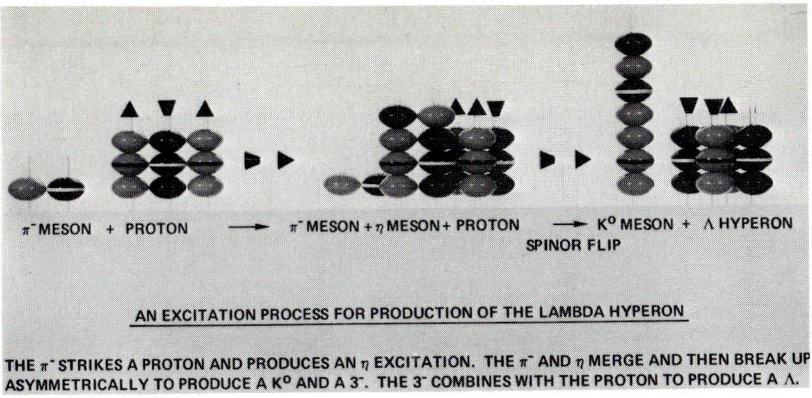

Figure 18.16. The Associated Production reaction $\pi^- + p \rightarrow \Lambda + K^0$. This reaction is the light-quark sequence $\pi + p \rightarrow \pi\{\eta\}p = \pi\{3\bar{3}\pi\}S\bar{S}S \rightarrow S\bar{S}S\bar{3} + 3\pi\pi \equiv \Lambda + K$, in which an η meson is first created and then broken apart asymmetrically: the $\bar{3}$ in the η combines with the proton to form a strangeness S = -1 hyperon; and the 3π in the η combines with the incident pion to form a strangeness S = +1 kaon. Thus the Production of an S = -1 particle is Associated with the Production of an S = +1 particle, and we have overall conservation of strangeness in this hadron production reaction.

manner in which the spinless cabers $\bar{3}$, $\bar{4}$, and $\bar{7}$ bind to an $S\bar{S}S$ spinor triplet to form hyperon resonances. The geometrical shape of the mass quantum M was deduced to be that of an oblate spheroid, with an equatorial radius of 0.59 fm and an axial length of 0.67 fm. This geometry was deduced for the spinning configuration of Fig. 18.1 with the aid of the systematics of Fig. 18.2; then, on the basis of the kaon mass systematics of Fig. 18.3 and the binding energy systematics of Fig. 18.6, this same geometry was assumed to apply to spinless configurations, as is illustrated in Figs. 18.4 and 18.5. Having obtained these results, we then demonstrated in Figs. 18.7 - 18.16 the accurate manner in which the geometrically-defined light-quark basis states of Fig. 18.5 account for the properties of the lowest-mass and stablest meson and baryon resonances. In all of these resonances, the light-quark basis states are shown clustered together in a side-by-side manner.

There is, however, an additional result which leads to a quite different picture of some of the light-quark excitations. This result involves the meson and baryon rotational moments of inertia, which we now discuss.

The meson and baryon resonances that we have shown in Figs. 18.3 - 18.16 of the present chapter are all rotationless ($\ell = 0$) S-state excitations. However, as we described in Chapters 15 and 16, there are rotational bands associated with many of these S states, and these rotational bands can also be used to obtain geometrical information. We can determine the energies of the levels in the rotational bands from experiment, and we can then use the equation

$$E_\ell = E_o + E_{rot} \times \ell(\ell+1) \equiv E_o + \hbar^2/2I \times \ell(\ell+1) \qquad (12.13)$$

to determine the average value for E_{rot} in each rotational band, and hence also the corresponding value for the moment-of-inertia $I = \hbar^2/2E_{rot}$. This empirical value for I in each rotational band then places a bound on the overall geometry of the rotating light-quark cluster which forms the bandhead of that rotational band. The geometrical bounds that we obtain by this procedure lead to some interesting implications about the manner in which light-quark excitations occur. (In the calculation of these geometrical bounds, the implicit assumption is made that the orbital motion of a rotating hadron is a rigid-body rotation of the entire light-quark structure. However, if this orbital motion is in fact due to the rotation of just a portion of the structure, as is conventionally assumed in quark models,[3] then we of course lose these geometrical bounds. In the present discussion we will not consider this latter possibility.) We now analyze the moments of inertia of the hadron rotational bands. Numerical values for the hadron rotational levels are summarized in Appendix H.

The nucleon is composed of a spinor triplet $S\bar{S}S$, and the rho meson is composed of a spinor pair $\ddot{S}S$. The baryon rotational bands of Chapter 15 are based on S-state bandheads which are excited states of the $S\bar{S}S$ triplet, and the meson rotational bands of Chapter 16 are based on S-state bandheads which are excited states of the $\ddot{S}S$ pair. Thus if we analyze the moments of inertia of the baryon and the meson rotational bands, we can obtain information about the geometries of the $S\bar{S}S$ and $\ddot{S}S$ excited states. We start by estimating the moments of inertia of the nucleon and of the $\ddot{S}S$ bandhead of the rho; these are obtained from the schematic representations shown in Fig. 18.17. The nucleon, which contains three cylindrically-shaped spinors S in a side-by-side cluster, has an estimated moment of inertia

$$I \simeq 3 \times (1/2\ MR^2 + 4/3\ MR^2) = 11/2\ MR^2 \qquad (18.13)$$

about its symmetry axis, where M and R are the mass and radius of the spinor S. The $\ddot{S}S$ rho meson bandhead has the corresponding moment of inertia

$$I \simeq 2 \times (1/2\ MR^2 + MR^2) = 3\ MR^2\ . \qquad (18.14)$$

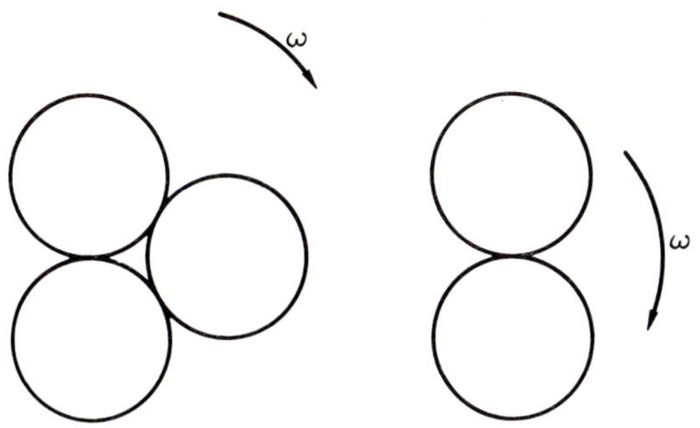

$I \simeq 3 \times (1/2\ MR^2 + 4/3\ MR^2)$ $I \simeq 2 \times (1/2\ MR^2 + MR^2)$

Figure 18.17. The three-spinor nucleon and the two-spinor rho meson as seen from above. These configurations can be used to make estimates for the moments-of-inertia I of these particles. The nucleon does not have identified rotational excitations, but the rho meson meson has a rotational energy that is in agreement with this estimate (see Fig. 13.1).

Inserting M = 313 MeV and R = 0.59 fm into Eq.(18.13) gives I = 559 MeV fm^2, which in turn gives E_{rot} = 32.5 MeV as the estimated rotational energy parameter [Eq. (12.13)] for the nucleon. Inserting M = 330 MeV (see Appendix D) and R = 0.59 fm into Eq. (18.14) gives I = 345 MeV fm^2, which leads to E_{rot} = 56.5 MeV as the estimated rotational energy parameter for the rho meson. There are no identified rotational levels for the nucleon that we can use to check the first estimate, but the rho meson has the experimental value E_{rot} = 57.5 MeV (Appendix H), in good agreement with the second estimate. The purpose in making these estimates is that we can use the geometrical models shown in Fig. 18.17 as the basis for extrapolating to the higher mass values of the observed meson and baryon rotational bands.

If the geometrical configuration of a rotating structure stays the same as its mass is increased, then its moment-of-inertia I will be proportional to its mass, and its rotational energy parameter E_{rot} will decrease in a manner that is inversely proportional to its mass. And if the geometrical size of a state increases with increasing mass, as is the case for example with the hyperon excitations shown in Fig. 18.13, the moment of inertia will increase more rapidly than 1/M, so that E_{rot} will decrease more rapidly than 1/M. However, this is not the result that is observed experimentally in the meson and baryon rotational bands. Figure 18.18 shows the E_{rot} values for the meson rotational bands (Table H5 in Appendix H) plotted against the meson bandhead masses, and it also shows a 1/M falloff curve that is normalized to the rho meson. As can be seen in Fig. 18.18, the experimental values for E_{rot} systematically fall above the 1/M line. This means that the experimental meson values for E_{rot} fall off more slowly than 1/M, so that the actual size of the rotating resonance structure (Fig. 18.17) must be decreasing as the S-state excitation energy (the bandhead mass) is increasing. Figure 18.19 shows the corresponding situation for the baryon rotational bands (Tables H1 - H4 of Appendix H), and here again the same situation occurs: the experimental values of E_{rot} fall off more slowly than 1/M, so that the dimensions of the rotor are decreasing as the mass of the rotor is increasing.

We saw in Chapters 15 and 16 that the S-state bandheads of these hadron rotational bands are formed by adding spinless excitation quanta to the spinor pair $\ddot{S}S$ (for mesons) and to the spinor triplet $S\bar{S}S$ (for baryons). The experimental results displayed in Figs. 18.18 and 18.19 indicate, if our identification of these hadron rotational bands is correct, that the spinless excitation quanta are "absorbed" into the spinors S (as vibrational excitations?) in a manner that not only does not increase their dimensions, but actually slightly decreases them. The decrease in these dimensions is brought out in Fig. 18.20, where we have plotted the values of the spinor R (based on the geometries shown in Fig. 18.17) which correspond to the observed moments of inertia. As can be seen in Fig. 18.20, if we assume that the spinor S has a radius R = 0.59 fm in the rho meson and in the nucleon, as seems to be experimentally correct, then this radius decreases to a value of about R = 0.5 fm in the more massive hadron excitations — a result which is true for both the meson and the baryon resonances.

If these meson and baryon rotational bands were our only examples of hadron resonances, we could logically inquire if it makes any sense to postulate the existence of spinless quanta: perhaps we are just observing vibrational excitations of the $\ddot{S}S$ spinor pair and the $S\bar{S}S$ spinor triplet which are quantized for some reason in multiples of 70 MeV. However, we do have the examples of the π, K, η, and (M,η') mesons, which are formed entirely of spinless quanta. Thus it is clear that spinless quanta M = 70 MeV do exist as entities in their own right. We have seen in Chapter 14 that spinless quanta can be converted into spinor pairs and back again via transitions of the type 333 $\leftrightarrow$ SS. What we are seeing in the present section (if our interpretation is correct) is that spinor pairs, once formed, can absorb even more spinless quanta and be excited into higher-energy states. The decrease in the geometries of these states may in part be a consequence of the need to conserve spin angular momentum.

The metastable hyperon resonances Λ, Σ, Ξ, and Ω do not appear to have associated rotational levels. It is clear from the present dicusssion that the geometries

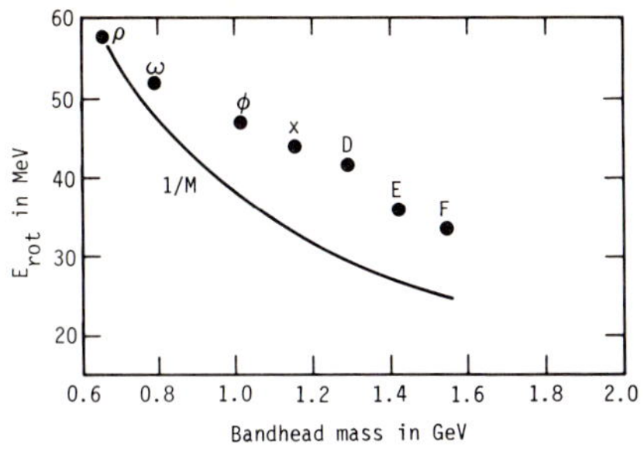

Figure 18.18. Experimental E_{rot} values for the meson rotational bands, as compared to a 1/M curve, where M is the mass of the bandhead. If the geometrical "size" of the bandhead remains invariant with increasing mass, then $E_{rot} \equiv \hbar^2/2I$ will fall off as 1/M. However, E_{rot} falls off more slowly than 1/M, which indicates that the size of the bandhead actually decreases as the mass is increased (if we assume that the rotational motion is a "rigid body rotation" of the entire resonance structure).

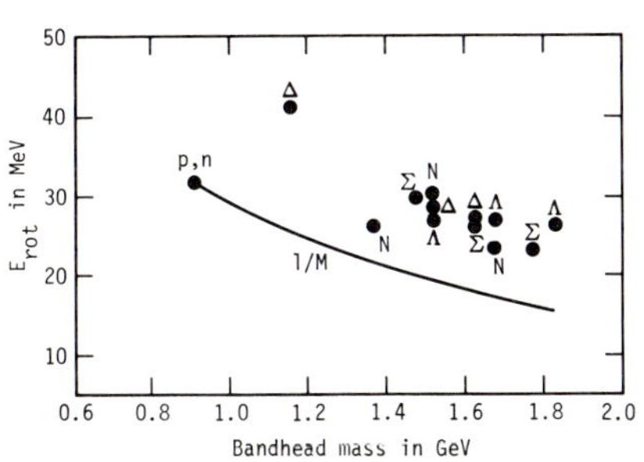

Figure 18.19. Experimental E_{rot} values for the baryon rotational bands, as compared to a 1/M curve. The same situation occurs here as occurred for the meson rotational bands of Fig. 18.8: the experimental values for E_{rot} fall off more slowly than 1/M as the mass of the bandhead is increased, thus indicating that the size of the bandhead decreases with increasing mass.

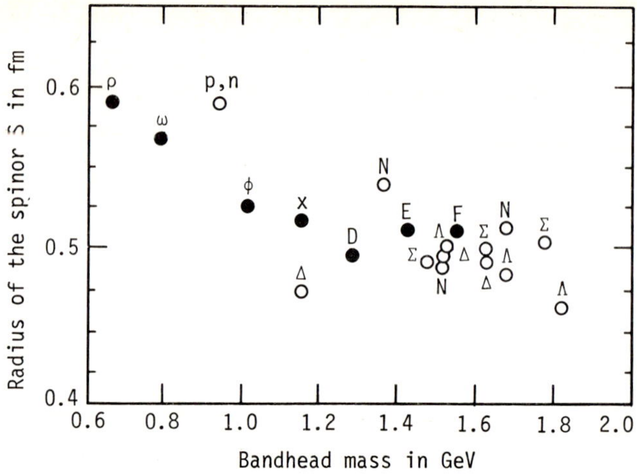

Figure 18.20. Calculated radii for the spinor S. If we assume that baryons and mesons all have the spinor geometries that are shown in Fig. 18.17, and if we use the experimental values for $E_{rot} \equiv \hbar^2/2I$ in Figs. 18.18 and 18.19 to obtain experimental values for I, then we can use Eqs. (18.13) and (18.14) to calculate values for the radius R of the spinor S as it appears in each of these resonances. These are the values shown in Fig. 18.20. If this interpretation is correct, then as spinless excitation energy is added to the rho spinor pair or to the nucleon spinor triplet, the radius of each spinor actually decreases, reaching a value $R \simeq 0.5$ fm for heavy particles as compared to the value $R \simeq 0.6$ fermi that we deduced for the lighter particles. This conclusion depends on the assumption that these are rigid body rotations.

shown in Fig. 18.13 for these hyperon states are much larger than the geometries of the rotating hyperon bandheads: the hyperon bandheads seem to be three-quark configurations of the type shown in Fig. 18.17. Thus the long-lived metastable hyperon resonances of Fig. 18.13 may actually represent a different type of geometry from that of the short-lived hyperon excitations.

The same result may also be occurring in the meson resonances. The spinless π, K, η, and (M,η') resonances, which have extended spatial geometries (see Figs. 18.10 and 18.11), do not have associated rotational levels. Also, the New Particle resonances, which are built on the $\bar{N}N$ light-quark cluster shown in Fig. 18.12, do not appear to have rotational excitations. It is only the mesons which contain the $\ddot{S}\ddot{S}$ spinor pair shown in Figs. 18.14 and 18.17 that have associated rotational bands.

We thus arrive at the same conclusion for both the baryon and meson resonances. In each of these systems, the lowest-mass resonances are spatially-extended structures which have long lifetimes and which do not give rise to rotational excitations. The higher-mass resonances in these systems seem to be excited states of the spinor clusters shown in Fig. 18.17, and they occur as 70 MeV-quantized S-states that have superimposed rotational excitations. The New Particles appear to constitute a third type of geometrical system built on the spatially-extended $\bar{N}N$ cluster, and they are manifested as a series of S-state resonances.

18.8 Summary

In Sections 18.1 through 18.6 of the present chapter, and then in Section 18.7, we have looked at the geometry of the nucleon quark from two different viewpoints, and we have thereby obtained two somewhat different geometries. If these viewpoints are logically consistent with one another, the different geometrical aspects should also be consistent. Whether or not this is actually the case is not clear from the present studies. In this connection, it should be noted that the duty of a physicist is not to present a fully self-consistent model (although that of course is the desired goal), but rather to present all of the information which can be brought to bear on the problem at hand – whether or not this information seems to contradict itself. In this respect the physicist differs from the trial lawyer – who introduces only the evidence that is favorable to his side of the argument. Historically, it is often the existence of recognized contradictions which leads to the next step in the development of a theoretical model.

In the present studies, we have not only set forth arguments which suggest a light-quark approach to hadron structure (in which all binding energies are small), but we have also pointed out from time-to-time various deficieicies in the SU(3) approach to this same problem. One reason we have done this is that when the multitude of papers on SU(3) and its applications are perused, the feeling seems to emerge (at least to the present author) that many of the proponents of SU(3) have an unconscious tendency to act more in the role of trial lawyers than of physicists. Many man-years of research work have been expended in marshaling the evidence which appears to be favorable to the SU(3) viewpoint, but the task of assembling the (numerous) deficiencies or contradictions in the SU(3) approach has received far less attention. From the standpoint of establishing a firm foundation for the use of SU(3) techniques in particle physics, these two tasks would seem to be of equal importance.

In the next chapter we conclude our basic studies of the hadron resonances with an examination of the electromagnetic aspects of these resonances.

References for Chapter 18

(1) See B. T. Feld, Models of Elementary Particles, Blaisdell, Waltham (1969), page 339, Eq. 18a'.

(2) T. G. Trippe et al., Review of Particle Properties, Rev. Mod. Phys. 48, No. 2, Part II, April 1976.

(3) See Ref. 1, Fig. 6.6 on page 382.

CHAPTER 19. ELECTROMAGNETIC EFFECTS AND PRECISION MASS CALCULATIONS

19.1 Do We Use Integral or Fractional Electric Charges?

In our study of the properties of hadrons, we have concentrated up to this point mainly on the "mechanical" properties of these particles – masses, angular momenta, hadronic binding energies, and geometrical shapes. When we finally include electromagnetic effects, as we do in the present chapter, we are confronted with two basic questions: (1) Do we assume integral or fractional internal electric charges? (2) How many internal charges do we assign? These two questions are illustrated, respectively, by the examples of the proton and the neutron. In nuclear and in atomic physics, the electric charge always seems to occur in integral multiples of the basic charge unit e. However, when we get down to the level of the proton, which has a total charge of +e, and which is composed of three equal-mass subunits, do we still use integral charges on the subunits; or should we, as mentioned in the first question, perhaps use 1/3 integral charges? In the case of the neutron, which has a total charge of zero, but which at the same time has a large (negative) magnetic moment, it is clear that both + and - internal charges are present; this leads us to the second question: how many of these internal + and - charges should we assume are present inside the neutron?

In the quark model of Gell-Mann and Zweig, the electric charges are assigned in a very clear-cut and definite manner; this has the phenomenological advantage that the electromagnetic properties of the particles are well defined, but it has the (potential) phenomenological disadvantage that a single counterexample suffices to invalidate the model. In the present light-quark model, on the other hand, we apparently have considerable freedom as to the choices we can make for the electromagnetic properties of the basis states; this means that the light-quark model, at least in its present formulation, has very little predictive ability with respect to electromagnetic effects, but it can be made to accommodate to a wide variety of experimental situations.

The Gell-Mann and Zweig quark model, in its usual formulation, features fractional electric charges: +2/3 e and -1/3 e for quarks; and -2/3 e and +1/3 e for antiquarks. The present light-quark model can be constructed with either integral or fractional electric charges. If we examine the electromagnetic aspects of the elementary particle carefully, the property which is most sensitive to the "integralness" of the electric charge is the magnetic moment. In Chapter 7 we constructed a model for the muon in which the magnetic moment arises from an equatorial current loop. We can similarly reproduce the magnetic moment of the nucleon quark — the spinor S — as an equatorial current loop, and there are a number of reasons which suggest this procedure; but if we do this, we are led to the conclusion (as we discuss in the next section) that these equatorial current loops contain one-third-integral electric charges. Thus the Gell-Mann and Zweig quark model is by no means the only model which suggests the possible existence of one-third-integral fractional charges. Gell-Mann and Zweig arrived at this conclusion by the phenomenological procedure of reproducing the total charges and the magnetic moment ratios of the elementary particles, whereas we obtain this conclusion by making a direct calculation of the magnetic moment of a quark. (In the Gell-Mann and Zweig approach, quark magnetic moments are assumed rather than calculated; furthermore, as we discussed in Chapter 4, the assumed quark magnetic moment values in the G-Z model are logically inconsistent with the implicit G-Z assumption that quarks are massive.)

In the present light-quark model, we have both spinning and nonspinning basis states. Thus we must decide between integral and fractional charge states not only for the spinor S, but also for the spinless cabers 1, 3, and 4. Furthermore, we construct the nucleon from two quarks and an antiquark, as in the Fermi-Yang model, rather than from three quarks as in the G-Z model. Thus, whereas Gell-Mann and Zweig must necessarily assume that quarks and antiquarks each contain both positive and negative charge states, it seems more logical in the light-quark model to assume that quarks carry only positive or zero charges, and antiquarks carry only negative or zero charges. Hence, the present light-quark basis states echo the properties of the spinless kaons (K^+,K^o) and antikaons ($\bar{K}^-$,$\bar{K}^o$), and also the properties of the spinning nucleons (p^+,n^o) and antinucleons ($\bar{p}^-$,$\bar{n}^o$).

In the electromagnetic calculations that are carried out in the present chapter we will assume that the light-quark basis states 1,3,4,7,S contain only integral internal electric charges, since no convincing experimental evidence has yet appeared which indicates the existence of fractional charges. In particular, we assume that the subquanta M which form the light-quark basis states (see Appendix C) occur in the charge states M^+, M^o, $\bar{M}^o$, $\bar{M}^-$ (see Fig. 18.4). Thus the spinor S and the cabers 3 and 4, which contain only subquanta M (see Figs. 18.7 and 18.8), are restricted to zero and positive charge states; and the spinor $\bar{S}$ and the cabers $\bar{3}$ and $\bar{4}$, which contain only subquanta $\bar{M}$, are restricted to zero and negative charge states. With these assumptions in mind, we now proceed to deduce the electromagnetic properties of the main hadron resonances. This discussion should be regarded as illustrative rather than definitive, since the experimental situation does not yet permit us to arrive at unique charge distributions for the light-quark basis states.

19.2 Magnetic Moments of the Nucleon and the Nucleon Quark

How many internal electric charges are contained in the proton and in the neutron? It turns out that by studying the magnetic moments of these particles, and by making use of the results developed in the preceding chapters, we can obtain an answer to this question. More specifically, we can deduce the minimum number of internal + and - charges that must be present in each of these nucleons.

We start with the following facts about the nucleon:

(1) the proton and neutron each have spin $S = 1/2\ \hbar$;
(2) the proton and neutron have total electric charges $Q_p = +1$ and $Q_n = 0$, in e.s.u.; (19.1a)
(3) the proton and neutron have magnetic moments $\mu_p \simeq +3$ and $\mu_n \simeq -2$, in nuclear magnetons;
(4) the neutron is more massive than the proton.

We then add the following assumptions:

(5) the proton and neutron are each composed of three equal-mass spin 1/2 quarks;
(6) these nucleon quarks carry integral electric charges;
(7) the quark magnetic moments arise from rotations of the quark charge distributions; (19.1b)
(8) charged quarks are somewhat more massive than neutral quarks, and doubly-charged quarks are more massive than singly-charged quarks.

With this information we are now prepared to answer our question about the number of internal charges in the proton and in the neutron. (In this discussion, we implicitly assume that we can work with just the z-components of the vector quantities – spins and magnetic moments.)

Let us start with the proton. Since the proton has a positive electric charge and a positive magnetic moment, it is clear that the magnetic moment must include a contribution from a positively-charged quark. As the simplest assumption, we first investigate the possibility that the intrinsic magnetic moment of a quark (the magnetic moment per unit charge) is $\mu_Q \simeq 3$ nuclear magnetons (this is the assumption that is made in the Gell-Mann and Zweig quark model[1]). We then construct the proton as the quark configuration

$$p = \uparrow^{+}\downarrow^{0}\uparrow^{0} \quad , \tag{19.2a}$$

which gives the correct spin and correct magnetic moment. However, with this choice for μ_Q, there is no way that we can reproduce the magnetic moment $\mu_n \simeq -2$ n. m. of the neutron; the quark magnetic moment is too large. (Gell-Mann and Zweig circumvent this difficulty by using fractional electric charges.) Thus, as a second choice, let us instead choose a smaller intrinsic magnetic moment for the quark: $\mu_Q \simeq 1$ n. m. With this choice, we now construct the proton as the quark configuration

$$p = \uparrow^{+}\downarrow^{-}\uparrow^{+} \quad , \tag{19.2b}$$

which again gives the correct spin and magnetic moment. The simplest configuration for the neutron is then

$$n = \uparrow^{+}\downarrow^{-}\downarrow^{0} \quad , \tag{19.3a}$$

which gives the correct spin and magnetic moment; but in view of Assumption (8) in Eq. (19.1) (which follows from our previous systematics), the neutron configuration of Eq. (19.3a) will have a smaller mass than the proton configuration of Eq. (19.2b) in disagreement with experiment. Hence we must instead construct the neutron as the quark configuration

$$n = \uparrow^{+}\downarrow^{=}\downarrow^{+} \quad , \tag{19.3b}$$

which, as we will demonstrate in Sec. 19.6, fulfills all of our requirements. Thus we finally conclude that the proton contains <u>three</u> internal charges (+-+), and the neutron contains <u>four</u> internal charges (+--+).

From these arguments, we have not only deduced the number of internal charges in the proton and in the neutron, but we have also deduced the intrinsic magnetic moment of the nucleon quark. Since the actual magnetic moments of the proton and neutron are

$$\mu_p = +2.79 \text{ n. m. and } \mu_n = -1.91 \text{ n. m.}, \tag{19.4}$$

the intrinsic magnetic moment of the nucleon quark must be

$$\mu_Q \simeq 1/3\ \mu_p \simeq 0.93 \text{ n. m.} \tag{19.5}$$

This value for μ_Q brings us to grips with a difficulty in the present phenomenology. In Chapter 7, where we constructed a model for the muon, we reproduced the magnetic moment of the muon by using an equatorial charge distribution (Fig. 7.3). Now, the spinor S that is shown in Fig. 7.3, and also in Fig. 18.1, has a radius $R \simeq 0.6$ fm for its constituent oblate spheroids (subquanta M). If we place an equatorial charge distribution on one of these spheroids, it gives rise to a calculated magnetic moment $\mu_Q = \mu_p = 2.79$ n. m. (see Fig. 18.2), which is just three times the value that we deduced in Eq. (19.5). One obvious way to correct this difficulty is to use one-third-integral charges for the equatorial current loops on the spinors S. However, it is not necessary to go to the assumption of fractional electric charges; since an equatorial charge distribution gives rise to the largest possible magnetic moment, our phenomenological difficulty with this factor of three can be corrected by choosing a smaller effective radius for the charge distribution – one that reduces the calculated magnetic moment by precisely a factor of three. But it is desirable to have a sound phenomenological basis for assigning this charge distribution, and thus far no reasonable suggestion has occurred to the author. In the remainder of the present discussion, we will continue to treat the spinors S as if they have equatorial current distributions, but we will at the same time assume that the intrinsic quark magnetic moment is given by Eq. (19.5); this inconsistent procedure causes no important difficulties in the present applications, but it is a situation which must ultimately be resolved. The discrepancy with the magnetic moments could simply mean that we have the wrong charge distribution in the spinor S, or it may be an indication that our very classical way of treating this problem (for example, by using only z-components), is leading us into essential difficulties.

On the basis of the above discussion, we have deduced the following spinning quark configurations:

$$p = \uparrow^{+}\downarrow^{-}\uparrow^{+} \quad ; \quad \mu_p \sim +3 \quad , \tag{19.6a}$$

and

$$n = \uparrow^{+}\downarrow^{=}\downarrow^{+} \quad ; \quad \mu_n \sim -2 \quad . \tag{19.6b}$$

A third spinning quark configuration, which occurs in the Λ hyperon, is

$$\Lambda = \uparrow^{+}\downarrow^{-}\downarrow^{+} \quad ; \quad \mu_\Lambda \sim -1 \quad . \tag{19.6c}$$

These three spinor configurations reproduce all of the baryon and hyperon magnetic moments which have been reported to date. In particular, we have[2]

$$\begin{aligned} &\mu_p = +2.793 \qquad && \mu_n = -1.913 \qquad && \mu_\Lambda = -0.606 \pm 0.034 \\ &\mu_{\Sigma^+} = +2.83 \pm 0.25 && \mu_{\Sigma^-} = -1.48 \pm 0.37 && \\ & && \mu_{\Xi^-} = -1.85 \pm 0.75 && \end{aligned} \tag{19.7}$$

as the measured hadron magnetic moments in nuclear magnetons.

19.3 Coulomb Effects

If nucleons contain multiple internal charges, as we have argued above, then these charges will have coulomb interactions that contribute to the total mass of the particle. In fact, we have already used the internal coulomb effects in the neutral kaon to deduce the geometry of the kaon (Sec. 18.4 of Chapter 18). The magnitude of these coulomb contributions to the mass can be estimated from the equation

$$\pm\, e^2/r \equiv \pm\, 1.44\ \text{MeV}/r \quad , \qquad (19.8)$$

where r is the separation distance between the charges in fermis. Eq. (19.8) applies only to point charges e; however, the corrections to this equation for the case of distributed charges are rather small ($\sim$20%). Since the separation distance r between two charges in a hadron is typically about one fermi, the coulomb interaction between these charges is roughly ±1.5 MeV, so that the error incurred by using Eq. (19.8) instead of an extended charge distribution is only about 0.3 MeV. Furthermore, the matrix of + and - charges in a hadron resonance tend to cancel each other out, so that the total coulomb potential inside a hadron doesn't exceed 2 or 3 MeV.

In making coulomb corrections for the light-quark geometries that we discuss later on in this chapter, we use the assumption of point charges [Eq. (19.8)] for the spinless cabers, and we assume ring geometries for the charges on the spinors S. The coulomb effects for adjacent rings are obtained by calculating the "point-ring" correction to the "point-point" interaction, and then doubling this correction; we thus assume that the coulomb interaction is not singular at the point of contact. As mentioned above, by making an allowance for extended charge distributions on the spinors S, we are only making small corrections to quantities which are rather small in the first place.†

The essential point to keep in mind about these hadronic coulomb corrections is that they amount in all cases to just a few MeV. This is a consequence of the fact that the overall charge on a metastable hadron resonance is usually either 0 or ±1. This situation in in marked contrast to the situation that occurs in nuclear physics[3].

---

†One example where the coulomb correction for an extended charge distribution becomes of real interest is in the case of the muon model of Chapter 7. If a spinning sphere is assigned a radius that is equal to its Compton wavelength, and if it is given an equatorial charge distribution, this charge distribution generates a series expansion for the coulomb potential that is in powers of $\alpha^2 = (e^2/\hbar c)^2$. This of course is the same result which emerges from the treatment of the muon in quantum electrodynamics.

wherein a large nucleus has an overall charge of up to 100 e. The total repulsive coulomb potential inside a large nucleus such as uranium can amount to a GeV, and this enormous repulsion not only distorts the nucleus into a prolate shape (in order to minimize the repulsion), but it also leaves the nucleus teetering on the knife edge of instability, so that the introduction of a neutron or other perturbation is sufficient to trigger the fission process. The energy that is released in fission is essentially just the coulomb energy that is liberated by the division of the nucleus into two smaller parts. This subject is discussed briefly in Chapter 20.

In addition to the coulomb binding energies that we have discussed above, the other coulomb effect that is important for hadron resonances is the self-energy carried by an electric charge. On phenomenological grounds, we have deduced that the charged quantum $M^{\pm}$ = 74.6 MeV is 4.6 MeV more massive than the neutral quantum M^{o} = 70.0 MeV. This coulomb self-energy of 4.6 MeV is also the mass difference between the $\pi^{\pm}$ and π^{o} mesons, and it is precisely equal to nine electron masses (which may or may not be a coincidence). The mass of a charged spinor $S^{\pm}$ is

$$S^{\pm} = 330.4 \text{ MeV} \tag{19.9}$$

(see Sec. 19.6; also see Appendix D), which includes the self-energy of a single spinning charge. From the mass relationship between the spinor $S^{\pm}$ = 330.4 MeV and its spinless counterpart $3^{\pm}$ = 214.6 MeV, we can roughly estimate the mass of a spinning charge to be 4.6 MeV x (330.4/214.6) = 7.1 MeV, which gives the mass of the doubly-charged spinor $S^{\pm\pm}$ as

$$S^{\pm\pm} = 337.5 \text{ MeV}. \tag{19.10}$$

The spinor $S^{\pm\pm}$ occurs in the neutron (Eq. 19.3b); and, as we will see in Sec. 19.6, the mass value shown in Eq. (19.10) is the proper $S^{\pm\pm}$ mass to reproduce both the absolute mass of the neutron and the n - p mass difference.

We now turn to the last important electromagnetic effect in hadrons — magnetic binding energies.

19.4 Magnetic Binding Energies

In Sec. 19.2 we deduced the value μ_Q = 0.93 n. m. as the magnetic moment of a singly-charged quark. If we calculate the magnetic dipole interaction energy between two of these quarks, it turns out to be quite small — about 0.1 MeV. However, if the magnetic field of the quark arises from a circular current distribution, and if two of these current distributions are placed adjacent to one another, the magnetic interaction can be greatly enhanced at the point of intersection, so that the total magnetic interaction energy is much larger than indicated by the quark magnetic moments, and also much shorter-ranged. Without assuming a specific cross-sectional geometry for the rotating charges, it is impossible to calculate this magnetic moment interaction. However, it turns out that we can deduce it phenomenologically.

The experiment that we use to deduce the quark-quark magnetic binding energy is the hyperon decay

$$\Sigma^{0} \to \Lambda + \gamma, \quad Q = 76.9 \text{ MeV} \quad . \qquad (19.11)$$

Phenomenologically, we conclude that the Σ^0 hyperon is the light-quark cluster $\Sigma^0 = S^+S^-S^+4^-$, and the Λ hyperon is the cluster $\Lambda = S^+S^-S^+3^-$. The important thing to notice about these two clusters is that they have essentially identical coulomb distributions. Thus their coulomb energies will be the same. Now, in the $\Sigma^0 \to \Lambda + \gamma$ decay shown in Eq.(19.11), a single mass quantum $M^0 = 70.0$ MeV is annihilated and a photon is emitted. But the energy liberated in the decay is Q = 76.9 MeV, and not 70.0 MeV. This difference of 6.9 MeV cannot be attributed to coulomb effects; hence it must be due to magnetic effects — that is, to the magnetic binding of adjacent internal quark-quark pairs.

From its observed magnetic moment [Eq. (19.7)], the Λ hyperon of reaction (19.11) is the light-quark configuration [see Eq. (19.6c)]

$$\Lambda = S^{+}_{\uparrow}S^{-}_{\downarrow}S^{+}_{\downarrow}3^{-} \quad . \qquad (19.12)$$

Unfortunately, the magnetic moment of the Σ^0 hyperon that appears in reaction (19.11) has not yet been measured, due to the short lifetime of the Σ^0 (see Table 10.1). However, we can provisionally assume that the Σ^0 must have one of the three spinor configurations p, n, or Λ shown in Eq.(19.6). But if it does, the Σ^0 cannot have the Λ configuration, since that is the configuration which occurs in the final-state particle of Eq. (19.11), and it cannot have the n configuration, since the n has the same magnetic interaction energy as the Λ; that is, if the Σ^0 has a Λ-type or an n-type spinor cluster, then the discrepancy of 6.9 MeV in the Q-value of Eq. (19.11) cannot be explained as a magnetic effect — in contradiction to our assertion above. Thus the Σ^0 must have a p-type spinor cluster, so that

$$\Sigma^{0} = S^{+}_{\uparrow}S^{-}_{\downarrow}S^{+}_{\uparrow}4^{-} \quad . \qquad (19.13)$$

This Σ^0 light-quark cluster, looking down from the top, has one or the other of the geometries (a) and (b) shown in Fig. 19.1; geometry (a) seems the most likely, since

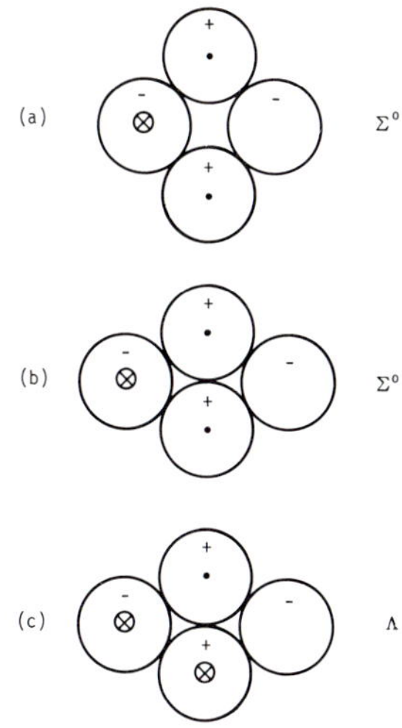

Fig. 19.1. Light-quark clusters for the Σ^O and Λ hyperons, as seen from above. The crosses and dots on the axes of the spinors S denote spin down and spin up, respectively, and the absence of a cross or dot indicates a spinless caber $\overline{3}$ or $\overline{4}$. The $\Sigma^O \equiv S\overline{S}S\overline{4}$ hyperon can occur conceptually in either of the geometries (a) or (b); the hadronic binding energy is the same in either case, but the short-ranged magnetic binding energy between the two S^+ spinors will be considerably different. Since the $S^+_\uparrow S^+_\uparrow$ magnetic binding energy in the Σ^O is repulsive, the magnetic interaction favors configuration (a). In the case of the Λ, however, the $S^+_\uparrow S^+_\downarrow$ magnetic binding energy is attractive, so that configuration (c) is the only logical cluster configuration.

the magnetic and coulomb interactions for the two positively-charged spinors (which have parallel spins) are both repulsive. The Λ hyperon, on the other hand, has the tightly-clustered spinor configuration of Fig. 19.1c, since the two positively-charged spinors (which have antiparallel spins) are attracted to one another magnetically by a force which is large enough (as we will show) to overcome the coulomb repulsion. If we say that the magnetic interaction operates only between adjacent spinors at their points of contact, then Fig. 19.1a represents two repulsive magnetic pairs; Fig. 19.1b represents three repulsive pairs; and Fig. 19.1c represents one attractive pair (that is, two attractive pairs and one repulsive pair). Hence the energy of 6.9 MeV in Eq. (19.11) that is attributed to magnetic interactions corresponds to the following magnetic pairing-energy (V_{mag}) terms:

$$+2V_{mag} - (-1)V_{mag} = +3V_{mag} = 6.9 \text{ MeV} \rightarrow V_{mag} = \pm 2.3 \text{ MeV} \qquad (19.14a)$$

for Figs. 19.1a and 19.1c;

$$+3V_{mag} - (-1)V_{mag} = +4V_{mag} = 6.9 \text{ MeV} \rightarrow V_{mag} = \pm 1.7 \text{ MeV} \qquad (19.14b)$$

for Figs. 19.1b and 19.1c. If the magnetic interaction is not completely short-ranged, then we might expect the actual situation to be intermediate between these two limits. Thus we choose

$$V_{mag} = \pm 2.0 \text{ MeV} \qquad (19.15)$$

as the empirically-determined magnetic interaction energy between adjacent singly-charged spinors $S^{\pm}$. We also assume that a doubly-charged spinor $S^{\pm\pm}$ and an adjacent singly-charged spinor $S^{\pm}$ have just twice this magnetic interaction energy.

As we will see in Sections 19.6 and 19.7, the magnetic interaction energy V_{mag} shown in Eq. (19.15) seems to be correct for the metastable baryon resonances. Also, as we demonstrate in Chapter 20, this same magnetic binding energy reproduces the binding energy in the alpha particle, which is made up of two protons and two neutrons.

Having evaluated the expected magnitudes of the electric and magnetic binding energies, we now apply these results to the calculation of the mass values for the metastable meson and baryon resonances.

19.5 The Spinless Mesons

Mass calculations for the spinless mesons π^o, $\pi^{\pm}$, $K^{\pm}$, K^o, η, M, and η' have already been discussed in Sec. 5.2 of Chapter 5 and in Sec. 9.2 of Chapter 9. We again stress the fact that these very-narrow-width mesons are the lowest-mass — and therefore the most fundamental — of all the hadron states. As such, they furnish perhaps the best test of theoretical models for hadrons.

Geometrical configurations for the spinless mesons are shown in Figs. 18.3 and 18.9 — 18.11 of Chapter 18. In assigning internal charge states to these particles, it seems to be a general rule that each hadronic state must contain at least one neutral subquantum M^o or $\bar{M}^o$. Thus the π^o meson is constructed as $\pi^o = M^o\bar{M}^o$ rather than $\pi^o = M^+M^-$ (see Fig. 18.9). It seems to be another general rule that spinors and spinless cabers tend to occur singly-charged. Hence the η, M, and η' mesons are constructed as the caber-anticaber bound state pairs 4^+4^-, 7^+7^-, and 7^+7^-, respectively, and the K^+ and K^- mesons are the unbound cabers 7^+ and 7^- (see Figs. 18.10 and 18.11). The K^o_L and K^o_S mesons (Fig. 18.3) each contain the substates 3^+ and 3^-, plus an additional subquantum M^o or $\bar{M}^o$, in a linear array.

The hadronic binding energies for these states have already been discussed in the previous chapter (see Figs. 18.6 and 18.11). The one new element that we include here is the addition of coulomb corrections to the hadronic binding energies. If we assume that the + and - charges in caber-anticaber bound-state pairs are adjacent to one another (which gives the largest coulomb correction), we obtain a coulomb contribution of -1.2 MeV to the calculated mass in the cases of the η, M, and η' mesons, since the + and - charges in the caber-anticaber pairs are separated by 1.2 fm [see Eq. (19.8)]. The neutral kaon configurations K^o_L and K^o_S shown in Fig. 18.3 each have a coulomb correction of -0.7 MeV due to the 2 fm spacing of the internal + and - charges.

The coulomb-corrected mass values for the spinless mesons are summarized in Table 19.1. As can be seen, this very simple light-quark model gives absolute mass values for these seven meson states to an average accuracy of ±0.07%, or ±0.4 MeV, where we are using a 70 MeV building block for all of these states. This compares to the 6% or 13% mass accuracy that is achieved in SU(3) for just the π, K, and η states. The simplicity and accuracy of the calculations shown in Table 19.1 speak for themselves. It is true that we have some freedom in the way that we assign the internal charge states for these mesons, but the choices shown in Table 19.1 are certainly as straightforward and plausible as any that we can make, and they are in general agreement with the choices that we make in the next two sections for the baryon and the hyperon states.

The main conclusion to be drawn from the mass calculations for the spinless mesons shown in Table 19.1 is the fact that it is possible to construct a quark model that reproduces these mass values, which are known experimentally to ±0.1%, to this same level of accuracy. This then becomes a standard against which other theoretical models should be tested. Moreover, as we discussed in Sec. 5.2 of Chapter 5, the light-quark configurations that are shown in Table 19.1 account for other properties of these particles (such as the strangeness quantum numbers, for example).

Table 19.1. Calculation of absolute mass values for the spinless meson and kaon resonances. The "intrinsic mass" in the table is the sum of the masses of the constituent cabers before binding energy corrections have been applied. With all corrections included, the absolute masses of these seven resonances are calculated to a precision of better than 0.1%.

Meson state	Caber array	Intrinsic mass (MeV)	Binding energies		Final mass values (MeV)		
			Hadr.	Coul.	Calc.	Exper.	Error
$\pi^\pm$	$1^o1^\pm$	144.6	-5.0		139.6	139.6	0.0
π^o	$1^o\bar{1}^o$	140.0	-5.0		135.0	135.0	0.0
η	4^+4^-	569.2	-20.0	-1.2	548.0	548.8	-0.8
M	7^+7^-	989.2	-35.0	-1.2	953.0	953	0.0
η'	7^+7^-	989.2	-30.0[a]	-1.2	958.0	957.6	+0.4
$K^\pm$	$7^\pm$	494.6	0		494.6	493.7	+0.9
$K^o,\bar{K}^o$	7^{+-}	499.2	0	-0.7[b]	498.5	497.7	+0.8

[a]See point <u>four</u> in Sec. 5.2 of Chapter 5.

[b]See point <u>five</u> in Sec. 5.2 of Chapter 5.

AVERAGE ABSOLUTE MASS ACCURACY FOR TABLE 19.1: ± 0.4 MeV .

AVERAGE PERCENTAGE MASS ACCURACY FOR TABLE 19.1: ± 0.07% .

19.6 The Proton, Neutron, and $\bar{p}n$ Bound State

"The proton-neutron mass difference problem, probably the oldest puzzle in hadron physics, has challenged and frustrated generations of theorists. The solution is still nowhere in sight."

A. Zee, Physics Reports 3C, 127 (1972)

The meson resonances of Table 19.1 have simple light-quark structures in that they consist of either one or a pair of spinless cabers formed from the spinless mass quantum $M \simeq 70$ MeV. The proton and neutron, on the other hand, have considerably more complicated structures in that they each contain three spinning quark states — the spinors $S \simeq 330$ MeV. This means that in the mass calculations for these nucleons, we will have to include magnetic as well as coulomb corrections to the hadronically-bound masses. However, with the same general light-quark systematics, we can obtain mass accuracies for the nucleons and the $\bar{p}n$ bound state which are comparable to the mass accuracies demonstrated above for the spinless mesons.

In this light-quark approach to hadron structure, we have considered the nucleon as a Fermi-Yang-type spinor triplet — $N = S\bar{S}S$ — in which two spinors S are bound to an antispinor $\bar{S}$. We have further postulated that the 4% hadronic binding energy operates just between the two spinor-antispinor pairs $S\bar{S}$ in the nucleon, and not between the spinors SS. This implies that the spinor triplet $S\bar{S}S$ is a "hinged" structure in which the spinors can line up either linearly or in a clustered manner.

$$N = S\bar{S}S \qquad \text{or} \qquad N = \bar{S}^{S}_{S} \quad . \tag{19.16}$$

If the spinors in the nucleon actually have this freedom of movement, then it is clear that the proton, from its electromagnetic interactions, will be a linear configuration; and the neutron, from its electromagnetic interactions, will be a clustered configuration. These are the configurations that we will assume in the present discussion, and they are the configurations that are shown geometrically in Fig. 19.2. However, it should be noted here that the assumption of a clustered configuration for the proton would have only a small effect on the present phenomenology. We know from electron scattering experiments that the proton and neutron both have the same rms magnetic moment radii, and it may be that this experimental result rules out a linear spinor configuration for the proton. In order to determine if this is in fact the case, we would have to average the experimental electron scattering results over all orientations of the target proton, and this analysis has not been carried out.

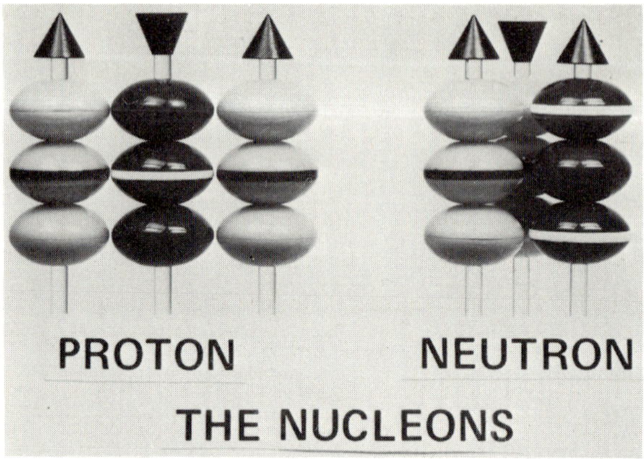

Figure 19.2. The proton and the neutron. From their different internal electromagnetic effects, the proton appears as a linear spinor configuration, whereas the neutron appears as a clustered spinor configuration.

The proton configuration shown in Fig. 19.2 is the linear cluster

$$p = S^{+}_{\uparrow}S^{-}_{\downarrow}S^{+}_{\uparrow} \quad . \tag{19.17}$$

In this cluster, the adjacent $S^{+}_{\uparrow}S^{-}_{\downarrow}$ and $S^{-}_{\downarrow}S^{+}_{\uparrow}$ pairs have magnetic interaction energies of +2.0 MeV each [Eq. (19.15)]. They also have coulomb interaction energies (assuming adjacent rings of charge) of -1.4 MeV each. And the two positively-charged spinors, which are widely separated, have zero magnetic interaction energy and +0.6 MeV coulomb interaction energy. Thus the proton mass, with electromagnetic effects removed, would be 938.3 - 2.0 - 2.0 + 1.4 + 1.4 - 0.6 = 936.5 MeV. Now, in Chapter 8 we estimated that the attractive $S\bar{S}$ binding energy [Eq. (8.5')] is roughly -26.4 MeV. It turns out phenomenologically that with coulomb and magnetic effects added in, a binding energy of -27.3 MeV is about right (this is the value shown in Fig. 18.6). Hence if we "unbind" the proton by separating the three singly-charged spinors, we then obtain a total unbound mass of 936.5 + 54.6 = 991.1 MeV, so that the mass of a singly-charged spinor $S^{\pm}$ is 991.1 ÷ 3 = 330.4 MeV [as shown in Eq.(19.9)].† The calculation of the proton mass, using the results that we have just described, is summarized in Table 19.2.

The neutron that is illustrated in Fig. 19.2 is the clustered configuration

$$n = S_{\downarrow}^{=} \begin{matrix} S_{\downarrow}^{+} \\ S_{\uparrow}^{+} \end{matrix} \quad . \tag{19.18}$$

The mass value for the doubly-charged spinor $S^{=}$ in the neutron is 337.5 MeV [see Eq. (19.10)]. As is shown in Table 19.2, the neutron has a total magnetic binding energy of -2.0 MeV and a total coulomb binding energy of -2.1 MeV. With these values, and with the same hadronic binding energy as for the proton, the neutron mass is accurately reproduced, and hence also the n-p mass difference.

Since the $S\bar{S}$ hadronic binding energy of -27.3 MeV and the $S^{\pm\pm} - S^{\pm}$ mass difference of 7.1 MeV appear here as somewhat adjustable (but phenomenologically plausible) parameters, it is obvious that we can accurately reproduce the mass values of both the proton and the neutron. However, an independent confirmation of the correctness of these calculations is obtained when we consider the mass of the $\bar{p}n$ bound state. The $\bar{p}n$ bound state, which is illustrated in Fig. 19.3, is the spinor configuration††

$$\bar{p}n = S_{\downarrow}^{+} \begin{matrix} S_{\uparrow}^{-} \\ S_{\uparrow}^{-} \end{matrix} \begin{matrix} S_{\downarrow}^{+} \\ S_{\uparrow}^{+} \end{matrix} S_{\uparrow}^{=} \quad . \tag{19.19}$$

The hadronic, magnetic, and coulomb corrections for this six-spinor cluster are shown in Table 19.2. As can be seen in Table 19.2, the same binding energies and spinor masses which accurately reproduce the proton and neutron masses also reproduce the $\bar{p}n$ bound state mass to a comparable accuracy. Furthermore, as we have discussed in the preceding sections, these parameter values are rooted in other facets

---

†The choices that we make for the $S\bar{S}$ binding energy and for the S and $\bar{S}$ masses are obviously related to one another, but the mass value of 330.4 MeV obtained here for the spinor $S^{\pm}$ is in close agreement with the mass values that we obtain from other independent determinations (see Appendix D).

††Equation (8.9) in Chapter 8 shows two different possible forms for this cluster; we have used the second form here. The answers are about the same in either case.

Table 19.2. Calculation of absolute mass values for the proton p, neutron n, and antiproton-neutron bound state $\bar{p}n$. Also calculated are the mass values for the spinor triplet "core states" $\hat{p}$, $\hat{\Lambda}$, $\hat{n}$, and $\hat{\Omega}$ that appear in the hyperon resonances of Table 19.3 [see Eq. (19.20)]. The "intrinsic mass" for the nucleon states is the sum of the masses of the constituent spinors before the various binding energy corrections have been applied. There is enough parameter freedom in the light-quark model to permit an accurate fit to the p and n masses, but these same parameters then give a precision fit to the $\bar{p}n$ mass and to the hyperon masses of Table 19.3 with no further adjustments being required.

Spinor state	Intrinsic mass (MeV)	Binding energies			Final mass values (MeV)		
		Mag.	Coul.[a]	Hadr.	Calc.	Exper.	Error
p	991.2	+4.0	-2.2	-54.6	938.4	938.3	+0.1
n	998.3	-2.0	-2.1	-54.6	939.6	939.6	0.0
$\bar{p}n$	1989.5	+2.0	-5.6	-191.1	1794.8	1794.5	+0.3
$\hat{p}$	938.3[b]	+1.0	+0.3		939.6		
$\hat{\Lambda}$	938.3[b]	-6.0	+0.8		933.1		
$\hat{n}$	939.6[b]	—	—		939.6		
$\hat{\Omega}$	939.6[b]	-4.0	—		935.6		

[a]Coulomb corrections are made by summing over all of the internal charge pairs.

[b]These are the experimental masses of the corresponding nucleon states; the treatment of these four basic spinor configurations is discussed in Sec. 19.7.

AVERAGE ABSOLUTE MASS ACCURACY FOR TABLE 19.2: ± 0.1 MeV .

AVERAGE PERCENTAGE MASS ACCURACY FOR TABLE 19.2: ± 0.01% .

of the phenomenology. Thus it seems apparent that the magnitudes of the hadronic and electromagnetic binding energies which we are using here must be correct.

Figure 19.3. The $\bar{p}n$ bound state. The same light-quark parameters which accurately reproduce the masses of the proton and neutron also accurately reproduce the mass of the $\bar{p}n$ bound state. The $\bar{p}n$ bound state has the largest binding energy (4.4 ± 0.1%) ever measured for any physical system.

At this point, we should summarize just what it is that we claim to have achieved with the mass calculations shown in Table 19.2. We started with a spinor mass $S^{\pm} \simeq 330$ MeV, and with an $S\bar{S}$ binding energy that is known to be about 4% (see Chapter 8). We then introduced a theoretically-calculated coulomb correction plus an empirically-deduced magnetic correction; we adjusted the $S\bar{S}$ binding energy and the mass of the spinor $S^{\pm}$ to match one another for the case of the proton when these electromagnetic effects are included; and we used a relativistic mass ratio to deduce a mass for the doubly-charged spinor $S^{\pm\pm}$. With the inclusion of these electromagnetic refinements, we are able to accurately reproduce the absolute masses of the proton and neutron, the n-p mass difference, and the $\bar{p}n$ bound state mass. We also reproduce the spin values and the magnetic moment ratio for the proton and neutron (these quantities are not yet known for the $\bar{p}n$ bound state). Now, by following this phenomenological procedure, we have not calculated the proton, neutron, and $\bar{p}n$ masses to an accuracy of a few tenths of an MeV, but we have reproduced these masses to this level of accuracy — and by using plausibly-selected parameters. This is a nontrivial achievement for two reasons: (1) The calculation of the n-p mass difference is commonly regarded as one of the great unsolved problems in particle physics (see the quotation by A. Zee at the beginning of this section), and a large number of papers have been written on this subject[4]; we can see from the results displayed in Table 19.2 that the n-p mass difference in fact arises as a delicate balance among the hadronic, coulomb, and magnetic binding energies — and in a manner that is readily understandable. (2) There has been no other quantitative explanation forthcoming as to why an antiproton and a neutron should bind together with a binding energy of -83.3 ± 1.4 MeV (which is the largest binding energy percentagewise ever observed for any physical system).

In addition to their ability to reproduce the spectroscopic properties of the nucleons, the geometric light-quark configurations that are shown in Fig. 19.2, with their equatorial charge distributions, have several other interesting properties: (1) the negative charges in the neutron are displaced slightly more from the center of the neutron than are the positive charges, so that electron scattering will reveal a slightly negative overall electric moment for the neutron; (2) the electric dipole moment with respect to the spin axis is zero for both the proton and neutron; (3) low-energy electron scattering off a current loop, where the wavelength is large as compared to the dimensions of the loop, appears as a dipole interaction; (4) high-energy electron scattering off a current loop, where the wavelength is small as compared to the dimensions of the loop, appears as a pointlike (parton) interaction. All four of these properties of the nucleon have been verified experimentally.

In Table 19.1, the present light-quark model was shown to be sufficient (and, in the opinion of the author, necessary) for the task of reproducing the spinless meson states; in Table 19.2 it was shown to be sufficient for the p, n, and $\bar{p}n$ nucleon states; and in Table 19.3 of the next section, we demonstrate that this same light-quark model also sufficies for the metastable hyperon states.

19.7 The Metastable Hyperons

The long-lived hyperon states Λ, Σ^+, Σ^0, Σ^-, Ξ^0, Ξ^-, and Ω^- are formed by adding spinless anticabers $\bar{3}$ and $\bar{4}$ to an $S\bar{S}S$ spinor core. These hyperons thus represent a combination of the spinless cabers of Sec. 19.5 and the spinor triplets of Sec. 19.6. The important fact to note here is that a straightforward extension of the phenomenology which worked so well for the absolute mass values of the spinless mesons of Table 19.1 and the nucleons of Table 19.2 also works for these basic hyperon states. However, the detailed spinor configurations that we assign to these states come only from an *a posteriori* fit to the masses and magnetic moments, and not from an *a priori* insight as to the way that these states might be expected to occur.

As we showed in Eq. (19.7), the measured baryon and hyperon magnetic moments seem to fall into three general groups: $\mu \lesssim +3$ n. m., $\mu \lesssim -2$ n. m., and $\mu \lesssim -1$ n. m. In detail, the two positively-charged particles, p and Σ^+, have $\mu \sim +3$; the neutron and the two negatively-charged hyperons, Σ^- and Ξ^-, have $\mu \sim -2$; and the neutral hyperon Λ has $\mu \sim -1$. Now, the magnetic moment values $\mu \sim +3$, -2, and -1 correspond to the spinor triplets p, n, and Λ, respectively, that were defined in Eq. (19.6). We will assume that these triplets and a spin 3/2 Ω^- triplet occur in the metastable hyperon resonances, and we give them the following notation: $\hat{p}$, $\hat{n}$, $\hat{\Lambda}$, $\hat{\Omega}$.

The mass values for the spinor triplets $\hat{p}$, $\hat{n}$, $\hat{\Lambda}$, and $\hat{\Omega}$ are calculated in the following manner. The spinor triplet $\hat{p} = S^+_\uparrow S^-_\downarrow S^+_\uparrow$ occurs as a linear structure in the proton [Eq. (19.17) and Fig. 19.2]. However, when a spinless caber $\bar{3}$ or $\bar{4}$ is added, which binds hadronically to both of the positively-charged spinors, the spinors will close into the clustered configuration shown in Fig. 19.1a. In this configuration, the magnetic potential energy is increased by 1.0 MeV†, and the coulomb potential energy is increased by 0.3 MeV. Thus the clustered configuration for p has an energy which is 1.3 MeV larger than the open configuration for $\hat{p}$, so that $\hat{p}_{clustered}$ = 939.6 MeV. The spinor triplet $\hat{n} = S^+_\uparrow S^=_\downarrow S^+_\downarrow$ which occurs only in a clustered configuration [Eq. (19.18) and Fig. 19.2], also has a mass of 939.6 MeV. The spinor triplet $\hat{\Lambda} = S^+_\uparrow S^-_\downarrow S^+_\downarrow$ has a magnetic potential energy of -2 MeV as against +4 MeV for the unclustered spinor triplet $\hat{p}$, and it has a coulomb potential energy that is +0.8 MeV higher (it corresponds to the cluster of Fig. 19.1c). Thus it has the mass value $\hat{\Lambda} = 938.3 - 6.0 + 0.8 = 933.1$ MeV. The spinor triplet $\hat{\Omega} = S^+_\uparrow S^=_\uparrow S^+_\uparrow$ is taken to be the $\hat{n}$ triplet with all spins aligned, and it has a magnetic potential energy of -4.0 MeV with respect to the $\hat{n}$, so that it has a mass of 935.6 MeV. On the basis of these electromagnetic binding energy calculations, we have the following triplet mass and magnetic moment values:

$$
\begin{aligned}
\hat{p}_{open} &= 938.3 \text{ MeV}, \quad \mu \sim +3;\\
\hat{p}_{clustered} &= 939.6 \text{ MeV}, \quad \mu \sim +3;\\
\hat{n}_{clustered} &= 939.6 \text{ MeV}, \quad \mu \sim -2;\\
\hat{\Lambda}_{clustered} &= 933.1 \text{ MeV}, \quad \mu \sim -1;\\
\hat{\Omega}_{clustered} &= 935.6 \text{ MeV}, \quad \mu \sim 0.
\end{aligned}
\qquad (19.20)
$$

---

†This value is obtained as the average of the two cases shown in Eqs. (19.14a) and (19.14b); this same average leads to the magnetic binding energy value shown in Eq. (19.15).

When we add spinless anticabers $\bar{3}$ and $\bar{4}$ to these spinor triplets, they bind hadronically (see Fig. 18.6) with a binding energy of -30 MeV for the first anticaber [two $S\bar{3}$ or $S\bar{4}$ hadronic bonds; see Eqs. (9.9) and (9.11) in Chapter 9]; -15 MeV for the second anticaber [one $S\bar{3}$ bond; see Eq.(9.13)]; and 0 MeV for the third anticaber [see the discussion of this point in connection with Eqs. (9.15)-(9.17) in Chapter 9]. If we finally include the coulomb corrections which occur when the spinless anticabers are charged (3^- or 4^-), we obtain the calculated hyperon mass values that are shown in Table 19.3.

In discussing the mass calculations shown in Table 19.3, consider first the Λ and Σ^+ hyperons. In the calculation of the Λ mass, which agrees very accurately with the experimental value, all of the binding energies and mass values, with the exception of the -30 MeV hadronic binding energy, have already been determined by the considerations discussed above. Thus, from the standpoint of the Λ mass calculation, we should possibly regard this hadronic binding energy of -30 MeV as an adjustable parameter. But then the calculation of the Σ^+ mass becomes a calculation which involves no adjustable parameters, and the calculated Σ^+ mass is also in very close agreement with experiment. The key point here is that the number of mass values we are calculating exceeds the number of adjustable parameters, so that the good agreement between the mass calculations and the experimental mass values suggests that these parameters must have physical significance.

Now consider the Σ^+, Σ^o, and Σ^- hyperons. From the arguments leading to Eq. (19.13), we concluded that the Σ^o hyperon must have the same $\hat{p}$-type spinor triplet core as the Σ^+ hyperon, and Table 19.3 shows that these assignments accurately reproduce the $\Sigma^o - \Sigma^+$ mass difference. However, the Σ^- hyperon, from its recently-measured magnetic moment of -1.48 n. m. [Eq. (19.7)], has an $\hat{n}$-type spinor triplet core. With this $\hat{n}$ spinor core for the Σ^-, we reproduce the absolute mass of the Σ^- quite accurately (to within 0.4%), but we do not reproduce the $\Sigma^- - \Sigma^o$ mass difference. However, if we compare the Σ^-, Ξ^-, and Ω^- mass calculations shown in Table 19.3, which are all based on $\hat{n}$-type spinor triplets (the Ω spinor triplet is an $\hat{n}$ with all spins aligned), we see that they are all too low by a few MeV. Thus the hadronic binding to an $\hat{n}$ spinor triplet may be somewhat different than to a $\hat{p}$ or $\hat{\Lambda}$ spinor triplet.

The Ξ^- hyperon shown in Table 19.3 clearly has an $\hat{n}$ spinor triplet core, since it has a magnetic moment of -1.85 n. m. [Eq. (19.7)]. However, the magnetic moment of the Ξ^o hyperon has not been measured. Table 19.3 shows calculated mass values for the Ξ^o which correspond to the spinor triplets $\hat{p}$, $\hat{\Lambda}$, and $\hat{n}$; the $\hat{\Lambda}$ triplet core reproduces the $\Xi^- - \Xi^o$ mass difference most accurately, but it is difficult to choose between these three possibilities just on the basis of the calculated mass values.

Table 19.3. Calculation of absolute mass values for the Λ, Σ^+, Σ^0, Σ^-, Ξ^0, Ξ^-, and Ω^- hyperons and the K^-p bound state. Three different choices for the Ξ^0 light quark configuration are shown in the table; in the absence of a measurement of the Ξ^0 magnetic moment, a choice among these three configurations cannot be made with any confidence. The baryon absolute mass calculations shown in Tables 19.2 and 19.3 have an overall average accuracy of about 0.1%, and thus are comparable in accuracy to the meson mass calculations shown in Table 19.1.

Hadronic state	Configuration	Constituent masses (MeV)	Binding energies: Coul.	Binding energies: Hadr.	Final mass values (MeV): Calc.	Final mass values (MeV): Exper.	Final mass values (MeV): Error
Λ	$\hat{\Lambda}3^-$	933.1 + 214.6	-1.8	-30.0	1115.9	1115.6	+0.3
Σ^+	$\hat{p}\bar{4}^0$	939.6 + 280.0		-30.0	1189.6	1189.4	+0.2
Σ^0	$\hat{p}4^-$	939.6 + 284.6	-1.6	-30.0	1192.6	1192.5	+0.1
Σ^-	$\hat{n}4^-$	939.6 + 284.6	-1.2	-30.0	1193.0	1197.4	-4.4
Ξ^0	$\hat{p}3^-\bar{3}^0$	939.6 + 424.6	-1.6	-45.0	1317.6	1314.9	+2.7
	$\hat{\Lambda}3^-\bar{3}^0$	933.1 + 424.6	-1.8	-45.0	1310.9	1314.9	-4.0
	$\hat{n}\bar{3}^0\bar{3}^0$	939.6 + 420.0		-45.0	1314.6	1314.9	-0.3
Ξ^-	$\hat{n}3^-\bar{3}^0$	939.6 + 424.6	-1.2	-45.0	1318.0	1321.3	-3.3
Ω^-	$\hat{\Omega}3^-\bar{4}^0\bar{4}^0$	935.6 + 774.6	-1.2	-45.0	1664.0	1672.2	-8.2
K^-p	$\hat{p}7^-$	939.6 + 493.7	-1.6	-30.0	1401.7	1402.4	-0.7

AVERAGE ABSOLUTE MASS ACCURACY FOR TABLE 19.3: ± 2.4 MeV .

AVERAGE PERCENTAGE MASS ACCURACY FOR TABLE 19.3: ± 0.18% .

The last two hyperon states shown in Table 19.3 are the Ω^- hyperon and the K^-p bound state [which is manifested as the $\Lambda(1402.4)$ resonance]. In the case of the Ω^-, we have used an ad hoc hadronic binding energy of -45 MeV rather than a value of -60 MeV that we might have more logically selected; thus we should only claim to have calculated the Ω^- mass to within an accuracy of about 20 MeV. The calculated K^-p bound state mass, on the other hand, is within 1 MeV of the experimental mass value, and this calculation features no adjustable parameters of any kind.

The important point to note with respect to the mass calculations of Table 19.3 is that the same absolute mass accuracy which was exhibited in Tables 19.1 and 19.2 is also exhibited here, and with the same general binding energy systematics. Thus the present light-quark model is capable of reproducing all of the fundamental low-mass meson and baryon states to an absolute mass accuracy of a few tenths of a percent, and in a self-consistent manner. However, it is not readily apparent from the systematics of Table 19.3 just why the $\hat{p}$, $\hat{n}$, $\hat{\Lambda}$, and $\hat{\Omega}$ spinor triplets occur in the manner that they do, or why the various hyperon states have the isotopic spins that they do. Hence, although this light-quark model does an excellent job in reproducing hyperon mass values, spins, magnetic moment ratios, and strangeness quantum numbers, it does not at this stage of its development qualify as a predictive model for these hyperon states.

19.8 The New Particles

In the preceding sections of this chapter, we have described the accurate manner in which the basic meson and baryon resonance mass values can be reproduced once we have properly taken into account the 4% hadronic binding energy and the $\sim$1% coulomb and magnetic binding energies (see Tables 19.1 - 19.3). In the present section, we briefly recapitulate some mass values that were obtained for the New Particle resonances of Chapter 17. These New Particle mass values are of interest here for three reasons: (1) they have calculated mass accuracies that are comparable to the examples cited above; (2) these accurately-calculated masses are obtained without applying any corrections for binding energies; and (3) these New Particle states were discovered after the light-quark basis-state mass values and binding energies had already been formulated and published.

The New Particle states that we consider here are the six states shown in Table 19.4. All six of these states are formed by adding either spinless excitation units in multiples of the mass quantum M = 70 MeV, or else spinor quartets $\dot{S}\dot{S}\dot{S}\dot{S}$ = 1310 MeV, to the fundamental 1795 MeV $\bar{N}N$ New Particle ground state. As is demonstrated in Table 19.4, we can, by simply adding together the masses of the $\bar{N}N$ bound state

Table 19.4. Calculation of absolute mass values for some New Particle resonances. In contrast to the meson and baryon states in Tables 19.1 - 19.3, which feature 4% hadronic binding energies, these New Particle states are formed by combining clusters with ∿0% binding energies. As can be seen, the mass accuracies that are obtained here are again at the level of ±0.1%.

Light-quark configuration	Calculated mass values (MeV)	New Particle state	Error (MeV)
$\bar{N}N\cdot M$	1794.5 + 70.0 = 1864.5	D^{o}(1863.3)	+1.2
$\bar{N}N\cdot\pi$	1794.5 + 140.0 = 1934.5	S(1935)	-0.5
$\bar{N}N\cdot 3$	1794.5 + 210.0 = 2004.5	D^{*o}(2006)	-1.5
$\bar{N}N\cdot\dot{S}\dot{S}\dot{S}\dot{S}$	1794.5 + 1310.0 = 3104.5	J/ψ(3098)	+6.5
$\bar{N}N\cdot(333)^3$	1794.5 + 1890.0 = 3684.5	ψ'(3684)	+0.5
$\bar{N}N\cdot(\dot{S}\dot{S}\dot{S}\dot{S})^2$	1794.5 + 2620.0 = 4414.5	ψ(4414)	+0.5

AVERAGE ABSOLUTE MASS ACCURACY FOR TABLE 19.4: ± 1.8 MeV .

AVERAGE PERCENTAGE MASS ACCURACY FOR TABLE 19.4: ± 0.07% .

and its associated excitation quanta, reproduce the masses of these six fundamental New Particle states to an accuracy of a couple of tenths of a percent; the zero binding energy between these light-quark basis states follows essentially from the fact that we are dealing with nonstrange excitations. (However, the "charmed" meson states D^o and D^{*o} shown in Table 19.4 each involve a strange subquantum M.)

These New Particle results were discussed in detail in Chapter 17. We have repeated them here in order to demonstrate the universal manner in which the same few light-quark basis states can be used to accurately reproduce the masses of all of the stablest, and therefore most fundamental, hadron resonances.

19.9 The Hadron Rotational Bands

In the preceding sections of this chapter we have discussed the calculation of mass values for the very-narrow-width meson and baryon resonances. In order to complete this survey of mass calculations, we should extend it to include the narrow-width S-state meson and baryon excitations and the broad-width rotational excitations. The mass values of the S-states, which constitute the rotors or bandheads for the rotational levels (see Chapters 15 and 16), are obtained by adding 70 MeV excitation quanta to the fundamental meson and baryon ground states $\dot{S}\dot{S}(659)$, N(939), $\Lambda(1116)$ and $\Sigma(1193)$. Thus we can evaluate the "mass accuracies" of the S-states by calculating the deviations of their mass values from 70 MeV mass grids that are erected above the masses of the corresponding ground states. In order to evaluate the mass accuracies of the rotational levels, we first obtain a universal moment-of-inertia curve (or, rather, the inverse of this curve) by combining all of the experimental E_{rot} values [Eq. (12.13)] for mesons and baryons in one graph, and then drawing a curve through the midpoints of these values; this universal E_{rot} curve is shown in Fig. 19.4. Next we insert the E_{rot} values from this universal curve together with the experimental S-state bandhead masses into Eq. (12.13), and thereby calculate the expected energies for the rotational levels.

The result of these calculations are summarized in Table 19.5. The 24 meson and baryon S-states that are included in Table 19.5 are on the average within 13 MeV of their 70 MeV-mass-grid values, and the average absolute mass accuracy for these states is at the 1% level. The 74 rotational levels included in Table 19.5 (see Tables H1 - H5 in Appendix H) are on the average within 23 MeV of their calculated values, and they have an average absolute mass accuracy of about 1.5%. Of course, many of these states are not well-pinned-down experimentally, so that the precise values we quote here for the mass accuracies are not really significant. However, the important point to note is that the average mass accuracy for all 98 of the meson and baryon states summarized in Table 19.5 is in the range of a few percent. Thus

Table 19.5. Calculation of mass values for 24 meson and baryon S-states and for 74 meson and baryon rotational levels. The S-state masses are calculated by erecting 70 MeV mass grids above the fundamental "ground state" masses. The rotational masses are calculated by combining the known bandhead masses with rotational energies as calculated from the universal E_{rot} curve shown in Fig. 19.4. Although not as precisely-determined as the masses listed in Tables 19.1 – 19.4, the masses summarized here still have an average calculated absolute mass accuracy of better than 2%. Tables 19.1 – 19.5 describe calculated mass values for almost all of the observed meson and baryon resonances.

ROTATIONLESS S-STATE RESONANCES			
Mesons	- 14 levels in Fig. 16.2	-	$\overline{\Delta M} \simeq 13$ MeV
Baryons	- 10 levels in Tables H1-H4	-	$\overline{\Delta M} \simeq 13$ MeV
Total	- 24 S-states	-	$\overline{\Delta M} \simeq 13$ MeV
ROTATIONAL RESONANCES			
Mesons	- 21 levels in Table H5	-	$\overline{\Delta M} \simeq 15$ MeV
Baryons	- 53 levels in Tables H1-H4	-	$\overline{\Delta M} \simeq 26$ MeV
Total	- 74 rotational levels	-	$\overline{\Delta M} \simeq 23$ MeV

AVERAGE PERCENTAGE MASS ACCURACY FOR THE 24 S-states of Table 19.5: $\sim$ 1%.

AVERAGE PERCENTAGE MASS ACCURACY FOR THE 74 ROTATIONAL LEVELS OF TABLE 19.5: $\sim$ 1.5%.

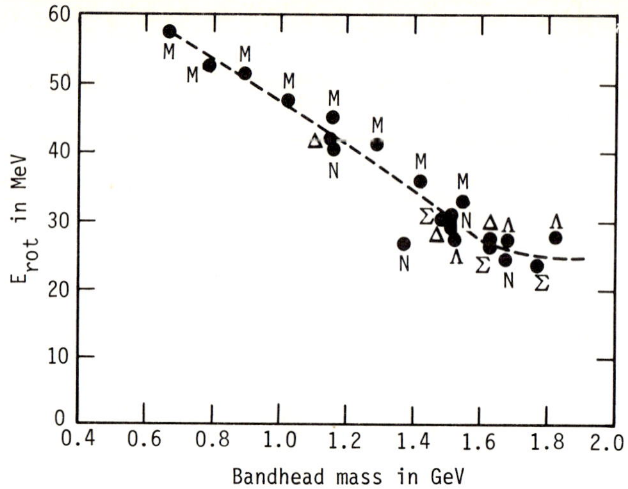

Figure 19.4. A universal E_{rot} curve (the dotted line). This curve is obtained as a rough average of the experimental meson (M) and baryon (N,Δ,Λ,Σ) values for E_{rot} shown in the figure. By using this universal curve plus the known bandhead masses, we can calculate the expected energies for the meson and baryon rotational levels. This parameter-free calculation gives the masses of 74 meson and baryon rotational levels to an average accuracy of better than 2% (±23 MeV), as summarized in Table 19.5. The significance of this result lies in the fact that the mass accuracies obtained for these same levels from SU(3) are typically on the order of 30% (±500 MeV).

the mass values for these broad-width states that are obtained from the light-quark model again represent an order-of-magnitude improvement over the mass values for these states which have emerged from calculations based on SU(3); the SU(3) mass calculations for these states are typically in error by about 30%[5].

19.10 Concluding Comments

With the addition of electromagnetic effects, we bring to a close our general study of the hadron resonances. What we have obtained from these studies is a light-quark set of basis states - 1,3,4,7,S — which are capable of reproducing the spectroscopic properties of the 185 or so hadron states that have been reported to date — including the New Particle states. This light-quark model is in a certain sense the opposite of the quark model of Gell-Mann and Zweig. The G-Z quarks have well-defined charges, but they do not have well-defined masses; accordingly, the G-Z quark model strikingly reproduces the observed meson and baryon charge states, but it does not work so well for the masses. The present light-quark basis states, on the other hand, have well-defined masses, but they do not (at least in the present formulation) have charge states that are completely delineated; accordingly, the present light-quark model gives very striking mass values, and it can reproduce the observed charge states, but it does not have the predictive properties of the G-Z model.

The predicted properties of the Gell-Mann and Zweig quark model, it should be noted, carry an associated responsibility: the predictions must turn out to be correct. In this regard, one of the most serious tests of the G-Z model, and of SU(3)-type symmetries in general, is that of correctly reproducing the magnetic moments of the baryons. Let us examine the calculation of baryon magnetic moments in the context of the Gell-Mann and Zweig quark model. If we start with the Gell-Mann and Zweig assumption that quark magnetic moments are proportional to quark charges, we are then faced with the problem of selecting quark wave functions that specify the averages to be taken over the quark spin orientations, since the G-Z quarks cannot reproduce the baryon magnetic moments if fixed spin orientations are assumed for the quarks. However, SU(6) provides these quark wave functions[6], and it leads to the predicted baryon magnetic moments[7] that are shown in Table 19.6, together with the experimental values[2] for these quantities. (Although the calculated values shown in Table 19.6 were originally derived on the basis of SU(6), they have also been obtained by other methods.) As can be seen in Table 19.6, the p, n, and Σ^+ magnetic moments are quite accurately reproduced, but the experimental value of the Λ magnetic moment is 35% smaller than its predicted value, and the Σ^- and Ξ^- absolute magnetic moments are about 50% to 100% larger than their predicted values (but with rather large experimental errors). It can be argued that the experimental Λ magnetic moment is smaller than the calculated value because higher-order correction terms are not included. But it must then be explained why it is that the Σ^- and Ξ^- magnetic moments are off in the other direction. Perhaps the most critical way of assessing the results shown in Table 19.6 is to note that SU(6) leads to the prediction that the Λ, Σ^- and Ξ^- hyperons all have the same magnetic moment, whereas the actual values are -0.61 ± 0.03 n.m., -1.48 ± 0.37 n.m., and -1.85 ± 0.75 n.m., respectively; these values represent a disagreement between SU(6) or the G-Z quark model and the experimental data which is well outside of the experimental errors. This is a critical result for SU(6) because the calculation of magnetic moments is one of the most difficult features to tamper with. As an example, a "broken

Table 19.6. Baryon and hyperon magnetic moments as calculated from SU(6) (Ref. 7) and as measured experimentally (Ref. 2). The calculated magnetic moments for the p, n, and Σ^+ are in agreement with experiment, but the calculated magnetic moments for the Λ, Σ^-, and Ξ^- are all outside of the experimental errors. This difficulty with the SU(6) calculation for the Λ, Σ^-, and Ξ^- magnetic moments is compounded by the fact that the SU(6) calculation is too high (in absolute magnitude) for the Λ, and too low for the Σ^- and Ξ^-, and no convincing theoretical argument has been advanced to explain this discrepancy. The recent increase in the "world average" value for the Σ^+ magnetic moment (Ref. 2) has intensified the discrepancy between the SU(6) calculations and experiment; and the Σ^+ magnetic moment value of 2.95 ± 0.31 n.m. reported by Doble et al. in Ref. 2 is even higher than the new "world average" that we have listed in Table 19.6.

Particle	Magnetic Moment Calculated from SU(6)	Experimental Value (in N.M.)
p	+2.79	+2.793
n	-1.86	-1.913
Λ	-0.93	-0.606 ± 0.034[a]
Σ^+	+2.79	+2.83 ± 0.25
Σ^0	+0.93	
Σ^-	-0.93	-1.48 ± 0.37
Ξ^0	-1.86	
Ξ^-	-0.93	-1.85 ± 0.75

[a]The new "world average" shown here for the Λ magnetic moment is from RPP78 (see Ref. 10), and it compounds the SU(6) difficulties with magnetic moment calculations even more. It seems clear now that the discrepancy between the SU(6) calculations and the experimental baryon magnetic moment data is beyond any possible range of experimental errors.

SU(6)" calculation is discussed by Feld[8] in which the three quark magnetic moments are arbitrarily adjusted to give a precision fit to the magnetic moment of the Λ hyperon; however, this calculation lowers the predicted Ξ^- magnetic moment from -0.93 n.m. to -0.65 n.m., so that it is in even more serious disagreement with experiment. Another modification, based on the use of Lagrangian field theory, was reported by Rédei[9], but his value for the Ξ^- magnetic moment (which was not known at that time) is even more in error than the "broken SU(6)" value. No satisfactory reason has yet been set forth which explains the inability of the Gell-Mann and Zweig quark model to provide accurate values for the baryon magnetic moments. Since a single well-established disagreement in the magnetic moment calculations is sufficient to invalidate the model, it seems that a reduction in the error limits for the magnetic moment of the Ξ^- should be a high priority item for elementary particle physicists.

In the next chapter, the last one in the book, we turn from the domain of the elementary particle to that of the atomic nucleus, and we attempt to make at least a partial repayment for the many concepts that we have borrowed from the nuclear physicists.

References for Chapter 19

1. See B. T. Feld, Models of Elementary Particles, Blaisdell, Waltham (1969), page 339, Eq. 18a'.
2. T. G. Trippe et al., Review of Particle Properties, Rev. Mod. Phys. 48, No. 2 Part II, April 1976. A more recent value for the Σ^- magnetic moment is quoted in G. Dugan et al., Nucl. Phys. A254, 396 (1975), which is in essential agreement with the value listed here. The value listed here for the Σ^+ magnetic moment is a new "world average," including new data, that is reported by N. Doble et al., Phys. Lett. 67B, 483 (1977); these authors point out that this large value, when combined with the values for the Λ and p magnetic moments, raises some real problems with the SU(3) calculations. (See Ref. 10 for the Λ magnetic moment.)
3. See M. H. MacGregor, Nuovo Cimento 36A, 113 (1976).
4. A Zee, Physics Reports 3C, 127 (1972); see the references that are cited in this review article on the problem of the n-p mass difference.
5. J. L. Rosner, Physics Reports 11C, 189 (1974); see page 219.
6. M. Beg et al., Phys. Rev. Lett. 13, 514 (1964).
7. See Table 15.10 on page 340 of Ref. 1.

8. See Table 15.11 on page 341 of Ref. 1.

9. L. B. Rédei, Phys. Rev. D2, 239 (1970).

10. The Λ magnetic moment value listed in Table 19.6 is the new "world average" as given in RPP78. It includes recent measurements by G. Bunce et al., Phys. Rev. Lett. 36, 1113 (1976), and by K.Heller et al., Phys. Lett. 68B, 480 (1977).

CHAPTER 20. IMPLICATIONS FOR NUCLEAR PHYSICS

> "The existence of large quadrupole moments and rotational spectra in certain nuclei forces one to the conclusion that these nuclei are appreciably aspherical. However, there is still the problem of <u>why</u> these nuclei acquire a non-spherical, and, in particular, an axially symmetric shape. Since the nuclear binding is due to forces between the nucleons themselves, it is clear that the nuclear shape must depend in an important way on the nucleon configuration, i.e., on the intrinsic structure."
>
> S. A. Moszkowski, Handbüch der Physik, Vol. 34, p. 498 (1957)

20.1 The Alpha Particle

In studying the properties of the hadronic elementary particles, we made use of some of the concepts of nuclear physics – for example, in the discussions of angular momentum barriers (Chapter 11) and rotational systematics (Chapters 12 and 13). We can now work in the other direction, and apply some of the information that we have learned about <u>nucleon</u> structure to the problem of <u>nuclear</u> structure. If the nucleus is constructed as a collection of closely-spaced nucleons, then it seems logical that the geometry of the nucleon should influence the geometry of the nucleus, as is suggested in the quotation by Moszkowski cited above. In particular, the wave aspects and the particle aspects of the atomic nucleus are of comparable dimensions, as we discussed in the Prologue to the present book, so that the nucleon geometry we deduced in Chapter 18 should have a substantial effect on the nuclear wave functions.

Protons and neutrons are the constituents of atomic nuclei, but the manner in which they combine together inside atomic nuclei seems to be mainly in the form of alpha particles or alpha-like nucleon correlations. This conclusion is based on the fact that a large and growing body of experimental evidence indicates that α-particle clusters are a prominent feature of the nuclear surface[1] and of the nuclear interior.[2] Thus, in applying nucleon structure to nuclear structure, our first task is to deduce the geometry of the α particle.

One of the rather surprising features about the geometry of the nucleon which emerged from Chapter 18 was the fact that the nucleon is composed of three cylindrically-shaped quarks which bind together in a side-by-side manner (Figs. 18.12 and

19.2). Furthermore, a nucleon and an antinucleon bind together in a side-by-side manner (Fig. 18.12), and the basic hyperon resonances are formed by adding spinless excitation quanta to a nucleon in a side-by-side manner (Fig. 18.13). Does this same result carry over to the structure of the α particle?

Conventionally, the spin zero α particle is pictured as a spherical object. However, the spin zero U^{238} nucleus was similarly thought to be a spherical object until measurements revealed that it has a markedly prolate shape. We can learn something about the geometry of the α particle by studying its binding energy. Empirically, the 4% hadronic binding energy which operates between particles and antiparticles, and between quarks and antiquarks, does not operate between the protons and neutrons in an atomic nucleus, since the proton and neutron are both particle states. Thus, if nuclear binding energies are related to the hadronic binding energies that we developed in the preceding chapters, the nuclear forces must be purely electromagnetic – that is, made up of the magnetic and coulomb binding energies which were defined in Chapter 19. One of the best tests of this assumption comes from the strongly-bound α particle, whose binding energy of 28 MeV has not been otherwise explained theoretically. Suppose we assume that the proton and neutron have the configurations that are shown in Fig. 19.2 and defined in Eqs. (19.17) and (19.18). Suppose further that the protons and neutrons which are inside of atomic nuclei have the coulomb and magnetic binding energies that are defined in Eqs. (19.8) and (19.15), respectively. If this is true, then, since the net <u>coulomb</u> force inside the α particle is both small and repulsive, the large attractive force that corresponds to the 28 MeV binding energy of the α particle must be supplied by <u>magnetic</u> interactions, which suggests that the two protons and two neutrons in the α particle must be bound together in a symmetric side-by-side manner.

To reproduce the α particle, we first write the proton and neutron quark configurations [see Eqs. (19.17) and (19.18)] as

$$p = S^{+}_{\uparrow}S^{-}_{\downarrow}S^{+}_{\uparrow} \qquad \text{and} \qquad n = S^{+}_{\uparrow}\begin{matrix} S^{=}_{\downarrow} \\ S^{+}_{\downarrow} \end{matrix} \quad , \tag{20.1}$$

which have the magnetic polarities

$$p = \mathrm{N\ N\ N} \qquad \text{and} \qquad n = \mathrm{N}\begin{matrix} \mathrm{NN} \\ \\ \mathrm{S} \end{matrix} \quad , \tag{20.2}$$

where N ≡ north and S ≡ south denote the magnetic dipole polarities as viewed (for example) from above, and where N and NN denote singly-charged and doubly-charged quarks, respectively. The α particle is contructed as the magnetic array

$$\alpha = \begin{array}{cccccccc} N & & N & & N & & & \\ & S & & SS & & S & & \\ & & N & & NN & & N & \\ & & & S & & S & & S \end{array} \quad . \tag{20.3}$$

This magnetic array contains six extranucleon pairs of the type N-S, two (net) pairs of the type NN-S, and one pair of the type NN-SS. If we assume that N-S = -2 MeV, NN-S = -4 MeV, and NN-SS = -8 MeV,[3] we then obtain -28 MeV as the calculated binding energy of the α particle. Adding in +1 MeV for the coulomb repulsion,[4] we obtain a net α particle binding energy of -27 MeV, which compares to the experimental α binding energy of -28 MeV. The assumptions made here are crude enough that this numerical agreement should not be taken too seriously. However, the following points should be noted: (1) the magnetic binding energy assumed here (±2 MeV per N-S pair) was obtained empirically [Eq. (19.15)] and is not an adjustable parameter; (2) this magnetic interaction is of the proper magnitude to bind the α particle, which is the crucial test for any proposed model of the nuclear force; (3) the magnetic interaction has the same tensor character as the tensor component which is observed in the nuclear force. This light-quark model of the α particle is shown in Fig. 20.1.

Figure 20.1 A light-quark configuration for the alpha particle. This twelve-quark cluster contains two protons and two neutrons in symmetric spin-up spin-down configurations. The quark electromagnetic binding energies that were deduced for baryons in Chapter 19 also reproduce the binding energy of the alpha particle (as an essentially magnetic effect). Furthermore, from the magnetic layering that occurs in the alpha particle [see Eq. (20.3)], α particles will tend to bind together "end-to-end" and in layers.

As we have just demonstrated, we are able to reproduce the binding energy of the α particle by using nucleon geometries (Chapter 18) and nucleon electromagnetic effects (Chapter 19) which were deduced on the basis of completely independent considerations. If we now combine α particles together so as to form an atomic nucleus, there are three key phenomenological features which emerge: (1) the α particle that is shown in Fig. 20.1 is not spherical, but is in the form of a "hexagonal brick"; (2) from the magnetic layering of the quarks that is shown in Eq. (20.3) and Fig. 20.1, α particles will bind together "end-to-end" in the form of two-dimensional magnetically-bound (Ising) layers,[5] and these layers will in turn bind together magnetically to form the nucleus; (3) since the "height" of a nucleon is two fermis (see Figs. 18.1 and 19.2), the thickness of a nucleon layer is two fermis, and we end up with a nucleus that is built up of a series of two-fermi-thick layers. As we will see in the next section, this third feature is precisely the result which is needed in order to account quantitatively for the observed properties of the fission process in large atomic nuclei.

Linus Pauling has long maintained[6] that the α particle is the main building block of atomic nuclei. We agree in essence with his conclusion, but we find that the α particles are arrayed with two-dimensional axial symmetry rather than with the three-dimensional spherical symmetry of the Pauling model. Since the nucleons in these two-dimensional layers are bound together magnetically, and since the magnetic axes of the nucleons are oriented perpendicular to the plane of the layers, these nucleon layers are closely analogous to the two-dimensional Ising layers which occur in solid state physics.[7] Hence α particles bind together in the form of two-dimensional "nuclear Ising layers", whose properties we now discuss.

20.2 Ising Layers and the Fission Process

One of the most remarkable aspects of nuclear physics is the asymmetric fission of the uranium nucleus. Heavy atomic nuclei have such powerful internal coulomb repulsion that they are just barely able to hold themselves together. Thus, when they are perturbed, or sometimes even spontaneously, they break apart into two separate pieces. Now, the interesting fact about a nucleus such as U^{238} is that when it fissions, it does not separate into two equal-sized parts; rather, it separates into a "heavy fragment" and "light fragment", with the heavy fragment containing on the average 42 more nucleons than the light fragment. The occurrence of this asymmetric fission process suggests that U^{238} has some kind of asymmetric internal structure, so that when it separates, it cleaves or scissions along an asymmetrically-located internal "fault line". Since the U^{238} nucleus is known to be in the shape of an axially-symmetric prolate spheroid, with a length of about 18 fm and a width at the girth of about 14 fm, one way to introduce some kind of a structured interior into this nucleus is to postulate that it has a series of internal "layers" which are arrayed perpendicular to the symmetry axis of the nucleus. The fission process then occurs as a fissure which opens up and propagates along the interface of two of these layers. This fission process is most likely to occur at the central layers, where the coulomb forces are the strongest; and, if the nucleus contains an odd number of layers, there will be a heavy fragment which contains one more layer than the light fragment. But if this picture of U^{238} fission is correct, then we can calculate exactly how thick one of these layers must be: the middle layer of the U^{238} nucleus contains 42 out of the 238 nucleons in the nucleus, and a straightforward volume calculation, using the known overall dimensions of the nucleus, shows that this layer must be precisely two fermis thick! This is the same thickness that we ascribed to the alpha particle in Sec. 20.1, which came in turn from the "height" of a nucleon in Chapters 18 and 19. Hence, from two completely independent arguments, we have deduced the existence of two-fermi-thick layers in the nucleus; these are formed from alpha particles bound together in a side-by-side manner (with the excess neutrons, as we will see in Sec. 20.5, wrapped around the outside of the nucleus in the form of a "neutron halo").

The picture we thus obtain for the U^{238} nucleus is that of a prolate spheroid formed from nine two-fermi-thick Ising layers arrayed perpendicular to the symmetry axis of the nucleus. This picture is shown schematically in Fig. 20.2. The magnetic forces which bind these nine Ising layers together in U^{238} are much shorter-ranged than the coulomb forces which tend to force them apart. Thus a "tear" that develops at an interface will propagate along the interface and lead to the scission of the nucleus. It is possible to calculate the coulomb forces which occur between the interfaces in Fig. 20.2, and also to estimate the relative magnitudes of the attractive magnetic forces that appear across these interfaces; these values are shown in Fig. 20.3. As can be seen in Fig. 20.3, the nucleus is most unstable at the central interfaces, and these are the interfaces where fissions usually occur. However, when uranium and bismuth nuclei are bombarded with high energy protons,[8] tantalum isotopes are produced; these correspond to fissions that take place at Ising interfaces which are one layer removed from the central interfaces in these nuclei. Thus the fission process can be induced at interfaces other than the central interface if the incident energy is large enough and localized enough.

If the asymmetric fission of uranium actually occurs in the manner that we have just described, then we can deduce an immediate consequence which follows from this laminar picture of the nucleus. If the uranium nucleus consists of <u>nine</u> two-fermi-thick Ising layers, and if the central layers each contain around 40 nucleons (Fig. 20.2), then when we move down in atomic mass A by about 20 nucleons, we should find nuclei which have <u>eight</u> Ising layers and hence fission symmetrically; and if we move up in A by about 20 nucleons, we should find nuclei which have <u>ten</u> Ising layers

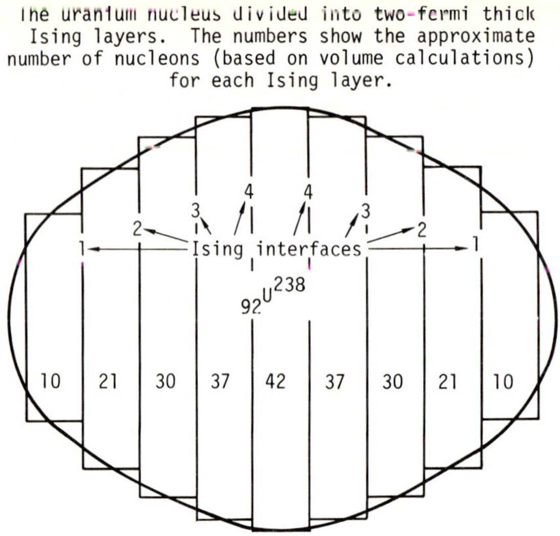

Figure 20.2. The uranium nucleus, pictured as a cluster of nine two-fermi-thick magnetically-bound layers of nucleons. The thickness of a layer comes from the two-fermi "height" of the nucleon (see Fig. 19.2). These magnetically-bound layers are in many respects similar to the "Ising" layers that occur in solid state physics, so that we can refer to them as nuclear Ising layers. From straightforward volume calculations, we can ascertain that the central Ising layer in U^{238} contains just about 42 nucleons, where the overall shape of the uranium nucleus shown in the figure is taken from a computer-generated plot that uses experimental deformation parameters. Fission occurs at a central Ising interface (see Fig. 20.3), so that the uranium nucleus divides into two unequal masses, with the "heavy fragment" having about 42 more nucleons than the "light fragment". This is precisely the result that is observed experimentally. Nuclei that are somewhat lighter (heavier) than uranium contain eight (ten) Ising layers, and hence fission symmetrically. This result is also observed experimentally. (Although we have given here a very literal "particle" description of the two-dimensional nucleon layers inside the nucleus, this nucleon layering can perhaps more realistically be thought of as a distortion of the nuclear wave functions caused by the nucleon-nucleon interactions inside the nucleus: cylindrically-shaped nucleons with side-by-side interactions tend to cluster in a manner that leads to laminar correlations in the nuclear wave functions.)

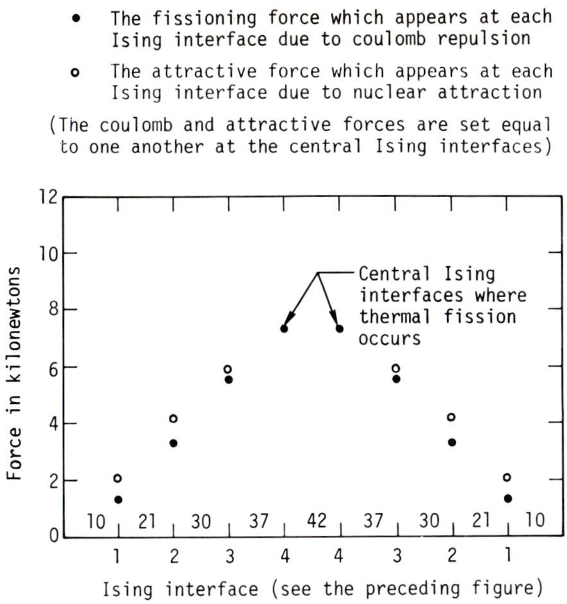

Figure 20.3. The forces which occur at the Ising layer interfaces in the uranium nucleus of Fig. 20.2. The fissioning force is due to coulomb repulsion, and it can be rather accurately calculated. The attractive force is due to magnetic interactions, and its relative magnitude can be estimated at each interface. By setting these two forces equal to one another at the central interfaces, we can determine that the outer interfaces are more stable than the inner interfaces. Thus thermal-neutron-induced or spontaneous fission will occur at a central Ising interface. Since a small perturbation affects the short-ranged attractive magnetic force more than it does the long-ranged repulsive coulomb force, a "tear" that develops at an interface propagates across the interface and results in the scission of the nucleus. The total fission energy of a heavy nucleus is essentially just the coulomb potential energy that is liberated in the fission process. The very large atomic nuclei all have prolate shapes in order to minimize the coulomb repulsion that appears across their central Ising interfaces.

and hence also fission symmetrically. This is precisely the situation which occurs experimentally: the nuclei A_u, Tl, Pb, and Bi, which fall in the range A = 197 to A = 209, all fission symmetrically; the nucleus $_{88}Ra^{226}$ fissions both symmetrically and asymmetrically (it is on the borderline between eight and nine layers); the nuclei from $_{90}Th^{230}$ to $_{100}Fm^{256}$ fission asymmetrically; the nucleus $_{100}Fm^{257}$ fissions asymmetrically for spontaneous fission and symmetrically for neutron-induced fission (it is on the borderline between nine and ten layers); and the nucleus $_{100}Fm^{258}$ fissions symmetrically. Thus the systematics of the fission process appears to be in quantitative experimental agreement with this concept of nuclear Ising layers.

One point to be noted here is that experimentally the mass distributions which correspond to symmetric fission are radically different from the mass distributions which correspond to asymmetric fission. Thus asymmetric fission does not gradually merge into symmetric fission as the atomic mass A = 257 is approached from below; fission symmetry is an either/or process. This is the result that we expect from the Ising layer point of view, but it is not a result which is to be expected from a liquid-drop type of potential-model approach.

In a recent experiment,[9] the element 105^{262} was observed to fission asymmetrically. Now, the atomic mass A = 262 is very close to the atomic mass A = 258 of the symmetrically-fissioning Fm^{258} nucleus, which seems to raise a question as to our Ising-layer interpretation of these events (since A = 257 is the nearby borderline between asymmetric and symmetric fission with increasing A). However, in order for a nucleus with an atomic number Z as large as Z = 105 to exist, it probably has to be in a very elongated geometry in order to minimize the repulsive coulomb potential. Thus the 105^{262} nucleus may contain <u>eleven</u> Ising layers. But more examples must be obtained before we can clarify the fission systematics of these very-high-Z nuclei.

There is one other fission result which follows from this Ising picture of the nucleus. It can be deduced (see Sec. 20.4) that the binding energies between α particles in an Ising layer are roughly 10 MeV. Thus a projectile which strikes a nucleus with sufficient energy will shatter these Ising layers, so that the nucleus will fission symmetrically. Hence we expect all atomic nuclei to fission symmetrically under high-energy bombardment, which is the situation that is observed experimentally.

In the present section we have used the systematics of the fission process as evidence for the existence of two-dimensional nucleon layers in the nucleus. In the next section we briefly describe some topological arguments which suggest this same conclusion.

20.3 The Topological Argument for Two-Dimensional Nuclear Layers

It has been determined experimentally that atomic nuclei occur in a variety of shapes, some of which are illustrated in the computer-generated plots[10] shown in Fig. 20.4. While some nuclei such $_{14}Si^{28}$ are oblate, the great majority of the

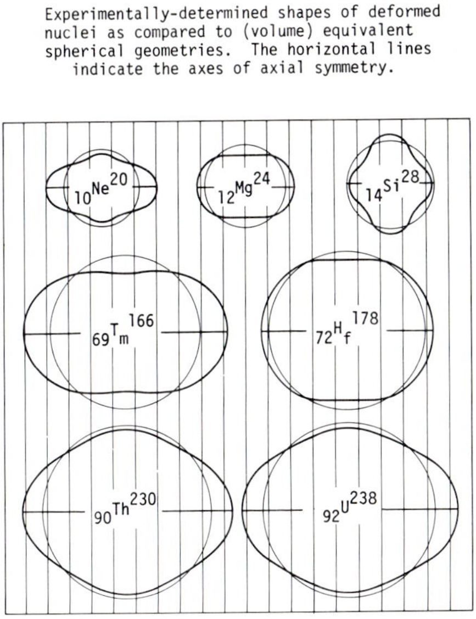

Figure 20.4 Computer-generated plots of some deformed nuclei, using experimental deformation parameters (Ref. 10). The circles denote volume-equivalent circular geometries for these nuclei. The vertical lines are spaced two fermis apart to suggest the appearance of Ising layers in these nuclei.

aspherical nuclei are prolate, and most of them appear to be axially symmetric. As Moszkowski has stressed in the quotation cited at the beginning of this chapter, the axial symmetry of the nucleus must be intimately related to the structures of the nucleon configurations inside the nucleus.

If a mathematical topologist were to be questioned as to the essential topological significance of a system of objects which exhibit axial symmetry (about the z axis, to be specific), he would reply that the system has one type of symmetry which is common to the x and y axes, and another type of symmetry with respect the z axis. Now one obvious topological shape which has these symmetry properties is a two-dimensional layer extending in the x and y directions. Thus the topological argument leads, though not inevitably, to the suggestion that these axially symmetric nuclei are composed of layers which are arrayed perpendicular to the z-axis of symmetry. In Fig. 20.4 the vertical grid lines, which are spaced two fermis apart (the thickness of a nucleon layer), are included so as to illustrate this topologically-suggested layering.

We can extend this line of topological reasoning and use it to resolve a paradox about atomic nuclei. The paradox is the following. From the successes of the liquid drop model in reproducing many observed features of nuclei, we know that the nucleus does behave in some respects as a rather mobile liquid. On the other hand, the existence of asymmetric low-energy fission suggests that the nucleus has a structured crystalline-like interior which tends to break up into unequal parts. To have both liquid and crystalline properties occurring in the same nucleus seems paradoxical. But this is where the topological argument comes to the rescue. The axially symmetric shapes (about the z axis) for nuclei suggest that the x,y symmetry is different from the z symmetry, as was discussed above. It can also follow that the structural fluidity of the nucleons is different in the x and y directions than it is in the z direction. In particular, if nucleons are bound together side-by-side into two-dimensional layers, then we can picture nucleon motion, nuclear surface waves, vibrational motion, and so forth as taking place within a two-dimensional layer and yet not strongly propagating from one layer to another, at least within certain time scales. Thus nuclear matter, if it is formed into two-dimensional layers, can behave as a fluid in the x and y directions and at the same time as a crystalline material in the z direction, which is precisely the property we require in order to account for these various facets of nuclear phenomenology. A situation which is somewhat analogous to this occurs in graphite, where the individual carbon clusters in the graphite are as hard as diamonds, but where the two-dimensional structure of these clusters enables them to slide freely past one another.

The topological argument for the existence of two-dimensional layering in the atomic nucleus is perhaps not as powerful as the argument based on fission systematics, but it is useful in that it suggests a bridge between the structureless liquid drop model of the nucleus and the more-structured nuclear shell model.

The existence of ordered layers in the nucleus indicates that nucleon clustering occurs not just at the surface of the nucleus, but also throughout the interior of the nucleus. We can investigate this idea by studying the binding energy systematics of atomic nuclei, which we do in the next section.

20.4 Binding Energy Systematics and Nucleon Clustering

Do nucleons cluster together in the interior of a nucleus? In order to answer this question, let us consider the simplest nuclear cluster model that we can devise: we will assume that (1) protons and neutrons combine together to form deuterons (D), He^3 nuclei, tritons (T), and He^4 nuclei (α particles); and (2) these D, He^3, T, and α clusters in turn combine together to form atomic nuclei. If this cluster picture has physical content, we should find a correlation between the sequential binding energies of atomic nuclei and the binding energies of the constituent internal clusters D, He^3, T, and α. The purpose of the present section is to demonstrate that this correlation does in fact exist.

In order to have a systematic way of building up atomic nuclei, we start with Z protons, and then add N neutrons — one at a time. As these N neutrons are added, they will combine with protons to form the clusters D, He^3, T, and α; these clusters have successively stronger binding energies, and the most strongly-bound cluster that can be formed in any situation is the one that is actually formed. Now, after a certain number of neutrons have been added, all of the Z protons will have combined with neutrons to form clusters, so that no further clustering can occur. It is useful to define this as the Cluster Saturation Point (CSP). When the proton number Z is even, the CSP occurs for N = Z; in this case all of the nucleons are formed into α particles. When Z is odd, the CSP occurs for N = Z + 1; in this case the nucleons are formed into α particles plus one triton. Thus we have the following definition for the Cluster Saturation Point:

$$\begin{aligned} \text{CSP} &\equiv (N = Z) \ , \qquad && Z \text{ even} \ ; \\ \text{CSP} &\equiv (N = Z + 1) \ , \qquad && Z \text{ odd} \ . \end{aligned} \tag{20.4}$$

If neutrons are added above the CSP (as actually occurs for all elements), we expect to find weaker binding energies, since no further strongly-bound clusters can be formed. Now let us examine the experimental data.

Figure 20.5 shows the total binding energies for the isotopes of the ten lightest elements, plotted as functions of the total atomic mass A of each isotope, with the isotope in each isotopic chain that corresponds to the CSP indicated by an arrow. As can be seen in Fig. 20.5, the binding energy curves, especially for the lightest elements, have points of inflection at the CSP, with the slopes above the CSP being considerably flatter than the slopes below the CSP. This indicates that the CSP has physical significance, so that nucleon clustering is occurring in these light atomic nuclei.

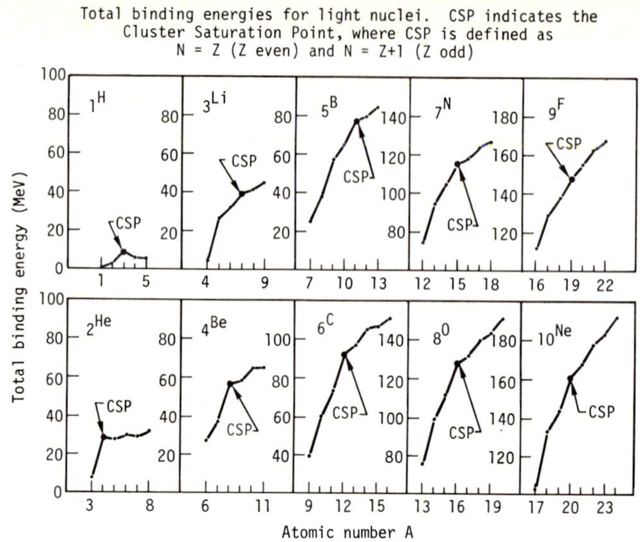

Figure 20.5. Total binding energies for light nuclei. If we assume that these nuclei are composed of D, T, He^3, and He^4 (alpha-particle) clusters, then the CSP denotes the Cluster Saturation Point [Eq. (20.4)] at which all of the available neutrons are combined into clusters. As can be seen in Fig. 20.5, there is a break in the binding energy curve at the CSP for each of these nuclei: neutrons added above the CSP are more weakly bound than are neutrons added below the CSP, which indicates that these internal clusters really do occur. These results are illustrated quantitatively in Fig. 20.6.

This conclusion is brought out clearly in Fig. 20.6, where the binding energy slopes for the data points of Fig. 20.5 are plotted. As is shown in Fig. 20.6, the slopes above the CSP are in every case flatter than the slopes below the CSP. The binding energies above the CSP, which correspond to the addition of excess unclustered neutrons, reveal a small n-n pairing effect that causes the data points to oscillate back and forth across the lines of constant average slope.

These binding energy results are extended to higher values of Z in Fig. 20.7. As the atomic number Z increases, the coulomb repulsion discriminates against proton-rich nuclei and pushes the CSP to the left of the isotopic chains in the figures. For

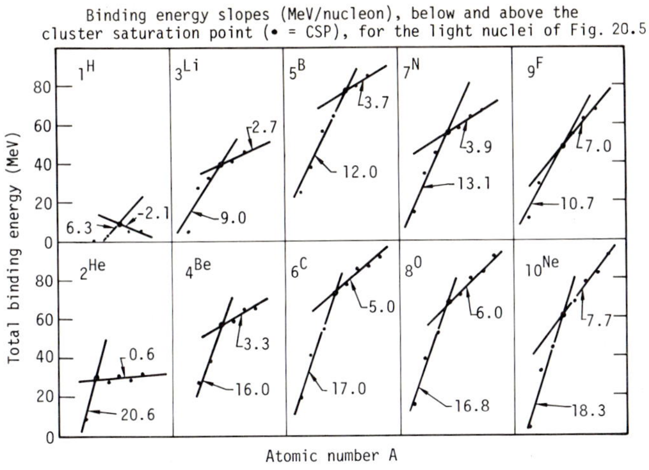

Figure 20.6. Binding energy slopes for the isotopic sequences of Fig. 20.5. In each case, the slope shows a discontinuity at the isotope that corresponds to the Cluster Saturation Point (CSP)

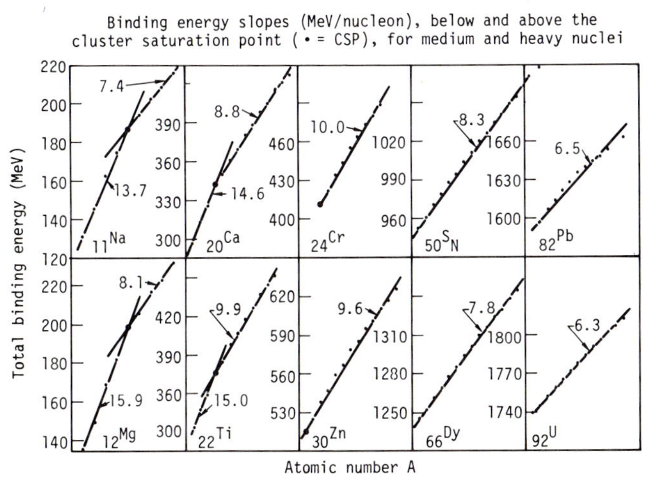

Figure 20.7. Binding energy slopes for the isotopic sequences of some higher-Z elements. As Z increases, the coulomb repulsion discriminates against proton-rich nuclei, which means that the CSP is pushed the left in the experimentally-observed isotopic sequence. For Z = 24 to Z = 30, the CSP corresponds to the lightest-observed isotope of the element; and for Z > 30, all of the stable isotopes occur above the CSP. Thus all of the heavy elements have smooth and featureless binding energy curves.

zinc (Z = 30), the CSP corresponds to the lightest nucleus that can be formed; and for the elements above zinc, all of the isotopes are above the CSP. Thus the elements above Z = 30 all exhibit uniform isotopic binding energies. These uniform binding energies above the CSP indicate that the excess neutrons which are added to these nuclei do not interact strongly with the clustered core region, but instead form as a neutron halo around the outside of the nucleus (see Sec. 20.5)

The difference between the nucleon binding energies below the CSP and the binding energies above the CSP is brought out quantitately in Fig. 20.8, which is a plot of the "average binding energy per last nucleon"; this average is taken over all

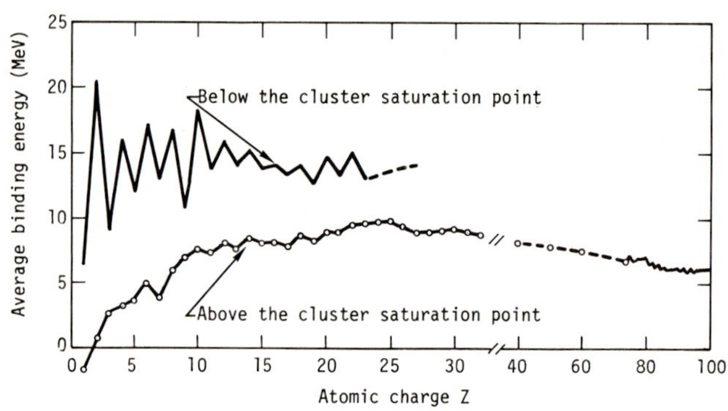

Figure 20.8. The "average binding energy per last nucleon" for isotopes below and above the CSP. This figure illustrates the dramatic decrease in binding energies that occurs once the internal clusters (mainly α particles) have been fully saturated. The significance of this result is that it indicates in a rather clear-cut manner that these internal α-particle clusters actually do exist.

of the isotopes which are below the CSP for a particular element (this includes only the elements with Z < 30), and also over all of the isotopes which are above the CSP for this same element (this includes all of the elements). As can be seen in Fig. 20.8, the average binding energy per last nucleon is in the range of 15 MeV for the isotopes below the CSP, but it is less than 10 MeV for the isotopes above the CSP. This figure graphically illustrates the sharp decrease in binding energies which occurs once the clusters inside the nucleus have been fully saturated (at the CSP).

The above results indicate that nucleon clustering inside the nucleus is an important phenomenon. We can make these results quantitative by comparing the actual nuclear binding energy intervals with the binding energies of the constituent clusters.

The clusters D, He^3, T, and α have the following binding energies:

$$D(2.2\text{ MeV}) \;;\quad He^3(7.7\text{ MeV}) \;;\quad T(8.5\text{ MeV}) \;;\quad \alpha(28.3\text{ MeV}) \quad . \qquad (20.5)$$

Thus, for example, when a neutron is added to a D cluster to form a T cluster, we expect to see a jump of about 6.3 MeV in the binding energy; and when a neutron is added to a He^3 cluster to form an α cluster, we expect to see a jump of about 20.6 MeV in the binding energy. These are very distinctive energy intervals, and they should be readily observable in the nuclear binding energies.

In order to make a quantitative comparison between the binding energies of the atomic nuclei and the binding energies of the constituent D, He^3, T, and α building blocks, we must define a nuclear cluster model for calculating absolute atomic masses. We actually define two such models, which are described in Table 20.1. In the First-Order Cluster Model that is defined in Rule 4a of Table 20.1, the entire binding energy of the nucleus is assumed to be the sum of the (experimental) binding energies of the constituent clusters; that is, we neglect cluster-cluster binding energies. In the Second-Order Cluster Model that is defined in Rule 4b of Table 20.1, we include a phenomenologically-determined cluster-cluster binding energy; and, for completeness, we add a small 1 MeV binding energy for unpaired protons (below the CSP), and binding energies of 6 and 8 MeV for unpaired and paired excess neutrons (above the CSP).

The binding energy results which are obtained with the simple first-order cluster model are shown in Fig. 20.9. As can be seen in Fig. 20.9, there is an unmistakable correlation between the calculated values (the o's) and the experimental values (the x's): every time that a strongly-bound α particle is formed, the experimental binding energy takes a decided jump. The importance of this result cannot be overemphasized. In order to realize its significance, we must understand how differently the nucleus appears from the viewpoint of the nuclear shell model. In the shell model, nucleons are assumed to move in single-particle orbitals around a mutually-formed potential-energy center. Since we know that nucleons interact strongly with one another, the only way that these single-particle orbitals can exist is if the Pauli Exclusion Principle prevents the nucleons from interacting with one another in the interior of the nucleus. But if this is the case, then the existence of nucleon clusters in the interior of the nucleus is strictly forbidden. Thus, from the standpoint of the nuclear shell model, there is no reason to expect to find correlations such as those shown in Fig. 20.9. Hence the correlations shown in Fig. 20.9 pose a real problem with respect to the conceptual foundations of the nuclear shell model.

The correlations shown in Fig. 20.9, while impressive, are still only qualita-

Table 20.1. A nucleon cluster model for reproducing the absolute masses of atomic nuclei. The rules defined here are sufficient to reproduce the masses of more than 1100 nuclides to an average absolute accuracy of better than 1%.

Rule 1. Protons and neutrons inside atomic nuclei combine into clusters, with the most strongly-bound clusters being dominant.

Rule 2. The relevant nucleon clusters are:

D: Z = 1, N = 1, B. E. = 2.2 MeV;
T: Z = 1, N = 2, B. E. = 8.5 MeV;
He^3: Z = 2, N = 1, B. E. = 7.7 MeV;
He^4: Z = 2, N = 2, B. E. = 28.3 MeV.

Useful mnemonic: these four clusters have a zero binding energy determinant:

$$ZN = \begin{vmatrix} 11 & 12 \\ 21 & 22 \end{vmatrix} = \begin{vmatrix} D & T \\ He^3 & He^4 \end{vmatrix} = \begin{vmatrix} 2.22\text{ MeV} & 8.48\text{ MeV} \\ 7.72\text{ MeV} & 28.30\text{ MeV} \end{vmatrix} \simeq \text{zero} .$$

Rule 3. The Cluster Saturation Point, CSP, occurs when all of the nucleons in the atomic nucleus are strongly bound into nucleon clusters: the CSP is at N = Z for even-Z nuclei, and at N = Z + 1 for odd-Z nuclei. Nuclei below the CSP contain unclustered protons, and nuclei above the CSP contain unclustered neutrons.

Rule 4a. First-order cluster model. The entire nuclear binding energy is taken to be the sum of the intra-cluster binding energies; that is, cluster-cluster and unpaired-nucleon binding energies are ignored. The results obtained with this first-order cluster model are illustrated in Fig. 20.9.

Rule 4b. Second-order cluster model. In addition to the intra-cluster binding energies of Rule 4a, the following extra-cluster binding energies are included: (1) cluster-cluster binding energies are calculated in accordance with Fig. 20.10, where T, He^3, and He^4 are counted as full clusters and D is counted as half a cluster; (2) unpaired protons are assigned binding energies of 1 MeV; (3) unpaired neutrons are assigned binding energies of 6 MeV (odd N) or 8 MeV (even N) as they are added above the CSP; (4) no coulomb corrections are applied (they are already implicitly contained in the curve of Fig. 20.10). The results obtained with this second-order cluster model are illustrated in Fig. 20.11.

(continued on next page)

Table 20.1 (continued).

Result A. The second-order cluster model of Rule 4b is sufficiently accurate that it reproduces the experimental binding energies of the 1135 reported isotopes between Li^{5} and Fm^{257} to an average absolute accuracy of better than 0.9%.

Result B. Evidence for the formation of nucleon clusters inside atomic nuclei is provided by the fact that the observed isotopic binding energy intervals accurately mirror the distinctive theoretically-calculated isotopic binding energy intervals which are obtained via the formation of successive clusters in the isotopic sequence. This result is illustrated in Figs. 20.9 and 20.11.

Result C. The empirically-determined cluster-cluster binding energy curve of Fig. 20.10 can be directly interpreted in terms of a linear nearest-neighbor type of cluster-cluster binding energy, as is shown in Fig. 20.13.

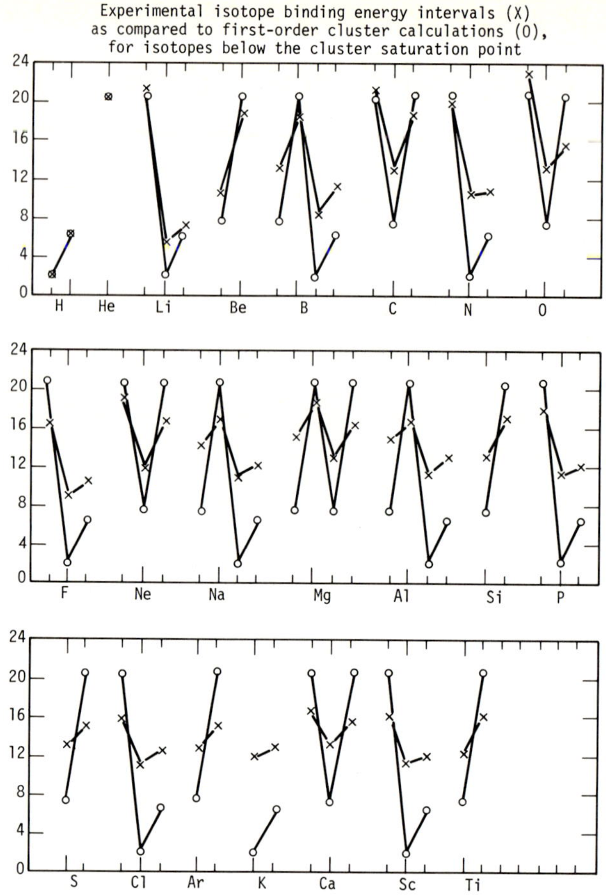

Figure 20.9. Experimental isotope binding energy intervals as compared to calculated intervals based on the First-Order Cluster Model of Rule 4a in Table 20.1. As can be seen, there is a clear-cut correlation between the experimental isotope binding energies and the internal cluster binding energies. In particular, every time that a strongly-bound internal α particle (He^4) is formed, the observed binding energy takes a large jump. On the basis of conventional shell-model systematics, there is no reason to expect to find this kind of a correlation.

tive. In order to improve on these results, we must include the binding energies between clusters, which certainly exist. This leads us to the second-order cluster model of Table 20.1. Instead of introducing an ad hoc cluster-cluster binding energy into our nuclear cluster model, we use the following phenomenological procedure: (1) starting with the masses of the 1144 measured atomic nuclei[11] as a data base, we calculate the masses of these 1144 isotopes in a computer calculation, using the binding energies of Rule 4b in Table 20.1; (2) we then determine the number of clusters C for each nucleus, and we select a total cluster-cluster binding energy E(C) for each value of C which optimizes the calculated fits to the masses of the nuclei which have this same number of clusters (there are roughly 20 different nuclei that correspond to each value of C, so that E(C) represents an average value which is obtained by fitting to these 20 nuclei). This empirically-determined cluster-cluster total binding energy curve, E(C), is shown in Fig. 20.10. As can be seen, it rises to a maximum of about 190 MeV for C =30, and then drops off due to the fact that the increasing coulomb repulsion is canceling out the cluster-cluster magnetic attraction.

The binding energy correlations that are obtained with this second-order cluster model are shown in Fig. 20.11. As can be observed in this figure, the correlations between the calculated and experimental binding energy intervals are now quite accurate, especially when the range of nuclei included is considered: Fig. 20.11 includes all elements which have more than one isotope below the CSP. This second-order cluster model gives calculated mass values for all of the 1135 nuclides between Li^5 and Fm^{257} to an average absolute accuracy of better than 0.9%.

The binding energy systematics that we have been discussing in connection with Figs. 20.9 and 20.11 are for isotopes which are below the CSP. In order to illustrate the strong contrast between the binding energy intervals observed for isotopic sequences below the CSP and those observed for isotopic sequences above the CSP, we show in Fig. 20.12 the experimental binding energy intervals for the isotopic chains in tin and in uranium. In each of these heavy elements, the isotopes are all well above the CSP; and, as can be seen in Fig. 20.12, the binding energy intervals, apart from an odd-even (O-E) neutron pairing correlation, are quite smooth and featureless. The systematics shown in Fig. 20.12 has a physical significance which goes well beyond just illustrating the contrast to Figs. 20.9 and 20.11: since the binding energy intervals of Fig. 20.12 do not reflect any nucleon cluster correlations, it seems clear that the sequential addition of these neutrons above the CSP does not disturb the basic core configurations of the Sn and U nuclei; hence these last-added neutrons must be arranged in a manner that is external to the clustered core region. This observation will take on added significance in the next section of the chapter, where we examine the evidence which points to the existence of a neutron halo in heavy nuclei.

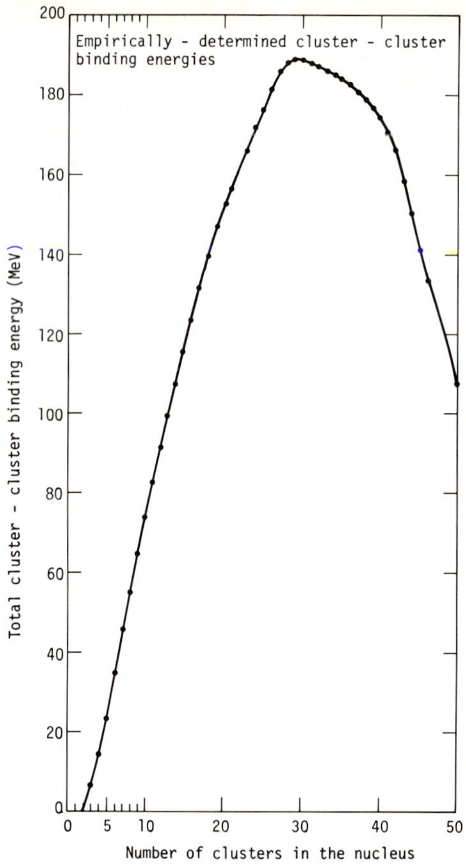

Figure 20.10. An empirically-deduced cluster-cluster binding energy curve. These cluster-cluster binding energies, which are between the internal α-particle clusters in the nucleus, are deduced from the computerized Second-Order Cluster Model of Rule 4b in Table 20.1, as described in the text. The curve shown here is for the total cluster-cluster binding energy in the nucleus. The fall-off of the curve for heavy nuclei is due to the increasing coulomb repulsion, which tends to cancel out the magnetic cluster-cluster attraction.

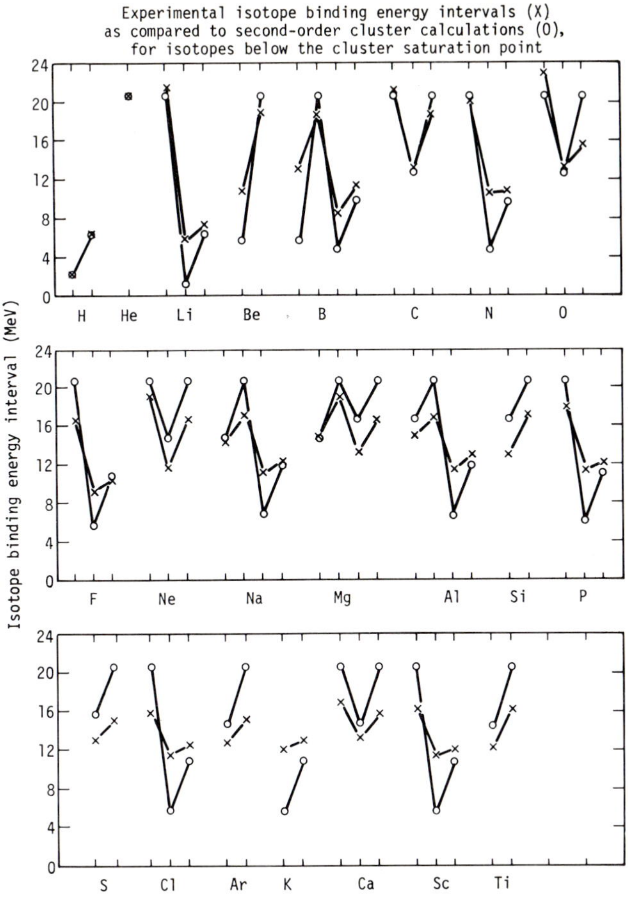

Figure 20.11. Experimental isotope binding energy intervals as compared to calculated intervals based on the Second-Order Cluster Model of Rule 4b in Table 20.1. Fig. 20.11 is the same plot as that of Fig. 20.9, but with cluster-cluster binding energies included (as obtained from Fig. 20.10), and with other small unpaired-nucleon effects added in. Figs. 20.9 and 20.11 both indicate clearly that internal D, T, He^3, and α clusters are being formed inside of these nuclei. The phenomenological significance of this result lies in the fact that this type of internal nucleon clustering is expressly forbidden in the formation of the nuclear shell model.

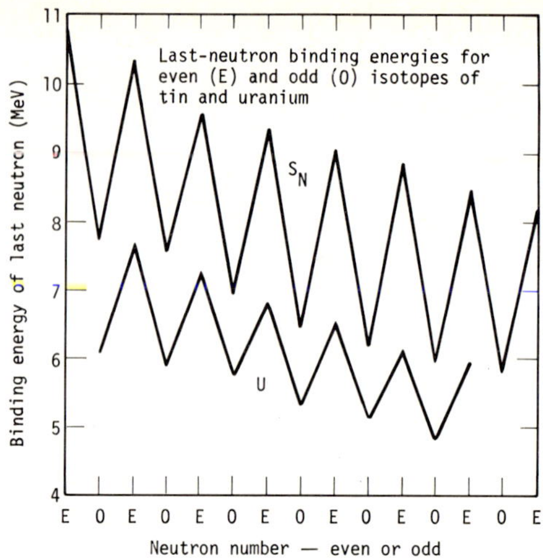

Figure 20.12. The last-neutron binding energies for even and odd isotopes of tin and uranium. Apart from the odd-even pairing energy of the neutron, these curves are completely smooth and featureless (contrast Fig. 20.12 with Figs. 20.9 and 20.11). This is an important result, because it means that the neutrons are added in a manner which does not disturb the clustered core region of the nucleus. Hence they must be combining in the form of a "neutron halo" that is wrapped around the outside of the clustered core region.

It is possible to give a physical interpretation to the empirically-determined inter-cluster binding energy curve of Fig. 20.10. If we remove coulomb effects (which cause the dip in the curve at high Z values), and if we divide by the number of clusters at each point on the abscissa so as to get the average inter-cluster binding energy per cluster, we obtain the solid curve shown in Fig. 20.13. Then, by giving each cluster a "cubic" geometry (so that it has at most six nearest neighbors), grouping the clusters together, counting the number of nearest-neighbor pairs for each number of clusters, and assigning a cluster-cluster binding energy of 10 MeV per adjacent pair (5 MeV if the pairing includes a deuteron), we obtain the calculated values shown by the circles in Fig. 20.13. The reasonable agreement between the solid curve and the circles in Fig. 20.13 indicates that both the intra-cluster and the inter-cluster binding energy systematics follow the linear nearest-neighbor type of interaction that we expect from this cluster model approach to nuclear binding

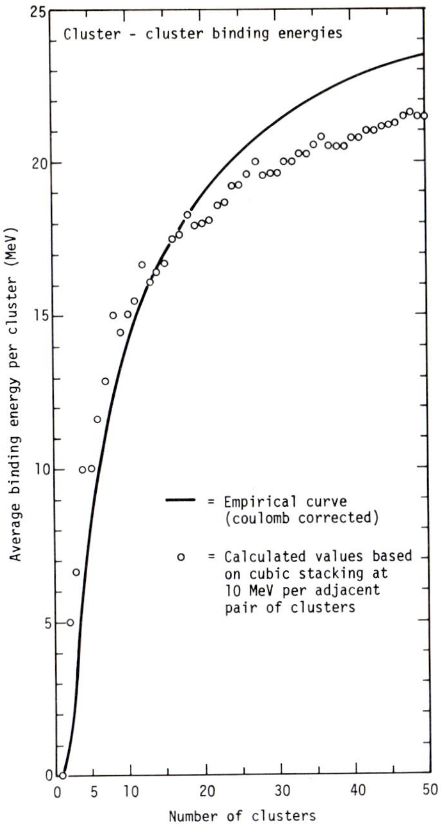

Figure 20.13. A comparison of empirical and calculated cluster-cluster binding energies. The solid curve is the empirically-deduced curve of Fig. 20.10 with coulomb effects removed, and with a division made by the number of clusters so as to obtain the average binding energy per cluster. The circles are calculated binding energy values obtained by stacking α particles together in a cubic manner (so that they have at the most six nearest neighbors), and assuming a binding energy of 10 MeV per adjacent α-α pair. As can be seen in the figure, this simple model rather closely matches the empirical curve, which suggests that this α-particle-model picture of the nucleus has physical content.

energies.

In Sec. 20.5 we discuss the experimental evidence for a neutron halo in heavy nuclei, and also the implications that can be drawn from the existence of this halo.

20.5 Neutron Halos in Heavy Nuclei

In the past two decades there have been numerous suggestions[12,13] that heavy nuclei contain "a nuclear skin of neutrons"[14]. Since the analysis of nuclear binding energies in the preceding section seems to mandate this result, it is worthwhile to briefly summarize some of the experimental evidence for this neutron halo.

One of the earliest experimental indications for a neutron halo came from electron and proton scattering off nuclei, where the rms radius for the nuclear force was found to be larger than the rms radius for the nuclear charge distribution,[15] as shown in Fig. 20.14. Recent determinations of this halo effect involve sub-coulomb pickup,[16] optical model calculations,[17] and K^- and $\bar{p}$ annihilation processes.[14,18] These experimental results indicate a nuclear neutron radius for heavy nuclei that is roughly one-half to one fermi larger than the corresponding proton radius.

Of these neutron halo experiments, the ones that probably give the most direct information are the K^- and $\bar{p}$ annihilation experiments, which have been analyzed theoretically by Leon and Seki.[19] When K^- or $\bar{p}$ antiparticles are captured in orbits around heavy nuclei, they cascade sequentially down to lower orbits by x-ray emission until they finally collide with nucleons in the nucleus and annihilate. Experimentally, these annihilations occur from orbits with large principal quantum numbers,[13] so that the annihilations take place from near-circular orbits and hence are logically associated with nucleons located at the periphery of the nucleus. Since heavy nuclei have more neutrons than protons, it is to be expected in any case that more neutron annihilations than proton annihilations will be observed. But it was found experimentally for heavy nuclei[14,18] that neutron annihilations predominate over proton annihilations by a factor which is considerably larger than the neutron-to-proton ratio in the nucleus, thus indicating in a straightforward manner that there is a concentration of neutrons near the surface of the nucleus.

As we described in detail in the last section, the binding energy systematics in atomic nuclei (specifically, the contrast between the isotopic binding energy intervals of Figs. 20.11 and 20.12) also point to the existence of this neutron halo. As evidence for nucleon clustering in atomic nuclei, the presence of a neutron halo is highly significant. A neutron halo in heavy nuclei is, from the standpoint of nuclear energetics, a very strange result. In a heavy nucleus the internal coulomb

Uranium nucleus as seen in electron scattering ($R_0 = 1.2\ A^{1/3}$) and in nucleon scattering ($R_0 = 1.4\ A^{1/3}$) experiments.

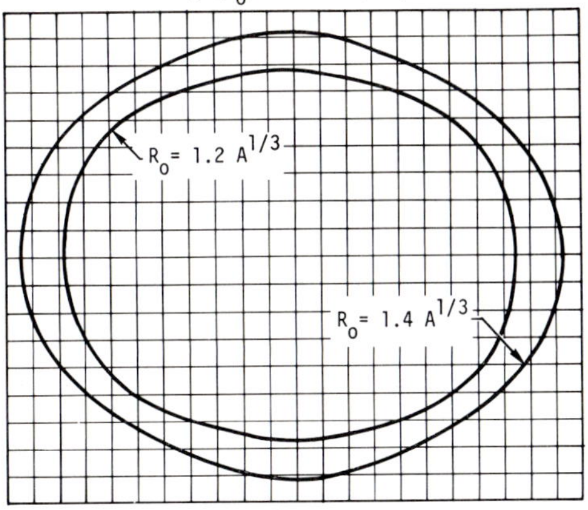

Figure 20.14. The uranium nucleus as seen in electron scattering experiments (which probe the charge radius) and in nucleon scattering experiments (which probe the hadronic matter radius). As can be seen, the charge radius appears experimentally to be smaller than the matter radius, which suggests that uranium has a "neutron halo" around the outside. Now, a neutron halo is a strange result to find in heavy nuclei, since coulomb effects are so large; the uranium nucleus in fact pays an "energy penalty" of about 150 MeV for having this neutron halo. The natural expectation is that we should find a "proton halo" in heavy nuclei. The absence of a proton halo can be explained by assuming that the protons in the nucleus are locked into strongly-bound α-particle clusters in the core of the nucleus, so that they are not free to move around and migrate to the surface of the nucleus under the action of their mutual coulomb repulsion.

forces are enormous, and we would intuitively expect to find a proton halo rather than a neutron halo. For example, the "energy penalty" paid by a uranium nucleus for having a neutron halo is about 150 MeV! Now, if the nucleons in a nucleus are free to move around, it seems inevitable that the protons will on the average be pushed towards the surface of the nucleus. In the liquid drop model, and also in the Fermi gas model (which forms the conceptual basis for the shell model orbitals), it is assumed that nucleons are in fact free to move around; thus on the basis of these models, it is difficult to account for neutron halos. The Pauli Exclusion Principle can be invoked to keep nucleons from crowding too closely together, but it doesn't prevent them from moving apart. And it certainly doesn't act in any way to screen out the coulomb forces among the protons.

The obvious implication to be drawn from the existence of a neutron halo in heavy nuclei is that the nucleons (or at least the protons) in these nuclei are not free to move around. But this is of course the result that follows from a cluster model viewpoint of the nucleus. Hence the existence of a neutron halo furnishes additional evidence for the kind of nuclear clustering that we have already been led to on the basis of binding energy systematics and of the observed symmetries in nuclear fission. If the protons in a nucleus are embedded in α particles, and if these α particles are bound together in the center of the nucleus (by 10 MeV per α-α pair, as we concluded from Fig. 20.13), then it becomes clear why the protons do not migrate to the surface of the nucleus to form a proton halo: the strong intra-α and inter-α binding energies are sufficient to counteract the repulsive coulomb forces between the protons.

20.6 Are Nuclear Cluster Effects Compatible with Shell Model Systematics?

The shell model plays a role in nuclear physics that is somewhat comparable to the role played by SU(3) in particle physics: the single-particle shell model, with its various collective-motion modifications, is generally accepted as the fundamental theory for atomic nuclei, just as SU(3), with its various quark model extensions, is generally accepted as the fundamental theory for hadronic particles. In the present chapter we have been summarizing evidence which indicates that nucleons pack together inside atomic nuclei in a very structured manner. Since this idea seems to be incompatible with the shell model idea that all of the nucleons in a nucleus are moving in orbits around a mutually-defined center, it is important to explore the relationship between these two different approaches to nuclear structure. We start by reviewing the results of the preceding sections.

In studying the systematics of the elementary particles, we came to the conclusion that the nucleon quark — the spinor S — has essentially a cylindrical geometry

(Fig. 18.8), so that the nucleon is constructed by clustering three of these spinors together in a side-by-side manner (Fig. 19.2). We also deduced phenomenologically that adjacent spinor pairs $S^{\pm}S^{\pm}$ have a (singular) nearest-neighbor magnetic interaction energy of about ±2 MeV [Eq.(19.15)], which is an order-of-magnitude larger than we would calculate from the magnetic moments of these spinors.† Since the large (4%) hadronic binding energy that operates between quark-antiquark substates and between nucleon-antinucleon states does not operate between nucleon-nucleon states, the attractive magnetic binding energy and the repulsive coulomb interaction energy emerge as the dominant factors in the binding together of nucleons in atomic nuclei.

In applying these results from elementary particle physics to nuclear physics, we started with the α particle. By clustering two protons and two neutrons together in a side-by-side manner (Fig. 20.1), and by assuming the same magnetic and coulomb binding energies that we have previously used for elementary particles (Chapter 19), we reproduced the 28 MeV binding energy of the α particle [Eqs. (20.1) - (20.3)], which is the first time that this result has been achieved from a theoretical model that has no adjustable parameters. Furthermore, we can see from the magnetic layering which occurs inside the α particle [(Eq.(20.3)] that these α particles (which have "cubic" rather than spherical geometries) bind together end-to-end in the form of two-fermi-thick nucleon Ising layers, and that these layers in turn bind together magnetically to form the atomic nucleus.

We next carried these results over to the fission process in heavy atomic nuclei. A large nucleus has between eight and ten Ising layers (see Figs. 20.2 and 20.4), and the central Ising layers each contain about 40 nucleons. Scission of the nucleus occurs at an Ising interface, where an incipient "tear" upsets the delicate balance between the short-ranged magnetic attraction and the long-ranged coulomb repulsion, and hence propagates along the interface. From a consideration of the forces involved (Fig. 20.3), it is apparent that low-energy-induced nuclear scisssion occurs at a central Ising interface, where coulomb forces are the strongest. Thus we are led naturally to the result that nuclei which contain an even number of Ising layers fission symmetrically, and nuclei which contain an odd number of Ising layers fission asymmetrically, with the "heavy fragment" in asymmetric fission having about 40 more nucleons than the corresponding "light fragment". This is precisely the result that is observed experimentally. The present theoretical model again provides the only parameter-free explanation that has yet been advanced to account for these experimental results.

---

†It is possible to calculate this singular magnetic interaction energy in terms of the cross-sectional radii of the equatorial current loops in these spinors (see Appendix B in Ref. 5).

The two-dimensional nature of these two-fermi-thick nucleon Ising layers means that the atomic nucleus can have quite different properties with respect to intra-layer nuclear excitations than it does with respect to inter-layer nuclear excitations. This possibility is embodied in the topological considerations that were discussed in Sec. 20.3, and it suggests a mechanism for bridging the conceptual gap that exists between the structureless liquid-drop model of heavy nuclei and the structured fission process which occurs in these same nuclei.

If two-dimensional nucleon layers do exist in atomic nuclei, then it seems apparent that nucleon clustering must be occurring throughout the interior of the nucleus. To demonstrate that this is in fact the case, we set up two simple versions of a cluster model in Sec. 20.4., and then used these models to show that the observed binding energies in light nuclei accurately mirror the binding energies of the constituent clusters which are formed inside the nucleus. From the binding energy characteristics of the isotopic chains, it is apparent that the excess neutrons in a nucleus [the neutrons which are added above the Cluster Saturation Point of Eq. (20.4)] do not interact with the clustered core region, but instead wrap around the outside of the nucleus in the form of a "neutron halo". Other experimental evidence for this neutron halo was summarized in Sec. 20.5. Independent evidence for the existence of nucleon clustering in the interior of the nucleus has also appeared recently from a variety of high-energy nuclear scattering experiments.[20]

All of the nuclear physics results that we have just cited follow as straightforward and self-consistent extensions of the light-quark elementary particle results which were obtained in the preceding chapters of the book. The question now is to reconcile these results with the known successes of the nuclear shell model. In this regard, there is one fact which is of paramount importance:

> due to the tendency of nucleons in a nucleus to pair off in self-canceling spin up - spin down pairs, the striking shell-model-like spin and magnetic moment regularities that are observed in atomic nuclei are due almost entirely to the outermost unpaired nucleons. (20.6)

In nuclei such as O^{17} and F^{17}, the outermost unpaired neutron and proton, respectively, seem to be clearly traveling in single-particle orbitals (see Table V in Ref. 5). However, this does not mean that the core region of the nucleus also consists of nucleons which are moving in single-particle orbitals, as is commonly assumed in the shell model. We can assume a clustered core region for the nucleus and still reproduce the essential shell-model features of these nuclei (see Tables III - V in Ref. 5). This observation is the central point of the present discussion. While single-particle orbital effects seem to be a feature of atomic nuclei, it does not appear to be either phenomenologically necessary nor theoretically justified to assume that the entire nucleus is composed of nucleons which are moving in single-particle orbitals; we can achieve essentially the same results by combining unpaired-nucleon single-particle orbitals with a clustered core region. Thus the answer to the question which serves as the title for this section is yes.

The theoretical difficulty with the shell model concept of the nucleus arises from the fact that there is no massive central core region, as there is in an atom (the relevance of this remark follows from the fact that the nuclear shell model was patterned after the atomic shell model). Thus shells can be formed in nuclei only by an overall collective motion in which the sum over the two-body nucleon-nucleon interactions serves to define an effective potential for the nucleus as a whole. But the concept of this collective motion has been seriously challenged by recent experimental results. For example, Lind et al.[21] bombarded Ca^{40} with 220 MeV pions and found that large cross sections were obtained for the production of gamma radiation from final-state nuclei which corresponded to the knocking out of 1, 2, 3, 4, or 5 α particles (or equivalent nucleons) from the Ca^{40} nucleus. Now, the simplest way to account for a process in which there is a high probability for ejecting several α particles is to assume that these α particles were present in the initial state; however, the shell model, due to the operation of the Pauli Exclusion Principle, does not permit the existence of these initial-state α-like correlations. Even more puzzling is the fact that a bombardment which results in the ejection of half of the particles from a Ca^{40} nucleus should lead to a chaotic transfer of momenta among the various single-particle orbitals, so that the residual nucleus is left in a highly-excited state, from which it must somehow regroup into the shell model orbitals which constitute Ne^{20} (that is, Ca^{40} minus five α particles), and then cascade down to the Ne^{20} ground state by gamma ray emission. But this is not the result that was observed experimentally: Lind et al.[21] found, in contrast to this shell model explanation, that the residual nucleus was left preferentially in its lower excited levels, and γ-ray transitions from the higher levels "were generally conspicuously absent". If Ca^{40} is composed in the first place as a cluster of ten α particles, then it seems possible to knock out 1, 2, 3, 4, or 5 α particles and still leave the nucleus in a state of low excitation; but if the Ca^{40} nucleus is a collection of carefully-balanced single-particle orbitals, then the experimental results of the Lind experiment seem inexplicable.

To point out one more recent example which indicates the presence of α particles inside nuclei, we mention the experiment of Głowacka et al.,[22] in which α particles were ejected from dysprosium nuclei by bombardment with 14 and 18 MeV neutrons. These authors concluded that "the experimental data can be described (by) assuming the existence of preformed α-clusters in target nuclei ...". Other α-knockout experiments are discussed in Section I.A of Ref. 5.

One conclusion we reach from the studies of the present chapter is that the geometry of the nucleon does affect the geometry of the nucleus, as was predicted by Moszkowski in the quotation given at the beginning of this chapter; in particular, it leads to a laminar two-dimensional substructure for the nucleus which is most clearly revealed in the process of nuclear fission. Another conclusion which seems to be indicated by these studies is that the fields of nuclear physics and elementary particle physics are intimately related – both experimentally and theoretically, as of course should be the case. To cite just one example, we found that in reproducing the electromagnetic properties of the nucleon (Chapter 18), we obtained results which seem to be in agreement with more-or-less classical electromagnetism – but not with the field-theory concept of the nucleon; and in studying the nuclear forces which bind the nucleus together, we found (in the present chapter) that these seem to have electromagnetic origins – and not the field-theory origins that correspond to the virtual exchange of pions. (The concept of the nuclear force as an electromagnetic phenomenon is discussed in Appendix B of Ref. 5.)

20.7 Final Remarks

In the present studies we have attempted to demonstrate, by means of direct calculations, that the microscopic properties of elementary particles can be understood, at least in part, in terms of concepts which were developed in the macroscopic world. In particular, we have tried to show that spin angular momenta and magnetic moments admit a classical representation, and that the empirically-invoked conservation laws relating to baryon number and strangeness are rooted in fundamental properties of the elementary particle basis states. Theoretically, we have pointed out the importance of relativistic equations as related to spinning particles, and we have used these equations to forge a link between spinless and spinning basis states. With the assumption of weak quark binding energies, we were inexorably led to the spinless mass quantum $M = 70$ MeV as a fundamental building block for all particles that are more massive than the electron.

We have concentrated our attention in this book on the strongly-interacting hadrons and on the muon – that is, on the particles that have masses in excess of 100 MeV/c^2. If these studies make sense, they must in some manner also relate to the lighter particles – the electron, neutrino, and photon. It is possible to make an extrapolation to include these particles. The model for the electron is the same as that for the muon, only scaled by the ratio of the electron and muon masses. With respect to the neutrino, the mechanical mass that we have introduced here to supply the inertial properties of the muon and electron necessarily has neutrino-like properties – that is, a spin angular momentum of 1/2 $\hbar$ and a very weak interaction with matter. This leaves the photon to be accounted-for. The photon, although it has zero rest mass, nevertheless has an angular momentum $\hbar$ that couples to the angular momenta of electrons, muons, hadrons, and atomic nuclei. Thus it seems that the photon should be explainable in the same terms that we use for these other particles. It turns out that the photon does have such an explanation,[23] provided that we are willing to introduce negative masses into our set of basis states. The properties of negative masses are defined by the Dirac equation, the Klein-Gordon equation, the equations of special relativity, and the equations of general relativity, and they are precisely the properties that are required to account for the special nature of the photon — namely, its combination of a large spin angular momentum and a small inertial mass. Furthermore, the $\psi^*\psi$ probability fields that furnish the "wave" aspects of the elementary particle can be reinterpreted as positive-mass, negative-mass perturbations of the vacuum state by means of the mathematically-correct identity[23] $\psi^*(+m)\psi(+m) \equiv \psi(-m)\psi(+m)$. These topics form an exciting new area of study, and perhaps someday they will lead to a second volume devoted to the Nature of the Elementary Particle.

> "All the fifty years of conscious brooding have brought me no closer to the answer to the question, 'What are light quanta?' Of course today every rascal thinks he knows the answer, but he is deluding himself."
>
> Albert Einstein, in 1951

<u>References for Chapter 20.</u>

The quotation by Albert Einstein is taken from R. H. Stuewer's book <u>The Compton Effect</u>, Science History Publications, New York (1975), page 332.

1. For example, see D. H. Wilkinson, <u>Proc. Rutherford Jubilee Conf.</u>, London (1961), p. 339; also see <u>Clustering Phenomena in Nuclei</u>, IAEA, Vienna (1969).

2. See references 4 through 12 in Ref. 5 of the present chapter.

3. This same assumption was also made in connection with Eq. (19.15).

4. This is the coulomb repulsion that is calculated from the configuration of Eq. (20.3) by using Eq. (19.8).

5. M. H. Mac Gregor, Nuovo Cimento 36A, 113 (1976).

6. L. Pauling, Phys. Rev. Lett. 15, 499 and 868 (1965); Phys. Rev. 182, 1357 (1969); Nature 208, 174 (1965); Science 150, 297 (1965).

7. See J. M. Ziman, Principles of the Theory of Solids, Cambridge (1972), p. 353.

8. S. K. Chang et al., Phys. Rev. C10, 2467 (1974).

9. C. E. Bemis, Jr., et al., Bull. Am. Phys. Soc. 22, 611 (1977).

10. The $_{10}Ne^{20}$, $_{12}Mg^{24}$, and $_{14}Si^{28}$ shapes are from the (p,p) measurements of R. de Swiniarski et al., Phys. Rev. Lett. 23, 317 (1969); the $_{69}Tm^{166}$ shape is from K. E. G. Löbner et al., Nuclear Data Tables A7, 495 (1970); the $_{72}Hf^{178}$ shape is from D. L. Hendrie et al., Phys. Lett. 26B, 127 (1968); the $_{90}Th^{230}$ and $_{92}U^{238}$ shapes are from F. K. Mc Gowan et al., Phys. Rev. Lett. 27, 1741 (1971).

11. A. H. Wapstra and N. B. Gove, Nuclear Data Tables 9, 267 (1971).

12. M. H. Johnson and E. Teller, Phys. Rev. 93, 357 (1954); P. B. Jones, Phil. Mag. 3, 33 (1958); R. G. Seyler and C. H. Blanchard, Phys. Rev. 131, 355 (1963); E. H. S. Burhop, Nucl. Phys. B1, 438 (1967); H. A. Bethe, Phys. Rev. 167, 879 (1968).

13. Y. N. Kim, Mesic Atoms and Nuclear Structure, North-Holland, New York (1971), especially Ch. 5, Sec. 2.

14. C. E. Wiegand, Phys. Rev. Lett. 23, 1235 (1969).

15. See R. D. Evans, The Atomic Nucleus, McGraw-Hill, New York (1955), p. 30.

16. H. J. Körner and J. P. Schiffer, Phys. Rev. Lett. 27, 1457 (1971).

17. G. W. Greenlees et al., Phys. Rev. 171, 1115 (1968).

18. D. H. Davis et al., Nucl. Phys. B1, 434 (1967); W. M. Bugg et al., Phys. Rev. Lett. 31, 475 (1973); W. M. Bugg et al., Nucl. Phys. B64, 29 (1973); W. M. Bugg et al., Phys. Rev. Lett. 35, 611 (1975).

19. M. Leon and R. Seki, Phys. Lett. 48B, 173 (1974) and Nucl. Phys. B74, 68 (1974).

20. Y. Iwasaki, Phys. Rev. Lett. 36, 1464 (1976); S. Frankel et al., ibid. 36, 642 (1976); M. de Llano and A. Plastino, ibid. 37, 556 (1976); J. Stevenson et al., ibid. 38, 1125 (1977). Other experiments are discussed in Ref. 5.

21. V. G. Lind et al., Phys. Rev. Lett. 32, 479 (1974).

22. L. Głowacka et al., Nucl. Phys. A284, 257 (1977). Also see A. G. Flowers et al., Phys. Rev. Lett. 40, 709 (1978).

23. M. H. Mac Gregor, LLL report UCRL-80467 (Rev. I), "Negative Masses, Photons, and $\psi^*\psi$ Probability Fields", December (1977).

APPENDIX A. A BRIEF OVERVIEW OF ELEMENTARY PARTICLE PHYSICS

The nature of the elementary particle is probably the central problem in the whole domain of physics. Hence it is a subject that is of interest to all physicists, and indeed to all scientists in general. The information about the elementary particle that we obtain from the studies of the present book is quite straightforward, both in its physical content and in its mathematics. Thus the main requirement that is needed in order to understand the material presented in this book is a familiarity with the basic concepts and the notation of elementary particle physics. Appendix A is written as a guide for the reader who is interested in learning about the elementary particle, but who is not familiar with the specialized literature on the subject, and it deals mainly with the aspects of the elementary particle which are of the most importance from the standpoint of the present studies.

The elementary particles that we are concerned with in the present book are the massive particles that have measured rest masses – namely, the electron, the muon, the mesons, and the baryons. Thus we will not discuss photons or neutrinos, although these are, properly speaking, part of the elementary particle family, and although neutrinos may in fact possess small rest masses (this is a question that has not as yet been resolved experimentally). Furthermore, we will concentrate our attention on the "heavy" particles, which we define here to mean the particles that are more massive than the electron; hence electrons are mentioned only incidentally in the text. Of the massive elementary particles, the only stable members are the electron e^- and proton p^+ and their corresponding antiparticles e^+ and p^-. The neutron, which is a particle that is closely related to the proton, is almost stable (it has a half life of 15 minutes), and neutrons are stable when bound together with protons in atomic nuclei. Physicists commonly regard the proton p and neutron n as two states of the same basic particle – the nucleon N. Apart from electrons and nucleons, all of the massive elementary particles are unstable, with half lives of a couple of millionths of a second or less. Thus, with the exception of occasional cosmic ray events, the unstable elementary particles for the physicist to examine must be produced by bombardments with high energy accelerators.

There are two general types of high energy accelerators that are used to produce unstable elementary particles – electron accelerators and proton accelerators; these can be either linear accelerators or circular accelerators. There are also two types of targets that are used – stationary targets and colliding-beam targets. In stationary-target accelerators the high energy electron or proton beam strikes a target made of some atomic material, and the kinetic collision energy is converted into excitation energy, which results in a shower of unstable particles. In colliding-beam accelerators the accelerated beam is trapped magnetically in two intersecting circular storage rings, and collisions are produced at the point of intersection of the two rings. The number of collision events is much smaller in colliding-beam accelerators than in fixed-target accelerators, but the center-of-mass energy of the colliding beams is much higher. Colliding-beam accelerators are also operated in a mode in which an initial accelerated electron beam strikes a converter and produces, via the resulting bremsstrahlung spectrum, a shower of electron-positron pairs; one storage ring is then filled with electrons and the other with positrons, so that high energy $e^- - e^+$ annihilations can be studied.

In addition to using the primary electron or proton beam directly in studying elementary particle reactions, physicists often use the primary beam to produce a stream of fairly-long-lived secondary mesons, which are then collimated and used in turn to produce other reactions at a secondary target location. These secondary beams are usually composed of pions or kaons, which have half-lives of about 10^{-8} seconds and thus travel many meters before they decay. The Fermilab proton accelerator — presently the world's largest — even has a useful secondary neutrino beam, something that would have been considered a practical impossibility a couple of decades ago (neutrinos have extremely small interaction cross sections, so that the probability of a neutrino collision with a target particle is almost vanishingly small).

The massive elementary particles fall into two general categories — strongly-interacting <u>hadrons</u> and weakly-interacting <u>leptons</u>. The <u>massive leptons</u> include just two particles, the 1/2 MeV[†] electron e^- and 105 MeV[†] muon μ^-, and their corresponding antiparticles e^+ and μ^+ (a third massive lepton, the 1820 MeV particle, may have recently been discovered). Since electron and muon interactions are purely electromagnetic, our main interest in the present studies centers on the hadrons — the strongly-interacting particles.[††] Hadronic interactions (the nuclear force, for example) are roughly a factor of α^{-1} stronger than electromagnetic interactions (where $\alpha \simeq 1/137$ is the fine structure constant), and it is from these strong hadronic interactions that we expect the main clues about elementary particle structure to appear. (However, as we demonstrate in Chapters 6 and 7, the leptonic muon

---

[†]In the present book we commonly use units in which the velocity of light c is set equal to unity, so that the terms <u>mass</u> and <u>energy</u> can be used interchangeably. The unit MeV (million electron volts) is an energy, and MeV/c^2 is a mass.

[††]The word "hadron" is derived from the Greek <u>hadros</u>, meaning stout or strong.

also provides us with some very basic information.) Hadrons occur in two general classes — baryons and mesons; and these two classes are in turn divided into strange and nonstrange categories. Strange baryons are denoted as hyperons, and strange mesons are denoted as kaons. We often use the term baryon to denote both baryons and hyperons, and we similarly use the term meson to include both mesons and kaons. We now delineate the physical ideas which underlie these definitions, and we introduce the basic quantum numbers that are used to categorize these hadron states.

When a nucleon — a proton or neutron — receives a violent impact, it can be transformed into one of a large number of excited states (about 100 such states are known at the present time). These excited states include both baryon and hyperon excitations. The remarkable fact about these nucleon excited states is that they invariably remember their origin — that is, they always decay back down again into nucleons. Physicists express this fact mathematically by assigning the additive baryon quantum number $B = +1$ to both baryons and hyperons, and then insisting that this quantum number remain invariant. Thus, for example, when an excited baryon state, which has the quantum number $B = +1$, decays back down to the nucleon ground state, which also has $B = +1$, it can emit other particles in the process, but these emitted particles are required to have baryon number $B = 0$. Antinucleons (antiprotons and antineutrons) are assigned the baryon quantum number $B = -1$. Hence when nucleon-antinucleon pairs are created (for example, by high energy $e^- - e^+$ collisions), the total baryon number of the system remains invariant. As far as is known experimentally, the baryon number B is absolutely conserved in any closed system. Thus the proton, which is the lightest of the baryons, is a stable particle; there is no lower state into which it can decay.

In a high-energy collision between elementary particles, kinetic energy is transformed into particle excitation energy — that is, energy is transformed into matter. This particle excitation energy can take two forms: (1) If nucleons are present, they can be excited into higher-mass baryon or hyperon states; these are the baryon excitations that we discussed in the preceding paragraph. (2) The kinetic energy can be directly transformed into meson excitations that do not involve a nucleon. Thus mesons are particles that have baryon number $B = 0$, and they ultimately decay into electrons, photons, and neutrinos, all of which have baryon number $B = 0$. As we will show in detail later on, the same elementary particle basis states that occur in baryons also occur in mesons, but with some characteristic differences in the ways that they are combined.

In addition to their difference with respect to the baryon quantum number B, baryons and mesons also differ characteristically in their spin angular momentum J: baryons have half-integral spins ($J = 1/2, 3/2, 5/2, \ldots$), whereas mesons have integral spins ($J = 0, 1, 2, \ldots$). This result follows from the facts that (1) protons and neutrons have spins J = 1/2, and (2) the excitation process in which kinetic energy is transformed into hadronic matter is a process that involves integral spins. Thus, when a half-integral-spin nucleon is excited, it is transformed into a half-integral-spin baryon or hyperon; and when kinetic energy is transformed directly into mesonic matter, an integral-spin meson or kaon is produced.

When two or more elementary particles have the same spin, same interaction strength, and approximately the same mass, and differ only in their electric charges, these are taken to be different charge states of the same basic particle. As an important example, the proton and neutron — which are both hadronically-interacting spin 1/2 particles, and which have masses of 938.3 and 939.6 MeV, respectively — are considered to be charged and neutral states of the same particle, the nucleon N. The formalism for handling these multiple charge states is taken from atomic physics and nuclear physics. In atomic physics, a state with a spin angular momentum J has J_z = -J, -J + 1, ..., +J = 2J + 1 different projections of the spin vector when it is placed in a magnetic field along the z direction. By analogy to this result, nuclear physicists assign an isotopic spin I to nuclear isotopes that differ only in their charge states, and they then treat I as a vector in "charge space". Elementary particle physicists, using this same formalism, assign (for example) isotopic spin I = 1/2 to the nucleon N, so that I_z = +1/2 corresponds to the proton, $p^+ \equiv N^+$; and I_z = -1/2 corresponds to the neutron, $n^o \equiv N^o$ (the antinucleons $\bar{N}$ have the charge states $\bar{p}^- \equiv \bar{N}^-$ and $\bar{n}^o \equiv \bar{N}^o$). As another example, the lowest-mass meson state, the pi meson π (which is commonly denoted as the pion), is assigned isotopic spin I = 1, corresponding to its charge states π^+, π^o, π^- (the pion is its own antiparticle, so that there is no antipion).

The baryon quantum number B applies uniquely to the domain of elementary particle physics, whereas the isotopic spin quantum number I applies also to the domain of nuclear physics, and the angular momentum quantum number J applies to the domains of nuclear, atomic, and molecular physics. Another quantum number that applies uniquely to elementary particles is the strangeness quantum number S. We can see the workings of the strangeness quantum number most clearly in the meson states. When kinetic energy is transformed directly into mesonic matter, the meson excitation that is produced necessarily has a symmetric internal particle-antiparticle structure; it is a "nonstrange" state that is assigned the strangeness quantum number S = 0. However, if this meson excitation, instead of decaying symmetrically, is separated into two unequal parts, then these two parts (which in this example constitute separate particles) can each have asymmetric internal particle-antiparticle structures (as we will see from the examples in the text). As a result of their asymmetric internal structures, these particles are prevented from having normal (fast) symmetric decays, so that they live for abnormally long lengths of time; hence they are "strange" particles, and they are denoted as kaons in order to distinguish them from the symmetric short-lived nonstrange mesons. (It should be noted here that the nonstrange pi meson also has a long lifetime; however, this is because the pion, as the lightest of the hadrons, has no lower-mass hadronic state to which it can decay.) There is a clear-cut empirical difference between kaons and mesons: the asymmetric kaons have antiparticles, whereas the symmetric mesons do not. Kaons are assigned the strangeness quantum number S = +1, and antikaons are assigned the strangeness quantum number S = -1. Kaons have isotopic spin I = 1/2, corresponding to the charge states K^+ and K^o (kaons) and $\bar{K}^-$ and $\bar{K}^o$ (antikaons).

The strangeness quantum number† S is conserved in the strong interactions that produce hadrons, which is why it is a useful concept.†† If an initial state in a production reaction has S = 0, then the particles that are produced must have total strangeness S = 0. However, if a kaon K, with strangeness S = +1, or an antikaon $\bar{K}$, with strangeness S = -1, is used as the bombarding particle, then the initial state has S = +1 or S = -1, respectively, and the final state must also have this same overall value for S. The strangeness characteristics of mesons and kaons are related to the strangeness characteristics of baryons and hyperons via the process of Associated Production, in which S is conserved. To see how Associated Production works, consider the reaction $\pi + p \rightarrow K + \Lambda$, where K is a kaon and Λ is a lambda hyperon. In the initial state of this reaction, the pion π and proton p each have strangeness S = 0, so that the initial-state total strangeness, and hence also the final-state total strangeness, is zero. But in the final state the kaon K has strangeness S = +1, so that the lambda hyperon Λ must have strangeness S = -1. Hence the production of a strangeness S = +1 particle is associated with the production of a strangeness S = -1 particle, as the name Associated Production suggests. Similarly, from the reaction $\bar{K} + N \rightarrow \Xi + K$, in which the initial-state antikaon $\bar{K}$ has S = -1 and the final-state kaon K has S = +1, we can deduce that the cascade particle Ξ has strangeness S = -2. The decay modes of the Ξ and Λ particles are $\Xi \rightarrow \Lambda + \pi$ and $\Lambda \rightarrow N + \pi$, so that strangeness is broken in the decays of these particles. However, the particles (both hyperons and kaons) which change in strangeness during their decays have very long lifetimes, so that strangeness is not easily broken.

---

†In the present book we will at various times use S to represent (1) the strangeness quantum number, or (2) the intrinsic spin of a particle, or (3) the spinor S (which is a 330 MeV spin 1/2 quark state); in each case, the context should make clear which quantity we are using.

††Strangeness is not conserved in the weak decays of these particles.

From the above discussion, we have identified the baryon number B, the angular momentum J,† the isotopic spin I, and the strangeness S as basic quantum numbers of elementary particles. Other quantum numbers include the electric charge Q, the parity P, the charge conjugation number C, and the G-parity G. The parity P characterizes the spatial symmetry properties of the particle: symmetric wave functions correspond to P = +, and antisymmetric wave functions correspond to P = –. The quantum numbers B, J, and Q are absolutely conserved in all reactions. In our present efforts to reproduce the observed spectrum of elementary particles, we will be mainly concerned with the quantum numbers B, J, and S, and we will make essentially no use of the quantum numbers C and G.

In terms of the quantum numbers B and S, the four categories of hadrons are:

baryons (B = +1, S = 0), hyperons (B = +1, S < 0);

mesons (B = 0, S = 0), kaons (B = 0, S = +1).

The antiparticles that correspond to these particles have the signs for B and S reversed. Since mesons have B = S = 0, there are no antimesons. The notation that is used for the observed families of baryons and hyperons is summarized in the following table:

Excitation	Strangeness	Isotopic spin	Charge states
N (or N*)	S = 0	I = 1/2	N^+, N^0
Δ (delta)	S = 0	I = 3/2	Δ^{++}, Δ^+, Δ^0, Δ^-
Λ (lambda)	S = -1	I = 0	Λ^0
Σ (sigma)	S = -1	I = 1	Σ^+, Σ^0, Σ^-
Ξ (cascade)	S = -2	I = 1/2	Ξ^0, Ξ^-
Ω (omega)	S = -3	I = 0	Ω^-

†Although we described J earlier as a spin angular momentum, J is actually the total angular momentum of the hadronic state. It can be decomposed into an "intrinsic" (quark) spin S and an orbital angular momentum ℓ, so that $\vec{J} = \vec{\ell} + \vec{S}$. The decomposition of J into its ℓ and S components is the subject that forms Part II of the present book.

Each of these families except the Ω contains a series of energy levels that have different values for the angular momentum and parity. These levels are characterized by the labels S, P, D, F, G, H, ..., where the parity P, total angular momentum J, and orbital angular momentum ℓ have the following values:

	S	P	D	F	G	H
$P^{\dagger}$	-	+	-	+	-	+
J	1/2	1/2,3/2	3/2,5/2	5/2,7/2	7/2,9/2	9/2,11/2
ℓ	0	1	2	3	4	5

In order to more completely specify these states, the mass or energy value is added in parentheses, and the isotopic spin I and angular momentum J are added as subscripts, with all of the half-integral values for I and J multiplied by 2. Thus, for example, $N(1514)D_{13}$ denotes an I = 1/2, J = 3/2 D-state N baryon with a mass of 1514 MeV; and $\Sigma(1774)D_{15}$ denotes an I = 1, J = 5/2 D-state Σ hyperon with a mass of 1774 MeV. Meson states are not sorted out according to families, but are denoted by a variety of symbols — π, η, ρ, ω, f, g, A, B, etc. If two or more meson states have the same quantum numbers, the higher-mass states are sometimes denoted by primes (e.g., η and η', ρ and ρ'). Kaon resonances are denoted by the letter K (e.g., K(494) and K*(892)), and also by other symbols (e.g., κ, Q, and L). As we discuss in Chapters 15 - 17, roughly 185 different baryon, hyperon, meson, and kaon states have been identified experimentally, and this list is continuing to increase as the experiments are extended.

There is one final aspect of elementary particle spectroscopy that we should mention here — namely, the use of the words particle and resonance. The lifetimes of the unstable elementary particles range from 2×10^{-6} seconds for the muon down to about 10^{-24} seconds for the shortest-lived states. Since the velocity of light is $c = 3 \times 10^{10}$ cm/sec, a particle that lives for less than $1/3 \times 10^{-10}$ sec will, apart from relativistic effects, travel less than 1 cm before it decays. The relativistic dilation of time can increase this path length by a couple of orders of magnitude or so, but in any event a particle that has a half life of less than (say) 10^{-15} sec has no measurable path length. More than 80% of all observed elementary particle states have extremely short lifetimes — less than 10^{-22} sec -, and hence can be identified

---

†The parity of these states is $P = -(-1)^{\ell}$, where $(-1)^{\ell}$ arises from the spatial Legendre wave function, and where the overall minus sign is from the presence of a negative-parity pion or kaon in the initial or final πN or KN state (the nucleon N has positive parity).

in a bubble chamber or counter experiment only on the basis of the long-lived particles that they emit when they decay. How then shall we characterize these very-short-lived states? Consider a reaction of the type $\gamma + N \to N^* + X$, where $X \to \pi\pi$ is a short-lived meson state that is produced by photon bombardment of a nucleon. By determining the energies of the two final-state pions, we can calculate the total center-of-mass energy (or mass) of the state X that produced the pions. If we have a large number of these events, and if X is a well-defined meson excitation, then the two final-state pions will have an energy distribution that peaks at a definite mass value and has a definite width Γ. If X represents the rho meson, for example, the $\pi\pi$ final states will have an energy distribution that peaks at 773 MeV and has a full width at half maximum Γ of about 150 MeV. These peaks in the energy or mass distributions have Breit-wigner resonance shapes of the same type as are found in nuclear physics, so that these short-lived particles are often denoted as resonances. The lifetime τ and width Γ of a particle state are related to one another by means of the equation $\tau = \hbar/\Gamma$, where $\hbar$ is Planck's constant divided by 2π; τ is the measurable quantity for the longer-lived particle states, whereas Γ is the measurable quantity for the short-lived resonances. As a terminology, we will generally refer to the longer-lived states as particles and the shorter-lived states as resonances, although these two terms are used somewhat interchangeably in the text.

As the final element in our notation for hadron resonances, we add the width Γ, so that we have, for example, the hadronic states $N(1514)D_{13}\Gamma 130$, $\Sigma(1774)D\Gamma 130$, $\rho(773)\Gamma 152$, and $\omega(783)\Gamma 10$. The width Γ turns out to be one of the most useful pieces of information we have for sorting out the characteristics of the various hadron resonances.

The concepts that have been introduced here will be amplified and extended in the main body of the text. In particular, we will discuss in considerable detail certain quantities such as the orbital angular momentum quantum number ℓ, whose interpretation for baryon and meson resonances has been somewhat obscure. Much of the currently-used elementary particle notation has been introduced in connection with SU(3)-type symmetries. In the present studies we will make very little use of these SU(3) symmetries, and we in fact cite a number of cases where the experimental data may be in conflict with the requirements of SU(3). However, for the sake of consistency we use this standard elementary particle notation.

APPENDIX B. THE EXPERIMENTAL ELEMENTARY PARTICLE DATA

B1. A Critique of the Review of Particle Properties

The field of elementary particle physics has been one of extraordinary activity — and extraordinary expenditures — during the past few decades. Perhaps the most enduring result to emerge from all of this effort is the compilation of data known as the Review of Particle Properties (RPP). Published every year or two by the Particle Data Group, it presents in an orderly manner the experimental data on elementary particles as collected from research laboratories all over the world.

Since the members of the Particle Data Group are physicists as well as compilers, they have a natural (and commendable) tendency to want to fit these data into some kind of theoretical framework. The one framework which has been offered to them by the theorists is that of SU(3) and its various derivatives. In spite of demonstrated shortcomings in the SU(3) approach (numerous examples are scattered throughout the pages of the present text), SU(3) has in general been accepted as "the theory" for elementary particles, and the pages of the Review of Particle Properties contain extensive descriptions of the various SU(3)-type symmetries which are believed to be applicable to elementary particle states.

The difficulty which is raised by this confrontation of experiment with theory is that the Review of Particle Properties is not "a collection of experimental data", but rather "a collection of experimental data as (implicitly) evaluated within the context of SU(3)". This procedure inevitably gives rise to a bias in the data selection: data which fit readily into SU(3) are quite promptly accepted, whereas data which do not fit into SU(3) tend to be treated with caution. Of course, this statement is somewhat of an exaggeration, and considerable efforts are in fact made by the members of the Particle Data Group to judge the data strictly on experimental grounds — using the criteria of reproducibility and of statistical significance. However, due to the incompleteness and the complexity of the data (see the quotations at the beginning of Chapters 15 and 16), an objective judgment is not always possible. There are a few important examples in the Review of Particle Properties[1] in which data that are not in agreement with SU(3) symmetries appear to have been discriminated against, and the purpose of the present section is to discuss these examples.

One important example is the M(953) meson. As we discuss in Sec. 5.2 of Chapter 5, the M(953) and $\eta'(958)$ are closely-related meson states that probably carry the same basic quantum numbers. However, there are three distinct differences between these states: (1) they have different production mechanisms [see Eqs. (5.8) and (5.9)]; (2) they have different decay mechanisms [see Eqs. (5.10) and (5.11)]; and (3) they have different mass values ($M_{\eta'} - M_M \simeq 5 \pm 2$ MeV). Now, the members of the Particle Data Group feel (with some justification) that the difference between the mass values of the M and the η' is not clear-cut (however, see the discussion of these mass values that is given on page 873 of Ref. 2). But the difference in the decay mechanisms is very clear-cut, as we see from the experimental results of M. Aguilar-Benitez et al.[3]:

$$M \rightarrow \rho^0\gamma/(\text{all } \pi\pi\gamma) = 0.05 \pm 0.10;$$

$$\eta' \rightarrow \rho^0\gamma/(\text{all } \pi\pi\gamma) = 0.94 \pm 0.20.$$

These radically-different decay modes constitute a clear experimental distinction between these two states, and they in fact formed the main basis for the initial claim[4] that the M meson is a new resonance. As one of the authors of the original experiment (V. E. Barnes) once commented to the present author in a discussion about the M meson, "One does not lightly set forth claims for a new resonance". An examination of the meson resonances which are listed at the front of RPP76[1] (the more-or-less established resonances) shows several whose experimental credentials are no better than those of the M. However, the M is still listed only at the back of RPP76. If SU(3) demanded the existence of both the M(953) and η'(958) meson states, it is a virtual certainty that the M would have long ago appeared at the front of RPP; but SU(3) has no place for the M, so it languishes along in the back pages.

Another, and somewhat more flagrant, example of bias is provided by the spin 2 η_N(1083) meson. This meson, like the M, overfills an established SU(3) nonet, so that SU(3) theorists have no place in which to put it. The Review of Particle Properties has for several years carried the quotation by Oh et al.[5] that this "bump is almost certainly the result of P-D interference", which is a remark that seems designed to dispose of this peak as a valid resonant state. However, this peak has been observed in a number of experiments[6], and the spin J = 2 seems quite well established. Furthermore, a careful reading of Oh's paper[5] shows that these workers considered several possibilities for this peak, and their favored explanation is that of a "D-wave resonance (loop or cusp) with other waves varying slowly". In Sec. 13.7 of Chapter 13, we give an explanation for the mysterious P-wave background that appears underneath the η_N(1083) (see Fig. 13.6).

One final meson resonance that does not fit into the SU(3) systematics, and which has had a hard time gaining acceptance in the Review of Particle Properties, is the $A_{1.5}$(1197) enhancement. Although this peak does not show up strongly, it has been[7] and continues to be[8] observed in a number of different experiments. Its spin and parity have been determined[8] as $J^P = 1^-$, which means that it also overfills an established SU(3) nonet.

The M, η_N, and $A_{1.5}$ meson resonances admittedly are not dominant mesons in the sense that the ρ, ω, and ϕ are dominant mesons. However, the evidence for their existence and their properties is as compelling as the evidence for some of the more-favored resonances which have been admitted to the front pages of the Review of Particle Properties. A statement is often encountered in the elementary particle conference summaries to the effect that "the systematics of SU(3) (or of the related Gell-Mann and Zweig quark model) is consistent with all of the well-established meson resonances". If one does not look too carefully at the experimental meson data, this is a correct statement. However, an examination of the meson data in more detail reveals that only by a somewhat cavalier definition of the phrase "well-established" can this statement be maintained.

When we turn to the baryon resonances, we face somewhat the same problem, but with an additional complication. The baryon resonances that are extracted from the data are already a mixture of experiment and theory, as K. C. Wali points out in the quotation at the beginning of Chapter 15. This mixing of theory with experiment is necessary because of the manifestly incomplete nature of the πN and $\bar{K}N$ scattering data, as we discuss in some detail in Chapter 15. Hence it is possible for an unconscious bias in favor of SU(3) to be built into the parametrizations that are used to represent the baryon scattering amplitudes as a function of energy.

One fact which does emerge clearly from the reported baryon and hyperon resonances is the relationship between the ratings assigned to the resonances and the manner in which they fit into SU(3) multiplets. Group I of Table 15.1 in Chapter 15 lists 45 baryon resonances (including the long-lived octet baryons p, n, Λ, Σ^+, Σ^0, Σ^-, Ξ^0, and Ξ^-) which have measured spin values and which have been rated as "well-established" by the Particle Data Group; Group II of Table 15.1 lists 24 baryon resonances with spin identification that are rated as "not-well-established" by the Particle Data Group (the S, P, and D levels above 1900 MeV are not included here, since the data are somewhat fragmentary). Of the 45 resonances rated as "well-established", 28 fit into "established" SU(3) multiplets (these are the multiplets with two or more identified members); however, of the 24 resonances rated as "not-well-established", none fit into "established" SU(3) multiplets. So here again we have a situation in which resonances that fit into SU(3) are accorded much better rankings than are resonances which do not fit into SU(3).

The situation that we have just described with respect to the ratings assigned to the meson and baryon resonances is not serious, but it does exist, and users of the Review of Particle Properties should be wary of this built-in SU(3) "glass slipper" effect. In general, the data compilation in the Review of Particle Properties is excellent, and it is from these data that the ultimate "truth" about elementary particles will emerge. As an aside, the present author is willing to wager that he has spent more time in studying these tables during the past eight years (probably by a factor of two) than any other physicist.

B2. The Experimental Data

Table B1 summarizes the units and conventions that are used in describing the baryon and meson resonances, and it defines the rating system that is used for these resonances. The ****, ***, **, and * labels for the baryon and hyperon resonances are the ratings that were assigned by the Particle Data Group in RPP76 (Ref. 1).

Tables B2, B3, B4, B5, and B6 list the N, Δ, Λ, Σ, and (Ξ,Ω) resonances, respectively; Tables B7 and B8 list the low-mass meson and kaon resonances; Table B9 lists the high-mass New Particle meson and kaon resonances; and Table B10 lists some possible long-lived New Particle baryon resonances. The resonances in each table are arranged in the order of increasing mass values.

The list of references at the end of Appendix B describes the manner in which the values for each resonance were obtained. Wherever possible, the RPP76-approved values were used. However, in some cases it was necessary to make rather arbitrary decisions in order to arrive at a set of values for a resonance; in these cases, we have documented the data used to make these decisions, although other workers might well choose different sets of values.

B2a. Review of Particle Properties — 1978 Edition (RPP78)

The 1978 edition of the Review of Particle Properties (RPP78) was being put together by the members of the Particle Data Group just as the present book was ready to be sent to the printer, and a copy of the data tables that appear at the front of RPP78 was received by the author after the present data tables in Appendix B had already been composed. In this section, we briefly summarize pertinent results from the RPP78 data tables that reflect on the data compilations which we give here in Appendix B. The author would like to thank the members of the Particle Data Group for making the RPP78 data tables available to him prior to publication.

In the present data compilation, we have distinguished between the high-mass New Particle baryon and hyperon states (Table B10) and the lower-mass Old Particle baryon and hyperon states (Tables B2 - B6), and also between the high-mass New Particle meson and kaon states (Table B9) and the lower-mass Old Particle meson and kaon states (Tables B7 and B8). In RPP78 this distinction is not made, and both the Old Particle and New Particle states are grouped together in the same tables.

With respect to the baryon and hyperon resonances, the following points can be noted in the RPP78 data tables. Table B2: N resonances. The new N(1540)P_{13} resonance is accorded a * rating; the N(1710)D_{13} state is promoted from ** to *** rating; the N(1989)F_{15} and N(2100)D_{15} states are promoted from * to ** ratings; and the N(2142)G_{19} resonance is accorded *** status. The ratings of the other N states are unchanged, and the determinations of masses and widths are pretty much as they were in RPP76. Table B3: Δ resonances. The new Δ(1550)P_{31} resonance is accorded * status; the Δ(1680)P_{33} state is promoted from * to *** rating; the Δ(1869)F_{35} and Δ(1900)P_{31} resonances are raised from *** to **** ratings; and the Δ(1889)D_{35} and Δ(2174)G_{39} resonances are promoted from ** to *** status (although the spin identification of the Δ(2174) level is listed as undetermined). Many of the Δ resonances still have wide variations in their reported mass values. Table B4: Λ resonances. The Λ(1573) P_{01} resonance is raised from * to ** status; a Λ(2325)D_{03} state is given a * rating; and the Λ(1827)S_{01} and Λ(2120)F_{05} states are raised from ** to *** ratings. The other Λ resonances are more-or-less unchanged from their RPP76 values. Table B5: Σ resonances. The Σ(1676)P_{11} resonance is raised from ** to *** status, and the other Σ resonances are essentially unchanged. Table B6: Ξ and Ω resonances. The Ξ(1956) state (or group of states) is changed from *** to ** status; the Ξ(2024) resonance is raised from ** to *** ranking; and the new Ξ(2121) level is awarded a * ranking. Table B10: New Particle baryon resonances. The Λ_c(2260) and Σ_c(2430) hyperon states are each accorded * status. These changes are noted in the Appendix B data tables.

The main conclusions which seem to emerge from a comparison of the baryon and hyperon resonances listed in RPP78 with those listed in RPP76 are that no major changes have occurred, and that the uncertainties in the determinations of the resonance parameters (masses and widths) have not appreciably decreased. This suggests that the currently-performed phase shift analyses are extracting most of the information content from the available experimental data. Thus, future improvements in these results must be based either on more complete experimental data (such as the very difficult R and A measurements), or else on theoretical revisions in the phase shift analyses (such as are involved in the energy dependences of the phase shifts). Most of the baryon and hyperon resonances that have received improved confidence-level ratings in RPP78 are states which fit naturally into the baryon rotational bands shown in Tables H1 - H4 of Appendix H.

With respect to the meson and kaon resonances, the following points can be noted in RPP78. Table B7: Low-mass meson resonances. The meson states that are shown in Table B7 with RPP76 A-ratings are still listed as such in RPP78, with one exception: the $F_1(1542)$ state is no longer listed at the front. Also, the RPP76 A-rated $\varepsilon(\sim 1200)\Gamma 600$ meson S-state (which we omitted from Table B7 because of its very broad width) now appears in RPP78 as the A-rated $\varepsilon(\sim 1300)\Gamma\sim 300$ S-state; and the new $\pi\pi(2150)3^-$ and $\pi\pi(2310)4^+$ states that are shown in Table B7 can be identified with the $T(2192)3^-$ and $U(2350)4^+$ states that are promoted to A rankings in RPP78. With these exceptions, the low-mass A-rated meson states listed in RPP78 are essentially the same as the corresponding states that are given in Table B7. Table B8: Low-mass kaon resonances. The Q(1368) resonance, which is denoted as $Q_2(1400)$ in RPP78, is considered as not being a well-established resonance; the $K_N(1785)3^-$ state has been promoted to an A ranking; and a broad $\kappa(\sim 1425)\Gamma\sim 250$ S-state is listed as a possible A-rated resonance. With these exceptions, the low-mass kaon data of Table B8 and of RPP78 are in substantial agreement. The RPP78 changes in rankings for the low-mass meson and kaon states of Tables B7 and B8 are noted in these tables. Table B9: High-mass New Particle meson and kaon resonances. These New Particle resonances have appeared recently enough that the quoted masses and widths are taken in most cases directly from the original research papers. The New Particle states which appear in the data tables at the front of RPP78 (and hence have the A rating of Table B1 in the present Appendix) are noted as A-ranked states in Table B9.

There is one other data change in RPP78 that should be noted here. The value for the magnetic moment of the Λ hyperon has been changed from -0.67 ± 0.06 n. m. to -0.606 ± 0.034 n. m. This change increases the difference between the magnetic moment of the Λ and the magnetic moments of the Σ^- and Ξ^- hyperons (which are all predicted by SU(6) to have the same value), and it heightens the difficulty for SU(6) that is discussed in connection with Table 19.6 in Chapter 19.

Table B1. Key to the notation and rating systems used in the baryon and meson data tables (Tables B2 - B10). The literature search for these tables was closed in March, 1978 (see Section B2a for final comments based on RPP78).

Notation (see Appendix A for a full description):

The mass is in MeV.

The width Γ is the full width at half maximum in MeV.

Baryon labels (e.g., P_{11}) denote ℓ_{IJ}, where ℓ = orbital angular momentum, I = isotopic spin, J = total angular momentum, and where half-integer values of I and J are multiplied by two.

Meson labels J^P denote total angular momentum J and parity P.

Ratings of the baryon resonances, as assigned in RPP76:

**** Good, clear, and unmistakable.

*** Good, but in need of clarification or not absolutely certain.

** Needs confirmation.

* Weak

Other baryon ratings used in the present tables:

† Data listed in RPP76, but not identified as a resonance.

new New data that appeared subsequent to RPP76.

"dead" Data declared "dead" by the members of the PDG in RPP76.

K76 Baryon states which are now recommended for *** status in the Karlsruhe 76 analysis (see Ref. 10).

Meson ratings used in the present tables:

A Listed at the front of RPP76.

B Listed at the back of RPP76, but not at the front.

C Not listed in RPP76, or controversial.

new New data that appeared subsequent to RPP76.

Table B2. The N baryon resonances.

Resonance	Rating	RPP78	References
p(938.3)Γ0	****		RPP76
n(939.6)Γ≈0	****		RPP76
N(~1230)P_{11}Γ~200	†		1
N(1413)P_{11}Γ187	****		2,3
N(1511)S_{11}Γ104	****		2,4
N(1517)D_{13}Γ130	****		2,4,10,11
N(1532)P_{11}Γ89	†		2,5
N(1540)P_{13}Γ200	new	*	6
N(1674)S_{11}Γ187	****		2,4
N(1674)D_{15}Γ159	****		2,4
N(1680)F_{15}Γ123	****		2,4
N(1710)D_{13}Γ100	**	***	2
N(1720)P_{11}Γ137	***		2,6,7,8
N(1722)P_{13}Γ132	***		2,6,7,8
N(1986)F_{17}Γ214	**,K76		9,10,11
N(1989)F_{15}Γ179	*	**	2
N(2029)D_{13}Γ116	**		2
N(2100)S_{11}Γ200	*,K76		12
N(2100)D_{15}Γ220	*,K76	**	2
N(2132)G_{17}Γ247	***		2,10
N(2142)G_{19}Γ297	†,K76	***	2,10
N(2183)H_{19}Γ364	***		2,10
N(2650)$I_{1,11}$Γ350	***		RPP76,10
N(3030)Γ400	***		RPP76

Table B3. The Δ baryon resonances.

Resonance	Rating	RPP78	References
$\Delta(1232)P_{33}\Gamma 115$	****		RPP76
$\Delta(1550)P_{31}\Gamma 110$	new	*	6
$\Delta(1623)S_{31}\Gamma 161$	****		2
$\Delta(1676)D_{33}\Gamma 207$	***		2,4,13
$\Delta(1680)P_{33}\Gamma 220$	*	***	12,14
$\Delta(1869)F_{35}\Gamma 255$	***	****	2
$\Delta(1889)D_{35}\Gamma 121$	**	***	2
$\Delta(1900)P_{31}\Gamma 200$	***	****	12,14
$\Delta(1928)F_{37}\Gamma 237$	****		2
$\Delta(2070)S_{31}\Gamma 100$	*,K76		11
$\Delta(2174)G_{39}\Gamma 205$	**	***	2
$\Delta(2200)G_{37}\Gamma 293$	new		11
$\Delta(2406)H_{3,11}\Gamma 289$	***		2,10
$\Delta(2850)\Gamma 400$	***		15
$\Delta(3230)\Gamma 440$	***		15

Table B4. The Λ hyperon resonances.

Resonance	Rating	RPP78	References
$\Lambda(1115.6)\Gamma\simeq 0$	****		RPP76
$\Lambda(1327)\Gamma<25$	"dead"		16
$\Lambda(1402.4)S_{01}\Gamma 38$	****		RPP76
$\Lambda(1519)D_{03}\Gamma 15$	****		RPP76
$\Lambda(1573)P_{01}\Gamma 147$	*	**	17
$\Lambda(1672)S_{01}\Gamma 33$	****		18
$\Lambda(1687)D_{03}\Gamma 30$	†		19,20
$\Lambda(1690)D_{03}\Gamma 60$	****		17,20
$\Lambda(1690)P_{01}\Gamma 22$	**		21
$\Lambda(1710)P_{03}\Gamma 20$	***		22
$\Lambda(1747)P_{01}\Gamma 120$	**		23
$\Lambda(1808)G_{09}\Gamma 27$	*		17,24
$\Lambda(1817)P_{01}\Gamma 12$	new		25
$\Lambda(1822)F_{05}\Gamma 81$	****		17
$\Lambda(1825)D_{05}\Gamma 94$	****		17
$\Lambda(1827)S_{01}\Gamma 150$	**	***	17,26
$\Lambda(1881)P_{03}\Gamma 101$	***		17,26,27
$\Lambda(2010)D_{03}\Gamma 130$	**		21
$\Lambda(2020)F_{07}\Gamma 160$	*		21
$\Lambda(2110)G_{07}\Gamma 250$	****		17
$\Lambda(2120)F_{05}\Gamma 170$	**	***	17
$\Lambda(2360)H_{09}\Gamma 120$	****		28
$\Lambda(2585)\Gamma 300$	***		29

Table B5. The Σ hyperon resonances.

Resonance	Rating	RPP78	References
$\Sigma^+(1189.4)\Gamma\simeq 0$	****		RPP76
$\Sigma^o(1192.5)\Gamma\simeq 0$	****		RPP76
$\Sigma^-(1197.4)\Gamma\simeq 0$	****		RPP76
$\Sigma(1385)P_{13}\Gamma 38$	****		RPP76
$\Sigma(1476)\Gamma 34$	*		30
$\Sigma(1582)D_{13}\Gamma 11$	**		31
$\Sigma(1623)S_{11}\Gamma 45$	**		22,26,32
$\Sigma(1670)D_{13}\Gamma 50$	****		17
$\Sigma(1676)P_{11}\Gamma 120$	**	***	17
$\Sigma(1680)P_{11}\Gamma 25$	new,†		33
$\Sigma(1697)S_{11}\Gamma 66$	†		34
$\Sigma(1760)P_{11}\Gamma 77$	*		35
$\Sigma(1770)S_{11}\Gamma 60$	***		17
$\Sigma(1774)D_{15}\Gamma 130$	****		17
$\Sigma(1840)P_{13}\Gamma 120$	*		26,36
$\Sigma(1920)D_{13}\Gamma 300$	***		17,20
$\Sigma(1920)F_{15}\Gamma 130$	****		17
$\Sigma(1925)P_{13}\Gamma 65$	†		34
$\Sigma(1945)D_{13}\Gamma 65$	†		37,20
$\Sigma(1954)P_{11}\Gamma 213$	**		38
$\Sigma(1955)S_{11}\Gamma 170$	*		17
$\Sigma(2040)F_{17}\Gamma 190$	****		17
$\Sigma(2070)F_{15}\Gamma 140$	*		39
$\Sigma(2090)G_{17}\Gamma 103$	*		21
$\Sigma(2100)P_{13}\Gamma 200$	**		40
$\Sigma(2210)G_{19}\Gamma 60$	****		41
$\Sigma(2455)\Gamma 220$	***		RPP76
$\Sigma(2620)\Gamma 175$	***		29
$\Sigma(3000)$	**		42

Table B6. The Ξ and Ω hyperon resonances.

Resonance	Rating	RPP78	References
$\Xi^0(1314.9)\Gamma\simeq 0$	****		RPP76
$\Xi^-(1321.3)\Gamma\simeq 0$	****		RPP76
$\Xi^0(1531.8)P_{13}\Gamma 9$	****		RPP76
$\Xi^-(1535.2)P_{13}\Gamma 10$	****		RPP76
$\Xi(1617)\Gamma 28$	**		RPP76
$\Omega(1672.2)\Gamma\simeq 0$	****		RPP76
$\Xi(1819)3/2\Gamma 19$	***		43
$\Xi(1871)\Gamma 17$	new		44
$\Xi(1956)\Gamma 43$	***	**	45
$\Xi(2024)\geq 5/2\Gamma 16$	**	***	46
$\Xi(2121)\Gamma 24$	†,new	*	47
$\Xi(2203)\Gamma<20$	new		44
$\Xi(2295)\Gamma<30$	*		48
$\Xi(2500)\Gamma 59$	**		49

Table B7. The low-mass meson resonances.

Resonance	J^P	Rating	RPP78	References
$\pi^0(135.0)\Gamma\simeq 0$	0^-	A		RPP76
$\pi^\pm(139.6)\Gamma\simeq 0$	0^-	A		RPP76
$\eta(548.8)\Gamma\simeq 0$	0^-	A		RPP76
$\pi\pi(658)\Gamma\simeq 15$		C		50
$\rho(773)\Gamma 152$	1^-	A		RPP76
$\omega(783)\Gamma 10$	1^-	A		RPP76
$M(953)\Gamma\leq 10$		B		51
$\eta'(957.6)\Gamma<1$	0^-	A		RPP76
$\delta^\pm(976)\Gamma 47$		A		RPP76
$S^*(993)\Gamma 40$	0^+	A		RPP76
$H(\sim 1000)$	1	B		RPP76,52
$\phi(1020)\Gamma 4$	1^-	A		RPP76
$\eta_N(1083)\Gamma 98$	2^+	B		RPP76
$A_1(1100)\Gamma\sim 300$	1^+	A		RPP76
$M^0(1148)\Gamma 15$		B		RPP76
$X^-(1153)\Gamma 16$		C		53
$A_{1.5}(1197)\Gamma<29$	1^-	B		54
$B(1230)\Gamma 125$	1^+	A		55
$\rho'(1250)\Gamma 130$	1^-	B		RPP76,56
$f(1271)\Gamma 180$	2^+	A		RPP76
$D(1285)\Gamma 21$	1^+	A		57
$A_2(1310)\Gamma 102$	2^+	A		RPP76
$A_2\text{dip}(1310)\Gamma\sim 10$	2^+	C		58
$X_0(1411)\Gamma 95$		B		59
$X_1(1415)\Gamma 30$		B		60
$E(1416)\Gamma 60$	1^+	A		RPP76
$f'(1522)\Gamma 62$	2^+	A		61
$F_1(1542)\Gamma 24$		A	B	62

(continued on next page)

Table B7 (continued).

Resonance	J^P	Rating	RPP78	References
$\rho'(1600)\Gamma\sim300$	1^-	A		56
$A_3(1640)\Gamma240$	2^-	A		63
$\omega(1667)\Gamma150$	3^-	A		RPP76
$X(1681)\Gamma64$		B		RPP76
$g(1690)\Gamma180$	3^-	A		RPP76
$A_4(1825)\Gamma90$		B		64
$h(2040)\Gamma193$	4^+	A		RPP76
$\pi\pi(2150)\Gamma200$	3^-	new	A	65
$\pi\pi(2310)\Gamma210$	4^+	new	A	65
$\pi\pi(2480)\Gamma280$	5^-	new		65

Table B8. The low-mass kaon resonances.

Resonances	J^P	Rating	RPP78	References
$K^{\pm}(493.7)\Gamma\simeq 0$	0^-	A		RPP76
$K^{0}(497.7)\Gamma\simeq 0$	0^-	A		RPP76
$\kappa(725)\Gamma 15$		C		66
$K^*(892)\Gamma 49$	1^-	A		RPP76
$Q(1234)\Gamma 188$	1^+	A		67
$K\pi\pi(1278)\Gamma 25$	1^+	B		68
$Q(1368)\Gamma 241$	1^+	A		67
$K_N(1421)\Gamma 108$	2^+	A		RPP76
$K_N(1673)\Gamma 77$		B		69
$K\pi\pi(1692)\Gamma 26$		new		70
$L(1765)\Gamma 138$	2^-	A		RPP76
$K_N(1785)\Gamma 100$	3^-	B	A	71
$K_N(1840)\Gamma 100$		B		72
$K_N(2115)\Gamma 300$	4^+	new		73

Table B9. The narrow-width high-mass New Particle meson and kaon resonances.

Resonance	J^P	RPP78	References
$\bar{p}n(1795)\Gamma\leq 8$			RPP76
$M(1817)\Gamma 30$			74
$D^{o}(1863.3)\Gamma<2.4$	0^-	A	75
$D^{+}(1868.3)\Gamma<2.4$	0^-	A	75
$S(1935)\Gamma\leq 4$		A	76
$D^{*o}(2006)\Gamma<2$	1^-	A	75,77
$D^{*+}(2008.6)\Gamma<2$	1^-	A	75,78
$\bar{p}p(2020)\Gamma 24$			79
$F(2030)$			80
$F^*(2140)$			80
$\bar{p}p(2204)\Gamma 16$			79
$(K\pi)^{o}(2290)$			81
$(K\pi)^{o}(2420)$			81
$(K\pi)^{o}_{R}(\sim 2430)$			82
$I(2600)\Gamma\leq 19$			83
$\bar{p}p(2660)$			81
$X(2830)$			RPP76S
$\bar{N}N(2850)\Gamma\leq 30$			84
$\bar{p}p\pi(2950)\Gamma<15$			85
$\bar{N}N(3050)\Gamma<20$			84
$J/\psi(3098)\Gamma 0.067$	1^-	A	RPP76
$\chi(3340)$			86
$\chi(3414)$	0^+	A	86
$\chi(3455)$			86
$\chi(3503)$	1^+?	A	86
$\chi(3551)$	2^+?	A	86
$\psi'(3684)\Gamma 0.228$	1^-	A	RPP76

(continued on next page)

Table B9 (continued).

Resonance	J^P	RPP78	References
$\psi''(3772)\Gamma 28$	1^-	A	87
$\psi(4028)$			RPP76S,88
$\psi(4100)$			RPP76
$\psi'''(4414)\Gamma 33$	1^-	A	RPP76
T(9400)	1^-	A	89
T(10010)			89
T(10400)			89

Table B10. The high-mass New Particle baryon and hyperon resonances.

Resonance	RPP designation	RPP78	Reference
$\bar{\Lambda}(2260)\Gamma\leq 40 \rightarrow \bar{\Lambda}\pi\pi\pi$	$\Lambda_c^+(2260)$	*	RPP76S
$\Lambda(2426) \rightarrow \Lambda\pi\pi\pi\pi$	$\Sigma_c(2430)$	*	RPP76S

References for Appendix B1.

1. Particle Data Group, "Review of Particle Properties", Rev. Mod. Phys. 48, No. 2, Part II, April (1976) — denoted here as RPP76.

2. M. H. Mac Gregor, Phys. Rev. D10, 850 (1974).

3. M. Aguilar-Benitez et al., Phys Rev. D6, 29 (1972)

4. M. Aguilar-Benitez et al., Phys. Rev. Lett. 25, 1635 (1970).

5. B. Y. Oh et al., Phys. Rev. D1, 2494 (1970).

6. M. H. Mac Gregor, Phys. Rev. D9, 1259 (1974). See the discussion of the η_N in Ref. 58 of this paper.

7. RPP76, Ref. 1, lists nine references to the $A_{1.5}$.

8. J. Böhm and B. R. French, CERN report CERN/D.Ph.II/PHYS 75-13, May (1975).

References for Appendix B2 data tables.

RPP76: Reference 1 above.

RPP76S: Particle Data Group, supplement to RPP76, Phys. Lett. 68B, 1 (1977).

1. The mass and width are a rough average of data given in H. Braun et al., Nucl. Phys. B95, 503 (1973), Fig. 3b; H. Braun et al., Phys. Rev. D8, 2765 (1973); and Z. Strachman et al., Nucl. Phys. B98, 120 (1975), Fig. 3d. The P_{11} assignment for this state is from D. Lissauer et al., Phys. Rev. D6, 1852 (1972).

2. AYED 74 in RPP76.

3. Although the AYED 74 values for the split Roper resonance are not confirmed in other phase shift analyses, this low value for the P_{11} phase shift agrees with a whole series of production peaks reported in RPP76.

4. BARBOUR 76 in RPP76.

5. Although this peak is not confirmed in other phase shift analyses, it fits in well with the systematics of Fig. 15.1 in Chapter 15.

6. R. S. Longacre and J. Dolbeau, Nucl. Phys. B122, 493 (1977).

7. Data from LBL-2636, as quoted in Longacre and Dolbeau (Ref. 6).

8. The mass values for this resonance vary widely in the different phase shift analyses, but the average of Refs. 2, 6, and 7 gives a representative value.

9. This is an average of six experiments in RPP76 (all except LEA 69) plus Refs. 10 and 11.

10. E. Pietarinen, Karlsruhe 76 πN phase shift analysis, as quoted in Ref. 82 of K. Lanius, PHE 76-16, rapporteur talk at the 18th International Conference on High Energy Physics, Tbilisi, USSR, July 1976.

11. CMU-LBL 76, R. E. Cutkosky et al., Ref. 84 in Ref. 10.

12. ALMEHED 72 in RPP76.

13. The average of Refs. 2 and 4 isn't too meaningful here, but it is in agreement with the other measurements in RPP76, and also with the average of Refs. 10 and 11.

14. We use Ref. 12 instead of Ref. 2 here because it agrees better with the other measurements in RPP76, and also with Refs. 10 and 11.

15. CITRON 66 in RPP76.

16. This has been labeled as a "dead" resonance. However, from the present systematics, we would expect it to exist (see the comments in Appendix E and at the end of Sec. 5.4 in Chapter 5).

17. RLIC 76 in RPP76.

18. This is an average of 13 measurements (years 1968 - 1976) in RPP76.

19. This is an average of BAXTER 73, ARMENTEROS-4 69, and BERTANZA 69 in RPP76.

20. These two resonances may be just one resonance with different reported widths, but we quote both here to illustrate the ambiguities that can arise when rotational and rotationless states occur at the same energy.

21. GALTIERI 70 in RPP76.

22. KIM 71 in RPP76.

23. This is an average of five measurements in RPP76 whose energies range from 1740 to 1755 MeV and whose widths range from 35 to 300 MeV. Three other measurements give masses in the range 1780 - 1853 MeV. Thus the resonance parameters for this state are not well-determined.

24. We list this resonance here although it is probably spurious. It appears near a $J = 1/2$ S-state and a very-narrow-width $J = 1/2$ P-state, and it may be some kind of discontinuity effect.

25. R. D. Tripp, Lawrence Berkeley Laboratory report LBL-5502, July 1976.

26. LANGBEIN 72 in RPP76.

27. HEMINGWAY 74 in RPP76.

28. DE BELLEFON 76 in RPP76; this experiment establishes $J^P = 9/2^+$.

29. ABRAMS 70 in RPP76.

30. This is an average of the three experiments in RPP76.

31. LICHTFIELD 74 in RPP76. Also seen by A. S. Carroll et al., Phys. Rev. Lett. 37, 806 (1976), and by M. Alston-Garnjost et al., LBL-6784 (1977).

32. BLUMENFELD 69 in RPP76.

33. This is an average of Refs. 25 and 32. It is interesting to note that the Blumenthal peak (Ref. 32) has a K_L^o + p final state, and that $K_L^o + \Sigma^+$ = 498 + 1189 = 1687 MeV, which compares to $\Sigma(1682)\Gamma 25$ for the Blumenthal peak.

34. VAN HORN 75 in RPP76.

35. This is an average of KANE 72, BAILLON 75, and RLIC 76 in RPP76.

36. A level has also been reported at 1925 MeV for this state.

37. LICHTFIELD 2 and LICHTFIELD 3 in RPP76.

38. This is the average of four experiments with masses between 1920 and 1985 MeV; however, several other experiments give lower masses, so that the mass of this state is not well-determined.

39. BERTHONI 70 in RPP76.

40. This is a rough average of the experiments listed in RPP76.

41. DE BELLEFON 76 in RPP76. This peak is seen in many experiments, but the De Bellefon measurement establishes $J^P = 9/2^-$.

42. EHRLICH 66 in RPP76.

43. This is an average of BADIER 65, SMITH 65, ROSS 73, and ACNO 76 in RPP76. The ACNO collaboration [as reported by R. A. Salmeron in Ref. 38 of M. Deutschmann et al., CERN/EP/PHYS 77-26 (Rev)] have determined the spin of this state to be J = 3/2.

44. J. M. P. Gago, C.N.R.S. — A.0.12.899, thesis, Pierre and Marie Curie University, June 29 (1976).

45. This is an average of ROSS 73, BMST 70(2), and BADIER 72 in RPP76.

46. R. J. Hemingway et al., Phys. Lett. 68B, 197 (1977). The ACNO collaboration [see Ref. 43] have determined a spin value J ≥ 5/2 for this state.

47. This is an average of Ref. 44 and of ACNO 76 in RPP76.

48. GOLDWASSER 70 in RPP76.

49. BARTSCH 69 in RPP76.

50. L. J. Gutay et al., Nucl. Phys. B27, 486 (1971) (see Fig. 8.1 in Chapter 8; also see Figs. 13.8 and 13.9 in Chapter 13).

51. M. Aguilar-Benitez et al., Phys. Rev. D6, 29 (1972).

52. The $H(\sim 1000) \rightarrow \pi^+\pi^-\pi^0$ enhancement may be related to the $\phi \rightarrow \pi^+\pi^-\pi^0$ decay mode of the $\phi(1020)$ meson. The spin 1 assignment for the H is from a paper by G. Goldhaber in Proceedings of the 13th International Conference on High Energy Physics, Berkeley, Calif. 1966 (Univ. of Calif. Press, Berkeley, 1967).

53. This is a CERN MMS^- peak that can be observed on page S88 of RPP71 [PDG, Rev. Mod. Phys. 43, S1 (1971)]. The quoted mass and width were obtained by fitting a Gaussian curve to the CERN data. This peak, unlike a nearby $X^-(1079)$ peak that is also seen in the CERN data on page S88, shows up clearly in various CERN preliminary reports that contain subsets of the final data.

54. See Ref. 8 in the references to Appendix B1.

55. The mass and width of the B meson are an average value of RPP76 and of S. M. Flatté et al., Phys. Lett. 64B, 225 (1976).

56. G. Bassompierre et al., CERN report, June 22 (1976).

57. The quoted mass of the D is from an average of eight experiments in RPP76 plus H. Grässler et al. and Ch. Defoix et al., papers submitted to the 18th International Conference on High Energy Physics, Tbilisi, USSR, July 1976. The quoted width is the RPP76 value.

58. See Fig. 8.1c in Chapter 8, and see Table D2 in Appendix D.

59. This is an average of BETTINI 66 and BARLOW 67 under $X_0(1430)$ in RPP76.

60. This is an average of FOLEY 71 and DEFOIX 73 under $X_1(1440)$ in RPP76.

61. F. Barreiro et al., Nucl. Phys. B121, 237 (1977). The broad width shown here for this resonance is indicated by several recent experiments.

62. This is an average of AGUILAR 69 and DUBOC 71 in RPP76. However, a recent experiment by A. M. Cooper et al., as described in a paper submitted to the Tbilisi Conference (see Ref. 57), failed to confirm this resonance. But P. Minnaert et al., Nucl. Phys. B132, 88 (1978), report that they have observed the state $F_1(1540)\Gamma 50$ in a different reaction, and with spin-parity $J^P = 2^-$.

63. KALELKAR 75 in RPP76.

64. This is an average of SALZBERG 72 and FRENCH 67 under $A_4(1900)$ in RPP76.

65. A. A. Carter et al., Phys. Lett. 67B, 117 (1977).

66. See the discussion of the $\kappa(725)$ meson in Section B1 of Appendix B.

67. This is Firestone 72 listed in the Q Region in RPP76.

68. This is from DAVIDSON 74 in the Q Region in RPP76; also, the existence of a narrow $K\pi\pi(1265)\Gamma\sim 10$ peak is reported by DAVIS 72.

69. This is the average of four experiments listed under $K_N(1700)$ in RPP76.

70. A. Etkin et al., Phys. Rev. Lett. 36, 1482 (1976).

71. R. Baldi et al., Phys. Lett. 63B, 344 (1976); S. U. Chung et al., Phys. Rev. Lett. 40, 355 (1978); D. D. Carmony et al., Phys. Rev. D16, 1251 (1977).

72. This is SPIRO 76 listed under K_N(1800) in RPP76, who also see the peak of Ref. 71.

73. D. D. Carmony et al., Phys. Rev. D16, 1251 (1977).

74. This is an average of M(1821)Γ31, M(1819)Γ24, and M(1812)Γ34, as quoted in B. Esposito et al., Phys. Lett. 68B, 389 (1977); C. Bacci et al., ibid. 393; and G. Barbiellini et al., ibid. 397.

75. I. Peruzzi et al., Phys. Rev. Lett. 39, 1301 (1977).

76. The width is from W. Brückner et al., Phys. Lett. 67B, 222 (1977); the mass is a weighted average of Brückner with CARROLL 72 and MARZANO 76 in RPP76.

77. The width is from G. J. Feldman, SLAC-PUB-2000, Aug. (1977).

78. The width is from G. J. Feldman et al., Phys. Rev. Lett. 38, 1313 (1977).

79. P. Benkheiri et al., Phys. Lett. 68B, 483 (1977).

80. R. Brandelik et al., Phys. Lett. 70B, 132 (1977).

81. M. A. Abolins et al., Phys. Rev. Lett. 37, 417 (1976); see the discussion of these data in Chapter 17.

82. G. Goldhaber, Lawrence Berkeley Laboratory report LBL-5534, Oct. (1976).

83. A. Apostolakis et al., Phys. Lett. 66B, 185 (1977). However, this peak was not seen in the experiment of G. W. van Apeldoorn et al., Phys. Lett. 72B, 487 (1978).

84. H. Braun et al., Phys. Lett. 60B, 481 (1976).

85. C. Evangelista et al., Phys. Lett. 72B, 139 (1977).

86. W. Tanenbaum et al., Phys. Rev. D (March 1, 1978).

87. P. A. Rapidus et al., Phys. Rev. Lett. 39, 526 (1977).

88. See the R-ratio curve in Fig. 17.2 of Chapter 17.

89. W. R. Innes et al., Phys. Rev. Lett. 39, 1240 (1977).

APPENDIX C. THE SYSTEMATICS OF THE LIGHT-QUARK MODEL

For pedagogical reasons, the present light-quark model has been developed in a step-by-step manner in the text. In Appendix C we gather the various facets of this light-quark model together in one place, so that it can serve as a guide to the reader. The light-quark model is presented here with the assumption that the light-quark substates contain only integral electric charges. The model can be formulated in terms of fractional electric charges, but until the existence of fractional charges is established experimentally, we prefer to remain with the integral-charge formulation given here.

C1. Mass Values and Strangeness Quantum Numbers

In this section we give a formal definition of the light-quark basis states. The fundamental hadronic building block is the mass quantum M = 70 MeV, which has the same basic quantum numbers as the kaon. Corresponding to its isotopic spin I = 1/2, the quantum M has the particle states M^+ and M^0 and the antiparticle states $\bar{M}^0$ and M^-, which have the mass values†

$$M^+ = 74.6 \text{ MeV}, \quad M^0 = 70.0 \text{ MeV}, \quad \bar{M}^0 = 70.0 \text{ MeV}, \quad M^- = 74.6 \text{ MeV}. \tag{1}$$

The mass quantum M appears in baryons and mesons in four basic spinless forms:

$$1 \equiv M, \quad 3 \equiv 3M, \quad 4 \equiv 4M, \quad 7 \equiv 7M. \tag{2}$$

The masses of the subquanta in these forms are additive, so that we have, for example,

$$\begin{aligned} 3^0 &= 210.0 \text{ MeV}, & 3^{\pm} &= 214.6 \text{ MeV}; \\ 4^0 &= 280.0 \text{ MeV}, & 4^{\pm} &= 284.6 \text{ MeV}; \\ 7^{\pm} &= 494.6 \text{ MeV}, & 7^{+-} &= 499.2 \text{ MeV}^{\dagger\dagger}. \end{aligned} \tag{3}$$

---

†The $M^{\pm} - M^0$ mass difference of 4.6 MeV reproduces the charge splitting of the $\pi^{\pm}$ and π^0 mesons exactly; however, it also gives accurate charge splittings for many other resonances. Phenomenologically, the $\pi^{\pm} - \pi^0$ mass difference is almost exactly equal to nine electron masses.

††The internal coulomb attraction lowers this mass value by 0.7 MeV (see Chapter 19).

The configurations 3 and 4 in Eq. (3) do not correspond to observed resonances, but the configurations $7^{\pm}$ and 7^{+-} can be identified with the $K^{\pm}$ and K^{0}, $\bar{K}^{0}$ mesons, respectively. The basis state 7 may in fact be a composite state, $7 \equiv 43$, and we sometimes treat it as such, but it appears geometrically to be a unique entity (see Chapter 18).

The quantum M also appears in one spinning form:

$$S \equiv (3M)_{spinning} \equiv 3_{spinning}, \tag{4}$$

in which the spinor S has spin angular momentum $J = 1/2\ \hbar$. S appears in baryon resonances in the charge states $S^{\pm}$ and $S^{\pm\pm}$, which have the mass values (see Chapter 19 and Appendix D):

$$S^{\pm} \simeq 330 \text{ MeV} \quad \text{and} \quad S^{\pm\pm} \simeq 337 \text{ MeV}; \tag{5}$$

the mass difference of 7 MeV between these spinning states is half again as large as the mass difference of 4.6 MeV between the spinless states 3^{0} and $3^{\pm}$.

The basis state 3 occurs in two unmixed forms,

$$3 \equiv MMM \quad \text{and} \quad \bar{3} \equiv \bar{M}\bar{M}\bar{M}, \tag{6}$$

and in two mixed forms,

$$\dot{3} \equiv M\bar{M}M \quad \text{and} \quad \dot{\bar{3}} \equiv \bar{M}M\bar{M}. \tag{7}$$

In dealing with the equations of Associated Production (Appendix I), it is sometimes useful to distinguish between the mixed forms $\dot{3}$ and $\dot{\bar{3}}$; however, in general we ignore this distinction, and we refer to them both as $\dot{3}$ states. The basis state $S \equiv (3M)_{spinning}$ similarly occurs in two unmixed forms,

$$S \equiv 3_{spinning} \quad \text{and} \quad \bar{S} \equiv \bar{3}_{spinning}, \tag{8}$$

and in two mixed forms,

$$\dot{S} \equiv \dot{3}_{spinning} \quad \text{and} \quad \dot{\bar{S}} \equiv \dot{\bar{3}}_{spinning}. \tag{9}$$

The unmixed forms occur in baryon resonances and in $N\bar{N}$ meson resonances as spinor triplets SSS and $\bar{S}\bar{S}\bar{S}$, whereas the mixed forms occur in meson resonances as spinor pairs $\dot{S}\dot{\bar{S}}$. Again, we do not usually distinguish between the forms $\dot{S}$ and $\dot{\bar{S}}$, and we refer to both of these as $\dot{S}$ spinors. The basis state 4 occurs in two unmixed forms,

$$4 \equiv MMMM \qquad \text{and} \qquad \bar{4} \equiv \overline{MMMM}, \tag{10}$$

and in at least one mixed form, which we denote here as

$$\dot{4} \equiv M\bar{M}M\bar{M}. \tag{11}$$

The mixed basis state $\dot{4}$ in Eq. (11) has balanced $M,\bar{M}$ internal symmetry, whereas the unmixed basis states $\dot{3}$ and $\dot{\bar{3}}$ in Eq. (7) do not; this fact seems related to the empirical result that resonances which contain $\dot{3}$-type excitation quanta have quite different symmetry characteristics than do resonances which contain $\dot{4}$-type excitation quanta, as we discuss in detail in Appendix E.

We now come to the assignment of the strangeness quantum numbers. It seems to be true in both the baryon and meson resonances that the <u>unmixed spinless excitations</u> carry the conventional strangeness quantum numbers. The spinless basis states of Eq (2) occur with the following assignments for the strangeness quantum number S:

S = +1	S = 0	S = -1
1		$\bar{1}$
3	$\dot{3}$	$\bar{3}$
4	$\dot{4}$	$\bar{4}$
7		$\bar{7}$

(12)

These strangeness quantum numbers are additive, so that in the baryon and hyperon resonances we have

Particle	Configuration	Strangeness
N	$S\bar{S}S$	0
Λ	$S\bar{S}S\bar{3}$	-1
Σ	$S\bar{S}S\bar{4}$	-1
Ξ	$S\bar{S}S\bar{3}\bar{3}$	-2
Ω	$S\bar{S}S\bar{3}\bar{4}\bar{4}$	-3

(13)

In analogy to the $S\bar{S}S\bar{3}$ and $S\bar{S}S\bar{4}$ excitations of Eq. (13), we also have $S\bar{S}S\bar{7} \equiv \Lambda(1402)$. And in the narrow-width $K\pi\pi$ kaon resonances of Fig. 5.3, we have the similar excitations $\pi\dot{S}\dot{S}K \equiv K\pi\pi(1278)$ and $\eta\dot{S}\dot{S}K \equiv K\pi\pi(1692)$, where $K \equiv 7$, and where the masses are additive.

The basis state 7 ≡ 7M does not seem to occur in the mixed form $\dot{7}$, although it would be difficult to distinguish $\dot{7}$ from the combination $\dot{3}\dot{4}$ (which appears, for example, in Fig. 5.7). We have identified the basis state 7 with the K meson, which itself has the following rather complicated behavior: the charged kaon states K^+ and K^- have strangeness values S = +1 and S = -1, respectively; the neutral kaon states K^0 and $\bar{K}^0$ also have strangeness values S = +1 and S = -1, but the neutral kaon states which are actually observed, K_L^0 and K_S^0, have mixed strangeness (they are thought of as being linear combinations of K^0 and $\bar{K}^0$).

We still have to consider the strangeness characteristics of the spinor S. As we have described above, the unmixed states 3 and $\bar{3}$ are strange, whereas the mixed state $\dot{3}$ is nonstrange. Since we construct the spinor S as S ≡ $3_{spinning}$, it seems logical to inquire if the strangeness assignments for 3, $\bar{3}$, and $\dot{3}$ carry over to the spinor S. There is an argument which suggests that they do. As we discuss in Section C2, there is a correlation between strangeness quantum numbers and binding energies: strange (unmixed) states have ∿4% binding energies, and nonstrange (mixed) states have ∿0% binding energies. Since the unmixed spinors S and $\bar{S}$ (as we will see) have 4% binding energies, they are, from this viewpoint, <u>strange</u>; however, from the conventional strangeness quantum number assignments, the nucleon N ≡ S$\bar{S}$S is nonstrange, so that the spinor S must be <u>nonstrange</u>. How can we reconcile these opposing requirements?

The dilemma about the strangeness of the spinor S which was posed in the preceding paragraph can be readily resolved by using the fact that the strangeness of a particle is <u>not</u> a property that can be directly identified <u>per se</u>; only the <u>change</u> in the strangeness of a particle state can be detected. For example, consider an Associated Production reaction in which a nucleon is bombarded by an antikaon, and a hyperon plus some final-state kaons are produced. By counting the number of strangeness-carrying kaons and antikaons in the initial and final states, we can determine the relative strangeness of the initial-state nucleon and the final-state hyperon. But from Eq. (13) we see that both the nucleon and the hyperon contain the same spinor triplet S$\bar{S}$S. Thus, regardless of what strangeness assignment we make for the spinor triplet S$\bar{S}$S (and hence also for the spinor S), it will not affect the relative strangeness of these two resonances, which is the only quantity that is determined in the Associated Production reaction. The spinor triplet S$\bar{S}$S is an invariant structure in baryon and hyperon resonances, and the conservation of S$\bar{S}$S triplets is in fact identical with the conservation of the baryon quantum number B. It should be noted that in hyperon decay processes, which can also give information about the strangeness of a resonance, the spinor triplet S$\bar{S}$S is again an invariant, so that we have the same situation as for production processes.

From the viewpoint of the present light-quark model, "strangeness" is a characteristic that is related to an internal particle-antiparticle asymmetry in the structure of the particle, and the spinors S and $\bar{S}$ are properly regarded in this sense as being "strange". However, the invariance of the spinor triplets $S\bar{S}S$ and $\bar{S}S\bar{S}$ (which are the only places in which the spinors S and $\bar{S}$ occur) effectively removes the spinors S and $\bar{S}$ from the bookkeeping of the process of Associated Production. In order to be in agreement with the conventional strangeness quantum number assignments, we should formally assign the strangeness quantum number S = 0 to these spinors. But the "strangeness" of the spinors S and $\bar{S}$ (that is, their internal particle-antiparticle asymmetry) is still manifested by their binding energies, as is described in the next section. Furthermore, the "strangeness" of the spinor triplet $S\bar{S}S$ has another observable consequence: the "positive strangeness" triplet $S\bar{S}S$ will combine with negative strangeness excitation quanta $\bar{3}$ and $\bar{4}$ to form the so-called Y* resonances, but it will not combine with positive strangeness excitation quanta 3 and 4 to form Z* resonances†. We can handle this situation within the context of the present light-quark model by recognizing two kinds of strangeness:

S_1 ≡ <u>first class strangeness</u>, which is relevant for both strangeness quantum numbers and binding energies, and which is carried by the spinless cabers;

(12a)

S_2 ≡ <u>second class strangeness</u>, which is relevant only for binding energies, and which is carried by the spinor S, and hence also by the spinor triplet $S\bar{S}S$.

These definitions will be useful in Appendix I, where we study Associated Production.

C2. Binding Energies

There are two different kinds of light-quark binding energies which must be considered — electromagnetic and hadronic. The electromagnetic binding energies in turn involve contributions from two different sources — coulomb interactions and magnetic interactions. These electromagnetic effects are quite small (<<1%), and they will not be discussed here; they are treated in Chapter 19. The hadronic binding energies, on the other hand, are the ∿4% and ∿0% binding energies that we have mentioned in the preceding section, and their influence on the stability of a resonance is much more important than their rather modest binding energy values would seem to indicate.

---

†Y* and Z* resonance formation is discussed in RPP76, "Review of Particle Properties", Particle Data Group, Rev. Mod. Phys. 48, No. 2, Part II, April 1976.

The "hadronic" binding energy is the unknown binding energy that holds the quark states together in an elementary particle resonance. Working phenomenologically, we can use the mass systematics of the meson and baryon resonances to figure out that two light-quark binding energies — ∿4% and ∿0% — must exist, and historically this deduction was made[1] before the appearance of the $\bar{p}n$ bound state resonance[2], which has a binding energy of 4.4%. This strong hadronic binding energy is of a much different nature than the conventional nuclear and electromagnetic binding energies. In the cases of nuclearly-bound and electromagnetically-bound systems, the application of a sufficiently large force results in the disruption of the bound system, with the subsequent ejection of some of the constituents. However, in the case of the hadronically-bound elementary particle, it seems to be impossible to break apart the system mechanically, no matter how large a force is applied. Thus, for example, a proton is stable even at ultra-high cosmic ray energies. It is difficult to envision a ∿4% binding energy as being that cohesive, much less a ∿0% binding energy, but even a 99% binding energy is not sufficient to maintain the stability of protons that come in from outer space. Hence it is apparent that the hadronic binding energy functions in a manner that is quite apart from our usual concept of potential-well interactions. This suggests that we should study hadron decay systematics in order to ascertain the mechanisms which break hadronic bonds.

We can obtain a clue as to the nature of hadron decay mechanisms by studying elementary particle lifetimes, which are discussed in Chapter 10. Now, in the present light-quark model, elementary particles are pictured as composed of subquanta M = 70 MeV that bind together hadronically in various combinations. From the systematics of the lifetimes shown in Chapter 10, it seems clear that the decay of an elementary particle involves the annihilation or transformation[†] of internal subquanta M, with the remaining subquanta being emitted in the form of final-state particles. Thus it appears that <u>hadronic bonds, once formed, can be broken only by the internal annihilation or transformation of some of the constitutent subquanta M</u>. In particular, the striking 1:2:4 ratios that appear in some of the elementary particle lifetimes (see Chapter 10) suggest that these particles have 4:2:1 independent subquanta M available to trigger the decay process.

Without inquiring further into the nature of the hadronic binding energy, we now give a formal definition of the hadronic bonds which operate between the light-quark basis states. Since the magnitude of the required binding energy depends in an essential manner on the mass values assumed for the basis states, it should be understood here that we are using the basis-state mass values defined in Sec. C1. As we

---

[†]These transformations involve the basis state transitions 333 ↔ SS (see Chapter 14).

indicated in Sec. Cl, the hadronic binding energy is related to the strangeness of the basis state.† Hadronic binding energies operate in accordance with the following set of definitions and rules:

Definition. Unmixed basis states are strange; these include both states and the corresponding antistates [see Eqs. (6),(8),(10),(12)]:

Strange basis states $1,3,4,7,S$ (14)

Strange basis antistates $\bar{1},\bar{3},\bar{4},\bar{7},\bar{S}$ (15)

Definition. Mixed basis states are nonstrange [see Eqs. (7),(9),(11),(12)]:

Nonstrange basis states $\dot{3},\dot{4},\dot{S};\dot{\bar{3}},\dot{\bar{S}}$ (16)

Rule A. Hadronic binding energies are very short-ranged, so that only adjacent basis states in a cluster bind together hadronically. (17)

Rule B. Matching strange state-antistate pairs bind together hadronically with ∿4% binding energies. (18)

Rule C. Nonstrange basis states bind hadronically to all adjacent basis states with ∿0% binding energies. (19)

Rule D. Strange state-state and antistate-antistate pairs do not bind together hadronically. (20)

Rule E. The light-quark basis states cluster together in a two-dimensional (side-by-side) manner. (21)

Rules A - E are empirical results which evolved from the attempt to reproduce the spectrum of elementary particle resonances by using a set of light-quark basis states. In particular, circumstances which lead to Rule E are discussed briefly in Chapter 8, Eqs. (8.8) and (8.9). The application of these rules in the remainder of the present Appendix will make clear both the phenomenological basis for the rules and the manner in which they operate.

---

†In the present section we consider the spinors S and $\bar{S}$ to be "strange" (see the discussion at the end of Sec. Cl).

In the spinless basis states, the 4% hadronic binding energies (Rule B) are:

$$1\bar{1} = -5.0 \text{ MeV},$$
$$4\bar{4} = -20.0 \text{ MeV}, \qquad (22)$$
$$7\bar{7} = -35.0 \text{ MeV}.$$

These binding energies are used in Eqs. (5.1) - (5.6) of Chapter 5. The fact that the binding energies in Eq. (22) are proportional to the number of subquanta M ($1 \equiv 1M$, $4 \equiv 4M$, $7 \equiv 7M$), and the fact (Rule A) that hadronic binding energies are very short-ranged, combine together to suggest that the basis states 4 and 7 are actually in the form of linear arrays of subquanta M. There are many other facts which lead to this same conclusion, and which suggest that the basis states 3 and S are also linear arrays of subquanta M (e.g., the $K^0 - K^{\pm}$ mass difference, the spectroscopic properties of the spinor S, and the binding energy systematics between spinless and spinning basis states). We have made implicit use of this linear structure for the basis states in the formulation of Rule E. The geometries of the light-quark basis states are discussed in detail in Chapter 18. As a generic term, we refer to the linear mass configurations 1, 3, 4, and 7 as "cabers"†; the spinor S is a spinning form of the caber 3. The cabers 1, 3, 4, and 7 and the spinor S are illustrated in Figure 18.5.

In the case of the spinors S and $\bar{S}$, the 4% hadronic binding energy (Chapter 8) is given in Eq. (8.5') as $S\bar{S} = -26.4$ MeV. When electromagnetic effects are included (Chapter 19), both the mass of the spinor S and the $S\bar{S}$ binding energy shift to slightly different values:

$$S^{\pm} = 330.4 \text{ MeV}, \quad S^{\pm\pm} = 337.5 \text{ MeV}, \quad S\bar{S} = -27.3 \text{ MeV}. \qquad (23)$$

As is shown in Eqs. (8.8) and (8.9), the combination of the -26.4 MeV $S\bar{S}$ binding energy of Eq. (8.5') and the 330 MeV $S^{\pm}$ spinor mass of Eq. (8.7) gives accurate mass values for both the nucleon N and the $\bar{N}N$ bound state. The inclusion of electromagnetic effects, Eq. (23), does not change this result appreciably, but it enables us to reproduce the n-p mass difference and to account more accurately for the hyperon mass values. These mass calculations are shown in Chapter 19.

In Eqs. (22) and (23) we have accounted for the hadronic binding energies of spinless pairs and spinning pairs. Now we must consider the binding energy between the spinor S and the spinless cabers 3, 4, and 7. This brings in an interesting

---

†A caber is a linear mass that is used in Scottish sporting events.

aspect of the phenomenology. The spinor

$$S \equiv (3M)_{\text{spinning}} \equiv M_s M_s M_s \tag{24}$$

has three spinning subquanta M_s in a linear configuration (Fig. 18.5). Hence the $S\bar{S}$ binding energy of -27.3 MeV in Eq. (23) corresponds to -9.1 MeV per matching $M_s\bar{M}_s$ pair. From Eq. (22), spinless cabers have a binding energy of -5.0 MeV per matching $M\bar{M}$ pair. Thus the $M\bar{M}_s = \bar{M}M_s$ binding energy should be somewhere between -5.0 MeV and -9.1 MeV. It turns out empirically that $M\bar{M}_s$ = -5.0 MeV gives accurate mass values. Hence we have $S\bar{3} = \bar{S}3$ = -15.0 MeV. But, as is shown in Fig. 18.6, the pairs $S\bar{4}$ and $S\bar{7}$ have the same binding energy as $S\bar{3}$, since no more than three of the subquanta $\bar{M}$ in the cabers $\bar{4}$ and $\bar{7}$ can pair off with the three subquanta M_s in the spinor S. Thus we have

$$\begin{aligned} S\bar{3} &= \bar{S}3 = -15.0 \text{ MeV}, \\ S\bar{4} &= \bar{S}4 = -15.0 \text{ MeV}, \\ S\bar{7} &= \bar{S}7 = -15.0 \text{ MeV}. \end{aligned} \tag{25}$$

These are precisely the binding energies that are required empirically in order to reproduce the mass values of the hyperon resonances [see Eqs. (9.10), (9.12), (9.14), and (9.19) in Chapter 9, and see Table 19.3 in Chapter 19].

For the final result in this Appendix, we illustrate the effect that clustering (Rule E) has on the binding energies. In the $\bar{N}N$ clusters shown in Eq. (8.9), one spinor S can bind hadronically to several adjacent spinors $\bar{S}$ at the same time. This result carries over to the spinless excitation quanta that are combined with spinor triplets $S\bar{S}S$ to form hyperon resonances [Eq. (13)]. The Λ hyperon is the caber cluster

$$\Lambda = \bar{S}{}^{S}_{S}\bar{3}. \tag{26}$$

In this cluster the spinless caber $\bar{3}$ combines with two spinors S, so that its total binding energy is -30 MeV. From Eq. (25), the binding energy for a caber $\bar{4}$ or $\bar{7}$ is also -30 MeV. The Ξ hyperon is the cluster

$$\Xi = \bar{S}{}^{S}_{S}\bar{3}_{\bar{3}}. \tag{27}$$

In this cluster the first caber $\bar{3}$ has a binding energy of -30 MeV, but the second caber $\bar{3}$ can only bind with one spinor S and hence has a binding energy of -15 MeV; thus the total $\bar{3}\bar{3}$ binding energy in Eq. (27) is -45 MeV. Finally, if we form the cluster [(see Eq. (13)]

$$\Omega = \bar{S}^{S_{\bar{3}}} {}_{S_{\bar{4}}}{}^{\bar{4}} \tag{28}$$

the total $\overline{3}\overline{4}\overline{4}$ binding energy is again -45 MeV.

The correlation between strangeness quantum numbers and hadronic binding energies, which is one of the most interesting and useful results to emerge from the present light-quark phenomenology, is illustrated graphically in Fig. 21 of Ref. 3.

The discussion in the present Appendix does not constitute a complete description of the present light-quark model. Instead, it is an attempt to put together in one place some of the more fundamental and interesting aspects of the model. Our goal in the present studies is not to obtain a fully-engineered theoretical model, but rather to develop a point-of-view for studying such models.

<u>References for Appendix C</u>.

1. See reference 7 in Chapter 4.
2. See reference 8 in Chapter 4.
3. M. H. Mac Gregor, <u>Phys. Rev</u>. <u>D10</u>, 850 (1974).

APPENDIX D. EXPERIMENTAL EVIDENCE FOR THE MASSES M^o = 70 MeV AND $S^{\pm}$ = 330 MeV

In this Appendix we summarize some experimental and theoretical determinations that can be used to obtain mass values for the fundamental light-quark mass quanta M^o = 70 MeV and $S^{\pm}$ = 330 MeV. Our goal here is not to obtain precision values for these masses, but rather to demonstrate, from as many viewpoints as possible, that these light-quark states or substates actually do exist. Thus the examples that are cited here should be judged not only on the basis of their consistency, but also on the basis of their universality. A notable feature is the manner in which the New Particle states have reinforced and extended the results that were previously obtained and published from the systematics of the Old Particle states. Since neither the quantum M^o nor the quantum $S^{\pm}$ (nor any other quark state) has been observed singly, we are forced to rely mainly on experimental mass differences in establishing values for these masses. We are aided here by the fact (as we argue throughout the book) that quark binding energies are very small (a few percent at most). And we also receive some help, as is discussed in Chapters 6 and 7, from the theory of (special or general) relativity as applied to rotating systems.

D1. The Mass Quantum M^o

Table D1 lists seven determinations of the mass quantum M^o that were published[1] prior to the appearance of the New Particles, plus eight determinations that are obtained from the systematics of the New Particles. Of the seven "Old Particle" determinations, the first three are based on accurately-known hyperon mass intervals (see Fig. 5.7), the fourth is taken from a meson mass interval (see Fig. 16.2), the fifth is obtained directly from the mass of the K^+ meson, the sixth is derived from the mass of the muon and the equations of relativity (Chapter 7), and the seventh is a purely phenomenological result based on the mass of the electron. The important point to note here is the universality of this evidence for the mass of the quantum M^o.

The New Particles were discovered subsequent to the publication of the present light-quark systematics, and there was no a priori reason (other than the present work) to expect to find any evidence for a 70 MeV mass quantization among the New Particle levels. However, as we have illustrated in Figs. 5.4 — 5.6, and have

Table D1. Experimental determinations of the fundamental mass quantum M^o.

MASS DETERMINATIONS PUBLISHED PRIOR TO THE APPEARANCE OF THE NEW PARTICLES

(1)	$\Lambda^o(1402.4) - \Sigma^o(1192.5) = 3M^o$	$\rightarrow M^o = 70.0$ MeV
(2)	$\Omega^-(1672.2) - \Xi^-(1321.3) = 5M^o$	$\rightarrow M^o = 70.2$ MeV
(3)	$\Lambda^o(1672) - \Lambda^o(1115.6) = 8M^o$	$\rightarrow M^o = 69.6$ MeV
(4)	$D^o(1285) - \ddot{S}\dot{S}(658) = 9M^o$	$\rightarrow M^o = 69.7$ MeV
(5)	$K^+(493.7) = 6M^o + M^+;\ M^+ = M^o + 4.6$ MeV	$\rightarrow M^o = 69.9$ MeV
(6)	muon: $M_\mu(105.66) = 3/2\ M^o$	$\rightarrow M^o = 70.4$ MeV
(7)	electron: $M_e(0.5110)/\alpha = M^o$	$\rightarrow M^o = 70.0$ MeV ($\alpha = e^2/\hbar c$)

MASS DETERMINATIONS BASED ON THE SYSTEMATICS OF THE NEW PARTICLES

(1)	$\psi(4028) - \psi'(3684) = 5M^o$	$\rightarrow M^o = 68.8$ MeV
(2)	$\psi(4028) - D^{*o}(2006) = 29M^o$	$\rightarrow M^o = 69.7$ MeV
(3)	$\psi(4028) - D^o(1863.3) = 31M^o$	$\rightarrow M^o = 69.8$ MeV
(4)	$\psi'(3684) - \bar{p}n(1795) = 27M^o$	$\rightarrow M^o = 70.0$ MeV
(5)	$\psi'(3684) - \chi(3551) = 2M^o$	$\rightarrow M^o = 66.5$ MeV
(6)	$\psi'(3684) - \chi(3414) = 4M^o$	$\rightarrow M^o = 67.5$ MeV
(7)	$\chi(3503) - J/\psi(3098) = 6M^o$	$\rightarrow M^o = 67.5$ MeV
(8)	$\chi(3455) - J/\psi(3098) = 5M^o$	$\rightarrow M^o = 71.4$ MeV

discussed in Chapter 17, the New Particle resonances are essentially universally-quantized in multiples of 70 MeV. Thus the discovery of this quantization must be regarded as a substantial reinforcement of the present light-quark viewpoint.

Of the eight M^o New Particle mass determinations shown in Table D1, the first and fourth are based on differences between New Particle resonance peaks, the second and third involve decays to the "charmed-meson" states D^{*o} and D^o, the fifth and sixth are radiative decay levels of the ψ', and the seventh and eighth are ψ' radiative levels that decay down to the J/ψ. Six of the eight New Particle determinations shown here are thus based on the use of intermediate-state or final-state decay levels. This is a unique feature of the New Particle resonances. In the decay of a low-mass Old Particle state, the annihilation process essentially destroys the original quark structure. The majority of the high-mass New Particle decays also occur in this manner; but the New Particles also contain a special class of "high-mass final-state" decays in which only a few of the subquanta M are annihilated, and the original quark structure either remains substantially intact ($\psi' \rightarrow \chi + \lambda$) or else splits in half to form two "charmed" mesons [$\Psi(4028) \rightarrow D + D^*$]. In the "high-mass final-state" decays, the quantization in multiples of 70 MeV is clearly in evidence both in the annihilation energies and in the mass values of the residual "high-mass" levels.

The New Particle determinations of M^o in Table D1 which involve large multiples of M^o (27 to 31 M^o) all place its mass in the range 69.7 – 70.0 MeV. However, the New Particle determinations which involve only a few quanta M^o (2 to 6 M^o) give values that range from 66.5 to 71.4 MeV. In these latter cases, it must be kept in mind that small (4%) binding energy effects and small electromagnetic effects (as shown by the charge splittings of the D and D^* mesons) can shift the level energies by a few MeV. Thus the key point to note in Table D1 is not the exact mass values that are obtained for M^o, but rather the fact that these values are all approximately equal to 70 MeV. On the basis of present-day SU(3)-type theories, there is no reason (except in the case of the D and D^* masses) to expect to find any mass intervals which are in multiples of 70 MeV.

In addition to the New Particle mass intervals shown in Table D1, there is a remarkable sequence of resonances which illustrates the 70 MeV quantization of the New Particle states in a particularly clear-cut manner. This is the sequence

$$\bar{p}n(1794.5), \quad D^o(1863.3), \quad \bar{p}p(1935), \quad D^{*o}(2006) \qquad (1)$$

that is shown in Fig. 5.4. All of these levels have very narrow widths, and they are known to mass accuracies of about ±2 MeV. If we take the differences between the matching levels in this sequence, we have:

$$\begin{array}{lll} \text{nonstrange levels} & \bar{p}p(1935) - \bar{p}n(1794.5) = 2M^{o} \rightarrow M^{o} = 70.3 \text{ MeV}, & \\ \text{strange levels} & D^{*o}(2006) - D^{o}(1863.3) = 2M^{o} \rightarrow M^{o} = 71.4 \text{ MeV}. & \end{array} \quad (2)$$

We also have the semi-independent pairings

$$\begin{array}{l} D^{*o}(2006) - \bar{p}n(1794.5) = 3M^{o} \rightarrow M^{o} = 70.5 \text{ MeV}, \\ D^{o}(1863.3) - \bar{p}n(1794.5) = M^{o} \rightarrow M^{o} = 68.8 \text{ MeV}, \end{array} \quad (3)$$

where the $\bar{p}n$ bound state, as we demonstrate in Chapter 17, serves as the "ground state" resonance for the New Particle excitations.

On the basis of all of the above evidence, we can pinpoint the mass M^{o} as being somewhere right around 70.0 MeV. The mass $M^{\pm}$ is set equal to M^{o} + 4.6 MeV in order to reproduce the $\pi^{\pm} - \pi^{o}$ mass difference. With these mass values, the $M\bar{M}$ binding energy must be -5.0 MeV (as described in Appendix C) in order to obtain precision mass values for the spinless meson resonances (Table 19.1). Of course, we can shift the mass M^{o} slightly and then readjust the $M\bar{M}$ binding energy to compensate for the shift, but the present systematics is probably accurate enough to pin down the allowable range of M^{o} masses to within a few tenths of an MeV.

D2. The Mass Quantum $S^{\pm}$

Although single spinors $S^{\pm}$ have not been observed experimentally, we can obtain some information about them from their magnetic moments, and also from their relationship, via the theory of relativity, to the spinless cabers $3^{\pm}$. There is, however, direct experimental evidence for spinor pairs, triplets, and quartets. We will discuss each of these types of information in turn.

(1). Single spinors S. (a) Magnetic Moments. As is pointed out in the quotation at the beginning of Chapter 8, the conventional quark model treatment of magnetic moments suggests a mass of 336 MeV for the nucleon quark (the spinor S). (b) Relative Theory. A spinless caber $3^{\pm}$ has a mass of 214.6 MeV (Appendix C). If it is given the geometry that is shown in Fig. 18.5, and if it is spun at the full relativistic limit, then from Eq. (6.16)(as evaluated on a computer), it has a calculated spinning mass of just about 330 MeV (see Fig. 18.2).

(2). Spinor pairs $\dot{S}\dot{S}$. The $\dot{S}^{+}\dot{S}^{-}$ spinor pair is manifested as an interference dip that is centered at 658 MeV (see Fig. 8.1b). Applying a correction of +3 MeV to allow for coulomb and magnetic effects (Chapter 19) places the center of the dip at 661 MeV, so that the mass of a spinor $\dot{S}^{\pm}$ is 330.5 MeV.

(3). Spinor Triplets $S\bar{S}S$. The proton is the spinor triplet $S^{+}S^{-}S^{+}$. We can calculate the mass of the spinor $S^{\pm}$ by unbinding the proton. The hadronic binding energy of an $S\bar{S}$ spinor pair is -27.3 MeV (see Fig. 18.6). The total binding energy in the proton (p = $S\bar{S}S$) is -54.6 MeV + 1.8 MeV = -52.8 MeV, where an electromagnetic correction of +1.8 MeV has been applied (see Table 19.2). Thus the mass of the unbound proton is 938.3 + 52.8 = 991.1 MeV, so that the mass of a single spinor $S^{\pm}$ is 991.1 ÷ 3 = 330.4 MeV.

(4). Spinor Quartets $\dot{S}\dot{S}\dot{S}\dot{S}$. The evidence from Old Particle data for spinor quartets comes from the interference effect in the A_2 meson (see Fig. 8.1c), which is discussed in Sec. 8.4 of Chapter 8; the experimental data on this A_2 interference effect are summarized in Table D2. As can be seen in Table D2, the "A_2 effect", with electromagnetic corrections included, places the mass of the $\dot{S}\dot{S}\dot{S}\dot{S}$ spinor quartet at about 1317 MeV. The evidence from New Particle data for spinor quartets comes from the mass differences among the excitations $\bar{p}n(1795) \equiv \bar{N}N$, $J/\psi(3098) \equiv \bar{N}N\cdot\dot{S}\dot{S}\dot{S}\dot{S}$, and $\psi(4414) \equiv \bar{N}N\cdot(\dot{S}\dot{S}\dot{S}\dot{S})^2$ (see Fig. 8.3 and Table 13.1). As can be seen in Table D2, the New Particle data place the electromagentically-corrected spinor quartet $\dot{S}\dot{S}\dot{S}\dot{S}$ at a mass of 1318 MeV, in good agreement with the value of 1317 MeV obtained from the Old Particle data. Thus once again, as in Table D1, we have a situation in which the New Particle data have substantiated a result that was already deduced[8] on the basis of the Old Particle data. If we set $\dot{S}\dot{S}\dot{S}\dot{S}$ = 1317.5 MeV, then $\dot{S}^{\pm}$ = 1317.5 ÷ 4 = 329.4 MeV.

Summarizing the above discussion, we have

(1a)	Single spinor (quark magnetic moment)	→	$S^{\pm}$ = 336 MeV
(1b)	Single spinor (relativity theory)	→	$S^{\pm} \simeq$ 330 MeV
(2)	Spinor pair $\dot{S}\dot{S}$	→	$S^{\pm}$ = 330.5 MeV
(3)	Spinor triplet $S\bar{S}S$	→	$S^{\pm}$ = 330.4 MeV
(4)	Spinor quartet $\dot{S}\dot{S}\dot{S}\dot{S}$	→	$\dot{S}^{\pm}$ = 329.3 MeV

We can see from the mass values in Eq. 4 that the strange spinor S and the nonstrange spinor $\dot{S}$ have essentially the same mass values, although they differ considerably in their binding energies (∿4% and ∿0%, respectively). The mass of a doubly-charged spinor $S^{\pm\pm}$ is about 337.5 MeV [Eq. (19.10)].

We have the same situation for the spinor S as we had above for the spinless quantum M^{o}: several different determinations all pinpoint the same mass value —

Table D2. Experimental determinations of the mass of the spinor quartet $\dot{S}\dot{S}\dot{S}\dot{S}$.

MASS DETERMINATIONS PUBLISHED PRIOR TO THE APPEARANCE OF THE NEW PARTICLES

Experiment	Location	Electromagnetic corr.	Corrected location
		Location of the A_2 dip	
Kienzle[2]	1310 ± 2 MeV	$(\dot{S}\dot{S}\dot{S}\dot{S})^-$ → +7 MeV	1317 ± 2 MeV
BCS[3]	1307 MeV	$(\dot{S}\dot{S}\dot{S}\dot{S})^0$ → +10 MeV	1317 MeV
BDNPT[4]	∿1310 MeV	$(\dot{S}\dot{S}\dot{S}\dot{S})^+$ → +7 MeV	∿1317 MeV
Barnham[5]	∿1310 MeV	$(\dot{S}\dot{S}\dot{S}\dot{S})^+$ → +7 MeV	∿1317 MeV
Bloodworth[6]	∿1305 MeV	$(\dot{S}\dot{S}\dot{S}\dot{S})^+$ → +7 MeV	∿1312 MeV
		Location of an A_2 peak	
Crennell[7]	1311 ± 2 MeV	$(\dot{S}\dot{S}\dot{S}\dot{S})^0$ +10 MeV	1321 ± 2 MeV
		Average location of the "A_2 effect"	
			≃1317 MeV

MASS DETERMINATIONS BASED ON THE SYSTEMATICS OF THE NEW PARTICLES

Experimental mass interval		Electromagnetic corr.	Corrected interval
$J/\psi(3098) - \bar{p}n(1795)$	= 1303 MeV	$(\dot{S}\dot{S}\dot{S}\dot{S})^+$ → +7 MeV	1310 MeV
$\psi(4414) - J/\psi(3098)$	= 1316 MeV	$(\dot{S}\dot{S}\dot{S}\dot{S})^0$ → +10 MeV	1326 MeV
	Average mass interval		
			1318 MeV

which in this case is $S^{\pm} \simeq 330$ MeV. The significance of this result lies not so much in the particular value that it gives for the $S^{\pm}$ mass, but rather in the evidence that it thereby provides for the reality of this light-quark state. Conventional Gell-Mann and Zweig quark modelists now occasionally consider nucleon quark masses that are in the range of 330 MeV (see the quotation at the beginning of Chapter 8, and see the discussion in Sec. 8.6 of Chapter 8), but at the time that this value was first set forth by the present author, in 1970[9], this was a highly heretical idea, since the quark mass was then commonly believed to be at least 5 GeV, and perhaps much larger[10].

References for Appendix D.

1. M. H. Mac Gregor, Phys. Rev. D10, 850 (1974), Table X.
2. W. Kienzle, "MMS, 1967", CERN report (1972)(unpublished).
3. M. Basile et al., Lett. Nuovo Cimento 4, 838 (1970).
4. Bonn-Durham-Nijmegen-Paris (E.P.) - Torino Coll., Nucl. Phys. B16, 221 (1969).
5. K. W. J. Barnham et al., Phys. Rev. Lett. 26, 1494 (1971).
6. I. J. Bloodworth et al., Nucl. Phys. B37, 203 (1972).
7. D. J. Crennell et al., Phys. Rev. Lett. 20, 1318 (1968).
8. Ref. 1, Table XI.
9. See M. H. Mac Gregor, Lett. Nuovo Cimento 4, 1043 (1970), Table I.
10. See B. T. Feld, Models of Elementary Particles, Blaisdell, Waltham (1969), page 364.

APPENDIX E. SYMMETRY CHARACTERISTICS OF 3-TYPE AND 4-TYPE EXCITATIONS

The fundamental hyperon "ground state" resonances are formed by adding strange spinless cabers $\bar{3}$ and $\bar{4}$ to an existing $S\bar{S}S$ nucleon spinor triplet, as is demonstrated in Eqs. (9.9)-(9.17) of Chapter 9 and in Table 19.3 of Chapter 19:

$$N \equiv S\bar{S}S, \quad \Lambda \equiv S\bar{S}S\bar{3}, \quad \Sigma \equiv S\bar{S}S\bar{4}, \quad \Xi \equiv S\bar{S}S\bar{3}\bar{3}, \quad \Omega \equiv S\bar{S}S\bar{3}\bar{4}\bar{4} \quad . \tag{1}$$

Excited states of these nuclei are in turn formed by adding nonstrange spinless cabers $\dot{3}$ and $\dot{4}$ to these basic ground states, as is shown in Figs. 5.7 and 15.1, which contain the excitations

$$\begin{gathered} N\dot{3},\ N\dot{3}\dot{3},\ N\dot{4}\dot{4},\ N\dot{3}\dot{3}\dot{4},\ \Lambda\dot{3},\ \Lambda\dot{4},\ \Lambda\dot{3}\dot{3},\ \Lambda\dot{4}\dot{4},\ \Lambda\dot{3}\dot{3}\dot{4},\ \Sigma\dot{3},\ \Sigma\dot{4},\ \Sigma\dot{3}\dot{3},\ \Sigma\dot{3}\dot{4}, \\ \Sigma\dot{4}\dot{4},\ \Xi\dot{3},\ \Xi\dot{4},\ \Xi\dot{3}\dot{4},\ \Xi\dot{4}\dot{4},\ \Xi\dot{3}\dot{3}\dot{4},\ \text{etc.} \end{gathered} \tag{2}$$

Now, in the case of the long-lived hyperon excitations of Eq. (1), the existence of the strange spinless excitations $\bar{3}$ and $\bar{4}$ was deduced both on the basis of the extremely accurate mass values that they give for these resonances (±0.1%), and also from the natural manner in which they account for the strangeness quantum numbers (see Eq. (13) in Appendix C). In the case of the short-lived baryon and hyperon excitations of Eq. (2), the existence of the nonstrange spinless excitations $\dot{3}$ and $\dot{4}$ was deduced on the basis of the observed mass values, and also from the manner in which these excitations account for the gaps that are observed in the energy level diagrams of Figs. 5.7 and 15.1. If these light-quark excitation mechanisms constitute a physically-significant procedure, and not just a phenomenological way of reproducing the resonances shown in Eqs. (1) and (2), then we might expect to find other physical consequences that are associated with these excitations. In particular, the excitations $\bar{3}$ and $\dot{3}$ contain an odd number of subquanta M and $\bar{M}$, so that they necessarily reflect a particle-antiparticle asymmetry; whereas the excitations $\bar{4}$ and $\dot{4}$ contain an even number of subquanta M and $\bar{M}$, so that they can reflect a more symmetric configuration (which may involve particle-antiparticle symmetries and/or parity symmetries). Thus, we might expect to find a symmetry difference between excitations which contain 3's and those which contain 4's. This difference does exist, and it is reflected in the following empirical set of rules:

(a) Excitations which contain 3's tend to have one or more of the following characteristics:

(1) They are difficult to observe directly as S-states:
(2) They are associated with γ-ray production and/or decay processes;
(3) They are associated with spinor-flip processes (e.g., they appear as narrow-width J = 3/2 resonances). (3a)

(b) Excitations which contain 4's usually do not have these characteristics (3b)

(c) This result applies to both the strange excitations $(\bar{3},\bar{4})$ and the nonstrange excitations $(\dot{3},\dot{4})$. (3c)

With respect to the nonstrange excitations, the rule of Eq. (3) is well-illustrated by the baryon and hyperon levels that are shown in the energy level diagrams of Figs. 5.7 and 15.1. The $N\dot{4}\dot{4}$, $\Lambda\dot{4}\dot{4}$, and $\Sigma\dot{4}\dot{4}$ levels in these figures correspond to the dominant and easily-observed S-state resonances N(1511)S, Λ(1672)S, and Σ(1770)S. By way of contrast, the $N\dot{3}\dot{3}$, $\Lambda\dot{3}\dot{3}$, and $\Sigma\dot{3}\dot{3}$ levels correspond to the unseen N(1359) bandhead of the N rotational band that is shown in Fig. 15.4 of Chapter 15 and in Table H1 of Appendix H; the narrow-width J = 3/2 Λ(1519)D resonance; and the controversial Σ(1623)S resonance (which was identified as an S-state on the basis of the present systematics long before it was thus identified experimentally[1]). (The Λ(1519)D resonance has γ-ray decays into both $\Lambda\gamma$ and $\Sigma^0\gamma$.) We also note that the $\Lambda\dot{4}$ excitation appears as the dominant and easily-observed Λ(1402)S resonance, and the $\Sigma\dot{4}$ excitation is observed as the Σ(1476) production peak†; whereas the $\Lambda\dot{3}$ excitation appears as the "dead" Λ(1327) resonance[2] (which has been observed only as a $\Lambda\gamma$ final state), and the $\Sigma\dot{3}$ excitation is the narrow-width J = 3/2 resonance Σ(1385)P.

The Λ(1327) resonance was declared "dead" in RPP76[2] because of the fact that it was not seen in several expected channels. However, from the systematics of Eq. (3), the difficulty in observing this $\Lambda\dot{3}$ excitation in all reaction channels is not surprising, and unless the several experiments in which the $\Lambda\dot{3}$ was observed[3] turn out to be irreproducible (which has not yet been demonstrated), it might be prudent to withhold judgment as to its demise.

The difficulty in observing the "dead" $\Lambda\dot{3}$(1327)S resonance is paralleled by the difficulty in observing the $N\dot{3}$(1149)S resonance, which is the unseen bandhead of the famous Δ(1232)P resonance (see Fig. 15.3). The Δ(1232)P resonance, which is the historic Δ_{33} enhancement (the first baryon resonance whose quantum numbers were determined), is an $N\dot{3}$ excitation in an $\ell = 1$ rotational state, and it has an unobserved S-state [Eq. (3a1)], a γ-ray decay mode [Eq. (3a2)], and a J = 3/2 spin

---

†From the present systematics (Fig. 15.5), the Σ(1476) peak is an S-State.

value [Eq. (3a3), but with a broad width].

Figure 5.7 contains three very-narrow-width J = 3/2 levels, Λ(1519)DΓ15, Σ(1583)DΓ11, and Ξ(1535)PΓ10. These are the excitations $S\bar{S}S\bar{3}\dot{3}\dot{3}$, $S\bar{S}S\bar{4}\dot{3}\dot{3}$, and $S\bar{S}S\bar{3}\bar{3}\dot{3}$, respectively, and they are examples of Eq. (3a3). The Ξ(1535) has the sequential decay modes $\Xi(1535) \rightarrow \Xi + \pi$, $\Xi \rightarrow \Lambda + \pi$, $\Lambda \rightarrow N + \pi$, which in the present light-quark notation correspond to the decay sequence

$$S\bar{S}S\bar{3}\bar{3}\dot{3} \rightarrow S\bar{S}S\bar{3}\bar{3} + \pi,\quad S\bar{S}S\bar{3}\bar{3} \rightarrow S\bar{S}S\bar{3} + \pi,\quad S\bar{S}S\bar{3} \rightarrow S\bar{S}S + \pi. \qquad (4)$$

In each of these decay steps, an excitation quantum $\bar{3}$ or $\dot{3}$ is transformed into a free pion by the annihilation of a subquantum M in the 3. The decay of the $\dot{3}$, which does not involve a change in strangeness, is a fast decay, but the decays of the $\bar{3}$'s, which do involve a change in strangeness, are slow decays.

The examples of Eq. (3) that we have discussed above involve the nonstrange excitations $\dot{3}$ and $\dot{4}$. However, we can establish a somewhat similar situation for the strange excitations $\bar{3}$ and $\bar{4}$ by studying the magnetic moments of the metastable baryon and hyperon resonances [Eq. (19.7)]. The light-quark configurations for these particles are shown in Table 19.3, and they have the general form:

particle	excitation form	magnetic moment	spinor configuration
p	p^{+}	$\mu \sim +3$	normal
Λ	$p^{+}\bar{3}^{-}$	$\mu \sim -1$	spinor-flipped
Σ^{+}	$p^{+}\bar{4}^{o}$	$\mu \sim +3$	normal
n	n^{o}	$\mu \sim -2$	normal
Σ^{-}	$n^{o}4^{-}$	$\mu \sim -2$	normal
Ξ^{-}	$n^{o}\bar{3}^{o}3^{-}$	$\mu \sim -2$	normal

(5)

In the case of the p, Λ, and Σ^{+} baryons, the $\bar{4}$ excitation leaves the $S\bar{S}S$ triplet unchanged [Eq. (3b)], but the $\bar{3}$ excitation causes the spinor-flip of one of the spinors [Eq. (3a3)], as is evidenced by the magnetic moments [see Eqs. (19.6a) and (19.6c)]. In the case of the n, Σ^{-}, and Ξ^{-} baryons, the $\bar{4}$ excitation leaves the $S\bar{S}S$ triplet unchanged, and the double $\bar{3}\bar{3}$ excitation also leaves the $S\bar{S}S$ triplet unchanged [it is not clear just how the Ξ^{-} result fits in with the systematics of Eq. (3)]. With the possible exception of the Ξ^{-}, the examples in Eq. (5) show that the symmetry differences between 3's and 4's apply to strange as well as to nonstrange spinless cabers. This is a significant result, because it indicates that the symmetry differences between 3's and 4's must involve a property that does not depend on strangeness, and the obvious candidate is the parity of the excitation.

The results that we have described here in Appendix E have an importance that goes well beyond just a small extension of the phenomenology. There is a general belief among physicists – as set forth, for example, by Richard Feynman in one of his television appearances – that if a valid discovery is made in one area of physics, it usually also provides answers to some questions in neighboring areas of physics. Similarly, a new theoretical idea, if it is correct, usually provides answers to several questions in addition to the question for which it was specifically devised. In the present instance, we originally deduced the existence of the excitation quanta 3 and 4 strictly on the basis of mass systematics; but we then discovered, after the fact, that the excitations 3 and 4 lead to resonances which have quite different symmetry properties. Results such as this add considerably to the plausibility of this light-quark approach to particle structure. The development of a theoretical model is similar in many respects to the working out of a crossword puzzle: the more cross-links that can be established, the more convinced we are as to the correctness of the answers.

References for Appendix E

1. M. H. Mac Gregor, Lett. Nuovo Cimento 4, 1043 (1970), Table III.

2. Particle Data Group, Rev. Mod. Phys. 48, No. 2, Part II, April (1976).

3. See Particle Data Group, Rev. Mod. Phys. 45, No. 2, Part II, April (1973), page S145. Also, see the comments on this resonance at the end of Sec. 5.4 in Chapter 5.

APPENDIX F. THE SPECIAL-RELATIVISTIC MASS EQUATION

In Section 6.3 of Chapter 6 we derived Eq. (6.2), which is the equation that gives the special-relativistic mass increase of a particle, in a manner that did not require the use of the Lorentz transformation equations. In the present Appendix we repeat this calculation, using an incremental-calculus approach. The interesting result we obtain here is not the mathematics per se, but rather the manner in which this mathematical approach brings out the underlying realities of the physical situation. In particular, it indicates that Eq. (6.2) has physical significance only for the case $v < c$, and not for the case $v > c$ that is studied in tachyon theory.

We start with a mass $m(v)$ that is initially at rest, with rest mass $m(0) \equiv m_o$, in the inertial frame of reference. We then impart a constant force F for a length of time Δt that suffices to bring the mass up to velocity v. During this time, the mass moves a distance Δx. The work that has been done on the mass is

$$W = F\Delta x \quad .$$

This work appears, according to the equation $E = mc^2$, as an increase in mass:

$$W = c^2\Delta m \quad .$$

The momentum that has been imparted to the mass is equal to the applied impulse:

$$F\Delta t = \Delta(mv) = (m_o + \Delta m)v \quad .$$

Eliminating F from the energy and momentum equations, we have

$$(m_o + \Delta m)\cdot v/\Delta t = c^2\Delta m/\Delta x \quad ,$$

which gives

$$\Delta m = m_o\cdot v/c^2\cdot\Delta x/\Delta t\cdot(1 - v/c^2\cdot\Delta x/\Delta t)^{-1}$$

$$\simeq m_o\cdot v/c^2\cdot\Delta x/\Delta t\cdot(1 + v/c^2\cdot\Delta x/\Delta t) \quad .$$

Thus we have

$$m = m_o + \Delta m$$

$$\simeq m_o(1 + v/c^2 \cdot \Delta x/\Delta t + (v/c^2 \cdot \Delta x/\Delta t)^2) \quad .$$

Now v is the final velocity and $\Delta x/\Delta t$ is the time-average of the velocity over the time interval Δt. We must express $\Delta x/\Delta t$ in terms of v. Since the force F is constant, we have

$$F = \text{const.} = d/dt(mv) = vdm/dt + mdv/dt \quad .$$

To an accuracy of order v^2/c^2, we can insert $\Delta x/\Delta t \simeq 1/2\ v$ into the above mass equation to give

$$m \simeq m_o(1 + 1/2\ v^2/c^2) \quad ,$$

so that a time differentiation gives

$$dm/dt \simeq m_o v/c^2 \cdot dv/dt \quad .$$

Inserting this value into the force equation gives

$$\text{const.} = m_o v^2/c^2 \cdot dv/dt + mdv/dt \quad .$$

Evaluating the constant at $t = 0$ gives

$$\text{const.} = m_o(dv/dt)_i \quad .$$

Thus

$$m_o(dv/dt)_i = (m_o v^2/c^2 + m) \cdot dv/dt \quad ,$$

$$\simeq m_o(1 + 3/2\ v^2/c^2) \cdot dv/dt \quad ,$$

which gives

$$dv/dt = (dv/dt)_i \cdot (1 + 3/2\ v^2/c^2)^{-1}$$

$$\simeq (dv/dt)_i \cdot (1 - 3/2\ v^2/c^2) \quad .$$

In terms of the acceleration a, we have

$$a_o \equiv (dv/dt)_i \quad .$$

Hence

$$a \simeq a_o(1 - 3/2\ v^2/c^2) \quad ,$$

and since $v \simeq a_o t$, we have to order v^2/c^2,

$$a \simeq a_o(1 - 3/2\ a_o{}^2 t^2/c^2) \quad .$$

This in turn gives

$$v = \int_0^t a dt$$

$$\simeq \int_0^t a_o(1 - 3/2\ a_o{}^2 t^2/c^2) dt$$

$$= a_o t - 1/2\ a_o{}^3 t^3/c^2 \quad ,$$

and

$$\Delta x = \int_0^t v dt$$

$$\simeq \int_0^t (a_o t - 1/2\ a_o{}^3 t^3/c^2) dt$$

$$= 1/2\ a_o t^2 - 1/8\ a_o{}^3 t^4/c^2 \quad .$$

Since $t \equiv \Delta t$, we have

$$\Delta x/\Delta t \simeq 1/2\ a_o \Delta t - 1/8\ a_o{}^3 \Delta t^3/c^2$$

and

$$v \simeq a_o \Delta t - 1/2\ a_o{}^3 \Delta t^3/c^2 \quad ,$$

so that we finally obtain

$$\Delta x/\Delta t \simeq 1/2\ v + 1/4\ a_o{}^3 \Delta t^3/c^2 - 1/8\ a_o{}^3 \Delta t^3/c^2$$

$$\simeq 1/2\ v + 1/8\ v^3/c^2 \quad .$$

Inserting this value into the mass equation gives

$$m \simeq m_o(1 + v/c^2 \cdot \Delta x/\Delta t + (v/c^2 \cdot \Delta x/\Delta t)^2)$$

$$\simeq m_o(1 + v/c^2 \cdot (1/2\ v + 1/8\ v^3/c^2) + (v/c^2 \cdot (1/2\ v + 1/8\ v^3/c^2))^2)$$

$$= m_o(1 + 1/2\ v^2/c^2 + 3/8\ v^4/c^4 + \ldots) \quad .$$

This expression for m is equal to the expression

$$m = m_o/\sqrt{1 - v^2/c^2}$$

only for the case $v < c$.

In Section 6.3 of Chapter 6, we have already demonstrated that the equation for the relativistic mass increase,

$$m = m_o/\sqrt{1 - v^2/c^2} \quad ,$$

is a direct consequence of the Einstein equation,

$$E = mc^2 \quad .$$

The reason that we have repeated this calculation here in a slightly different form is to demonstrate that the fundamental equation for the relativistic mass increase is actually the equation

$$m = m_o(1 + 1/2\ v^2/c^2 + 3/8\ v^4/c^4 + \ldots) \quad ,$$

and not the equation

$$m = m_o/\sqrt{1 - v^2/c^2} \quad .$$

These two equations are equal if $v < c$ and if the positive square root is taken, but otherwise they are not equal. The physical significance of this result is that when a particle is given an increase in velocity, it also receives an increase in mass, and this increase occurs without limit. Thus the particle can never be given a velocity $v > c$. Hence the mathematical extension of the equation $m = m_o(1 - v^2/c^2)^{-1/2}$ to the case where $v > c$, which is a procedure that is followed in tachyon theory (tachyons are hypothetical particles that move faster than the velocity of light), may be an interesting theoretical exercise, but it bears no relationship to the physical ideas that are anchored in the special theory of relativity.

APPENDIX G. ELEMENTARY PARTICLE LIFETIME PHASE SPACE CORRECTIONS

In Chapter 10 we discussed the systematics of elementary particle lifetimes. The striking lifetime systematics displayed in Figs. 10.1 - 10.9 were obtained by working directly with the experimental lifetimes, and not with phase-space-corrected lifetimes. Although phase space corrections are probably important for the short-lived "resonance" lifetimes (see Sec. 10.3 in Chapter 10), the lifetime regularities noted for the long-lived "particles" in these figures indicate very directly that phase-space corrections to the "particle" lifetimes must be small (∿20% or less). This is a fortunate circumstance, because in general it is impossible to correct "particle" total widths (lifetimes) for phase space effects, due to the different dimensionalities of the various final states. In the present Appendix, we first sketch the manner in which phase space corrections are applied, and we then show the results of applying (or trying to apply) these corrections to the "particle" lifetimes of Fig. 10.4 and Table 10.1.

G1. Phase Space Formalisms

Discussions of phase space calculations are given, for example, in books by Feld[1] and Källén[2]. The transition probability

$$\lambda_{fi} \tag{1}$$

for going from a given initial state <u>i</u> of a particle to a given final state <u>f</u> is inversely proportional to the lifetime of the particle,

$$\lambda \propto 1/\tau, \tag{2}$$

and it is given by Fermi's "golden rule"[1]:

$$\lambda_{fi} = 2\pi/\hbar \cdot |M_{fi}|^2 \cdot dN_f/dE. \tag{3}$$

In cases where the transition matrix M_{fi} is slowly-varying, the phase-space density dN_f/dE becomes the dominant factor in determining the distribution of the decay products <u>f</u>. Since particle wave functions, and hence particle transition matrix elements, are unknown, we are forced to use very general phenomenological considerations when we attempt to apply Eq. (3) to actual experimental situations. In particular, the accuracy of the assumption that the matrix element M_{fi} is constant over the region spanned by the final state phase space can only be ascertained by the results to which this assumption leads in specific situations.

If the final state in Eq. (3) contains only two particles, the phase space factor dN/dE has the simple form (in the barycentric system)[1]

$$dN(2)/dE = V/2\pi^2 \cdot p\varepsilon_1\varepsilon_2/E, \tag{4}$$

where V is the normalization volume, p is the linear momentum of either particle, ε_1 and ε_2 are the particle energies, and E is the energy of the decaying resonance. If the decay products are all strongly-interacting hadrons, then the normalization volume is sometimes replaced by the short-ranged interaction volume υ, where $\upsilon \propto (\hbar/mc)^3$, with m being a characteristic mass for the reaction. This gives

$$dN(2)/dE \propto (\hbar/mc)^3/2\pi^2 \cdot p\varepsilon_1\varepsilon_2/E \tag{5}$$

for the two-particle final state of Eq. (4). If the final state contains three particles, then dN/dE can be handled analytically only in the cases where all three particles are either non-relativistic or else highly relativistic. These two limiting cases give[1]

$$dN(3)/dE = 4\pi^3 \cdot V^2/(2\pi)^6 \cdot (m_1m_2m_3/(m_1 + m_2 + m_3))^{3/2} \cdot Q^2 \tag{6}$$

and

$$dN(3)/dE = 7\pi^2/240 \cdot V^2/(2\pi)^6 \cdot E^5, \tag{7}$$

respectively, where m_i are the masses of the nonrelativistic final state particles, Q is the reaction energy, and E is the total energy of the resonance. Eq. (6) is applicable to $K \rightarrow \pi\pi\pi$ decays, and Eq. (7) is applicable to the $\mu \rightarrow e + \nu + \nu$ decay.

By comparing Eq. (4) with Eqs. (6) and (7), we can see an immediate difficulty if we attempt to apply phase space corrections to total widths, and hence to lifetimes. The η meson, for example, decays 38% of the time into $\gamma\gamma$, which is a two-particle final state, and 54% of the time into $\pi\pi\pi$, which is a three-particle state. The phase space correction for the $\eta \rightarrow \gamma\gamma$ decay involves the normalization volume V, and the correction for the $\eta \rightarrow \pi\pi\pi$ decay involves V^2. Thus the ratio of these two corrections depends on the normalization volume V, which of course is a completely unphysical situation. The only way out of this dilemma is to assign to V a specific value that comes from physical considerations, and the only logical choice here is to replace V by the interaction volume υ, as we did in going from Eq. (4) to Eq. (5). However, the effective interaction radius is not known to better than a factor of two or three (for example, on p. 68 of Ref. 1 Feld suggests the pion Compton wavelength; and on p. 158 of Ref. 1 he suggests a radius $\hbar/mc$, with m = 335 MeV), so that the effective interaction volume υ is not known to better than an order of magnitude. Hence there is no meaningful way that we can correct lifetimes for phase space effects if the final states are of different dimensionalities. Furthermore, even if two final states have the same dimensionality, they may still have quite different effective interaction radii (e.g., $K^{\pm} \to \mu + \nu$ and $K^{\pm} \to \pi\pi$). The final states may also contain different spin factors. As a result of these intrinsic difficulties, the only way to judge the significance of making phase space corrections to lifetimes (total widths) is the operational procedure of applying the corrections and seeing what happens.

In addition to the difficulties just discussed, there is some arbitrariness in the choice of equations to be used for representing phase space effects. The transition amplitudes can be written in the form of Eqs. (4) - (7) above, which are not covariant and hence must be used only in the center-of-mass system[1], or they can be written in a covariant form, as described for example in RPP76[3]. Also, there is some choice as to whether certain energy-dependent factors should be left in the transition matrix elements or else removed and added to the phase space factors. If we go over to a covariant form for the transition matrix elements, the factors d^3p_i are replaced by d^3p_i/ε_i in the density-of-states term, which removes the factors ε_1 and ε_2 from Eq. (4). Also, in RPP76 a factor of E^{-1} is removed from the transition matrix element. Thus, for <u>meson</u> resonances, the RPP76 two-body transition probabilities are of the form

$$\lambda \sim |M|^2 \cdot p/E^2, \tag{8}$$

in the notation of Eqs. (3) and (4). For baryon decays to baryon plus meson, the spin terms supply an extra factor of E, so that for the <u>baryon</u> decays, the RPP76 two-body transition probabilities are of the form

$$\lambda \sim |M|^2 \cdot p/E. \tag{9}$$

The above discussion has been centered on the phase space factor dN_f/dE in Eq. (3). However, the matrix element M_{fi} in Eq. (3) also merits some discussion. If we study Eq. (3) from the point-of-view of the experimental particle decay rates, it seems clear that the large differences observed between the very slow (inhibited) decays and the fast (strong) decays must be due to fundamentally different decay mechanisms, and not to phase space factors. Since phase space factors dN/dE can change by factors of 10 - 100 in going from one resonance to another, we can account for variations in λ by factors of 10 - 100 as arising from phase space effects[4], but we cannot account for variations by factors of 10^{10} or so in this manner. For example, the $J^P = 1^-$ meson resonances $\rho(770) \rightarrow \pi\pi$ and $\omega(783) \rightarrow \pi\pi\pi$ have lifetimes $\tau \sim 4 \times 10^{-24}$ sec and $\tau \sim 6 \times 10^{-23}$ sec, respectively, and the factor of 15 difference in their lifetimes can be partly attributed to the difference between 2π and 3π final-state phase space volumes. But if we now compare $\rho(770) \rightarrow \pi\pi$ with $K_S^0(498) \rightarrow \pi\pi$, we find lifetimes $\tau \sim 4 \times 10^{-24}$ sec and $\tau \sim 1 \times 10^{-10}$ sec, respectively, which is a lifetime ratio of more than 10^{13}. This large a difference in the lifetimes cannot be attributed to phase space or spin effects. The K_S^0 decay is inhibited, and the inhibiting factor (strangeness and/or parity) must be in the matrix element M_{fi} and not in the phase space factor dN/dE. From this viewpoint, it seems reasonable that the matrix elements M for all of the "resonances" in the "resonance region" of Fig. 10.5 (to the right of the arrow) do not contain inhibiting factors, and the matrix elements M for all of the "particles" in the "particle region" of Fig. 10.5 (the region $X \leq 11$) do contain (electromagnetic) inhibiting factors. In particular, the ψ and ψ' resonances, which stand at the boundary between these two regions, are clearly inhibited, since otherwise their large mass values would lead to very rapid decays. If the matrix elements M differ discontinuously in these two regions, then the validity of the way in which phase corrections can be applied may also differ. Empirically, as we have already discussed, the total widths of the "resonances" appear to be responsive to the dictates of phase space, whereas the total widths of the "particles" are not. This leads to the conclusion that the "inhibiting factors" in the "particle" matrix elements are essentially independent of phase space considerations.

There is another viewpoint that we can use in studying phase space corrections. These corrections are based in essence on the Gibbsian assumption that the hadronic interaction has a statistical distribution of final state momenta. Now the ω meson, for example, has a lifetime $\tau \sim 6 \times 10^{-23}$ sec, so that the ω resonance persists for a distance of about 18 fermis (where we ignore any special-relativistic distortions). Thus the ω stays together for a distance that is about ten times the "length" of the interacting particles, and then breaks up. Hence it seems plausible that this relatively brief existence will be strongly influenced by statistical considerations. This result is even more true for the ρ meson, which persists for a distance of about one fermi; the ρ breaks up essentially as soon as it is formed, and statistical considerations must be all-important in such a case. However, the $\pi^{\pm}$ meson, for example, which persists for a distance of about 8 meters, is a stable particle in comparison to the ρ and ω mesons, and there is no reason to assume that the factors which lead to its eventual decay bear any resemblance at all to the Gibbsian statistical production and decay mechanisms which obtain for the ρ and ω mesons.

G2. Phase Space Applications

The seven particles clustered near X = 6 in Fig. 10.4 are the only particles which decay primarily into two-hadron final states; hence these are the only particles to which phase space corrections can be applied with any degree of confidence. The $\pi^{\pm}$, Σ^{0}, and π^{0} particles in Fig. 10.4 also have decays into two-particle final states, but the final-state particles for these decays are not both hadrons, so that the effective volumes for the decay interactions may vary from decay to decay. The other particles in Fig. 10.4 either have three-particle final states or else have a mixture of two-particle and three-particle final states, so that phase space corrections for the total widths of these particles either are difficult to compare with the corrections for the two-particle decays just described or else are essentially impossible to apply at all.

If we confine our attention to the particles in Fig. 10.4 which have two-body decays, then we should use a particle such as the Σ^{-} for a reference lifetime, since the neutron has a three-body decay mode. Figure G1 shows the results of applying phase space corrections to these particles. We first apply Eq. (4), with the normalization volume V kept the same for all decays. As is shown in the middle row in Fig. G1, these phase space corrections have only a small effect on the lifetimes near the logarithm X = 0 (which was the logarithm X = 6 in Fig. 10.4), but they shift the $\pi^{\pm}$ and π^{0} lifetimes so as to destroy the scaling in powers of α. Since these last two lifetimes are for decays which include particles that are not hadrons, the interaction volumes for these decays may be different from the interaction volumes for the particles near X = 0. To investigate this possibility, we next apply Eq. (5) to these same lifetimes, with the interaction volume now set proportional to $(\hbar/mc)^{3}$, where m is the mass of the decaying particle. Again, the scaling of lifetimes in powers of α is destroyed. (In these last results, we have also included a corrected lifetime for the $\mu^{\pm}$ meson, which has a three-body final state, by using Eq. (7) for the correction.) Finally, if we apply the covariant correction factors of Eqs. (8) and (9), we obtain results which are very similar to the results obtained by using Eq. (4); these results are partially illustrated by the small circles shown in the middle row of Fig. G1. The main difference between using Eq. (4) and using Eqs. (8) and (9) is that in the latter case the K_S^0 meson, due to its different energy dependence (Eq. 8) from that (Eq. 9) of its X = 0 neighbors, is shifted to the right by about one power of α. Once again, applying these phase space correction factors destroys the observed scaling of lifetimes in powers of α.

From a visual inspection of Fig. 10.4, it seems intuitively clear that the observed correlation between particle lifetime ratios and the fine structure constant $\alpha = e^2/\hbar c$ must be physically significant. If this conclusion is correct, then the fact that the phase space corrections shown in Fig. G1 destroy this scaling in powers of α must mean that these corrections were not properly applied. We can investigate this possibility by studying in detail the seven particles in Fig. 10.4 which are grouped together near X = 6, since these are the particles for which phase space corrections can be most reliably applied. The effect of applying phase space

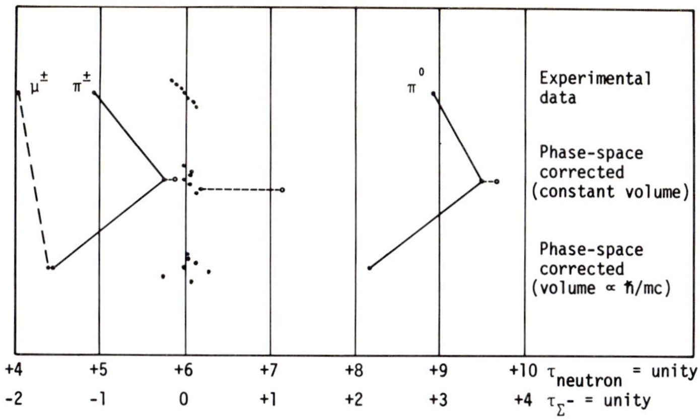

Figure G1. Phase space corrections as applied to the long-lived "particles" of Fig. 10.4 which have two-body decays; these particles have the same phase space dimensionality and hence can be intercompared. The Σ^- lifetime is used as the reference lifetime. The $\mu^{\pm}$ is included for comparison purposes. As is discussed in the text, the seven particles near X = 0 (X = 6), which all have two-hadron final states, are only mildly affected by the phase space corrections, whereas the lifetimes of the other particles, which have nonhadronic final states, are shifted radically so that the scaling in α is destroyed.

corrections to these seven particles, which was shown on a compressed scale in Fig. G1, is shown on an expanded scale in Fig. G2 (which is a plot analogous to that of Fig. 10.9). Phase space corrections are applied with most reliability to the Σ^-, Σ^+ and Ξ^0, Ξ^- pairs in Fig. G2, and it can be seen there that the ratios of these pairs of lifetimes remain essentially constant as corrections based on Eqs. (4), (5), and (9) are successively applied. The Λ/Ω^- lifetime ratio, on the other hand, changes somewhat with the various phase space corrections. (In making the Λ/Ω^- corrections, a $\Omega^- \rightarrow \Xi\pi$ and $\Omega^- \rightarrow \Lambda K$ branching ratio of 1/3 to 2/3 was assumed, and spin differences were ignored.) The Ξ^-/K_S^0 phase space correction is expected to be the most uncertain since the Ξ^- is a spin 1/2 baryon and the K_S^0 is a spin 0 meson, and as can be seen in Figs. G1 and G2 the Ξ^-/K_S^0 lifetime ratio varies wildly as Eqs. (4) and then (5) and then (8) and (9) are successively applied.

If we now compare Fig. G2 with Fig. 10.9, we can by inference draw some rather powerful general conclusions about phase space corrections for "particles". In the

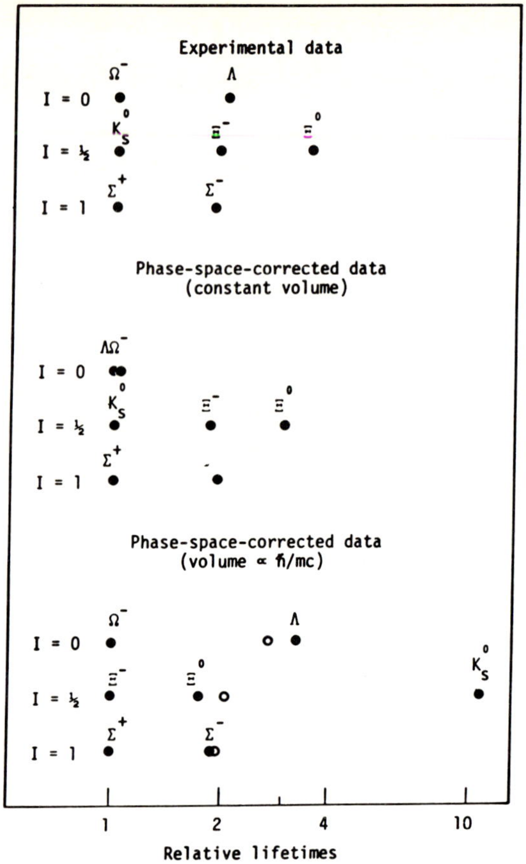

Figure G2. The effect of applying phase space corrections to the seven particles shown near X = 6 in Fig. 10.4, all of which have two-hadron final states. The experimental data shown at the top of Fig. G2 are plotted as in Fig. 10.9 in order to reveal the striking factor-of-two lifetime ratios that occur for these particles. As is discussed in the text, the observed stability of the (Ξ^-,Ξ^0) and (Σ^+,Σ^-) pairs indicates that phase space is not responsible for these factor-of-two lifetime ratios. If we turn this argument around, we can conclude that the observed factors of two indicate that the phase space corrections for all of the particles shown in Fig. G2 must be small. The open circles correspond to the covariant Eqs. (8) and (9).

only two cases in Fig. G2 where the phase space corrections are known to be reliable, namely in the lifetime ratios Σ^-/Σ^+ and Ξ^0/Ξ^-, the corrections turn out to be small. In particular, if we apply the most reliable phase space correction to these ratios — namely Eq. (9), which is covariant and contains the proper spin factors — , we obtain the following results:

Lifetime ratio	Experimental	Corr. by Eq. (9)	
Σ^-/Σ^+	1.85	1.90	(10)
Ξ^0/Ξ^-	1.75	2.00	

These results are crucial to the present discussion, because when they are applied to the systematics of Figs. G2 and 10.9, they lead to the following chain of conclusions:

(1) the factors of two shown in Eq. (10) for the phase-space-corrected Σ^-/Σ^+ and Ξ^0/Ξ^- lifetime ratios must be physically significant;

(2) these factors of two must be inherent in the (covariant) transition matrix elements $|M|^2$, and not in the phase space factors dN/dE, so that the observed factor-of-two lifetime ratios are not a phase space effect;

(3) if the factors of two observed for the Σ^-/Σ^+ and Ξ^0/Ξ^- lifetime ratios in Fig. 10.9 are physically significant, then all of the other factors of two shown in Fig. 10.9 are probably physically significant, and they should likewise be attributed to factors in the matrix elements rather than to phase space factors;

(4) conclusion (3) suggests in turn that the phase space corrections to all of the particles in Fig. 10.9, if properly applied, must be small, since the experimental lifetime ratios are very nearly factors of two;

(5) if we carry the factor-of-two line of reasoning of Fig. 10.9 over to Fig. 10.4 and apply it to the observed factors of α, we can at least tentatively conclude that phase space corrections for all of the "particles" of Fig. 10.4 must be small (which implies that the "electromagnetic lock" on the decay modes of each of these "particles" is independent of phase space considerations);

(6) the above results indicate that the large phase space corrections shown in Figs. G1 and G2 for the $\mu^\pm$, $\pi^\pm$, Ω^-, K_S^0, Σ^0, and π^0 lifetimes are not meaningful. In particular, the large fluctuations observed for the phase-space-corrected lifetimes of these particles as we go from Eqs. (4) to (5) to (8) and (9) are an indication that the magnitude of the correction depends in an essential way on (unknown) normalization factors, and conclusions (1) - (5) suggest that with properly chosen normalizations the phase space corrections would in fact be small.

Hence, for the "particles" of Fig. 10.4, we are forced back to experimental lifetimes as the most meaningful quantities to study from the point-of-view of the present systematics.

References for Appendix G.

1. B. T. Feld, Models of Elementary Particles, Blaisdell, Waltham (1969), Chapter IV.

2. G. Källén, Elementary Particle Physics, Addison-Wesley, Reading (1964).

3. Particle Data Group, Rev. Mod. Phys. 48, No. 2, Part II, April (1976).

4. Ref. 1, Sec. 7.4.

APPENDIX H. SUMMARY OF HADRON ROTATIONAL BANDS

This Appendix contains a listing of the baryon rotational bands that were defined in Chapter 15 and the meson rotational bands that were defined in Chapter 16. The mass, spin, and width of each resonance is given, together with the confidence-level rating assigned to the resonance and the empirical value for the rotational energy parameter $E_{rot} \equiv \hbar^2/2I$. The average value of E_{rot} for the whole rotational band is given at the bottom of each rotational band. Tables H1 - H4 contain the N, Δ, Λ, and Σ baryon rotational bands, respectively, and Table H5 contains the meson and kaon rotational bands. The N rotational bands of Table H1 and the Σ rotational bands of Table H4 each contain some high-spin levels which have been entered in two different rotational bands to indicate the ambiguous nature of these assignments. The baryon confidence ratings that are given in parentheses (***) are ratings that have recently been upgraded by the Karlsruhe group (see Table B1 in Appendix B). The confidence ratings of a few of the baryon and hyperon states listed in Tables H1 - H4 have recently been upgraded in RPP78; these upgraded ratings are denoted in the tables by bracketed ratings placed next to the old (RPP76) ratings.

Table H1. The N baryon rotational bands.

Band	Level	Resonance	Rating	E_{rot}	$<E_{rot}>$
$N\dot{3}$	S	(1149)	--		
	P	N($\sim$1230)$P_{11}\Gamma\sim200$	--	40.5	<40.5>
$N\dot{3}\dot{3}$	S	(1359)			
	P	N(1413)$P_{11}\Gamma187$	****	27.0	
	D	N(1517)$D_{13}\Gamma130$	****	26.3	
	F	N(1680)$F_{15}\Gamma123$	****	26.8	
	G				
	H	N(2183)$H_{19}\Gamma364$	***	27.5	<26.9>
$N\dot{4}\ddot{4}$	S	N(1511)$S_{11}\Gamma104$	****		
	P	N(1540)$P_{13}\Gamma200$	new [*]	(14.5)	
	D	N(1674)$D_{15}\Gamma159$	****	27.2	
	D	N(1710)$D_{13}\Gamma100$	** [***]	33.2	
	F				
	G†	N(2132)$G_{17}\Gamma247$	***	31.0	
	G†	N(2142)$G_{19}\Gamma297$	(***) [***]	31.6	
	H				
	I†	N(2650)$I_{1,11}\Gamma350$	***	27.1	<30.0>
$N\dot{3}\ddot{3}\dot{4}$	S	N(1674)$S_{11}\Gamma187$	****		
	P	N(1720)$P_{11}\Gamma137$	***	23.0	
	P	N(1722)$P_{13}\Gamma132$	***	24.0	
	D				
	F	N(1986)$F_{17}\Gamma214$	(***)	26.0	
	F	N(1989)$F_{15}\Gamma179$	* [**]	26.3	
	G†	N(2132)$G_{17}\Gamma247$	***	22.9	
	G†	N(2142)$G_{19}\Gamma297$	(***) [***]	23.4	
	H				
	I†	N(2650)$I_{1,11}$ 350	***	23.2	<24.1>

†These resonances, with very broad widths, are each entered in two rotational bands.

Table H2. The Δ baryon rotational bands.

Band	Level	Resonance	Rating	E_{rot}	$\langle E_{rot} \rangle$
$N\dot{3}$	S	(1149)			
	P	$\Delta(1232)P_{33}\Gamma 115$	****	41.5	<41.5>
$N\dot{4}\dot{4}$	S	(1511)			
	P	$\Delta(1550)P_{31}\Gamma 110$	new [*]	(19.5)	
	D	$\Delta(1676)D_{33}\Gamma 207$	***	27.5	
	F	$\Delta(1869)F_{35}\Gamma 255$	*** [****]	29.8	<28.7>
$N\dot{3}\dot{3}\dot{4}$	S	$\Delta(1623)S_{31}\Gamma 161$	****		
	P	$\Delta(1680)P_{33}\Gamma 220$	* [***]	28.5	
	D				
	F	$\Delta(1928)F_{37}\Gamma 237$	****	25.4	
	G	$\Delta(2174)G_{39}\Gamma 205$	** [***]	27.6	
	G	$\Delta(2200)G_{37}\Gamma 293$	new	28.9	
	H	$\Delta(2406)H_{3,11}\Gamma 289$	***	26.1	<27.3>

Table H3. The Λ hyperon rotational bands.

Band	Level	Resonance	Rating		E_{rot}	$\langle E_{rot} \rangle$
$\Lambda\dot{3}\dot{3}$	S	$\Lambda(1519)D_{03}\Gamma 15$	****			
	P	$\Lambda(1573)P_{01}\Gamma 147$	*	[**]	27.0	
	D	$\Lambda(1690)D_{03}\Gamma 60$	****		28.5	
	F	$\Lambda(1822)F_{05}\Gamma 81$	****		25.3	
	G	$\Lambda(2110)G_{07}\Gamma 250$	****		29.6	
	H	$\Lambda(2360)H_{09}\Gamma 120$	****		28.0	
	I	$\Lambda(2585)\Gamma 300$	***		25.4	<27.3>
$\Lambda\dot{4}\dot{4}$	S	$\Lambda(1672)S_{01}\Gamma 33$	****			
	P	$\Lambda(1747)P_{01}\Gamma 120$	**		(37.5)	
	D	$\Lambda(1825)D_{05}\Gamma 94$	****		25.5	
	F	$\Lambda(2020)F_{07}\Gamma 160$	*		29.0	<27.3>
$\Lambda\dot{3}\dot{3}\dot{4}$	S	$\Lambda(1827)S_{01}\Gamma 150$	**	[***]		
	P	$\Lambda(1881)P_{03}\Gamma 101$	***		27.0	
	D	$\Lambda(2010)D_{03}\Gamma 130$	**		30.5	
	F	$\Lambda(2120)F_{05}\Gamma 170$	**	[***]	24.4	<27.3>

Table H4. The Σ hyperon rotational bands.

Band	Level	Resonance	Rating	E_{rot}	$<E_{rot}>$
$\Sigma\dot{4}$	S	$\Sigma(1476)\Gamma34$	*		
	P				
	D	$\Sigma(1670)D_{13}\Gamma50$	****	32.3	
	F				
	G	$\Sigma(2090)F_{17}\Gamma103$	**	30.7	
	H				
	I	$\Sigma(2620)\Gamma175$	***	27.2	<30.0>
$\Sigma\dot{3}\dot{3}$	S	$\Sigma(1623)S_{11}\Gamma45$	**		
	P	$\Sigma(1676)P_{11}\Gamma120$	** [***]	25.5	
	D	$\Sigma(1774)D_{15}\Gamma130$	****	24.8	
	F	$\Sigma(1920)F_{15}\Gamma130$	****	24.6	
	G†	$\Sigma(2210)G_{19}\Gamma60$	****	29.4	
	H†	$\Sigma(2455)\Gamma220$	***	27.7	<26.4>
$\Sigma\dot{4}\dot{4}$	S	$\Sigma(1770)S_{11}\Gamma60$	***		
	P	$\Sigma(1840)P_{13}\Gamma120$	*	(35.0)	
	D	$\Sigma(1920)D_{13}\Gamma300$	***	25.5	
	F	$\Sigma(2040)F_{17}\Gamma190$	****	22.5	
	F	$\Sigma(2070)F_{15}\Gamma140$	*	25.0	
	G†	$\Sigma(2210)G_{19}\Gamma60$	****	22.0	
	H†	$\Sigma(2455)\Gamma220$	***	22.8	<23.5>

†These resonances are each entered in two rotational bands.

Table H5. The meson and kaon rotational bands. Matching meson and kaon bands are paired together. When two meson resonances occur at the same level, an average mass is used.

Band	Level	Meson resonances	<Mass>	Kaon resonances	E_{rot} Mesons	E_{rot} Kaons
$\dot{S}\dot{S}$	S	$\pi\pi(658)\Gamma{\simeq}15$	658			
	P	$\rho(773)\Gamma152$	773		57.5	
$\dot{S}\dot{S}\pi$	S	$\omega(783)\Gamma10$	783	(783)		
	P			$K^*(892)\Gamma49$		54.5
	D	$\eta_N(1083)\Gamma98$	1083			50.0
$\dot{3}\dot{S}\dot{S}$	S	(873)	(873)			
	P	$\delta^{\pm}(976)\Gamma47$[a]	976		51.5	
$\dot{3}\dot{S}\dot{S}\pi$	S	$S^*(993)\Gamma40,\phi(1020)\Gamma4$	<1006>			
	P	$A_1(\sim1100)\Gamma{\sim}300$	1100		47.0	
	D	$f(1271)\Gamma180,A_2(1310)\Gamma102$	<1290>		47.3	
$\dot{3}\dot{S}\dot{S}\dot{4}$	S	$M^o(1148)\Gamma15$[a]$,X^-(1153)\Gamma16$[a]	<1150>	(1150)		
$(K\dot{S}\dot{S})$	P	$B(1230)\Gamma125,\rho'(1250)\Gamma130$	<1240>	$Q(1234)\Gamma188$	45.0	42.0
	D	$X_0(1411)\Gamma95$[a]	1411	$K_N(1421)\Gamma108$	43.5	45.2
	F	$\omega(1667)\Gamma150,g(1690)\Gamma180$	<1678>	$K_N(1673)\Gamma77$[a]	44.0	43.6
	G	$h(2040)\Gamma193$	2040		44.5	
	H	$\pi\pi(2480)\Gamma280$	2480		44.3	
$\dot{3}\dot{3}\dot{3}\dot{S}\dot{S}$	S	$D(1285)\Gamma21$	1285	$K\pi\pi(1278)\Gamma25$[a]		
$(K\dot{S}\dot{S}\pi)$	P			$Q(1368)\Gamma241$		45.0
	D	$f'(1522)\Gamma62$	1522		39.5	
	F			$K_N(1785)\Gamma100$		42.3
	G			$K_N(2115)\Gamma300$		41.8
$\dot{3}\dot{3}\dot{3}\dot{S}\dot{S}\pi$	S	$E(1416)\Gamma60,X_1(1415)\Gamma30$	<1416>	(1416)		
$(K\dot{S}\dot{S}\dot{4})$	P					
	D	$A_3(1640)\Gamma240$	1640		37.3	
	F	$A_4(1825)\Gamma90$[a]	1825	$K_N(1840)\Gamma100$[a]	34.1	35.3

(continued on next page)

[a]No spin identification.

Table H5 (continued).

Band	Level	Meson resonances	<Mass>	Kaon resonances	E_{rot} Mesons	E_{rot} Kaons
$\dot{3}\dot{3}\dot{3}\dot{S}\dot{S}\dot{4}$	S	$F_1(1542)\Gamma 24$	1542	(1542)		
$(\dot{3}\dot{3}K\dot{S}\dot{S})$	P	$\rho'(1600)\Gamma\sim 300$	1600		29.0	
	D			$L(1765)\Gamma 138$		37.2

APPENDIX I. EQUATIONS FOR ASSOCIATED PRODUCTION

In the present hadronic light-quark model, all of the basis states are formed from the single mass quantum $M \sim 70$ MeV. By writing down equations for hadron production and decay reactions in terms of the quantum M, it is possible to trace these reactions through in some detail. In this Appendix, we briefly discuss a number of Associated-Production reactions — not from the viewpoint of providing a comprehensive treatment, but rather in the sense of sketching the manner in which the systematics of these reactions might be worked out.

I1. Basis States and Rules for Associated Production

Definition of Strangeness

S_1 = first-class strangeness — carried by spinless cabers — relevant for both binding energies (Fig. 18.6) and strangeness quantum numbers.

S_2 = second-class strangeness — carried by spinor triplets — relevant only for binding energies [see Eq. (12a) in Appendix C].

Excitation Basis States

π	$\equiv M\bar{M}$	$S_1 = 0$	(Fig. 18.9)
3	$\equiv 3, \dot{3}, \dot{\bar{3}}, \bar{3}$	$S_1 = +1,0,0,-1$	(Fig. 18.7)
4	$\equiv 4, \dot{4}, \bar{4}$	$S_1 = +1,0,-1$	
K	$\equiv 3\pi\pi,\ \bar{K} \equiv \bar{3}\pi\pi$	$S_1 = +1,-1$	(Fig. 18.3)
333	$\equiv 3\bar{3}\dot{3}$	$S_1 = +1-1 = 0$	Basic 3-triplet
$\overline{333}$	$\equiv 3\bar{3}\dot{\bar{3}}$	$S_1 = +1-1 = 0$	Basic 3-antitriplet

Spinor Core Configurations

$(\hat{p}) = S^{+}_{\uparrow}S^{-}_{\downarrow}S^{+}_{\uparrow}$ - strangeness $S_2 = +1$.

$(\hat{n}) = S^{+}_{\uparrow}S^{=}_{\downarrow}S^{+}_{\downarrow}$ - strangeness $S_2 = +1$. [Eq. (19.6)]

$(\hat{\Lambda}) = S^{+}_{\uparrow}S^{-}_{\downarrow}S^{+}_{\downarrow}$ - strangeness $S_2 = +1$.

Spinless Mass Operators

{ } = creation operator - creation of quanta M.

[ ] = annihilation operator - annihilation of quanta M.

< > = transformation operator - transformation of quanta M or $\bar{M}$ into their antistates: $\langle M\rangle \rightarrow \bar{M}$; $\langle\bar{M}\rangle \rightarrow M$.

Rules for the Mechanism of Associated Production

(R.1) Strangeness $S_2 = +1$ (-1) spinor cores combine with strangeness $S_1 = -1$ (+1) spinless cabers to form Y* resonances; strangeness $S_2 = +1$ (-1) spinor cores do not combine with strangeness $S_1 = +1$ (-1) spinless cabers to form Z* resonances.

(R.2) The conventional strangeness quantum number of a resonance is the sum of the strangeness quantum numbers S_1 carried by the spinless cabers in the resonance cluster.

(R.3) The dominant excitations produced by the creation operator { } in Associated Production are the same excitations which are dominant in the meson production reactions shown in Chapter 16. These excitations, which are the straight conversion of kinetic energy into hadronic matter, are necessarily nonstrange.

(R.4) The transformation operator < > occurs in conjunction with the annihilation operator [ ], since the annihilation process is associated with a geometrical rearrangement of the resonance structure. Also, the transformation operator < > seems to be responsible for the K_S^0 regeneration process ($K_L^0 = 3\bar{3}M \rightarrow 3\bar{3}\langle M\rangle = 3\bar{3}\bar{M} = 3\bar{4} = K_S^0$), and it suggests that the physical K_L^0 secondary collision which causes the regeneration of K_S^0 mesons leads to the reorientation of the odd quantum M in the linear K_L^0 structure (see Fig. 18.3). These results both suggest that the distinction between the particle state M^0 and the antiparticle state $\bar{M}^0$ resides in its spatial orientation with respect to the resonance structure.

(R.5) The asymmetric collision process $\pi + N$, in which a strangeness $S_2 = +1$ spinor core is involved, leads to the production of the excitation sequence $\{\overline{333}\}$, $\{\overline{333}\cdot 333\}$, $\{\overline{333}\cdot 333\cdot\overline{333}\}$, etc., but not to the production of excitations such as $\{333\}$, $\{\overline{333}\cdot\overline{333}\}$, etc.

I2. Λ Hyperon Production

An Associated Production reaction for the Λ hyperon is

$$\pi + p \rightarrow \pi\{3\bar{3}\pi\}(\hat{p}) \rightarrow (\hat{\Lambda})\bar{3} + 3\pi\pi = \Lambda + K \quad . \tag{1}$$

This reaction is shown pictorially in Fig. 18.16. The excitation $3\bar{3}\pi = 4\bar{4}$ depicted in (1) is the η meson. The spinor-flip transition $(\hat{p}) \rightarrow (\hat{\Lambda})$ is characteristically associated with an $I = 0$ excitation channel, which may account for the isotopic spin $I = 0$ of the Λ. From Rule (R.1) the $(\hat{\Lambda})$ spinor core will combine with a $\bar{3}$ to form a Y* resonance (the Λ), but it will not combine with a 3 to form a Z* resonance. Typical decay processes for the final states of (1) are:

$$\Lambda = (\hat{\Lambda})\bar{3} = (\hat{\Lambda})\bar{M}\bar{M}\bar{M} \rightarrow (\hat{p})[\bar{M}]<\bar{M}>\bar{M} = (\hat{p}) + M\bar{M} = p + \pi \quad .$$

$$\rightarrow (\hat{n})[\bar{M}]<\bar{M}>\bar{M} = (\hat{n}) + M\bar{M} = n + \pi \quad .$$

$$K = 3\pi\pi = MMM\pi\pi \rightarrow [M]<M>M\pi\pi = \bar{M}M + \pi + \pi - \pi + \pi + \pi \quad .$$

$$\rightarrow [3]\pi\pi = \pi + \pi \quad .$$

I3. Σ Hyperon Production

An Associated Production reaction for the Σ hyperon is

$$\pi + p \rightarrow \pi\{\overline{333}\}(\hat{p}) = \pi 3\bar{3}\dot{\bar{3}}(\hat{p}) \rightarrow (\hat{p})\bar{4} + 3\pi\pi = \Sigma + K \quad . \tag{2}$$

The excitation $\overline{333}$ is the excitation which appears in meson resonances as the rho meson ($\overline{333} \rightarrow \dot{S}\dot{S}$). Since no core spinor-flip is involved in reaction (2), the Σ is produced with isotopic spin I = 1. Typical decay processes of the Σ are:

$$\Sigma = (\hat{p})\bar{4} = (\hat{p})\bar{M}\bar{M}\bar{M}\bar{M} \rightarrow (\hat{p})<\bar{M}>\bar{M}<\bar{M}>\bar{M} \rightarrow (\hat{p})[M\bar{M}]M\bar{M} = p + \pi \quad .$$

$$\rightarrow (\hat{\Lambda})[\bar{M}]\bar{M}\bar{M}\bar{M} + \gamma = \Lambda + \gamma \quad .$$

The gamma-producing decay of course occurs only in the $I_Z = 0$ mode.

Since $\Lambda = S\bar{S}S\bar{3}$ and $\Sigma = S\bar{S}S\bar{4}$, the only other possible strangeness $S_1 = -1$ configuration is $S\bar{S}S\bar{7}$. This configuration actually occurs, in the form of the Λ(1402) resonance:

$$\bar{K} + p \rightarrow (\hat{p})\bar{3}\pi\pi = (\hat{p})\bar{7} = \Lambda(1402)S \quad .$$

The Λ(1402) has a single decay mode:

$$\Lambda(1402) = (\hat{p})\bar{7} = (\hat{p})\bar{3}\bar{M}\bar{M}\bar{M}\bar{M} \rightarrow (\hat{p})\bar{3}\bar{M}[M]\bar{M}M = (\hat{p})\bar{4} + \bar{M}M = \Sigma + \pi \quad .$$

I4. Ξ Hyperon Production

We have seen above that the strangeness $S_1 = -1$ resonances occur in all possible combinations:

$$S\bar{S}S\bar{3} = \Lambda, \; S\bar{S}S\bar{4} = \Sigma, \; S\bar{S}S\bar{7} = \Lambda(1402) \quad .$$

The main question with regard to mapping the strangeness $S_1 = -2$ Ξ resonances is to explain why they occur only in the combination

$$S\bar{S}S\bar{3}\bar{3} = \Xi \quad , \tag{3a}$$

and not in the combinations

$$S\bar{S}S\bar{3}\bar{4} = \Xi' \qquad \text{or} \qquad S\bar{S}S\bar{4}\bar{4} = \Xi'' \quad . \tag{3b}$$

As we show below, the Ξ' and Ξ'' resonance forms can be ruled out on the basis of the unnatural excitation processes that would be required.

Two Associated Production reactions for the Ξ are:

$$\pi + p \rightarrow \pi\{333\cdot\overline{333}\}(\hat{p}) = \pi3\bar{3}\dot{3}3\bar{3}\dot{\bar{3}}(\hat{p}) \rightarrow (\hat{\Lambda})\bar{3}\bar{3} + 3\pi\pi + 3\pi\pi = \Xi + K + K \quad . \tag{4}$$

$$\bar{K} + p \rightarrow \bar{3}\pi\pi\{3\bar{3}\}(\hat{p}) \rightarrow (\hat{\Lambda})\bar{3}\bar{3} + 3\pi\pi = \Xi + K \quad . \tag{5}$$

In order to produce the Ξ' and Ξ'' hyperons, which are not observed experimentally, we would have to have Associated Production reactions of the types shown below:

$$\pi + p \rightarrow \pi\{\overline{333}\pi\cdot3\bar{3}\pi\}(\hat{p}) = \pi3\bar{3}\dot{\bar{3}}\pi3\bar{3}\pi(\hat{p}) \rightarrow (\hat{\Lambda})\bar{3}\bar{4} + 3\pi\pi + 3\pi\pi = \Xi' + K + K \quad . \tag{6}$$

$$\pi + p \rightarrow \pi\{\overline{333}\pi\cdot\overline{333}\}(\hat{p}) = \pi3\bar{3}\dot{\bar{3}}\pi3\bar{3}\dot{\bar{3}}(\hat{p}) \rightarrow (\hat{\Lambda})\bar{4}\bar{4} + 3\pi\pi + 3\pi\pi = \Xi'' + K + K \quad . \tag{7}$$

$$\bar{K} + p \rightarrow \bar{3}\pi\pi\{\dot{3}\bar{3}\pi\}(\hat{p}) \rightarrow (\hat{\Lambda})\bar{4}\bar{4} + 3\pi\pi = \Xi'' + K \quad . \tag{8}$$

The excitations $\{\overline{333}\pi\cdot3\bar{3}\pi\}$ and $\{\overline{333}\pi\cdot\overline{333}\}$ in (6) and (7), respectively, are both mixed excitations, and the experience with excitations in the meson resonances suggests† that these mixed excitations occur with very small cross sections. Also, the excitation $\{\overline{333}\pi\cdot\overline{333}\}$ in (7), which contains two $\overline{333}$ combinations, is not allowed by Rule (R.5). The excitation $\{\dot{3}\bar{3}\pi\}$ in (8) has non-zero strangeness, and hence is not allowed by Rule (R.3). Thus we can see that the nonobservance of the cascade resonances Ξ' and Ξ'' of Eq. (3b) is simply explained in terms of the requisite excitation mechanisms.

The dominant Ξ decay mode is

$$\Xi = S\bar{S}S\bar{3}\bar{3} = S\bar{S}S\bar{3}\bar{M}\bar{M}\bar{M} \rightarrow S\bar{S}S\bar{3}[\bar{M}]\langle\bar{M}\rangle\bar{M} = S\bar{S}S\bar{3} + M\bar{M} = \Lambda + \pi \quad .$$

---

†See M. H. Mac Gregor, Phys. Rev. D9 1259 (1974), Table XXIII.

The decay mode

$$\Xi = S\bar{S}S\bar{3}\bar{M}\bar{M}\bar{M} \rightarrow S\bar{S}S\bar{3}\bar{M}\langle\bar{M}\rangle\bar{M} = S\bar{S}S\bar{4} + M\bar{M} = \Sigma + \pi$$

is not allowed, because the transformation $\langle\bar{M}\rangle \rightarrow M$ can only take place in cases where an annihilation has occurred (e.g., [M] or [3]), or where a secondary particle scattering has occurred (K_S^0 regeneration).

APPENDIX J. PUBLICATIONS, PREDICTIONS AND FULFILLMENTS

The conceptual ideas behind the present light-quark model were initiated in the Summer of 1969. A large number of the experimental results that are now contained in the Review of Particle Properties (RPP76) have appeared subsequent to that time. Hence it is of interest to ascertain if the experiments are or are not deveoping in the manner that we would expect from the viewpoint of the light-quark model. The answer to this question, in the opinion of the present author, is yes. The most spectacular example of this is in the systematics of the New Particles, which reveals their 70 MeV mass quantization and their small binding energies very clearly. But there are also interesting examples among the Old Particles – most notably, in the 4% binding energy of the $\bar{p}n$ bound state, which was anticipated on the basis of the mass systematics of the Old Particles, and which, for the record, is the largest binding energy ever measured experimentally for any physical system.

In order to trace the development of the present ideas and the corresponding developments with respect to the experimental situation, we list in this Appendix the series of papers and reports that have been written on this work, and we briefly mention some predictions and fulfillments that are associated with these papers. Some of the speculations that were contained in these papers have not turned out to be correct, and there is no real purpose in reviewing them here. However, the main thrust of the work – that quarks are light, and quark binding energies are very small – seems to be vindicated, and this has led to a number of successful predictions, particularly in the identification of controversial baryon and meson resonances.

Predictions and fulfillments of the present studies up to the year 1974 are summarized in Section XI of the article in Phys. Rev. D9, 1259 (1974) (which we will refer to in the present Appendix as Paper I), and also in the Note Added in Proof to the article in Phys. Rev. D10, 850 (1974) (which we will refer to in the present Appendix as Paper II). We will mention some, but not all, of those results here.

We now summarize the series of published papers (plus a few unpublished reports) that have been written by the present author on the subject of the light-quark model. These papers all have Malcolm H. Mac Gregor as the sole author.

UCRL-71842 (Rev.) Part I, Aug. 8 (1969) (unpublished)

This paper introduced the idea that quark binding energies are very small ($\sim$1%), and it identified the muon mass and pion mass as fundamental light-quark building blocks. The necessity for having a neutral muon mass μ^o = 105 MeV was pointed out. (With the subsequent development, a few months later, of the relativistic equations for a spinning sphere, the neutral muon mass devolved into the neutral spinless mass quantum M^o = 70 MeV.)

UCRL-71842 (Rev.) Part II, Aug. 8 (1969)(unpublished)

This paper introduced the idea that baryons and mesons occur in nuclear-physics-type rotational bands, with an $\ell(\ell+1)$ splitting of the levels. The meson rotational bands that were identified in this 1969 paper differ substantially from the present-day (1978) assignments, but the baryon rotational bands are very similar. Table XXXII of Paper I shows a detailed comparison between the baryon rotational bands as they appeared experimentally in 1969 and in 1974. During those five years, four new levels appeared – in particular, the predicted S states N(1508)S and Σ(1620)S – , and the quantum numbers of two other rotational levels were clarified. Also, the P_{11} Roper resonance, which was predicted in 1969 to be a split resonance (on the basis of these baryon rotational bands), was finally identified as such in the 1974 phase shift analysis of AYED and coworkers (see RPP74: Particle Data Group, Physics Letters 50B, No. 1, April 1974).

The ideas that were discussed in UCRL-71842 (Rev.), Parts I and II, are documented in the following abstracts: Bull. American Physical Society 14, 1178 (1969); ibid., 14, 1206 (1969).

UCRL-72128, Nov. 5 and 14 (1969)

This report contains the equations for a relativistically-spinning sphere, which are applied to the calculation of the spin angular momentum and magnetic moment of the muon and electron. In particular, the spinless 70 MeV mass quantum M^o is related to the spinning 105 MeV muon. This report was submitted for publication in December 1969, and it was published in Lett. Nuovo Cimento 4, 211 (1970). UCRL-72187, Dec. 10 (1969), is a later version of UCRL-72128; and UCRL-72287, Jan. 27 (1970) and UCRL-72288, Jan. 29 (1970) contain extensions of the above ideas.

Nuovo Cimento Letters 4, 211 (1970)

This paper contains the material described above in UCRL-72128.

UCRL-72495 (Rev. I), July 17 (1970)(unpublished)

This report identifies the meson interference dips at $\sim$650 MeV and $\sim$1300 Mev with the spinor excitations $\ddot{S}\ddot{S}$ and $\ddot{S}\ddot{S}\ddot{S}\ddot{S}$, respectively (see Sec. 8.4 in Chapter 8).

<u>UCRL-72646, Aug. 14 (1970)(unpublished)</u>

This paper contains a comprehensive treatment of the baryon rotational bands, but it has a now-discarded method for handling the meson excitations.

<u>UCRL-72736, Aug. 31 (1970)</u>

This paper contains a discussion of the structure of the nucleon as seen from the present light-quark viewpoint, and it was presented as part of an invited talk (on another subject) that was given at the "Third International Symposium on Polarization Phenomena in Nuclear Reactions," Madison, Wisconsin, Aug. 31 (1970).

<u>Nuovo Cimento Letters 4, 1043 (1970)</u>

This paper contains the light-quark systematics for the <u>baryon</u> resonances, including the S-state energy level diagrams and the baryon rotational bands.

<u>Nuovo Cimento Letters 4, 1249 (1970)</u>

This paper contains the light-quark systematics for the <u>meson</u> resonances, including the S-state energy level diagrams; and it gives, for the first time, the presently-used method for sorting the meson resonances into rotational bands.

<u>Nuovo Cimento Letters 4, 1309 (1970)</u>

This paper is an attempt to apply the ideas of Kemmer symmetric coupling to meson decays in the context of the light-quark model.

<u>Lectures from the Coral Gables Conference on Fundamental Interactions at High Energy, January 20 - 22, 1971, M. Dal Cin, G. J. Iverson, and A. Perlmutter, editors, Gordon and Breach, publishers, New York (1971), Vol. 3, Invited Papers, pages 75 - 154</u>.

This is an article on "Hadron Spectroscopy" written by the present author in connection with his invitation to participate in this closed conference on problems in high energy physics. Page 96 of this article contains the following statement: "In both the mesons and baryons, there seem to be two types of resonant structure, one that occurs right on the mass shell (a resonance), and one that has about a 3% binding energy (a bound state)." This observation was made prior to the appearance, in the Summer of 1971, of the $\bar{p}n(1795)$ bound state with its 4% binding energy; that is, this observation was made at a time when no binding energy larger than 1% had ever been observed for any physical system.

<u>Nuovo Cimento Letters 1, 427 (1971)</u>

This paper containa a discussion of the systematics of nuclear physics as applied to hadron rotational levels.

Nuovo Cimento Letters 1, 437 (1971)

This paper discusses possible ΛN dibaryon resonances that have baryon number $B = 2$.

Nuovo Cimento Letters 1, 759 (1971)

This paper shows the universal manner in which the fine structure constant $\alpha = e^2/\hbar c \simeq 1/137$ appears as a scaling factor in particle masses, particle lifetimes, and stellar densities. The periodicity of metastable elementary particle lifetimes in powers of α is displayed for the first time, and a phenomenological relationship is deduced between the mass of the muon and the gravitational constant G. The scaling of lifetimes in powers of α was extended several years later by the appearance of revised experimental lifetimes for the η and η' mesons and by the discovery of the J/ψ and ψ' mesons.

UCRL-73193, May 10 (1971)(unpublished)

This report identifies possible nonadiabatic rotations in very light atomic nuclei that seem to be analogous to the nonadiabatic rotations observed in hadron excitations.

Particles and Nuclei 1, 473 (1971)

This paper discusses the light-quark explanation for the small interference dip observed in the A_2 meson.

UCRL-73328, July 30 (1971)

This paper is the original version of the two long articles that were published in The Physical Review in 1974. The three year delay between the original submission of this paper and its final publication was due mainly to continual self-imposed revisions and extensions by the author, and not to any action or inaction on the part of the editors of The Physical Review.

Nuovo Cimento Letters 2, 846 (1971)

This paper discusses rotation invariance as applied to hadron rotational bands.

UCRL-73585, Dec. 8 (1971)(unpublished)

This paper describes the accurate manner in which the linear spinless cabers reproduce the mass values of the spinless π, K, η, M, and η' mesons.

Particles and Nuclei 2, 358 (1971)

This paper describes some nearby peaks which seem to be produced in coincidence with the interference dip that is observed in the A_2 meson.

Nuovo Cimento Letters 3, 197 (1972)

This paper gives, for the first time, the presently-used binding energy systematics and mass splittings of the light-quark basis states, and it contains a calculation of the mass of the newly-reported $\bar{p}n$ bound state.

Nuovo Cimento 8A, 235 (1972)

This is essentially the same paper that was published in the Lectures from the Coral Gables Conference in 1971. It contains a long four-page Note Added in Proof at the end which describes a number of recent experimental and theoretical developments.

"A relationship between the gravitational constant and the masses of elementary particles," April 8 (1972)

This is an essay that was given an Honorable Mention award in the competition of the Gravity Research Foundation, New Boston, N. H., June 1, 1972.

Particles and Nuclei 3, 173 (1972)

This paper contains a discussion of the invariant strangeness of the nucleon spinors in the context of the light-quark model.

Nuovo Cimento Letters 5, 135 (1972)

This paper sketches the relationship between the $\bar{p}n$ bound state and the mass of the nucleon quark.

Particles and Nuclei 5, 69 (1973)

This paper uses the $\bar{p}n$ bound state to establish the Fermi-Yang nature of nucleon quarks.

UCRL-73328 (Rev. I), March 15 (1973)

This is a revision of the July 1971 report that was published in 1974 in the form of two long articles in The Physical Review. This version of the report received wide circulation as a preprint. It contains the first use of vacuum-formed plastic models to represent light-quark geometries. It also contains, on page 227, the meson excitation series $(\dot{3}\dot{3}\dot{3})^n$ that serves as the generator of the high-mass meson resonances. At the time that this report was issued, terms in this series up to n = 3 were known. The J/ψ New Particle peak subsequently appeared as the n = 5 term in the series, followed shortly after by the ψ' peak as the n = 6 term and the ψ(4414) peak as the n = 7 term.

Nuovo Cimento 18A, 169 (1973)

This nineteen-page paper is devoted to a study of the rho meson. In particular, it explores the consequences of the idea that the rho is a spin $J = 1$ resonance only at the peak of the resonance, and it has $J < 1$ below the peak and $J > 1$ above the peak, as is of course required if the distribution of masses in the rho is correlated with a corresponding distribution of angular momentum values.

Nuovo Cimento 20A, 471 (1974)

This paper points out the significance of resonance widths as a key identification symbol for distinguishing between rotationless and rotational excitations, and it also discusses in detail the systematics of elementary particle lifetimes. When this paper was originally issued in preprint form, the preprint librarian at CERN wrote to the author and said that he had received 65 requests for copies of the paper.

Physical Review D9, 1259 (1974)

"Light-quark hadron spectroscopy: experimental systematics and angular momentum systematics," 71 pages, denoted as Paper I.

Physical Review D10, 850 (1974)

"Light-quark hadron spectroscopy: a geometric quark model for S-states," 34 pages, denoted as Paper II.

These two papers are the published version of the report on the light-quark model that was initially submitted to The Physical Review in July 1971. Because of its great length, the original document was split into two parts. Even so, the first of these two papers is probably the longest paper ever accepted for publication in The Physical Review, and certainly the longest paper by a single author. These two papers contain a complete account of all the work done by the author on the light-quark model between the years 1969 and 1974. The work since that time has centered mainly on extensions in the area of nuclear physics, and on the incorporation of the subsequently-discovered New Particles into the systematics of the model.

Physical Review D12, 1492 (1975)

This paper identifies the J/ψ and ψ' New Particles as $\bar{N}N$ excitations, and it shows that they are the $n = 5$ and $n = 6$ terms in the previously-published $(\dot{3}\dot{3}\dot{3})^n$ meson excitation series. It also mentions the fact that the lifetime of the J/ψ fits into the scaling in powers of α which was previously established on the basis of the lifetimes of the long-lived Old Particles.

Physical Review D13, 574 (1976)

This paper completes the study of elementary particle lifetimes by extending the span of lifetimes to include the short-lived "resonances" as well as the long-lived "particles". A detailed discussion is given as to the validity or nonvalidity of applying phase space corrections to these elementary particle lifetimes.

Physical Review D14, 1323 (1976)

This paper points out the fact that the centrifugal barrier concept used by elementary particle physicists to make decay rate corrections is probably invalid. As a semi-related topic, the paper also discusses the question of the spin of the Ω^- hyperon.

Physical Review D14, 1463 (1976)

This paper extends the systematics of the New Particles within the framework of the light-quark model: the $\psi(4414)$ and other New Particle peaks are incorporated; the decay modes of the ψ' are analyzed; and the striking "charmed meson" sequence of levels $\bar{p}n(1795)$, $D(\sim1870)$, $\bar{p}p(1935)$, $D^*(\sim2010)$ is pointed out.

UCRL-78679, Sept. 11 (1976)(unpublished)

This report demonstrates the 70 MeV quantization of the excitation mode and "charmed-meson" decay modes of the $\psi(4028)$ peak.

Nuovo Cimento 36A, 113 (1976)

This is a long 58 page article that describes the applications of the light-quark model in the area of nuclear physics. In particular, these studies suggest that atomic nuclei contain two-dimensional laminar layers of (mostly) α particles; these layers are manifested experimentally by the systematics of symmetric and asymmetric fission in heavy nuclei.

UCRL-80467, Dec. 27 (1977), "Negative Masses, Photons, and $\psi^*\psi$ Probability Fields"

This paper explores the concept of "negative masses" as they are defined (for example) by the Dirac equation. The peculiar properties of negative masses are precisely the properties that are required in order to reproduce the "particle" aspects of the photon. The mathematical identity $\psi^*(+m)\psi(+m) \equiv \psi(-m)\psi(+m)$ is also explored as a means of providing a physical interpretation for the $\psi^*\psi$ probability fields of quantum mechanics.

UCRL-80734, Feb. 6 (1978), "The Two Lifetimes of the Ω^-"

This paper contains the results that are displayed in Table 10.2 of Chapter 10. On the basis of these results, it is suggested that the two lifetimes which have been measured for the Ω^- probably both exist.

In addition to these research papers, the author has also given invited talks on the light-quark model at the following universities: UCLA, Berkeley (twice), Florida State, U. of Florida, Minnesota, Wisconsin, Purdue, Arizona, Virginia,

Michigan, Ohio State, Ohio U., Wayne State, Notre Dame, Arizona State, U. of New Mexico, U. of Western Ontario, Virginia Polytechnic, Oregon, Oregon State, Rice, and U. C. Davis. Thus this light-quark model has not only passed under the close scrutiny of the professional journal referees, but it has also been exposed in seminars and colloquia to a substantial segment of the elementary particle physics community — a process which is usually quite successful in eliminating any demonstrably-wrong aspects of the model.

The purpose of the present book is to make this material available in its entirety to interested readers, so that they can ascertain for themselves the simplicity and at the same time the accuracy and comprehensiveness of this light-quark approach to elementary particle structure.

> "If a man does not keep pace with his companions, perhaps it is because he hears a different drummer. Let him step to the music he hears, however measured or far away."
>
> Henry David Thoreau, Walden

Selected Issues from

Lecture Notes in Mathematics

Vol. 431: Séminaire Bourbaki – vol. 1973/74. Exposés 436–452. IV, 347 pages. 1975.

Vol. 433: W. G. Faris, Self-Adjoint Operators. VII, 115 pages. 1975.

Vol. 434: P. Brenner, V. Thomée, and L. B. Wahlbin, Besov Spaces and Applications to Difference Methods for Initial Value Problems. II, 154 pages. 1975.

Vol. 440: R. K. Getoor, Markov Processes: Ray Processes and Right Processes. V, 118 pages. 1975.

Vol. 442: C. H. Wilcox, Scattering Theory for the d'Alembert Equation in Exterior Domains. III, 184 pages. 1975.

Vol. 446: Partial Differential Equations and Related Topics. Proceedings 1974. Edited by J. A. Goldstein. IV, 389 pages. 1975.

Vol. 448: Spectral Theory and Differential Equations. Proceedings 1974. Edited by W. N. Everitt. XII, 321 pages. 1975.

Vol. 449: Hyperfunctions and Theoretical Physics. Proceedings 1973. Edited by F. Pham. IV, 218 pages. 1975.

Vol. 458: P. Walters, Ergodic Theory – Introductory Lectures. VI, 198 pages. 1975.

Vol. 459: Fourier Integral Operators and Partial Differential Equations. Proceedings 1974. Edited by J. Chazarain. VI, 372 pages. 1975.

Vol. 461: Computational Mechanics. Proceedings 1974. Edited by J. T. Oden. VII, 328 pages. 1975.

Vol. 463: H.-H. Kuo, Gaussian Measures in Banach Spaces. VI, 224 pages. 1975.

Vol. 464: C. Rockland, Hypoellipticity and Eigenvalue Asymptotics. III, 171 pages. 1975.

Vol. 468: Dynamical Systems – Warwick 1974. Proceedings 1973/74. Edited by A. Manning. X, 405 pages. 1975.

Vol. 470: R. Bowen, Equilibrium States and the Ergodic Theory of Anosov Diffeomorphisms. III, 108 pages. 1975.

Vol. 474: Séminaire Pierre Lelong (Analyse) Année 1973/74. Edité par P. Lelong. VI, 182 pages. 1975.

Vol. 484: Differential Topology and Geometry. Proceedings 1974. Edited by G. P. Joubert, R. P. Moussu, and R. H. Roussarie. IX, 287 pages. 1975.

Vol. 487: H. M. Reimann und T. Rychener, Funktionen beschränkter mittlerer Oszillation. VI, 141 Seiten. 1975.

Vol. 489: J. Bair and R. Fourneau, Etude Géométrique des Espaces Vectoriels. Une Introduction. VII, 185 pages. 1975.

Vol. 490: The Geometry of Metric and Linear Spaces. Proceedings 1974. Edited by L. M. Kelly. X, 244 pages. 1975.

Vol. 503: Applications of Methods of Functional Analysis to Problems in Mechanics. Proceedings 1975. Edited by P. Germain and B. Nayroles. XIX, 531 pages. 1976.

Vol. 507: M. C. Reed, Abstract Non-Linear Wave Equations. VI, 128 pages. 1976.

Vol. 509: D. E. Blair, Contact Manifolds in Riemannian Geometry. VI, 146 pages. 1976.

Vol. 515: Bäcklund Transformations. Nashville, Tennessee 1974. Proceedings. Edited by R. M. Miura. VIII, 295 pages. 1976.

Vol. 516: M. L. Silverstein, Boundary Theory for Symmetric Markov Processes. XVI, 314 pages. 1976.

Vol. 518: Séminaire de Théorie du Potentiel, Proceedings Paris 1972–1974. Edité par F. Hirsch et G. Mokobodzki. VI, 275 pages. 1976.

Vol. 522: C. O. Bloom and N. D. Kazarinoff, Short Wave Radiation Problems in Inhomogeneous Media: Asymptotic Solutions. V, 104 pages. 1976.

Vol. 523: S. A. Albeverio and R. J. Høegh-Krohn, Mathematical Theory of Feynman Path Integrals. IV, 139 pages. 1976.

Vol. 524: Séminaire Pierre Lelong (Analyse) Année 1974/75. Edité par P. Lelong. V, 222 pages. 1976.

Vol. 525: Structural Stability, the Theory of Catastrophes, and Applications in the Sciences. Proceedings 1975. Edited by P. Hilton. VI, 408 pages. 1976.

Vol. 526: Probability in Banach Spaces. Proceedings 1975. Edited by A. Beck. VI, 290 pages. 1976.

Vol. 527: M. Denker, Ch. Grillenberger, and K. Sigmund, Ergodic Theory on Compact Spaces. IV, 360 pages. 1976.

Vol. 532: Théorie Ergodique. Proceedings 1973/1974. Edité par J.-P. Conze and M. S. Keane. VIII, 227 pages. 1976.

Vol. 538: G. Fischer, Complex Analytic Geometry. VII, 201 pages. 1976.

Vol. 543: Nonlinear Operators and the Calculus of Variations, Bruxelles 1975. Edited by J. P. Gossez, E. J. Lami Dozo, J. Mawhin, and L. Waelbroeck, VII, 237 pages. 1976.

Vol. 552: C. G. Gibson, K. Wirthmüller, A. A. du Plessis and E. J. N. Looijenga. Topological Stability of Smooth Mappings. V, 155 pages. 1976.

Vol. 556: Approximation Theory. Bonn 1976. Proceedings. Edited by R. Schaback and K. Scherer. VII, 466 pages. 1976.

Vol. 559: J.-P. Caubet, Le Mouvement Brownien Relativiste. IX, 212 pages. 1976.

Vol. 561: Function Theoretic Methods for Partial Differential Equations. Darmstadt 1976. Proceedings. Edited by V. E. Meister, N. Weck and W. L. Wendland. XVIII, 520 pages. 1976.

Vol. 564: Ordinary and Partial Differential Equations, Dundee 1976. Proceedings. Edited by W. N. Everitt and B. D. Sleeman. XVIII, 551 pages. 1976.

Vol. 565: Turbulence and Navier Stokes Equations. Proceedings 1975. Edited by R. Temam. IX, 194 pages. 1976.

Vol. 566: Empirical Distributions and Processes. Oberwolfach 1976. Proceedings. Edited by P. Gaenssler and P. Révész. VII, 146 pages. 1976.

Vol. 570: Differential Geometrical Methods in Mathematical Physics, Bonn 1975. Proceedings. Edited by K. Bleuler and A. Reetz. VIII, 576 pages. 1977.

Vol. 572: Sparse Matrix Techniques, Copenhagen 1976. Edited by V. A. Barker. V, 184 pages. 1977.